Guanidino Compounds in Biology and Medicine

Guanidino Compounds in Biology and Medicine

Edited by P.P. De Deyn, B. Marescau, V. Stalon and I.A. Qureshi

1992

British Library Cataloguing in Publication Data

Guanidino Compounds in Biology and Medicine
    I. Deyn, Peter de
  574.19

ISBN: 0 86196 330 X

Published by

John Libbey & Company Ltd, 13 Smiths Yard, Summerley Street, London SW18 4HR, England.
Telephone: 081-947 2777: Fax 081-947 2664
John Libbey Eurotext Ltd, 6 rue Blanche, 92120 Montrouge, France.
John Libbey - C.I.C. s.r.l., via Lazzaro Spallanzani 11, 00161 Rome, Italy

Printed in Great Britain by Hartnolls Ltd, Bodmin, Cornwall, U.K.

Table of Contents

Foreword

This book contains selected papers presented at the 3rd International Symposium on Guanidino Compounds in Biology and Medicine that was held in September 1991 in Antwerp, Belgium. The first International Symposium on Guanidino Compounds in Biology and Medicine that was held in 1983 in Tokyo, Japan, coincided with the sixth Annual Guanidino Compound Research Meeting of the Japanese Guanidino Compound Research Association. The second International Symposium on Guanidino Compounds in Biology and Medicine took place in Susono City, Schizuoko, Japan in 1987. These meetings were realized under the inspiring and powerful impulse of Professor Akitani Mori and the respective presidents of the Japanese Guanidino Compound Research Association.

The two-hundred-fifty authors who contributed to this book reflect the world-wide distribution of Guanidino Compound research. The book offers a comprehensive updated 'State-of-the-Art' information on natural guanidino compounds and contains contributions of experts involved with several research aspects of guanidino compounds in microorganisms, plants, invertebrates and mammals, including man. It illustrates the increasingly multidisciplinary approach to research on these natural substances.

The selection of papers covers biochemical, nutritional, metabolic, (patho)physiological, clinical, therapeutical, diagnostic, electrophysiological, pharmacological, toxicological and analytical aspects of natural guanidino compounds.

Guanidino compounds and/or their synthetic or catalytic enzymes with their molecular genetics in microorganisms, plants, invertebrata and man are covered. The diagnostic, therapeutic and prognostic value of guanidino compound studies in a variety of human diseases such as inborn errors of metabolism, epilepsy, metabolic encephalopathy, renal failure, cerebrovascular, cardiovascular and neuromuscular diseases is presented as well.

We believe that this volume will further contribute to our aim of promoting a concerted action between all scientists involved with Guanidino Compound studies.

Finally, we would like to acknowledge the following institutions and companies that financially and logistically contributed to the realization of the symposium and this volume: Antwerp O.C.M.W. Medical Research Foundation, Born-Bunge Foundation, Byk Belga, Japanese Guanidino Compound Association, NATO, University of Antwerp.

Dr Bart Marescau
Laboratory of Neurochemistry and Behaviour
Department of Medicine
Born-Bunge Foundation
University of Antwerp
Universiteitsplein 1
2610 Wilrijk-Antwerp, Belgium

Dr Peter De Deyn
Laboratory of Neurochemistry and Behaviour
Department of Medicine
Born-Bunge Foundation
University of Antwerp
Universiteitsplein 1
2610 Wilrijk-Antwerp,
Belgium;
Department of Neurology
General Hospital Middleheim
Lindendreef 1, 2020 Antwerp, Belgium

Section I

Guanidino compounds in microorganisms,
plants and invertebrates

Guanidino Compounds in Biology and Medicine, eds. by P.P. De Deyn, B. Marescau, V. Stalon and
I.A. Qureshi. ©1992 John Libbey & Company Ltd., pp. 3–12.

Chapter 1

Novel mechanism of guanidino compound transport in bacteria: arginine:ornithine exchange

Arnold J.M. DRIESSEN, Hans C. VERHOOGT and Wil N. KONINGS

Department of Microbiology, State University of Groningen, Kerklaan 30, 9751 NN Haren, The Netherlands

Summary

The arginine deiminase pathway is widely distributed among bacteria, and allows growth with arginine as sole energy source. Conversion of arginine into ornithine, CO_2, and two molecules of ammonia is coupled to the production of ATP from ADP. Ornithine is excreted into the growth medium. Arginine uptake and ornithine excretion are coupled processes mediated by a novel transport system, the arginine:ornithine antiporter. Transport is driven by the concentration gradients of both substrates which are maintained by the metabolism. As no additional energy is required for the translocation of these solutes, a high efficiency of energy conversion is accomplished. By this mechanism an optimal energy conversion is possible when cells are exposed to conditions of limited energy supply.

Introduction

Bacteria have developed various strategies to utilize the nitrogen-rich amino acid, arginine[1,8,38]. The arginine deiminase (ADI) pathway is widely distributed among bacteria, and involves only three cytoplasmic enzymes (Fig. 1). Hydrolysis of the guanidino group of arginine is catalysed by *arginine deiminase* (EC 3.5.3.6) (ADIase) yielding the ureido-compound citrulline and NH_3. This reaction is essentially irreversible. The catabolic *ornithine carbamoyltransferase* (EC 2.1.3.3) (cOTCase) catalyses the reaction of citrulline with inorganic phosphate to produce ornithine and carbamoyl phosphate. The equilibrium of this reaction favours the formation of citrulline[36,37]. *Carbamate kinase* (EC 2.7.2.2) (CKase) transfers the phosphate group of carbamoyl phosphate to ADP to produce ATP, CO_2 and ammonia. The equilibrium of this reaction is in favour of ATP formation[23]. Some bacteria are also able to metabolize the decarboxylated derivative of arginine, agmatine[8]. Agmatine is converted via the agmatine deiminase (AgDI) pathway into putrescine. Although this metabolic route shares many similarities with the ADI pathway, unique enzymes are involved. The pathway comprises the enzymes *agmatine deiminase* (EC 3.5.3.12), a catabolic *putrescine carbamoyltransferase* (EC 2.1.3.6.), and a CKase different from the one belonging to the ADI pathway[34].

The ADI and AgDI pathways provide a source of ATP derived from the catabolism of arginine and agmatine, respectively. In the presence of small amounts of carbohydrate, arginine can serve as a sole energy source in the lactic acid bacteria *Lactococcus lactis* and *Enterococcus faecalis*[9,20,35]. The latter organism is also able to utilize agmatine[9,32,34,35]. The ADI pathway allows *Pseudomonas* and *Halobacterium* species to survive anaerobiosis when a suitable electron-acceptor or light is absent[8]. Most bacteria utilize only the guanidino-group of arginine, and ornithine is stoichiometrically

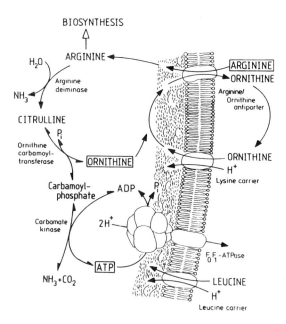

Fig. 1. Schematic representation of the arginine deiminase pathway in lactococci. From reference[15] with permission.

excreted into the medium. The maximal ATP yield (mole of ATP produced per mole of arginine consumed) of the ADI pathway equals one. The net gain, however, depends on the energetic costs of arginine uptake and ornithine excretion.

Energetic limitations on the excretion of ornithine

Bacterial cells maintain a proton motive force (Δp) across the cytoplasmic membrane which consists of an electrical potential ($\Delta\psi$), inside negative, and a pH gradient (ΔpH), inside alkaline. Δp allows solutes to accumulate against their chemical gradient. The pK of the δ-amino group of ornithine is 10.8 (pI 9.70) and the pK of the guanidino group of arginine is 12.5 (pI 11.15). At neutral pH values these molecules carry one net positive charge. Presuming a secondary transport mechanism, $\Delta\psi$ may function as a driving force for the uptake of arginine. Excretion of ornithine takes place against the $\Delta\psi$, and cells need a mechanism by which they prevent the massive accumulation of this amino acid. Examples of excretion mechanisms are shown in Fig. 2. When ornithine is excreted through a 'uniport' mechanism (Fig. 2A), expulsion will only be possible when the chemical gradient of ornithine exceeds $\Delta\psi$. These are conditions at which the direction of the ornithine flow reverses. This outwardly directed flux of ornithine contributes to the generation of an $\Delta\psi$. Efficient excretion of ornithine is possible when this process is coupled to the inward movement of protons as depicted in Fig. 2B. Ornithine:H$^+$ antiport is driven by ΔpH and allows excretion against a concentration gradient. When the number of protons translocated is greater than one, excretion depends on $\Delta\psi$ as well. However, this process presents a major energetic sink and reduces the net energy gain of the ADI pathway. An efficient excretion mechanism is the exchange between intracellular ornithine and extracellular arginine by an arginine:ornithine (Arg:Orn) antiporter (Fig. 2C). When the stoichiometry of exchange equals one, no net charge is translocated. Thus, transport will be solely driven by the concentration gradients of both amino acids. Arg:Orn exchange is of major advantage when the metabolic energy gain is low as *in vivo* translocation is primarily dictated by the metabolism.

Molar growth yield studies in lactic acid bacteria suggest that arginine metabolism is efficiently

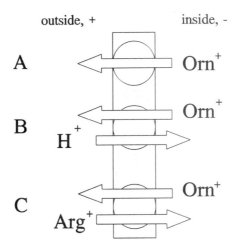

Fig. 2. Putative ornithine excretion mechanisms. A. Ornithine uniport; B. Ornithine:H^+ antiport; and C. Arginine:ornithine exchange.

coupled to growth with the production of 1 mole of ATP per mole of arginine utilized[3,7,27]. These results are consistent with recent findings that uptake of arginine and excretion of ornithine are coupled processes mediated by a unique antiport transport system[10]. Arg:Orn exchange has been studied in detail in *L. lactis*[10,11,29,39] and *P. aeruginosa*[42]. In this paper we will discuss the energetic and biochemical properties of both systems in relation to their metabolic functions.

Arginine:ornithine antiporter of *Lactococcus lactis*

In vitro studies

Membrane vesicles derived from galactose-arginine grown cells of *L. lactis* mediate rapid Arg:Orn exchange activity (Fig. 3)[10,12]. These vesicles are devoid of ADIase and thus provide a suitable model system for detailed studies of kinetics, specificity, and the mechanism of arginine uptake and ornithine. Without external arginine, ornithine release by membrane vesicles is characterized by a biphasic process, i.e. a rapid initial phase, and a slow second phase[12]. The initial burst of ornithine release represents half a turnover of the carrier in which an amount of ornithine is released equal to the number of ornithine-binding sites present in the membrane. This phenomenon is absent in membrane vesicles derived from cells which lack the Arg:Orn antiporter. These vesicles display only the slow phase of ornithine release which is not related to antiport activity. Ornithine efflux is accelerated by several orders of magnitude when arginine is present on the outer surface of the membrane. Under those conditions, release of ornithine is coupled to uptake of arginine with a one-to-one stoichiometry. The carrier thus operates as a strict antiport mechanism. The Arg:Orn antiporter of *L. lactis* does not catalyse net uptake of arginine or ornithine. The exchange reaction is electrically silent at neutral pH, conditions at which both substrates bear one net positive charge[10,12]. The pH dependency of Arg:Orn exchange indicates that monovalent positively charged arginine and ornithine are the translocated species (A.J.M. Driessen, unpublished work). Kinetic properties of Arg:Orn exchange have been studied in detail in membrane vesicles of *L. lactis* using an inhibitor stop assay with a time resolution of about 1 s[12]. Comparison of exchange kinetics in right side-out and inside-out vesicles suggests that the Arg:Orn antiporter operates by a kinetically symmetric mechanism. Arginine and ornithine compete for the same binding site(s), and are transported with almost identical V_{max}. The apparent K_t values of arginine and ornithine are different (Table 1), but comparable for the inner- and outer surface of the membrane. The V_{max} and K_t of arginine uptake increase with increasing ornithine concentration on the inside of the vesicles. These kinetic parameters

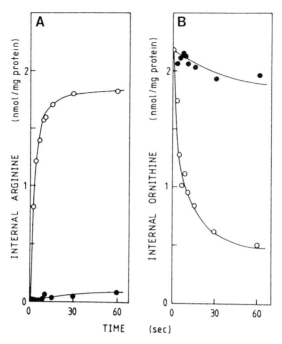

Fig. 3. Time-course of arginine uptake and ornithine release by membrane vesicles of L. lactis. *Uptake of arginine (A) and release of ornithine (B) by induced (○) and noninduced (●) membrane vesicles. External arginine and internal ornithine concentrations were 20 μM and 500 μM, respectively. Uptake was assayed at pH 6.5. From reference[12] with permission.*

reach their half-maximum value at an internal ornithine concentration of 27–35 μM. There is little variation in the V_{max} to apparent K_t ratio with internal ornithine. These kinetic data are consistent with a ping-pong type of mechanism. This mechanism is characterized by the formation of a binary complex. The substrate binding site is presented alternately to the two faces of the membrane (Fig. 4). The positively charged forms of arginine and ornithine are translocated. $\Delta\psi$ has no effect on the net flow of ornithine during the initial burst of ornithine release. The rate by which net flow occurs may be too fast to detect an effect of $\Delta\psi$. The measurement of the number of substrate binding sites indicate a turnover of 6 s^{-1} at 18°C.

Structural analogues of arginine and ornithine have been used to gain insight in the specificity of the substrate binding site of Arg:Orn antiporter. The system is specific for L- and D-enantiomers of amino acids or analogues with a guanidino- or NH_2 group at the C^5- or C^6-position of the side-chain[12,39]. Citrulline, a metabolic intermediate of the ADI pathway, is not transported. In this respect, the lysine carrier of *L. lactis* which catalyses Δp-dependent uptake of lysine is only specific for amino acid analogues with a guanidino- or NH_2 group at the C^6-position[13]. Arg:Orn exchange exhibits a sharp optimum at pH 5.0. Exchange is inhibited by an acidic internal pH, and an alkaline external pH. The rate of arginine uptake increases sixfold when the temperature is increased from 2 to 36 °C. The activation energy of the exchange reaction is 49.3 kJ/mol[12]. The Arg:Orn antiporter contains reactive sulphhydryl groups and is reversibly inactivated by organomercurials such as p-chloromercuriben-zene sulphonic acid (pCMBS)[14]. The reactive groups are exposed to the outer surface of the membrane at or near the substrate binding site. Substrates protect the carrier against pCMBS inactivation, and inhibition results in a reduction of the V_{max} of exchange while the K_t remains unchanged.

The Arg:Orn antiporter has been partially purified from cholate-extracted membranes using *n*-octyl-β-D-glucopyranoside as detergent (B. Tolner and A.J.M. Driessen, unpublished work). For functional reconstitution, the protein requires acidic phospholipids and the osmolyte glycerol during the solubilization step[19a]. Further purification by ion-exchange column chromatography and FPLC gelfiltration leads to a major enrichment of proteins with an apparent molecular mass in the range of 50–60 kD (Fig. 5A). Proteoliposomes reconstituted with the partially purified carrier exhibit a highly enriched Arg:Orn exchange activity (i.e. rates of up to 2 µmol/mg of protein·min) (Fig. 5b). Further purification is hampered by the low abundance of the antiporter in native membranes and non-specific interactions with column resins. Purification of the antiporter will facilitate molecular cloning of this transport protein.

In vivo studies

The Arg:Orn antiport is found in many *Lacto-coccus*, *Enterococcus* and *Streptococcus* species[29,39] and absent in *L. lactis* subsp. *cre-moris*[14]. The regulatory mechanisms controlling expression of the ADI pathway is complex and differs among these bacteria[8]. In *L. lactis*, enzymes of the ADI pathway are coordinately induced by a high arginine concentration in the medium[7,12,29], except the constitutively ex-pressed CKase[7]. CKase is possibly involved in other metabolic routes besides the ADI pathway. Galactose and arginine are concurrently metabo-lized. Glucose acts as a repressor and cAMP functions as an antagonist[7,29]. Glucose re-pression is not relieved by cAMP in *S. sanguis*[16], while glucose and arginine are simultaneously metabolized in *Ec. faecalis*[35]. Barotolerant vari-ants of *S. sanguis*[16] and *Ec. faecalis*[5,25] are con-stitutive for the ADI pathway and exhibit a reduced sensitivity to glucose catabolite re-pression. Growth conditions which promote the energy status of *Ec. faecalis* lead to reduction of the levels of the ADI pathway enzymes[11,35]. Induction of the ADI pathway in *L. lactis* re-quires external arginine. Many oral streptococci are able to hydrolyse arginine containing pep-tides at the cell surface[18,32]. Released arginine is immediately converted into ornithine by the ADI pathway[19]. The proteolytic proficiency and ADI activity allows these bacteria to survive lethal acidification resulting from the fermentation of carbohydrate (*see* Chapter 4)[6,24].

Mutants of the ADI pathway in *Ec. faecalis* are available[34] with a defect in the Arg:Orn antipor-ter (H. Verhoogt, unpublished work). The gene for the antiporter is located on the bacterial chro-mosome (H. Verhoogt, unpublished work). The ADIase (*arcA*), cOTCase (*arcB*) and CKase (*arcC*) genes of *S. sanguis* have been cloned and

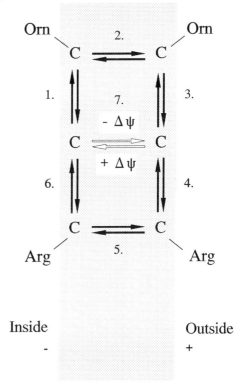

Fig. 4. Scheme of the kinetic mechanism of Arg:Orn exchange. Ornithine binds to the carrier on the inner surface of the membrane (Step 1) to form a binary complex which reorients (Step 2) followed by release of ornithine (Step 3) and the restoration of the unloaded carrier with its binding site facing outward. Arginine binds to the carrier on the outer surface of the membrane (Step 4), the binary complex reorients (Step 5) and arginine dissociates from the carrier on the inner membrane surface (Step 6). Reorientation of unloaded carrier (Step 7) allows net transport of arginine or ornithine as observed with the Arg:Orn exchanger of P. aeruginosa. This step does not exist in the catalytic cycle of the Arg:Orn antiporter of L. lactis.

7

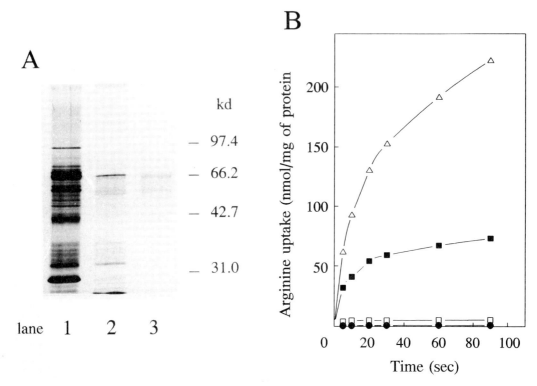

Fig. 5. Partial purification of the Arg:Orn antiporter of L. lactis. *A. Fractions with equal Arg:Orn antiport activity were analysed on silver-stained SDS-PAGE. Lane 1, octylglucoside extract; lane 2, DEAE Sephacel; lane 3, FPLC Superose 12 gelfiltration. B. Arg:Orn exchange activity of purified fractions reconstituted into proteoliposomes. (●) vesicles derived from arginine/galactose grown cells; (▢) octylglucoside extract, (■) DEAE Sephacel flow-through; and (Δ) FPLC Superose 12 gelfiltration. (B. Tolner and A.J.M. Driessen, unpublished work).*

expressed in *Escherichia coli*[4]. Insertional mutagenesis data suggests that the genes are clustered and possibly transcribed as an operon, with the gene order being *arcCAB*. Whether the gene coding for the Arg:Orn antiporter is also organized in this gene cluster remains to be established.

L. lactis cells grown in the presence of galactose and arginine maintain a high intracellular ornithine pool[29]. The ornithine level drops rapidly when cells are supplied with arginine while intracellular citrulline increases up to levels exceeding 20 mM. These conditions direct the thermodynamic unfavourable reaction catalysed by cOTCase towards the formation of ornithine and carbamoyl phosphate. A high turnover of the ATP pool in growing cells may further facilitate the cOTCase reaction as it depletes the cytosol for available carbamoyl phosphate. Ionophores that dissipate the Δp stimulate the ADI pathway activity[29]. Δp has no effect on the Arg:Orn antiporter[12] and affects the ADI pathway indirectly by its effect on the turnover of the cellular ATP pool. Conditions which lower ATP consumption, i.e. inhibition of F_0F_1-ATPase by *N,N'*-dicyclohexylcarbodiimide (DCCD) or high (internal) pH, decrease ADI activity, while protono- and ionophores which promote the hydrolysis of ATP increase the activity. The perturbation of the Δp may affect the ADI pathway activity also via intracellular pH effects. A major share of these regulatory effects may result from complex allosteric interactions of ATP with CKase and cOTCase[28].

Arginine:ornithine exchanger of *Pseudomonas aeruginosa*

P. aeruginosa is well adapted to conditions of limited oxygen supply. In the absence of oxygen and nitrate, cells are able to utilize arginine as a sole energy source to allow slow growth and motility on rich media[40]. Ornithine is excreted into the growth medium during anaerobic growth on arginine[8]. The enzymes of the ADI pathway are organized in the *arcDABC* operon[21,31]. These genes code for the Arg:Orn exchanger[42], ADIase, cOTCase and CKase, respectively[21,22]. The ADI pathway is repressed under aerobic conditions[26]. The *arcDABC* operon is under control of a positive regulatory protein, ANR, a FNR homologue of *P. aeruginosa*[17,40]. ANR is thought to bind to the consensus FNR-binding site centred around position –41 from the transcriptional start site of the *arcDABC* operon. ANR activates transcription of the *arc* genes during oxygen limitation. The *arcD* gene, located at the proximal part of the *arc* operon encodes a hydrophobic protein with a predicted molecular mass of 52 kD[22]. Hydropathy analysis[22] and topological studies (J.P. Borodineaud *et al.*, submitted) suggest that ArcD protein transverses the membrane 13 times with its amino- and carboxyl-terminus exposed to the cytoplasm and periplasm, respectively. Growth and complementation studies suggest that ArcD functions as a transport protein[22,41]. The *arcD* gene has been cloned and expressed in *E. coli*[22]. High level expression of *arcD* in *E. coli* allowed *in vitro* studies on the function of the protein. The parental strain used was defective in the uptake of arginine. Proteoliposomes have been reconstituted with membrane-extracted proteins from a strain expressing the ArcD protein. These proteoliposomes catalyse a rapid exchange of arginine and ornithine (Fig. 6) with a stoichiometry close to one[42]. This system resembles the Arg:Orn antiporter of *L. lactis* in many respects (Table 1), but differs in its ability to mediate a slow, but significant rate of Δp-driven uptake of arginine and ornithine. Δp-driven uptake is biased towards the Δψ, albeit a role of protons can not definitely be excluded. Although the kinetic mechanism of Arg:Orn exchange by the ArcD protein has not been studied in detail, the mechanistic explanation for this phenomenon must reside in the ability to relocate the unloaded carrier (Fig. 3, step 7). With H⁺:solute symport systems, reorientation of the unloaded carrier is often Δψ-dependent and rate-limiting when Δψ is low or absent. Reorientation of the binary complex is usually an order of magnitude faster allowing rapid exchange. Exchange mediated by the ArcD protein is several orders of magnitude faster than the rate of Δp-driven uptake[42]. Exchange is favoured by a high intracellular ornithine concentration which forces a rapid reoccupation of the binding site when arginine is released. Importantly, exchange rates are optimal when both the intra- and extracellular ornithine and arginine concentration are saturating. A strictly coupled antiport reaction will not lead to ornithine accumulation. Another mechanism

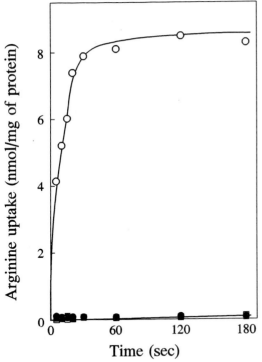

Fig. 6. Reconstitution of the Arg:Orn exchanger of P. aeruginosa. *Proteoliposomes were prepared from octylglucoside-extracted membranes of* E. coli JC182–5 *harbouring plasmid pME3719 (squares) carrying the P.* aeruginosa arcD *and* arcA *genes, or plasmid pKSII⁺ (circles) without insert. Proteoliposomes were loaded with 500 μM ornithine (open symbols) or not loaded (closed symbols) and diluted 50-fold in a buffer containing 1.5 μM L-[*¹⁴*C-]arginine. Uptake was assayed at 25 °C.*

is required to maintain a high intracellular concentration of ornithine. Aside from biosynthetic solutions to this problem, Δp-driven uptake may serve as an alternative route for *P. aeruginosa* to acquire ornithine or arginine. Arginine may be used for biosynthetic purposes, or converted into ornithine to allow accumulation of this amino acid within the cell to maximize the exchange rate. In *L. lactis*, ornithine can be replaced by lysine which is accumulated via a separate Δp-dependent transport system.

Table 1. Characteristics of arginine:ornithine exchange

		L. lactis	*P. aeruginosa*
Exchange stoichiometry[a]		one-to-one	one-to-one
Kinetic mechanism		ping pong	ND
K_t (µM)[b]	Arginine	6	11
	Ornithine	38	60
	Lysine	193	ND
Δp-dependent		No	Yes
Inhibitors		pCMBS, pCMB, mersalyl DCCD, DEPC	–

[a]Arg:Orn exchange; [b]K_t values for exchange with saturating ornithine on the inside; DEPC = diethylpyrocarbonate; ND = not determined.

Antiport of other guanidine compounds

Ec. faecalis can use the decarboxylated derivative of arginine, agmatine as a sole energy source for growth[8,9,33–35]. Agmatine is metabolized via a pathway analogous but not similar to the ADI pathway. The end product putrescine is excreted into the medium. Uptake of agmatine and the excretion of putrescine in *Ec. faecalis* is mediated by an agmatine:putrescine antiporter (Agm:Put antiporter)[11]. Agmatine-grown cells rapidly accumulate agmatine and putrescine. The intracellular putrescine pool in these cells is high, and rapidly drops in the presence of agmatine. These cells harbour only a low level of Arg:Orn antiporter activity. Agm:Put exchange and Put:Put homoexchange is electroneutral and not affected by the Δp. Putrescine is accumulated with a K_t of 20 µM. Putrescine uptake is competitively inhibited by agmatine (K_i values of 7 and 58 µM) and not by arginine. Conclusive evidence for an Agm:Put antiporter was obtained by the use of membrane vesicles. Uptake of agmatine was only observed with putrescine-loaded membrane vesicles. As *Ec. faecalis* lacks arginine decarboxylase activity, induction of the antiporter[11] and the other enzymes of the AgDI pathway[34] depends on the presence of agmatine in the medium.

In vivo, agmatine may commence from the decarboxylation of arginine by certain lactobacilli[8]. Decarboxylation of amino acids by these bacteria is associated with a growth advantage, while the biogenic amines accumulate in the medium. Transport of arginine and agmatine may be mediated by an Arg:Agm antiport. This reaction differs from the systems discussed thus far as it creates a charge imbalance thereby generating a Δψ, inside negative. Such exchange reactions have been found for the decarboxylation of certain anions[2,30].

Epilogue

Antiporters for cationic guanidino compounds are unique to bacteria, and are only expressed when the cells are grown in the presence of these compounds. These efficiently designed transport systems make a significant contribution to the overall energy-production as the concentration gradients are continuously maintained by the metabolic process. Cells may acquire further energetic advantages when ammonia and HCO_3^- are excreted in symport with protons.

References

1.   Abdelal, A.T. (1979): Arginine catabolism by microorganisms. *Annu. Rev. Microbiol.* **33**, 139–168.

2.   Anantharam, V., Allison, M.J. & Maloney, P.C. (1989): Electrogenic oxalate:formate exchange, the basis of energy coupling in *Oxalobacter formigens. J. Biol. Chem.* **264**, 7244–7250.

3.   Bauchop, T. & Elsden, S.R. (1960): The growth of micro-organisms in relation to their energy supply. *J. Gen. Microbiol.* **23**, 457–469.

4.   Burne, R.A., Parsons, D.T. & Marquis, R.E. (1989): Cloning and expression in *Escherichia coli* of the genes of the arginine deiminase system of *Streptococcus sanguis* NCTC 10904. *Infect. Immun.* **57**, 3540–3548.

5.   Campbell, J., III, Bender, G.R. & Marquis, R.E. (1985): Barotolerant variant of *Streptococcus faecalis* ATCC 9790 with reduced sensitivity to glucose catabolite repression. *Can. J. Microbiol.* **31**, 644–650.

6.   Casiano-Colón, A. & Marquis, R.E. (1988): Role of arginine deiminase system in protecting oral bacteria and an enzymatic basis for acid tolerance. *Appl. Environ. Microbiol.* **54**, 1318–1324.

7.   Crow, V.L. & Thomas, T.D. (1982): Arginine metabolism in lactic Streptococci. *J. Bacteriol.* **150**, 1024–1032.

8.   Cunin, R., Glansdorff, N., Pierárd, A. & Stalon, V. (1986): Biosynthesis and metabolism of arginine in bacteria. *Microb. Rev.* **50**, 314–352.

9.   Deibel, R.H. (1964): Utilization of arginine as an energy source for the growth of *Streptococcus faecalis. J. Bacteriol.* **87**, 989–992.

10.   Driessen, A.J.M., Poolman, B., Kiewiet, R. & Konings, W.N. (1987): Arginine transport in *Streptococcus lactis* is catalyzed by a cationic exchanger. *Proc. Natl. Acad. Sci. USA* **84**, 6093–6097.

11.   Driessen, A.J.M., Smid, E. & Konings, W.N. (1988): Transport of diamines by *Enterococcus faecalis* is mediated by an agmatine:putrescine antiporter. *J. Bacteriol.* **170**, 4522–4527.

12.   Driessen, A.J.M., Molenaar, D. & Konings, W.N. (1989): Kinetic mechanism and specificity of the arginine-ornithine antiporter of *Lactococcus lactis. J. Biol. Chem.* **264**, 10361–10370.

13.   Driessen, A.J.M., van Leeuwen, C. & Konings, W.N. (1989): Transport of basic amino acids by membrane vesicles of *Lactococcus lactis. J. Bacteriol.* **171**, 1453–1458.

14.   Driessen, A.J.M. & Konings, W.N. (1990): Reactive exofacial sulfhydryl groups on the arginine-ornithine antiporter of *Lactococcus lactis. Biochim. Biophys. Acta* **1015**, 87–95.

15.   Driessen, A.J.M. & Konings, W.N. (1990): Energetic problems of bacterial fermentations: extrusion of metabolic end products. In *The bacteria, a treatise on structure and function*, Vol. XII, ed. T.A. Krulwich, pp. 449. San Diego: Academic Press.

16.   Ferro, K.J., Bender, G.R. & Marquis, R.E. (1983): Coordinately repressible arginine deiminase system in *Streptococcus sanguis. Curr. Microbiol.* **9**, 145–150.

17.   Galimand, M., Gamper, M., Zimmermann, A. & Haas, D. (1991): Positive FNR-like control of anaerobic arginine degradation and nitrate respiration in *Pseudomonas aeruginosa. J. Bacteriol.* **173**, 1598–1606.

18.   Hiraoka, B.Y., Mogi, M., Fukasawa, K. & Harada, M. (1986): Coordinate repression of arginine aminopeptidase and three enzymes of the arginine deiminase pathway in *Streptococcus mitis. Biochem. Int.* **12**, 881–887.

19.   Hiraoka, B.Y., Harada, M., Fukasawa, K. & Mogi, M. (1987): Intracellular location of the arginine deiminase pathway in *Streptococcus mitis. Curr. Microbiol.* **15**, 81–84.

19a. In 't Veld, G., de Vrije, T., Driessen, A.J.M. & Konings, W.N. (1992): Acidic phospholipids are required during solubilization of amino acid transport systems of *Lactococcus lactis. Biochim. Biophys. Acta* **1104**, 250–256.

20.   Konings, W.N., Poolman, B. & Driessen, A.J.M. (1989): Bioenergetics and solute transport in lactococci. *CRC Crit. Rev. Microbiol.* **16**, 419–476.

21.   Lüthi, E., Mercenier, A. & Haas, D. (1986): The *arcABC* operon required for fermentative growth of *Pseudomonas aeruginosa* on arginine: Tn*5-751* -assisted cloning and localization of structural genes. *J. Gen. Microbiol.* **132**, 2667–2675.

22.   Lüthi, E., Baur, H., Gamper, M., Brunner, F., Villeval, D., Mercenier, A. & Haas, D. (1990): The *arc* operon for anaerobic arginine catabolism in *Pseudomonas aeruginosa* contains an additional gene, *arcD*, encoding a membrane protein. *Gene* **87**, 37–43.

11

23. Marshall, M. & Cohen, P.P. (1970): Carbamate kinase (*Streptococcus faecalis*). *Methods Enzymol.* **17a**, 229–234.

24. Marquis, R.E. & Bender, G.R. (1980): Isolation of a variant of *Streptococcus faecalis* with enhanced barotolerance. *Can. J. Microbiol.* **26**, 371–376.

25. Marquis, R.E., Bender, G.R., Murray, D.R. & Wong, A. (1987): Arginine deiminase system and bacterial adaptation to acid environments. *Appl. Environ. Microb.* **53**, 198–200.

26. Mercenier, A., Simon, J.-P., Vander Wauven, C., Haas, D. & Stalon, V. (1980): Regulation of enzyme synthesis in the arginine deiminase pathway of *Pseudomonas aeruginosa*. *J. Bacteriol.* **144**, 159–163.

27. Pandey, V.N. (1980): Interdependence of glucose and arginine catabolism in *Streptococcus faecalis* ATCC 3043. *Biochem. Biophys. Res. Commun.* **96**, 1480–1487.

28. Pandey, V.N. & Prodhan, D.S. (1981): Reverse and forward reactions of carbamyl phosphokinase from *Streptococcus faecalis* R. Participation of nucleotides and reaction mechanisms. *Biochem. Biophys. Acta* **660**, 284–292.

29. Poolman, B., Driessen, A.J.M. & Konings, W.N. (1987): Regulation of arginine-ornithine exchange and the arginine deiminase pathway in *Streptococcus lactis*. *J. Bacteriol.* **169**, 5597–5604.

30. Poolman, B., Molenaar, D., Smid, E.J., Ubbink, T., Abee, T., Renault, P.P. & Konings, W.N. (1991): Malolactic fermentation: electrogenic malate uptake and malate/lactate antiport generate metabolic energy. *J. Bacteriol.* (in press).

31. Rella, M., Mercenier, A. & Haas, D. (1985): Transposon insertion mutagenesis of *Pseudomonas aeruginosa* with a Tn5 derivative: application to physical mapping of the *arc* gene cluster. *Gene* **33**, 293–303.

32. Rogers, A.H., Zilm, P.S., Gully, N.J. & Pfennig, A.L. (1988): Response of a *Streptococcus sanguis* strain to arginine-containing peptides. *Infect. Immun.* **56**, 687–692.

33. Roon, R.J. & Barker, H.A. (1972): Fermentation of agmatine in *Streptococcus faecalis*: occurrence of putrescine transcarbamoylase. *J. Bacteriol.* **109**, 44–50.

34. Simon, J.P. & Stalon, V. (1982): Enzymes of agmatine degradation and the control of their synthesis in *Streptococcus faecalis*. *J. Bacteriol.* **152**, 676–681.

35. Simon, J.P., Wargnies, B. & Stalon, V. (1986): Control of enzyme synthesis in the arginine deiminase pathway of *Streptococcus faecalis*. *J. Bacteriol.* **150**, 1085–1090.

36. Stalon, V. (1972): Regulation of the catabolic ornithine carbamoyl transferase of *Pseudomonas fluorescens*: a study of the allosteric interactions. *Eur. J. Biochem.* **29**, 36–46.

37. Stalon, V., Ramos, F., Pierárd, A. & Wiame, J.M. (1972): Regulation of the catabolic ornithine carbamoyltransferase in *Pseudomonas fluorescense*: a comparison with the anabolic transferase and with a mutationally modified catabolic transferase. *Eur. J. Biochem.* **29**, 25–35.

38. Stalon, V. (1985): Evolution of arginine metabolism. In *Evolution of prokaryotes*, eds. K.H. Schleifer & E. Stackebrandt, pp. 277. London: Academic Press.

39. Thompson, J. (1987): Ornithine transport and exchange in *Streptococcus lactis*. *J. Bacteriol.* **169**, 4147–4153.

40. Zimmermann, A., Reimmann, C., Galimand, M. & Haas, D. (1991): Anaerobic growth and cyanide synthesis of *Pseudomonas aeruginosa* depend on *anr*, a regulatory gene homologous with *fnr* of *Escherichia coli*. *Mol. Microb.* **5**, 1483–1490.

41. Vander Wauven, C., Pierárd, A., Kley-Raymann, M. & Haas, D. (1984): *Pseudomonas aeruginosa* mutants affected in anaerobic growth on arginine: evidence for four-gene cluster encoding the arginine deiminase pathway. *J. Bacteriol.* **160**, 928–934.

42. Verghoogt, H.J.C., Smit, H., Abee, T., Gamper, M., Driessen, A.J.M., Haas, D. & Konings, W.N. (1992): *arcD*, the first gene of the *arc* operon for anaerobic arginine catabolism in *Pseudomonas aeruginosa*, encodes an arginine-orinithine exchanger. *J. Bacteriol.* **174**, 1569–1573.

Guanidino Compounds in Biology and Medicine, eds. by P.P. De Deyn, B. Marescau, V. Stalon and I.A. Qureshi. ©1992 John Libbey & Company Ltd., pp. 13–17.

Chapter 2

L-arginine catabolism in the cyanobacterium *Anacystis nidulans*

Matthias KUHLMANN and Elfriede K. PISTORIUS

Lehrstuhl für Zellphysiologie, Fakultät für Biologie, Universität Bielefeld, 4800 Bielefeld 1, Germany

Summary

The cyanobacterium *Anacystis nidulans* (*Synechococcus* PCC6301) contains an L-amino acid oxidase with high specificity for basic L-amino acids, such as L-arginine. A number of previous results indicated that this flavoprotein might be the precursor form of the water oxidizing enzyme in photosystem II. The results presented here show that this enzyme also participates in L-arginine degradation, when the cells are grown on L-arginine as the nitrogen source. Furthermore, it could be demonstrated that a complex interrelationship exists in *A. nidulans* between the activation state of photosystem II and the induction of a pathway leading to degradation of 4-guanidinobutyrate (formed by the thylakoid bound L-amino acid oxidase when photosystem II is damaged) to succinate. These results suggest that L-arginine can be used as nitrogen and carbon source when photosystem II is damaged, but that L-arginine is only used as nitrogen source when photosystem II is intact.

Introduction

The cyanobacterium *Anacystis nidulans* contains a constitutive L-amino acid oxidase (L-AOX) with high specificity for basic L-amino acids (L-Arg > L-Lys > L-Orn > L-His)[7]. In cells broken by French press treatment, the enzyme is partly present in the soluble fraction and partly tightly associated with the thylakoid membrane. The enzyme isolated from the soluble fraction of the cell extracts was shown to consist of two subunits of 50 kDa and to contain FAD as well as modified flavin of unknown structure(s)[8]. The L-AOX activity of this flavoprotein is strongly inhibited by cations ($M^{3+} > M^{2+} > M^+$) and less strongly by anions[6].

A number of results[2,7] indicated that this flavoprotein might be the precursor form of the long-sought water oxidizing enzyme in photosystem II (PS II)[3]. We could show that an antiserum raised against the isolated L-AOX protein detected a 36 kDa peptide which we believe to be a modified form of the soluble L-AOX protein (possibly associated with lipids), in highly purified O_2-evolving PS II complexes from *A. nidulans*[2]. Our initial reason to suspect that this protein might be involved in photosynthetic water oxidation was based on the observation that $CaCl_2$ has antagonistic effects on the two examined reactions: $CaCl_2$ stimulates photosynthetic water oxidation but inhibits the L-AOX activity detectable in *Anacystis* thylakoid membranes[7]. These results indicate that L-AOX activity has to become suppressed before water oxidation can occur. Based on these and additional results we suggested that the water oxidizing enzyme in PS II evolved from an L-arginine dehydrogenase/oxidase which originally mediated electron flow from L-arginine to the plastoquinone pool of the electron transport chain in cyanobacteria. Our model further predicts that during evolution this flavoprotein became modified with the inorganic cofactors Mn, Ca^{2+} and Cl^-, all of them required for photosynthetic water oxidation. The combination of this modified flavoprotein with the D_1/D_2/cytochrome b559 complex which in our model only catalyses photochemical charge separation, became the present-day water plastoquinone oxidoreductase[2,7].

Since the L-AOX activity of this flavoprotein is relatively high and not completely suppressed in favour of the suggested water oxidizing activity in *A. nidulans*, we investigated whether this enzyme might also have a function in L-arginine degradation.

Materials and methods

Anacystis nidulans (*Synechococcus* PCC6301) was grown on nitrate or L-arginine as previously described[7] with slight modifications. Regular cation concentration in the nutrient medium corresponded to 1.5 mM $MgSO_4$ and 0.15 mM $CaCl_2$, and the reduced cation concentration corresponded to 0.1 mM $MgSO_4$ and 0.05 mM $CaCl_2$. Preparation of French press extracts and measurements of the photosynthetic O_2 evolving activity in *Anacystis* cells were performed as previously described[7]. All other enzyme activities were determined in French press extracts (100 µl cells/ml). L-AOX activity was determined as NH_4^+ formation from L-arginine, and 4-guanidinobutyrase activity was measured as NH_4^+ production from 4-guanidinobutyrate (urease present)[7]. 4-Aminobutyrate transaminase and succinate semialdehyde dehydrogenase activities were determined via succinate formation[4] from 4-aminobutyrate (or via NAD reduction). Reaction mixtures for the succinate formation contained in a total volume of 3 ml: 20 mM tricine-NaOH, pH 9.5, 10 mM 4-aminobutyrate, 10 mM 2-ketoglutarate, and 10 mM NAD, and 200 µl French press extract.

Results and discussion

As previously shown[7], *A. nidulans* is able to grow on L-arginine as sole nitrogen source. So far, our results indicate that in *A. nidulans* the L-AOX is the only L-arginine degrading enzyme and that no other L-arginine degrading enzymes become induced after transfer of cells from nitrate to L-arginine-containing medium (results not shown). Originally[7], we thought that an arginase is induced when cells were grown on L-arginine, but the activity measured as NH_4^+ production from L-arginine in the presence of $MnCl_2$ in crude extracts (urease present) was most likely also associated with the L-AOX enzyme (unpublished results). Therefore, since the major (and possibly only) L-arginine degrading enzyme in *Anacystis* cells seems to be the constitutive L-AOX, we were interested to investigate whether the initial major product of this reaction was further metabolized. The initial products of the L-AOX reaction (utilizing molecular O_2 as electron acceptor) are either 2-ketoarginine and NH_4^+ when catalase is present to degrade the formed H_2O_2, or 4-guanidinobutyrate and NH_4^+ in the absence of catalase. Under such conditions the initially formed 2-ketoarginine is degraded nonenzymatically to 4-guanidinobutyrate by the H_2O_2. Results in Table 1 show that a 4-guanidino-butyrase, a 4-aminobutyrate transaminase and a succinate semialdehyde dehydrogenase are being induced when cells are grown on L-arginine. These enzymes will degrade 4-guanidinobutyrate to succinate as shown in Fig. 1. However, these enzymes only become induced when *Anacystis* cells are grown on L-arginine in a nutrient medium containing a reduced Mg^{2+} and Ca^{2+} concentration, while the enzymes are absent in cells grown on L-arginine in a medium with the regular cation concentration. As previously shown[9], reduction of the Mg^{2+} and Ca^{2+} concentrations in the nutrient medium leads to inactivation of photosynthetic water oxidation in *A. nidulans*. The results of Table 2 show that induction of the 4-guanidinobutyrase (and subsequent enzymes) coincides with reduction of the photosynthetic O_2-evolving activity. This reverse correlation between O_2 evolution and induction of enzymes degrading 4-guanidinobutyrate to succinate suggests that under conditions where CO_2 fixation becomes limiting due to reduction of the PS II activity (causing a deficiency in NADPH and ATP), L-arginine can be used as nitrogen and carbon source in *A. nidulans*.

Since the induction of this pathway in *A. nidulans* depends on the presence of L-arginine in the nutrient medium and on the activation state of PS II, the question is which signal leads to induction of this pathway, since it can not be L-arginine. We think that most likely the first substrate for the subsequent

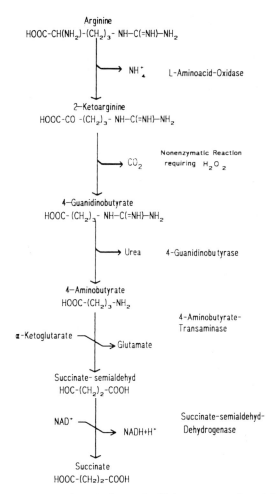

Fig. 1. Tentative L-amino acid oxidase pathway in Anacystis nidulans *grown on L-arginine as nitrogen source in a nutrient medium with reduced cation concentration.*

reaction sequence, 4-guanidinobutyrate, might be the inducer. This would imply that in *A. nidulans* 4-guanidinobutyrate will only be formed from L-arginine under conditions of PS II damage. A likely explanation could be that under such conditions the suggested L-AOX protein in PS II might become accessible for the hydrophilic substrate L-arginine and that part of the L-arginine is degraded by the membrane bound L-AOX. Probably, the products of this reaction 2-ketoarginine, NH_4^+ and H_2O_2 are released into the intrathylakoid space (H_2O_2 not being accessible to catalase). In consequence, 2-ketoarginine will be degraded to 4-guanidinobutyrate by H_2O_2 in a nonenzymatic reaction. A tentative model for the interrelation of the activation state of PS II and degradation of L-arginine to succinate is given in Fig. 2.

When cation concentration in the nutrient medium is sufficiently high, PS II not being damaged, the L-AOX protein in PS II is not accessible for L-arginine, since PS II is optimized for water oxidation. Under such conditions only the L-AOX protein present in the soluble fraction (not incorporated into the thylakoid membrane) is available for L-arginine degradation, thus leading to 2-ketoarginine as final product, since in the soluble fraction catalase effectively degrades the formed H_2O_2 – preventing nonenzymatic degradation of 2-ketoarginine to 4-guanidinobutyrate. Under such conditions 2-keto-

15

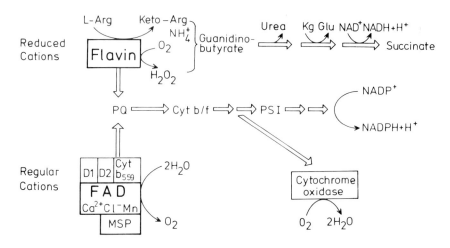

Fig. 2. Hypothetical model describing the interrelationship between photosynthetic O_2 evolution and L-arginine degradation to succinate in A. nidulans. This model is based on our hypothetical model of PS II[2,7] which suggests that the water-oxidizing enzyme has evolved from an L-arginine dehydrogenase/oxidase. As previously shown[9], under cation deficiency various PS II peptides (especially the D_1 peptide) are degraded. Under such conditions the L-AOX protein in the thylakoid membrane most likely becomes accessible for the hydrophilic substrate L-arginine leading to 4-guanidinobutyrate formation from L-arginine (with participation of H_2O_2) and induction of a pathway leading to succinate formation from 4-guanidinobutyrate.

Abbreviations: Flavin (FAD) referring to the flavoprotein with L-AOX activity; D1/D2/Cyt b559, reaction center complex of PS II; MSP, manganese stabilizing protein of PS II; PQ, plastoquinone; Cyt b/f, cytochrome b/f complex; PS I, photosystem I.

arginine is most likely excreted into the medium, since enzymes, such as 2-ketoarginine decarboxylase and 4-guanidinobutyraldehyde oxidoreductase, have so far not been detected in *Anacystis* cell extracts. Therefore, it seems that induction of the 4-guanidinobutyrate degrading pathway leading to succinate formation depends on the participitation of the thylakoid bound L-AOX in L-arginine degradation, because only under such conditions 4-guanidinobutyrate can be formed from L-arginine by the L-AOX (with participation of H_2O_2).

Table 1. L-amino acid oxidase, 4-guanidinobutyrase, and 4-aminobutyrate transaminase/succinate semialdehyde dehydrogenase activities in French press extracts of *Anacystis nidulans* cells grown on L-arginine in a nutrient medium with regular or reduced cation concentration

	Cation concentration	
Anacystis nidulans grown on L-arginine:	Regular	Reduced
	(μmol NH_4^+ or succinate formed/ml French press extract $\times$ h)	
L-amino acid oxidase activity	12.8	8.0
4-Guanidinobutyrase activity	0.4	7.2
4-Aminobutyrate transaminase/succinate semialdehyde dehydrogenase activity	0	4.9

Anacystis cells were grown for 4 days on L-arginine as N-source in a nutrient medium containing the regular or reduced cation concentration. The enzyme activities were determined as described under Materials and Methods. The L-AOX and the 4-guanidinobutyrase activity (urease present) were determined as NH_4^+ formation from L-arginine and 4-guanidinobutyrate, respectively, and the 4-aminobutyrate transaminase/succinate semialdehyde dehydrogenase activities were determined as succinate production from 4-aminobutyrate in the presence of 2-ketoglutarate and NAD.

Table 2. Photosynthetic O$_2$-evolving activity and 4-guanidinobutyrase activity in *Anacystis* cells grown for 4 days on L-arginine as N-source in a nutrient medium with a reduced cation concentration

	Photosynthetic O$_2$ evolution (μmol O$_2$ evolved per mg chlorophyll $\times$ h)	4-Guanidinobutyrase activity (μmol NH$_4^+$ formed per ml French press extract $\times$ h)
2. Day	435	0.4
3. Day	166	0.9
4. Day	88	17.5

Photosynthetic O$_2$-evolving activities and 4-guanidinobutyrase activities were determined as described in Materials and Methods

The reaction sequence leading to succinate formation from 4-guanidinobutyrate is comparable to the corresponding reactions of the L-AOX pathway converting L-arginine to succinate in *Pseudomonas putida*[1,5]. However, in *Pseudomonas putida*[5] the whole pathway including the L-arginine oxidase, 2-ketoarginine decarboxylase and 4-guanidinobutyrate oxidoreductase become induced when cells are grown on L-arginine. On the other hand, in *A. nidulans* only enzymes converting 4-guanidinobutyrate to succinate become induced, while the L-AOX is constitutive. As mentioned above, a number of results[2,7] have indicated that the main function of the L-AOX in *A. nidulans* is not in L-arginine degradation, but in photosynthetic water oxidation after the flavoprotein has been modified and incorporated into PS II. The results presented here suggest that a complex interrelationship exists between photosynthetic water oxidation and L-arginine catabolism in the cyanobacterium *A. nidulans* and that the activation state of PS II determines whether L-arginine is used by the cells only as a nitrogen source or as a carbon source as well.

Acknowledgements

The financial support of the Studienstiftung des Deutschen Volkes and Deutsche Forschungsgemeinschaft is gratefully acknowledged.

References

1. Cunin, R., Glansdorff, N., Piérard, A. & Stalon, V. (1986): Biosynthesis and metabolism of arginine in bacteria. *Microbiol. Rev.* **50**, 314–352.

2. Gau, A.E., Wälzlein, G., Gärtner, S., Kuhlmann, M., Specht, S., & Pistorius, E.K. (1989): Immunological identification of polypeptides in photosystem II complexes from the cyanobacterium *Anacystis nidulans*. *Z. Naturforsch.* **44c**, 971–975.

3. Hansson, Ö. & Wydrzynski, T. (1990): Current perceptions of photosystem II. *Photosynth. Res.* **23**, 131–162.

4. Michal, G., Beutler, H.-O., Lang, G. & Güntner, U. (1976): Enzymatic determination of succinic acid in foodstuff. *Z. Anal. Chem.* **279**, 137–138.

5. Miller, D.L. & Rodwell, V.W. (1971): Metabolism of basic amino acids in *Pseudomonas putida*. *J. Biol. Chem.* **246**, 5053–5058.

6. Pistorius, E.K. (1985): Further evidence for a functional relationship between L-amino acid oxidase activity and photosynthetic oxygen evolution in *Anacystis nidulans*. Effect of chloride on the two reactions. *Z. Naturforsch.* **40c**, 806–813.

7. Pistorius, E.K., Kertsch, R. & Faby, S. (1989): Investigations about various possible functions of the L-amino acid oxidase in the cyanobacterium *Anacystis nidulans*. *Z. Naturforsch.* **44c**, 370–377.

8. Wälzlein, G., Gau, A.E. & Pistorius, E.K. (1988): Further investigations about the flavin in the L-amino acid oxidase and a possible flavin in photosystem II complexes from the cyanobacterium *Anacystis nidulans*. *Z. Naturforsch.* **43c**, 545–553.

9. Wälzlein, G. & Pistorius, E.K. (1991): Inactivation of photosynthetic O$_2$ evolution in the cyanobacterium *Anacystis nidulans*: influence of nitrogen metabolites and divalent cation concentration. *Z. Naturforsch.* **46c**, 1024–1032.

Guanidino Compounds in Biology and Medicine, eds. by P.P. De Deyn, B. Marescau, V. Stalon and I.A. Qureshi. ©1992 John Libbey & Company Ltd., pp. 19–28.

Chapter 3

Catabolism of the guanidino compounds nopaline, octopine, and L-arginine in *Agrobacterium tumefaciens*: enzymes, genes, and regulation

Joachim SCHRÖDER, Johannes von LINTIG and Hans ZANKER

Institut für Biologie II, Biochemie der Pflanzen, Universität Freiburg, Schänzlestrasse 1, D-7800 Freiburg, Germany

Summary

Plant cells transformed with Ti plasmids of *Agrobacterium tumefaciens* synthesize unusual substances (opines) which are catabolized by the bacteria. The utilization of the guanidino opines nopaline (N^2-(1,3-D-dicarboxypropyl)-L-arginine) and octopine (N^2-(1-D-carboxyethyl)-L-arginine) requires inducible functions in the *noc* and the *occ* region of nopaline and octopine Ti plasmids, respectively. The *noc* region is split into two parts (each with one nopaline-induced promoter), while the *occ* region is one stretch of DNA with one octopine-induced promoter. Both regions code for a related set of functions: opine transport system (four proteins), opine oxidase, ornithine cyclodeaminase, and a constitutively expressed positive regulator (nocR, occR) for activation of the promoters in presence of the opines. Only the *noc* region contains an arginase gene. The similarities of the *noc* and *occ* region in organization and function indicate a common evolutionary origin.

Introduction

*A*grobacterium tumefaciens, a member of the Rhizobiaceae, causes a neoplastic disease in a wide range of gymnosperms and dicotyledonous angiosperms. In nature, the tumours are usually found at the root–stem interface (crown gall). Tumour formation requires the presence of the very large tumour-inducing (Ti) plasmids in the bacteria. The infection of the plants needs the presence of a wound area, certain signal molecules synthesized in the plant cells, low pH, and temperatures below 28 °C. Under these conditions, the bacteria initiate a complex sequence of events during which a part of the Ti plasmids, the T-region, is transferred from the bacteria to the cells where it is stably integrated into the plant DNA (see[43] for review). The transformed cells possess two important new properties:

(1) They form tumours, often with more or less well differentiated shoot-like structures. The cells are able to grow in sterile culture without the addition of auxin and cytokinins. A number of studies have shown that the DNA transferred from the agrobacteria codes for enzymes of auxin and cytokinin biosynthesis (reviewed in[58]). The activities of these genes appear to be the major factors in the hormone independence and in the neoplastic growth pattern.

(2) The tumour cells synthesize and secrete unusual substances (opines), most often amino acid or

19

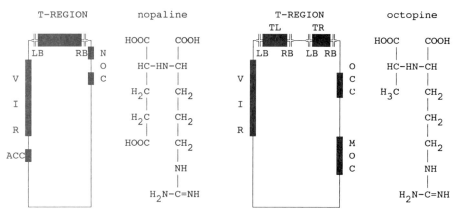

Fig. 1. Schematic overview of some functional units in nopaline and octopine Ti plasmids. NOC and OCC: noc and occ region, respectively. T-region: the part of the Ti plasmids transferred to the plant cells; RB and LB represent the right and left border sequences defining the extent of the transferred DNA. TL and TR: left and right part of the T-region in octopine plasmids. VIR: vir region with the genes responsible for transfer of the T-region to plant cells; ACC: agrocinopine catabolism[21,22]; MOC: mannityl opine catabolism[8,10,16].

sugar conjugates. They are not catabolized by the plants, whereas the tumour-inducing bacteria are able to utilize them as sole sources of carbon and nitrogen. The type of opines synthesized in the plant cells and their catabolism in *Agrobacterium* depend on the specificity of the Ti plasmids. According to the 'opine concept'[19,42,56], the benefit of opine utilization was the driving force in the evolution of the interaction with plants, and the induction of neoplastic growth appears to be a method to ensure the rapid proliferation of opine producing cells.

Nopaline (N^2-(1,3-D-dicarboxypropyl)-L-arginine) and octopine (N^2-(1-D-carboxyethyl)-L-arginine) (Fig. 1) are two guanidino opines which are typical for tissues transformed with nopaline and octopine type Ti plasmids, respectively. They are synthesized in the plant cells by a reductive condensation of L-arginine with 2-oxoglutarate (nopaline) and pyruvate (octopine). Like with all other opines, utilization in *Agrobacterium* requires Ti plasmid functions which are induced by the opines. Octopine, but not nopaline[12,14], has a second regulatory function[32,33], because it also activates the conjugal mechanisms of the Ti plasmids which ensure that the majority of the plant-colonizing agrobacteria contain a copy of the Ti plasmid. It also has been proposed that opines have additional roles in the activation of the *vir* genes for efficient transfer of the T-region into plant cells[37,57]. Interestingly, many soil-borne *Pseudomonas* strains are able to catabolize these opines ([1,45] and references therein), but neither the pathways nor the genetics are understood at present.

We summarize our present knowledge of the Ti plasmid encoded catabolic utilization of nopaline (plasmid pTiC58) and octopine (plasmid pTiAch5).

Results and discussion

The *noc* region

In typical nopaline-type Ti plasmids (Fig. 1A) the catabolic functions for this opine are located directly to the right of the T-region, in a stretch of approximately 17 kbp which is called the *noc* region[28,30,48]. The analysis of mutants also suggested that the region is split into two parts which are separated by several kbp DNA of unknown function[28,30]. The present state of knowledge is summarized in Fig. 2.

NOC REGION IN NOPALINE PLASMID pTiC58

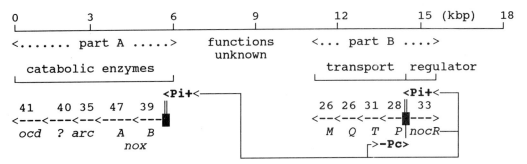

Fig. 2. Organization, functions, and regulation of the noc region in nopaline Ti plasmid pTiC58. The lines with arrows represent the protein coding regions and the direction of transcription. The names of the genes are given below, and the numbers on top indicate the approximate size of the proteins in kDa. ocd: L-ornithine cyclodeaminase; ?: protein of unknown function; arc: arginase; noxA and noxB: two proteins necessary for oxidative cleavage of nopaline; M, Q, T, P: transport proteins; nocR: regulator protein. < Pi+: promoters activated by nocR in presence of nopaline; –Pc >: constitutive promoter for expression of the regulator protein; the minus sign indicates that the promoter is negatively autoregulated by the protein expressed from nocR. The data are taken from [38,46,47,50] and from unpublished results (H. Zanker, J. von Lintig, J. Schröder, submitted for publication).

Catabolic enzymes

Part A contains the genes for nopaline oxidase (EC 1.5.1.19; *nox*, two proteins), arginase (EC 3.5.3.1; *arc*), a 40 kDa protein of unknown function, and L-ornithine cyclodeaminase (EC 4.3.1.12; *ocd*)[47]. The functions were identified with the proteins expressed in *E. coli*. Together, the enzymes represent a catabolic pathway which leads from nopaline via L-arginine and L-ornithine to L-proline which is then further catabolized by enzymes encoded in the *Agrobacterium* chromosome.

Several features are interesting in this pathway. The oxidative cleavage of nopaline to L-arginine and 2-oxoglutarate requires two different polypeptides, but their precise functions and the mechanism of the reaction are not understood. The proteins also cleave octopine, although apparently with somewhat lower efficiency. With the proteins expressed in *E. coli*, no cofactor appears to be necessary[47].

The presence of an arginase is unusual, since most other bacteria utilize other pathways for L-arginine degradation[5]. The gene and the properties of the enzyme have been characterized[50]. The protein sequence is related to that of arginases in eucaryotic cells (28–33 per cent identity), and the enzyme properties, including a very high K_m, are similar although some differences were noted[50].

The last enzyme in the pathway, L-ornithine cyclodeaminase, is unusual even for bacteria, because otherwise it has been identified only in some anaerobic bacteria and in some *Pseudomonas* strains. The reaction converts L-ornithine with release of ammonia directly into L-proline. The gene has been sequenced and the protein properties were characterized [46]. Although the overall reaction involves no oxidation, the activity is highly stimulated by NAD+. It seems likely that the cofactor plays a catalytic rather than a co-substrate role in the reaction, presumably as a transient electron acceptor. The activity is regulated by L-arginine which has pronounced effects on the optima for pH and temperature, and on the K_m for L-ornithine. It has been proposed that the effects of L-arginine reflect a part of the mechanisms which regulate the metabolic flow through the pathway[46]. The 40 kDa polypeptide remains an enigma. The gene is part of the operon containing all of the other known catabolic enzymes. It would seem likely that the protein performs some reaction related to opine catabolism, but the precise function remains unclear. The protein sequence deduced (A. Schrell, J. Schröder, unpublished) reveals no significant similarity with amino acid sequences in data libraries.

Promoters and regulator protein

The transcriptional analysis for promoter activities were performed in *Agrobacterium* with subfragments of the *noc* region and the reporter gene *lacZ*. The results[38] are also presented in Fig. 2.

Part A of the *noc* region contains one promoter which is activated in the presence of the opine (Pi). It was identified in approximately 460 bp at the extreme right end of part A, and it seems likely that it governs the expression of all downstream genes, i.e. for the catabolic enzymes described above.

A second nopaline induced promoter was detected close to the right end of part B. It reads leftwards into sequences which, as deduced from previous mutant studies, belong to the *noc* region, but the precise functions were unknown at the time of the identification of the promoter. The same region revealed a constitutive promoter activity (Pc >) reading rightwards.

All of these transcriptional analyses were not only performed in the presence of a resident Ti plasmid containing the *noc* region (pGV3850)[60], but also with an *Agrobacterium* strain free of any plasmids (APF2)[29]. Constructs containing the inducible promoter from part A were inactive under all conditions in APF2, indicating that the Ti plasmid contains regulatory elements, and that the regulation is by activation, not repression. In most cases, the plasmids with the promoter from part B showed the same result, but constructs with the right end of this part also revealed nopaline inducible activities in the absence of any other Ti plasmid sequences. This indicated that the cloned fragments contain elements which are necessary and sufficient for induction by the opine.

The DNA sequence analysis of the part responsible for the regulation revealed that the constitutive promoter reads into a gene coding for a protein (*nocR* in Fig. 2) with significant similarity to a family of positive regulators in other bacteria[2,3,4,41,51,53,59]. All of these proteins contain the lysR signature, an amino acid sequence characteristic for this type of regulatory protein[23], and the signature is also present in the polypeptide encoded in the *noc* region. All of the available evidence indicates that *nocR* is the only Ti plasmid encoded factor which is necessary for the activation of both inducible promoters by nopaline. It is expressed from a promoter which is active in absence of the opine (Pc >). Recent experiments (J. von Lintig, J. Schröder, unpublished) indicate that the promoter is autoregulated, i.e. the regulatory protein from *nocR* down-regulates the activity.

The *occ* region

The catabolic functions for octopine (*occ* region, see Fig. 1B) have been located to a single stretch of approximately 9 kbp[7,17,35,52]. The region maps between coordinates 33 and 45 kbp on the octopine-

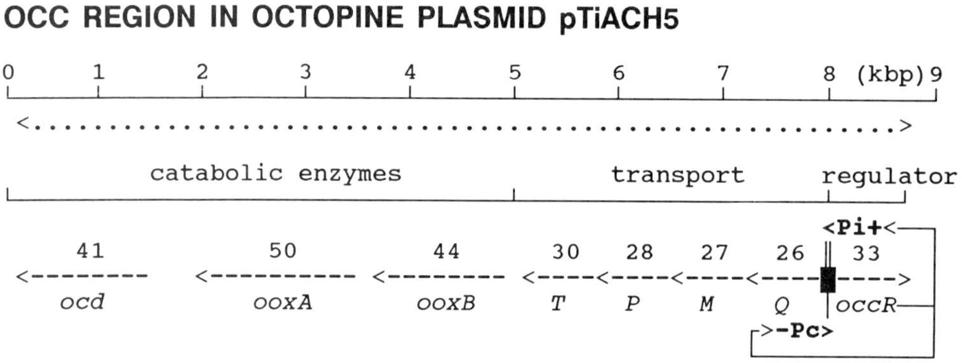

OCC REGION IN OCTOPINE PLASMID pTiACH5

Fig. 3. Organization, functions, and regulation of the occ region in octopine plasmid pTiAch5. ocd: L-ornithine cyclodeaminase; ooxA and ooxB: two proteins necessary for oxidative cleavage of octopine; T.P.M.Q. transport proteins; occR: regulator protein. < Pi+: promoter activated by occR in presence of octopine. See legend to Fig. 2 for other explanations. The data are taken from [38,49] and from unpublished results (H. Zanker, J. Schröder, unpublished; H. Zanker, J. von Lintig, J. Schröder, submitted for publication).

type Ti plasmid map which uses as reference a T-region *Sma*I site conserved in nopaline and octopine type Ti plasmids[11]. The organization of the *occ* region is shown in Fig. 3.

Catabolic enzymes

As with nopaline oxidase, octopine oxidase activity (*oox*, reaction products: L-arginine and pyruvate) requires two different proteins (H. Zanker and J. Schröder, unpublished results). In contrast to nopaline oxidase, the octopine oxidase cleaves only octopine, but not nopaline.

Table 1. Relationship between proteins encoded in the *noc* region (pTiC58) and the *occ* region (pTiAch5)

Gene name	Protein function	Identity (%)
ocd	L-ornithine cyclodeaminase	69
R	Positive regulator	36
M	Transport	51
Q	Transport	50
T	Transport	48
P	Transport	58

The other catabolic activity identified in the *occ* region is L-ornithine cyclodeaminase (*ocd*)[9,15], and the gene, its precise position, and the enzyme have been characterized[49]. It is related to the corresponding gene and polypeptide from the *noc* region (Table 1), and the enzyme properties are similar, but not identical. Differences were detected in the regulation of the activity by L-arginine (e.g. no stimulation of the *occ* region enzyme by the amino acid), and in the response to varying ratios of NAD+/NADPH[49]. Interestingly, no arginase gene was discovered in the *occ* region (see later), and a gene for a 40 kDa polypeptide corresponding to that in the *noc* region is also missing. The absence of the latter in the *occ* region suggests that the function of the protein may be specific for the catabolism of nopaline.

Promoters and regulator protein

The transcriptional analyses of octopine plasmid pTiAch5[38] showed that the *occ* region contained, close to the right end, a promoter induced in presence of octopine (< Pi, Fig. 3). This is the only octopine-induced promoter that could be identified so far. It reads leftward and may be responsible for induced expression of all downstream sequences. This was of considerable interest at the time of discovery, because the role of the 3.5 kbp DNA sequences between the promoter and *oox*B in octopine utilization was unknown. The finding of the inducible promoter in fact prompted the analysis which led to the identification of the transport functions. The same region contains a promoter which reads rightwards and is active in absence of octopine (Pc > in Fig. 3). Interestingly, the experiments with several subfragments from the *occ* region revealed the same phenomenon as observed in the *noc* region: the inducible promoter by itself is inactive in an *Agrobacterium* strain in the absence of a resident octopine Ti plasmid (pGV2260)[6], and it is not activated by octopine. Induction is possible, however, with subfragments which contain sequences from the right end of the *occ* region which are under control of the constitutive promoter. This clearly indicated that this part contains elements which are necessary and sufficient for induction of < Pi. The analysis of the DNA sequences revealed that the region expressed from the constitutive promoter codes for a protein (*occR* in Fig. 3) that is related to *nocR* (Table 1) and also contains the lysR signature. This is the only Ti plasmid encoded protein necessary for activation of the inducible promoter in the *occ* region. The positive regulator has also been analysed from a different octopine plasmid (pTiA6)[20]. The expression is autoregulated, and the protein sequence is identical with that described for Ti plasmid pTiAch5.

Transport functions in the *occ* and *noc* regions

The transcriptional analysis indicated that both regions contain opine induced promoters reading into sequences of poorly defined function in opine catabolism. This was of interest, since the catabolic enzymes identified so far appeared to be sufficient for conversion of the opines into components that are easily accommodated in general metabolism. We therefore sequenced these regions, and the results (H. Zanker, J. von Lintig, J. Schröder, submitted for publication) are also summarized in Figs. 2 and 3.

Both regions contain open reading frames for four different proteins. The polypeptides in the *occ* and *noc* region are related (Table 1), but the arrangement is different (*occQ*→*occM*→*occP*→*occT* *vs* *nocP*→*nocT*→*nocQ*→*nocM*). Most importantly, all of the deduced polypeptides are closely related to proteins in other bacteria which are involved in transport of amino acids[18,24-27,36,39,40].

The highest similarity was obtained with the components for the transport of L-histidine and L-arginine in *Salmonella typhimurium* (summarized in[25]), and therefore most of the terminology established in that system was adopted for the *occ* and *noc* region. This binding-protein-dependent transport system consists of four components which all have their counterparts in the Ti plasmid proteins: (i) two membrane proteins (hisM and hisQ) which are related to occM/occQ and nocM/nocQ, (ii) one protein (hisP) which carries the characteristic ATP/GTP binding site motif A and the ATP binding 'active transport' family signature, and these motifs are also present in occP and nocP, and (iii) one periplasmic protein which is responsible for the amino acid specificity (argT for L-arginine, hisJ for L-histidine). Characteristic for these polypeptides is a protein transport signal at the N-terminal end, and this is also found in occT and nocT.

The sum of these results indicates that the two catabolic regions each contain a complete set of proteins for a binding-protein-dependent transport system of the opines. The deduced functions were also confirmed by uptake studies with radioactive octopine and nopaline. These experiments also confirmed that the *noc* region functions transported nopaline *and* octopine, whereas the *occ* region proteins were active with octopine, but not nopaline[31,55]. Opine 'permeases' have been postulated before, but it was not known which and how many polypeptides are involved[31,48].

It appears likely that the pronounced similarity with transport systems for basic amino acids is not accidental, because both octopine and nopaline are L-arginine derivatives. In this context it is of interest that octopine tumours also synthesize opines with other basic amino acids[54]: lysopine (N^2-(1-D-carboxyethyl)-L-lysine), octopinic acid (N^2-(1-D-carboxyethyl)-L-ornithine), and histopine (N^2-(1-D-carboxyethyl)-L-histidine). Likewise, nopaline type tumours do not only produce nopaline, but also nopalinic acid (N^2-(1,3-D-dicarboxypropyl)-L-ornithine). These are also inducers of opine uptake in *Agrobacterium tumefaciens*[55], and presumably they are transported by the same permease function[31]. It appears therefore likely that the transport functions identified in the catabolic regions possess a broad substrate specificity.

Comparison of the two catabolic regions

The overall organization of the *occ* and *noc* regions appears to be very similar: (i) the principal arrangement of the genes in the DNA sequence, (ii) a single, autoregulated regulator gene; the protein product is necessary and sufficient for induction of the other functions, (iii) a transport system consisting of four proteins, and (iv) basically similar catabolic functions to convert the opines into substances which are further metabolized in the general metabolism of *Agrobacterium*[9]. The similarity is also obvious in the homology of the proteins (Table 1). All of these findings suggest that the two catabolic regions share a common evolutionary origin.

However, there are also differences which reflect to some extent the different evolution of octopine and nopaline Ti plasmids. The *noc* region is split into two parts (each with one inducible promoter), while the *occ* region represents a single block (one inducible promoter).

Another difference is that agrobacteria with octopine plasmids can utilize octopine, but not nopaline, whereas nopaline plasmids under suitable conditions confer utilization of both. The prerequisite for octopine usage is that a trace of nopaline must be present, because only this opine can activate the genes in the *noc* region[31,38]. The transport functions in the *noc* region also import octopine (H. Zanker, J. von Lintig, J. Schröder, submitted for publication), and, as noted before, nopaline oxidase also

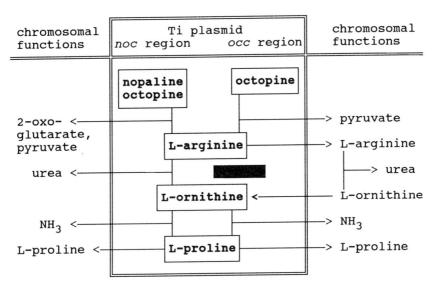

Fig. 4. Comparison of the Ti plasmid encoded catabolic reactions for nopaline and octopine.

cleaves octopine. These two properties appear to be essential for the usage of the 'foreign' opine. The inability of agrobacteria with octopine Ti plasmids to utilize nopaline is most likely due to the fact that the *occ* region functions lack these properties (no uptake of nopaline by the transport functions and no cleavage by octopine oxidase).

With respect to the catabolic pathways of the guanidino opines, the lack of arginase in the *occ* region is of particular interest (see Fig. 4 for a comparison of the Ti plasmid encoded reactions). The presence of L-ornithine cyclodeaminase indicates that the catabolic pathway proceeds via L-ornithine, and therefore the arginase activity must be provided by a chromosomal gene. Arginase genes are indeed present in all *Agrobacterium* strains tested, and it has been reported that octopine induces the enzyme activity in strains with octopine plasmids[9]. It also has been described for nopaline strain C58 and its Ti plasmid-free derivatives that L-arginine degradation with chromosomal genes is under the control of the central nitrogen regulation (*ntr*) system[44]. The interaction of gene activities in Ti plasmids and the chromosome deserves further investigation.

All agrobacteria can utilize L-arginine as N-source *via* arginase, but the use as sole C-source requires octopine or nopaline Ti plasmids and induction by the opine; in that sense, arginine catabolism is a function of the Ti plasmids[13]. The presence of arginase in all strains excludes the possibilty that this activity is the limiting factor. The available evidence indicates that its efficient use as C-source requires the activity of ornithine cyclodeaminase, and, with exception of a few octopine type strains[9], genes for this enzyme are present only on the Ti plasmids. This conclusion is supported by the finding that the requirements for the use of L-ornithine as C-source are identical[9]. The main catabolism of L-arginine and L-ornithine in *Agrobacterium* apparently proceeds through L-proline, and ornithine cyclodeaminase has a key role in the pathway.

Concluding remarks

The summary of these data indicates that many aspects of the catabolic utilization of the guanidino compounds nopaline and octopine in *Agrobacterium* are understood. The precise mechanisms of regulation, however, deserve further attention.

Earlier studies led to the proposal that the *occ* region is under negative regulation[32,33,34]. This conclusion was based on the analysis of constitutive mutants that were recessive to wild-type genes. The more recent investigations clearly suggest a positive control[20,38], and this apparent contradiction needs to be resolved. This should be possible with the regulatory genes and the inducible promoters which are now available.

The presently available data and the similarity of the proteins encoded in *nocR* and *occR* with other regulators already allow some predictions on the mechanisms of gene regulation. It seems likely that the proteins bind to the inducible promoters regardless of the presence or absence of the opines, that the opines binding to the regulators change the binding of nocR and occR to the DNA, and that this leads to activation of the promoters. Another interesting finding is that the constitutive, autoregulated regulatory gene and the inducible operon are transcribed from divergent promoters which are either very close or overlapping. These aspects will be studied in future investigations.

Acknowledgement

This work was supported by Deutsche Forschungsgemeinschaft (SFB206) and Fonds der Chemischen Industrie.

References

1. Bergeron, J., MacLeod, R.A. & Dion, P. (1990): Specificity of octopine uptake by *Rhizobium* and *Pseudomonas* strains. *Applied Environ. Microbiol.* **56**, 1453–1458.

2. Bölker, M. & Kahmann, R. (1989): The *Escherichia coli* regulatory protein *OxyR* discriminates between methylated and unmethylated states of the phage Mu *mom* promoter. *EMBO J.* **8**, 2403–2410.

3. Campbell, J.I.A., Scahill, S., Gibson, T. & Ambler, R.P. (1989): The phototrophic bacterium *Rhodopseudomonas capsulata* sp108 encodes an indigenous class A β-lactamase. *Biochem. J.* **260**, 803–812.

4. Chang, M., Hadero, A. & Crawford, I.P. (1989): Sequence of the *Pseudomonas aeruginosa trpI* activator gene and relatedness of *trpI* to other procaryotic regulatory genes. *J. Bacteriol.* **171**, 172–183.

5. Cunin, R., Glansdorff, N., Piérard, A. & Stalon, V. (1986): Biosynthesis and metabolism of arginine in bacteria. *Microbiol. Rev.* **50**, 314–352.

6. Deblaere, R., Bytebier, B., De Greve, H., Deboeck, F., Schell, J., Van Montagu, M. & Leemans, J. (1985): Efficient octopine Ti plasmid-derived vectors for *Agrobacterium*-mediated gene transfer to plants. *Nucleic Acids Res.* **13**, 4777–4788.

7. De Greve, H., Decraemer, H., Seurinck, J., Van Montagu, M. & Schell, J. (1981): The functional organization of the octopine *Agrobacterium tumefaciens* plasmid pTiB6S3. *Plasmid* **6**, 235–248.

8. Dessaux, Y., Guyon, P., Petit, A., Tempé, J., Demarez, M., Legrain, C., Tate, M.E. & Farrand, S.K. (1988): Opine utilization by *Agrobacterium* spp.: octopine-type Ti plasmids encode two pathways for mannopinic acid degradation. *J. Bacteriol.* **170**, 2939–2946.

9. Dessaux, Y., Petit, A., Tempé, J., Demarez, M., Legrain, Ch. & Wiame, J.-M. (1986): Arginine catabolism in *Agrobacterium* strains: role of the Ti-plasmid. *J. Bacteriol.* **166**, 44–50.

10. Dessaux, Y., Tempé, J. & Farrand, S.K. (1987): Genetic analysis of mannityl opine catabolism in octopine-type *Agrobacterium tumefaciens* strain 15955. *Mol. Gen. Genet.* **208**, 301–308.

11. De Vos, G., De Beuckeleer, M., Van Montagu, M. & Schell, J. (1981): Restriction endonuclease mapping of the octopine tumor-inducing plasmid pTiAch5 of *Agrobacterium tumefaciens*. *Plasmid* **6**, 249–253.

12. Ellis, J.G., Kerr, A., Petit, A. & Tempé, J. (1982): Conjugal transfer of nopaline and agropine Ti-plasmids–the role of agrocinopines. *Mol. Gen. Genet.* **186**, 269–274.

13. Ellis, J.G., Kerr, A., Tempé, J. & Petit, A. (1979): Arginine catabolism: a new function of both octopine and nopaline Ti plasmids of *Agrobacterium*. *Mol. Gen. Genet.* **173**, 263–269.

14. Ellis, J.G., Murphy, P.J. & Kerr, A. (1982): Isolation and properties of transfer regulatory mutants of the nopaline Ti-plasmid pTiC58. *Mol. Gen. Genet.* **186**, 275–281.

15. Farrand, S.K. & Dessaux, Y. (1986): Proline biosynthesis encoded by the *noc* and *occ* loci of *Agrobacterium* Ti plasmid. *J. Bacteriol.* **167**, 732–734.

16. Farrand, S.K., Tempé, J. & Dessaux, Y. (1990): Localization and characterization of the region encoding catabolism of mannopinic acid from the octopine-type Ti plasmid pTi15955. *Mol. Plant-Microbe Interactions* **4**, 259–267.

26

17. Garfinkel, D.J. & Nester, E.W. (1980): *Agrobacterium tumefaciens* mutants affected in crown gall tumorgenesis and octopine catabolism. *J. Bacteriol.* **144**, 732–743.

18. Gowrishankar, J. (1989): Nucleotide sequence of the osmoregulatory *proU* operon of *Escherichia coli. J. Bacteriol.* **171**, 1923–1931.

19. Guyon, P., Chilton, M.-D., Petit, A. & Tempé, J. (1980): Agropine in 'null-type' crown gall tumors: evidence for generality of the opine concept. *Proc. Natl. Acad. Sci. USA* **77**, 2693–2697.

20. Habeeb, L.F., Wang, L. & Winans, S.C. (1991): Transcription of the octopine catabolism operon of the *Agrobacterium* tumor-inducing plasmid pTiA6 is activated by a lysR-type regulatory protein. *Mol. Plant-Microbe Interactions* **4**, 379–385.

21. Hayman, G.T. & Farrand, S.K. (1988): Characterization and mapping of the agrocinopine-agrocin 84 locus on the nopaline Ti plasmid pTiC58. *J. Bacteriol.* **170**, 1759–1767.

22. Hayman, G.T. & Farrand, S.K. (1990): *Agrobacterium* plasmids encode structurally and functionally different loci for catabolism of agrocinopine-type opines. *Mol. Gen. Genet.* **223**, 465–473.

23. Henikoff, S., Haughn, G.W., Calvo, J.M. & Wallace, J.C. (1988): A large family of bacterial activator proteins. *Proc. Natl. Acad. Sci. USA* **85**, 6602–6606.

24. Higgins, C.F. & Ames, G.F.L. (1981): Two periplasmic transport proteins which interact with a common membrane receptor show extensive homology: complete nucleotide sequences. *Proc. Natl. Acad. Sci. USA* **78**, 6038–6042.

25. Higgins, C.F., Haag, P.D., Nikaido, K., Ardeshir, F., Garcia, G. & Ames, G.F.-L. (1982): Complete nucleotide sequence and identification of membrane components of the histidine transport operon of *S. typhimurium. Nature* **298**, 723–727.

26. Higgins, C.F., Hiles, I.D., Whalley, K. & Jamieson, D.J. (1985): Nucleotide binding by membrane components of bacterial periplasmic binding protein-dependent transport systems. *EMBO J.* **4**, 1033–1040.

27. Hogg, R.W. (1981): The amino acid sequence of the histidine binding protein of *Salmonella typhimurium. J. Biol. Chem.* **256**, 1935–1939.

28. Holsters, M., Silva, B., Van Vliet, F., Genetello, C., De Block, M., Dhaese, P., Depicker, A., Inzé, D., Engler, G., Villarroel, R., Van Montagu, M. & Schell, J. (1980): The functional organization of nopaline *Agrobacterium tumefaciens* plasmid pTiC58. *Plasmid* **3**, 212–230.

29. Hynes, M.F., Simon, R. & Pühler, A. (1985): The development of plasmid-free strains of *Agrobacterium tumefaciens* by using incompatibility with a *Rhizobium meliloti* plasmid to eliminate pAtC58. *Plasmid* **13**, 99–105.

30. Inzé, D. (1984): Een genetische studie van het pTiC58 plasmide van *Agrobacterium tumefaciens*: de constructie en karakterisatie van T-DNA mutanten en de identificatie van functies voor ornithine catabolisme. *PhD thesis* Belgium: Rijksuniversiteit Gent.

31. Klapwijk, P.M., Oudshoorn, M. & Schilperoort, R.A. (1977): Inducible permease involved in the uptake of octopine, lysopine and octopinic acid by *Agrobacterium tumefaciens* strains carrying virulence-associated plasmids. *J. Bacteriol.* **102**, 1–11.

32. Klapwijk, P.M., Scheulderman, T. & Schilperoort, R.A. (1978): Coordinated regulation of octopine degradation and conjugative transfer of Ti-plasmid in *Agrobacterium tumefaciens*: evidence for a common regulatory gene and separate operons. *J. Bacteriol.* **136**, 775–785.

33. Klapwijk, P.M. & Schilperoort, R.A. (1979): Negative control of octopine degradation and transfer genes of octopine Ti plasmids in *Agrobacterium tumefaciens. J. Bacteriol.* **141**, 424–431.

34. Klapwijk, P.M. & Schilperoort, R.A. (1982): Genetic determination of octopine degradation. In *Molecular biology of plant tumors* eds. G. Kahl & J. Schell, pp. 475–495. New York: Academic Press.

35. Knauf, V.C. & Nester, E.W. (1982): Wide host range cloning vectors: a cosmid clone bank of an *Agrobacterium* Ti plasmid. *Plasmid* **8**, 45–54.

36. Kraft, R. & Leinwand, L.A. (1987): Sequence of the complete P protein gene and part of the M protein gene from the histidine transport operon of *Escherichia coli* compared to that of *Salmonella typhimurium. Nucleic Acids Res.* **15**, 8568.

37. Krishnan, M., Burgner, J.W., Chilton, W.S. & Gelvin, S.B. (1991): Transport of nonmetabolizable opines by *Agrobacterium tumefaciens. J. Bacteriol.* **173**, 903–905.

38. Lintig, J. von, Zanker, H. & Schröder, J. (1991): Positive regulators of opine-inducible promoters in the nopaline and octopine catabolism regions of Ti plasmids. *Mol. Plant-Microbe Interactions* **4**, 370–378.

39. Nohno, T., Saito, T. & Hong, J.-S. (1986): Cloning and complete nucleotide sequence of the *Escherichia coli* glutamine permease operon (*glnHPQ*). *Mol. Gen. Genet.* **205**, 260–269.

40. Nonet, M.L., Marvel, C.C. & Tolan, D.R. (1987): The *hisT-purF* region of the *Escherichia coli* K-12 chromosome. *J. Biol. Chem.* **262**, 12209–12217.

41. Ostrowski, J., Jagura-Burdzy, G. & Kredich, N.M. (1987): DNA sequences of the *cysB* regions of *Salmonella typhimurium* and *Escherichia coli*. *J. Biol. Chem.* **262**, 5999–6005.

42. Petit, A., Dessaux, Y. & Tempé, J. (1978): The biological significance of opines. I. A study of opine catabolism by *Agrobacterium tumefaciens*. In *Proceedings of the 4th International Conference on Plant Pathology and Bacteriology* ed. M. Ride, pp. 143–152. Angers, France: Institut National de Recherche Agronomique.

43. Ream, W. (1989): *Agrobacterium tumefaciens* and interkingdom genetic exchange. *Ann. Rev. Phytopathol.* **27**, 583–618.

44. Rossbach, S., Schell, J. & De Bruijn, F.J. (1987): The *ntrC* gene of *Agrobacterium tumefaciens* controls glutamine synthetase (GSII) activity, growth on nitrate and chromosomal but not Ti-encoded arginine catabolism pathways. *Mol. Gen. Genet.* **209**, 419–426.

45. Rossignol, G. & Dion, P. (1985): Octopine, nopaline, and octopinic acid utilization in *Pseudomonas*. *Can. J. Microbiol.* **31**, 68–74.

46. Sans, N., Schindler, U. & Schröder, J. (1988): Ornithine cyclodeaminase from Ti plasmid C58: DNA sequence, enzyme properties, and regulation of activity by arginine. *Eur. J. Biochem.* **173**, 123–130.

47. Sans, N., Schröder, G. & Schröder, J. (1987): The *noc* region of Ti plasmid C58 codes for arginase and ornithine cyclodeaminase. *Eur. J. Biochem.* **167**, 81–87.

48. Schardl, C.L. & Kado, C.I. (1983) A functional map of the nopaline catabolism genes on the Ti plasmid of *Agrobacterium tumefaciens* C58. *Mol. Gen. Genet.* **191**, 10–16.

49. Schindler, U., Sans, N. & Schröder, J. (1989): Ornithine cyclodeaminase from octopine Ti plasmid Ach5: identification, DNA sequence, enzyme properties and comparison with gene and enzyme from nopaline Ti plasmid C58. *J. Bacteriol.* **171**, 847–854.

50. Schrell, A., Alt-Mörbe, J., Lanz, T. & Schröder, J. (1989): Arginase of *Agrobacterium* Ti plasmid C58 – DNA sequence, properties, and comparison with eucaryotic enzymes. *Eur. J. Biochem.* **184**, 635–641.

51. Shearman, C.A., Rossen, L., Johnston, A.W.B. & Downie, J.A. (1986): The *Rhizobium leguminosarum* nodulation gene *nodF* encodes a polypeptide similar to acyl-carrier protein and is regulated by *nodD* plus a factor in pea root exudate. *EMBO J.* **5**, 647–652.

52. Stachel, S.E., An, G., Flores, C. & Nester, E. W. (1985): A Tn3 *lacZ* transposon for the random generation of β-galactosidase gene fusions: application to the analysis of gene expression in *Agrobacterium*. *EMBO J.* **4**, 891–898.

53. Stragier, P. & Patte, J.-C. (1983): Regulation of diaminopimelate decarboxylase synthesis in *Escherichia coli*. III. Nucleotide sequence and regulation of the *lysR* gene. *J. Mol. Biol.* **168**, 333–350.

54. Tempé, J. & Goldmann, A. (1982): Occurrence and biosynthesis of opines. In *Molecular biology of plant tumors* eds. G. Kahl & J. Schell, pp. 427–449. New York: Academic Press.

55. Tempé, J. & Petit, A. (1982): Opine utilization by *Agrobacterium*. In *Molecular biology of plant tumors* eds. G. Kahl & J. Schell, pp. 451–459. New York: Academic Press.

56. Tempé, J. & Petit, A. (1983): La piste des opines. In *Molecular genetics of the bacteria–plant interaction* ed. A. Pühler, pp. 14–32. Berlin: Springer Verlag.

57. Veluthambi, K., Krishnam, M., Gould, J.H., Smith, R.H. & Gelvin, S.B. (1989): Opines stimulate induction of the *vir* genes of *Agrobacterium tumefaciens* Ti plasmid. *J. Bacteriol.* **171**, 3696–3703.

58. Weiler, E.W. & Schröder, J. (1987): Hormone genes and crown gall disease. *Trends Biochem. Sci.* **12**, 271–275.

59. Wek, R.C. & Hatfield, G.W. (1986): Nucleotide sequence and *in vivo* expression of the *ilvY* and *ilvC* genes in *Escherichia coli* K12. *J. Biol. Chem.* **261**, 2441–2450.

60. Zambryski, P., Joos, H., Genetello, C., Leemans, J., Van Montagu, M. & Schell, J. (1983): Ti plasmid vector for the introduction of DNA into plant cells without alteration of their normal regeneration capacity. *EMBO J.* **2**, 2143–2150.

Guanidino Compounds in Biology and Medicine, eds. by P.P. De Deyn, B. Marescau, V. Stalon and I.A. Qureshi. ©1992 John Libbey & Company Ltd., pp. 29–34.

Chapter 4

Role of the arginine deiminase system in the acid–base physiology of oral streptococci

R.E. MARQUIS, A.C. CASIANO-COLÓN and R.A. BURNE

Departments of Microbiology & Immunology and of Dental Research, University of Rochester Medical Center, Rochester, NY 14642, USA

Summary

The arginine deiminase system can act to protect dental plaque streptococci from acid damage by production of ammonia reactive with cytoplasmic protons to raise the intracellular pH value. The system seems well designed for this protective function in terms of its regulation and in terms of its acid tolerance. This tolerance can be related in part to inherent or molecular acid tolerance, primarily for carbamate kinase, but also in part to tolerance which appears to depend on aggregated states of the enzymes, especially for ornithine carbamyl transferase. The major importance of the system in dental plaque may be in allowing less acid tolerant bacteria, such as *S. sanguis*, to survive during periods of acidification, to pH values as low as 4, associated with glycolytic activities of more acid tolerant organisms such as *S. mutans*. Thus, the system would serve to reduce the cariogenicity of plaque and to promote remineralization of acid damaged teeth.

Introduction

Arginine appears to play important roles in the ecology of dental plaque, and as a consequence in dental caries, which is perhaps the most prevalent infectious disease in the developed world. Although the disease is coming under control, at least in part because of widespread use of agents such as fluoride, caries reductions in most countries are currently only about 50 per cent. Moreover, the disease pattern is changing from one in which those in their teens and twenties are primarily affected to one of a life-long problem affecting more and more older people who now retain their teeth for life. Also, a highly susceptible subpopulation has been identified.

The world picture for caries is one of considerably more disease as people in developing countries switch to more cariogenic diets.

The virulence of cariogenic bacteria such as *Streptococcus mutans* depends in part on acidogenicity[13]. The pH value in plaque can be reduced to 4 or somewhat lower as a result of production of acid from dietary carbohydrates[14]. This acidification, especially if prolonged, results in demineralization of teeth, which may not be fully remineralized during the alkalinization phase of the acid–base cycle in plaque. Thus, the pathology of caries is due to imbalance in the normal acidification-demineralization and alkalinization-remineralization cycles of plaque.

Virulence of cariogenic bacteria depends also on acid tolerance, specifically, the ability to carry out glycolysis at low pH values. Acid tolerance for organisms such as *S. mutans* depends on F-ATPases able to move protons out of the cell coupled to ATP hydrolysis[2,3]. The more acid tolerant organisms in plaque have higher levels of F-ATPase activity, and their enzymes have lower pH optima for activity. The glycolytic enzymes of the cell are acid sensitive, and for example, enolase is severely inhibited at pH values much below about 6.5. Thus, the capacity of the bacterium to carry out glycolysis at low pH values depends on its ability to maintain ΔpH across the cell membrane with the cytoplasm relatively alkaline compared with the environment.

Bacteria such as *Streptococcus sanguis* are considered to be desirable organisms in dental plaque. *S. sanguis* is only moderately acid tolerant, has low cariogenic potential, and moreover, has the arginine deiminase system (ADS). Catabolism of arginine by bacteria such as *S. sanguis* results in alkali production in plaque and is thought to contribute to the pH rise seen after exhaustion of carbohydrate following a sugar challenge. The maximal pH in plaque is greater than the salivary pH value, presumably because of base production by plaque bacteria. Other ADS-positive organisms commonly in plaque include: *Streptococcus rattus, S. gordonii, S. milleri, S. mitis* biovar 2, *S. anginosus, Actinomyces naeslundii* and *Lactobacillus fermentum*. Levels of ammonia in plaque fluid may be as high as 43 mM[11]. Part of this ammonia may derive from ureolysis and from Stickland fermentation reactions, but the ADS is generally considered to be a major contributor to plaque ammonia[15]. The major sources of arginine for plaque bacteria are the diet and salivary peptides which can be hydrolysed by various plaque bacteria.

The arginine deiminase system in oral streptococci is highly regulated. It requires induction by arginine and is subject to catabolite repression. It may also be under the influence of a positive regulator such as the Anr protein described recently by Zimmermann *et al.*[16] because expression of the system is reduced by aeration[6]. The genes for the system of *S. sanguis* NCTC10904 have been cloned and transferred to *Escherichia coli* and *S. mutans*[4,5]. The cloning work is part of a project aimed at developing implantable strains of oral streptococci, either strains of *S. mutans* to which ADS genes have been transferred or strains of *S. sanguis* altered to be hyperproducers of ammonia. The project has required that we come to terms with the functions of the ADS in oral streptococci, and as reviewed here, one of the functions appears to be to protect acid sensitive organisms such as *S. sanguis* from acid damage in plaque acidified by acid-tolerant, cariogenic organisms such as *S. mutans*.

Materials and methods

Bacteria

S. sanguis NCTC10904 and *Streptococcus rattus* FA-1 were maintained by weekly transfer on tryptic-soy agar (Difco Laboratories, Detroit, MI) and as stocks frozen in 25 per cent glycerol solution at –70 °C. Both are arginine-deiminase positive. Cells were grown routinely in static culture at 37 °C in complex medium with excess sugar so that growth was acid-limited.

Acid killing

Cells from cultures in the early stationary phase of growth were prepared as described previously[8]. For determinations of colony-forming units per ml (CFU/ml), samples were diluted in 1 per cent peptone broth (Difco Laboratories, Detroit, MI). 0.1-ml samples of the diluted suspensions were then spread over the surfaces of plates of tryptic-soy agar, and the plates were incubated at 37 °C for at least 48 h before final counts of colonies were made.

Enzyme isolation and determination of pH profiles for activity

Enzymes were isolated as described by Casiano-Colón[9], basically with ammonium-sulphate fraction-ation followed by chromatographic separations with Sephadex G-200 (Pharmacia Corp., Piscataway, NJ) and DEAE-BioGel A (BioRad Corp., Richmond, CA). Cells were permeabilized by subjecting them to one freeze–thaw cycle in mercaptoethanol buffer (5 mM 2-mercaptoethanol, 5 mM $MgCl_2$, 50 mM potassium phosphate, pH 7.0) and addition of 0.025 ml toluene per ml of cell suspension.

Arginine deiminase (ADase) activity was assayed in terms of production of citrulline determined by the method of Archibald[1]. Ornithine carbamyl transferase (OTCase) activity was assayed by measuring ammonia produced from citrulline in the presence of 500 mM arsenate buffer with an ammonia electrode (Orion Research, Inc., Cambridge, MA). Carbamate kinase (CKase) activity was assayed in terms of ATP production by the enzymatic method involving hexokinase, glucose-6-phosphate dehydrogenase and NADP.

Results and discussion

Protection against acid killing

The data presented in Fig. 1 show that 20 mM L-arginine fully protected cells of *S. rattus* FA-1 against acid killing at pH values of 3.5 or 4.0 but that 20 mM NH$_4$Cl had only minor protective effects. Acid killing of this organism did not occur during the experimental period in suspensions at pH values of 4.5 or 7.0. Organisms without the ADS system, such as *S. mutans* GS–5, were not protected against acid killing by arginine. Protection of organisms such as *S. rattus* FA-1 or *S. sanguis* NCTC10904 could be achieved with addition of as little as 1 mM arginine to suspensions initially with approximately 10^9 CFU/ml. *S. rattus* FA–1 is more acid tolerant than *S. sanguis* NCTC10904, and for example, minimum pH values at which the oganisms can carry out glycolysis in dense suspensions with 1 mM MgCl$_2$ and 20 mM potassium phosphate were, respectively, 3.7 and 4.1. This difference was apparent also in terms of acid killing. *S. sanguis* NCTC10904 cells were found to be killed rapidly at pH 4.0 with a D value (time required to kill 90 per cent of the population) of some 0.5 h. Moreover, the persisting population after 4 h of acidification was low, only about 10 CFU/ml from an initial population of some 10^9 CFU/ml. As shown in the Fig., killing of *S. rattus* FA–1 at a pH of 4.0 had an average D value of more than 1 h, and a persisting population after 4 h of some 10^6 CFU/ml. However, *S. rattus* FA–1 was killed rapidly at a pH of 3.5 with a D value of about 0.25 h.

The protective effect of arginine against acid killing did not require pH rise in the suspensions associated with ammonia production but could be demonstrated in highly buffered suspensions or suspension to which acid was added to maintain the pH value. However, arginolysis by intact cells was found to be highly acid-tolerant. For example, cells of *S. rattus* FA-1 with 20 mM L-arginine at an initial pH value as low as 3.1 could carry out glycolysis to raise the pH above neutrality in less than 4 h. Arginolysis by intact cells of *S. sanguis* NCTC10904 was somewhat less acid tolerant, but still, the organism could carry out arginolysis in suspensions with initial pH values as low as 4.0 to raise the pH above neutrality within 4 h. Thus, the ADS is sufficiently acid tolerant to allow for arginolysis even at minimum pH values in dental plaque of about

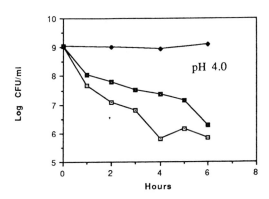

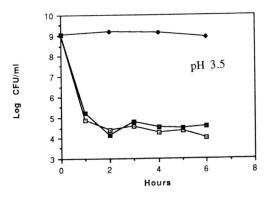

Fig. 1. Protection against acid killing by arginine and NH$_4$Cl. Cells of Streptococcus rattus *FA–1 were suspended in 20 mM phosphate buffer with 1 mM MgCl$_2$, suspension pH values were adjusted to 3.5 or 4.0, and samples were taken at intervals to determine colony-forming units (CFU)/ml. Data are shown for suspensions with no other additions (□), with 20 mM NH$_4$Cl (■) or with 20 mM L-arginine (◆).*

31

4. The protective effect of the system appeared to depend on production of ammonia (pK_a = 9.25.) within the cytoplasm, where it would associate with protons to raise the cytoplasmic pH. Ammonium ion is relatively non-toxic for bacteria and can even spare K^+ requirements[7]. Thus, the protective effect of the ADS is primarily for arginolytic organisms. However, the general rise in environmental pH during arginolysis would provide secondary aid to other organisms. As reviewed by Kleiner[12], bacterial cell membranes are highly permeable to NH_3, which can move between the cell and its environment in response to ΔpH across the membrane to the more acid compartment.

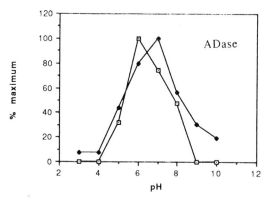

Nature of acid tolerance of the arginine deiminase system

Previous work with organisms such as *S. rattus* FA–1 indicated that the acid tolerance of the ADS does not depend on an intact cell membrane able to maintain ΔpH[8]. This organism can carry out arginolysis at pH values as low as about 2.5, well below the minimum pH for glycolysis or for activity of F-ATPases. Arginolysis by intact cells of *S. sanguis* NCTC10904 is less acid tolerant, but as indicated above, the organism still can carry out arginolysis at the lowest pH values commonly occurring in dental plaque.

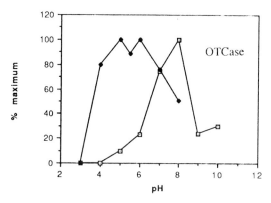

At least part of the acid tolerance of the ADS is enzymatic, as shown by data presented in Fig. 2[9]. CKase isolated from *S. sanguis* NCTC10904 was found to have an optimal pH value for activity of about 4, when working in the direction of ATP synthesis, and remarkably, the enzyme was still active at a pH value of 2, at some 60 per cent of the level at the optimal pH. Thus, the enzyme has inherent or molecular acid tolerance. The figure shows also the results of assays carried out with permeabilized cells. There is little difference in pH-activity profiles between the isolated enzyme or that in cells with damaged membranes.

Similarly, there was little difference in pH profiles for citrulline production by isolated ADase or the enzyme in permeabilized cells, although an apparent shift in pH optimum from 7 to 6 was evident.

ADase appeared to be the enzyme that may largely determine acid tolerance of the ADS. Certainly, it had significantly less inherent acid tolerance than CKase.

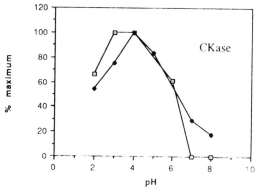

Fig. 2. Acid sensitivities of enzymes of the arginine deiminase system of Streptococcus sanguis *NCTC10904 in isolated form (□) or in permeabilized cells (◆).*

OTCase activity, assayed in terms of ammonia production from citrulline, showed a striking difference in pH profile depending on whether assays were carried out with purified enzyme or with the enzyme in permeabilized cells. The enzyme in the permeabilized cells was much more acid tolerant but still not as tolerant as CKase. Still at a pH value of 4, it was nearly fully active in permeabilized cells. In contrast, isolated, purified CKase was totally inactive at a pH value of 4.

Isolation and purification of the individual ADS enzymes proved to be very difficult because of aggregation and coelution. Moreover, even commercial preparations of CKase from cells of *Enterococcus faecalis* were found to be contaminated with ADase and OTCase. HPLC separation of enzymes from salt-lysed spheroplasts of *S. sanguis* NCTC10904 indicated that the most abundant aggregate had an apparent molecular weight of 670,000 with both ADase and OTCase activities. Similarly, isolations involving initial ammonium-sulphate fractionation and column chromatographic separations again indicated that the bulk of the enzyme was in aggregates from which ADase and OTCase could be separated only with difficulty. CKase could be somewhat more easily separated.

At pH values much below 4, cell membranes of oral streptococci tend to have reduced barrier function, and so the arginine/ornithine antiporter present in many lactic acid bacteria[10] may not be important under conditions of severe acid stress for entry of arginine into cells[8]. However, under less stressful conditions, the acid sensitivity of the antiporter system could play a role in the acid tolerance of the ADS.

Our overall conclusion is that the acid tolerance of the ADS in oral streptococci depends largely on molecular properties of the enzyme aggregates rather than on the capacities of the cells to maintain ΔpH across the cell membrane. CKases appear to have inherent acid tolerance. However, especially for OTCase, tolerance appears to depend also on the aggregated states of the ADS enzymes. In fact, in the cell, the enzymes may function best in the aggregated state rather than as a set of separated molecules. The type of acid tolerance shown by the ADS seems well suited to an enzyme system designed to protect acid-sensitive organisms against acid damage in dental plaque. Thus, even when short-term acid damage results in reversible inactivation of the glycolytic system and F-ATPases, NH_3 can still be produced from arginine to raise the pH value in the cytoplasm and in plaque so that membrane function could be restored. The ADS thus would function to reduce the cariogenicity of plaque and to promote remineralization repair.

Acknowledgements

Our work is supported by grants P50 DE07003 (Rochester Caries Research Center) and R01 DE06127 from the US National Institute of Dental Research.

References

1. Archibald, R.M. (1944): Determination of citrulline and allantoin and demonstration of citrulline in blood plasma. *J. Biol. Chem.* **156**, 121–142.

2. Bender, G.R., Sutton, S. & Marquis, R.E. (1986): Acid tolerance, proton permeability, and membrane ATPases of oral streptococci. *Infect. Immun.* **53**, 331–338.

3. Bender, G.R. & Marquis, R.E. (1987): Membrane ATPases and acid tolerance of *Actinomyces viscosus* and *Lactobacillus casei*. *Appl. Environ. Microbiol.* **53**, 2124–2128.

4. Burne, R.A., Parsons, D.T. & Marquis, R.E. (1989): Cloning and expression in *Escherichia coli* of the genes of the arginine deiminase system of *Streptococcus sanguis* NCTC10904. *Infect. Immun.* **57**, 3540–3548.

5. Burne, R.A., Parsons, D.T. & Marquis, R.E. (1990): Stable integration of arginine deiminase genes into *Streptococcus mutans*. *J. Dent. Res.* **69**, 325.

6. Burne, R.A., Parsons, D.T. & Marquis, R.E. (1991): Environmental variables affecting arginine deiminase expression in oral streptococci. In *Genetics and molecular biology of streptococci, lactococci and enterococci*, eds. G.M. Dunny, P.P. Cleary & L.L. McKay, pp. 276–280. Washington DC: American Society for Microbiology.

7. Buurman, E.T., Pennock, J., Tempest, D.W., Joost Teixeira de Mattos, M. & Neijssel, O.M. (1989): Replacement of potassium ions by ammonium ions in different micro-organisms grown in potassium-limited chemostat culture. *Arch. Microbiol.* **155**, 391–395.

8.   Casiano-Colón, A. & Marquis, R.E. (1988): Role of the arginine deiminase system in protecting oral bacteria and an enzymatic basis for acid tolerance. *Appl. Environ. Microbiol.* **54**, 1318–1324.

9.   Casiano-Colón, A.C. (1990): The arginine deiminase system and base production by oral streptococci. *PhD Thesis.* Rochester, NY: University of Rochester.

10. Driessen, A.J.M. & Konings, W. (1990): Energetic problems of bacterial fermentations: extrusion of metabolic end products. In *Bacterial energetics*, ed. T.A. Krulwich, pp. 449–478. San Diego: Academic Press, Inc.

11. Edgar, W.M. & Higham, S.M. (1990): Plaque fluid as a bacterial milieu. *J. Dent. Res.* **69**, 1332–1336.

12. Kleiner, D. (1985): Bacterial ammonium transport. *FEMS Microbiol. Rev.* **32**, 87–100.

13. Loesche, W.J. (1986): Role of *Streptococcus mutans* in human dental decay. *Microbiol. Rev.* **50**, 353–380.

14. Schachtele, C.F. & Jensen, M.E. (1982): Comparison of methods for monitoring changes in the pH of human dental plaque. *J. Dent. Res.* **61**, 1117–1125.

15. Wijeyeweera, R.L. & Kleinberg, I. (1989): Acid-base pH curves *in vitro* with mixtures of pure cultures of human oral microorganisms. *Arch. Oral Biol.* **34**, 55–64.

16. Zimmermann, A., Reimmann, C., Galimand, M. & Haas, D. (1991): Anaerobic growth and cyanide synthesis of *Pseudomonas aeruginosa* depend on *anr*, a regulatory gene homologous with *fnr* of *Escherichia coli*. *Molec. Microbiol.* **5**, 1483–1490.

Guanidino Compounds in Biology and Medicine, eds. by P.P. De Deyn, B. Marescau, V. Stalon and
I.A. Qureshi. ©1992 John Libbey & Company Ltd., pp. 35–37.

Chapter 5

Metabolism of arginine in *Euglena gracilis* Z

Bong Sun PARK, Aiko HIROTANI [1] and Yoshihisa NAKANO[1]

*Chungang University, Seoul, Korea; [1]Department of Agricultural Chemistry, University of Osaka Prefecture, Sakai,
Osaka 591, Japan*

Summary

Euglena gracilis Z accumulated intracellular arginine, when grown in nitrogen-rich medium, to reach at 190 mM and
addition of carbon source, such as glucose and ethanol, resulted in the rapid decrease with a new cell growth. The
catabolism of arginine was shown to be operated by arginine dihydrolase pathway. The subcellular location and
properties of arginine deiminase and citrullinase related to the pathway were determined.

Introduction

In *Euglena gracilis* arginine does not support growth either as the nitrogen or carbon source[2], while
when grown on glutamate *Euglena gracilis* accumulates a great amount of arginine as a free amino
acid or dipeptide. We describe how arginine is synthesized and degraded and what physiological
role it plays in *Euglena*.

Materials and methods

Reagents

L-[U-^{14}C]-Arginine (327 mCi/mM) and L-[carbamoyl-^{14}C]-citrulline (54 mCi/mmol) were purchased from New England Nuclear. All other chemicals were obtained from Nacalai Tesque and were
of reagent grade.

Organism and culture

E. gracilis Z was cultured under illumination (2500 lux) at 27°C as described elsewhere[3].

Analysis of amino acid pool

Euglena cells (400 mg wet wt) were used for the analysis of amino acids in trichloroacetic acid soluble
fraction[3]. Amino acids in the aqueous extract before and after hydrolysing by 6 N HCl at 107 °C for
24 h were analysed with a Hitachi KLA–5 amino acid autoanalyser.

Enzyme assays[3]

Arginase (EC 3.5.3.1) activity was determined by measuring ^{14}C-urea released from ^{14}C-arginine by
enzyme reaction. Glycine amidinotransferase (EC 2.1.4.1) and arginine deiminase (EC 3.5.3.6) were
assayed by determining guanidinoacetate formed from arginine and glycine, and citrulline formed by

release of ammonia from arginine, respectively. Citrullinase (EC 3.5.1.20) was determined by measuring labelled CO_2 released from citrulline.

Subcellular fractionation of *Euglena* cell homogenate

A cell homogenate was obtained by partial trypsin digestion of pellicle followed by mild mechanical disruption[7], and fractionated by differential centrifugation according to Shigeoka *et al.*[6] and Tokunaga *et al.*[7].

Purification of arginine deiminase and citrullinase

Arginine deiminase was homogeneously purified 56-fold with the techniques of ultracentrifugation, and DEAE-cellulose, P-cellulose and Sephacryl S–200 chromatographies with recovery of 24 per cent[4].

Citrullinase was also homogeneously purified 64-fold with ultracentrifugation, and DEAE-cellulose and Sephacryl S–200 column chromatographies[5].

Results and discussion

Arginine was not taken up by *Euglena gracilis* and accordingly did not support the growth. However, it was accumulated in both free and peptide forms as a major nitrogen reserve in the cells, when an excess of a nitrogen source was supplied into the medium. Its content was greatest in the early exponential phase of growth, reaching a concentration as high as 220 µmol/10^9 cells or 190 mM in the cell volume. In the mid-exponential phase it decreased and accumulated again in the late-exponential phase. Addition of ethanol or glucose to a culture of *E. gracilis* in the late-exponential phase rapidly eliminated the accumulated arginine with renewed cell division, and the further addition of ammonium salts or amino acid caused renewed accumulation of arginine. The degradation of accumulated arginine involved *de novo* synthesis of enzymes related to arginine metabolism[3]. Analysis of metabolic products after incubation of labelled arginine in cell-free extracts and assay of related enzymes showed that in *Euglena* arginine is catabolized to ornithine via citrulline by the so-called arginine dihydrolase pathway, releasing ammonia successively[3]. The enzymes involved in this pathway, arginine deiminase and citrullinase, were detected in mitochondrial fraction, as judged from the distribution of activities of marker enzymes, as shown in Table 1. Sonication of the isolated mitochondrial fraction released the deiminase and citrullinase as well as succinate semialdehyde dehydrogenase into the $100,000 \times g$ supernatant, indicating the location in mitochondrial matrix[4,5].

Table 1. Distribution of arginine deiminase, citrullinase and marker enzymes in subcellular fractions of *E. gracilis*

	Enzyme activity[a]			
	Cytosol	Mitochondria	Microsomes	Chloroplasts
Arginine deiminase	7.4 (0.03)[b]	78.0 (1.5)[b]	3.9 (0.1)[b]	13.6 (0.5)[b]
Citrullinase	6.8 (2.7)[b]	84.2 (33.4)[b]	0.9 (0.4)[b]	5.1 (2.0)[b]
Glutamate dehydrogenase	94.1	0	0	0
Succinate semialdehyde dehydrogenase	5.1	86.9	0	3.5
Glucose-6-phosphatase	0	0	27.3	0
Ribulose-1,5-bisphosphate carboxylase	7.5	9.9	3.7	76.0

[a]per cent against the activity in the crude extract; [b]Specific activity, µmol/mg protein/min.

The highly purified arginine deiminase required Co^{2+} for the enzyme reaction with the K_m value of 0.23 mM, and its optimum pH was 9.7 to 10.3. The molecular weight of the native enzyme protein was 87,000 by gel filtration, and SDS-acrylamide gel electrophoresis showed that the enzyme consisted of two identical subunits with molecular weight of 48,000. *Euglena* arginine deiminase was

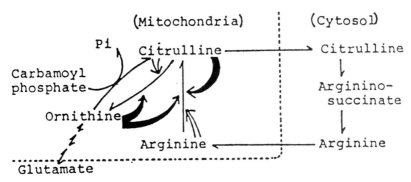

Fig. 1. Arginine metabolism in Euglena.

inhibited by sulphhydryl inhibitors, indicating that a sulphhydryl group is involved in the active centre of the enzyme. It exhibited positive cooperativity in binding with arginine. L-α-amino-β-guanidino-propionate, D-arginine and L-homoarginine strongly inhibited the enzyme while β-guanidinopropionate, γ-guanidinobutyrate and guanidinosuccinate did not. Considerable inhibition was observed with citrulline and ornithine[4].

The molecular weight of homogeneously purified citrullinase protein was calculated to be 94,000 by gel filtration and 31,000 by SDS-acrylamide gel electrophoresis, indicating that the citrullinase consists of three identical subunits. The maximum activity was obtained at 30 °C and pH 7.0 to 7.2. The enzyme was stable in the pH range of 6.0 to 7.0 when treated at various pHs for 10 min at 50 °C before enzyme reaction. *Euglena* citrullinase obeyed Michaelis–Menten kinetics toward citrulline and K_m value was 4.55 mM[5], while *Tetrahymena* enzyme did not follow normal Michaelis–Menten kinetics[1]. *Euglena* enzyme was strongly inhibited by sulphhydryl reagents such as mersaryl acid and $HgCl_2$ at 1 mM and the inhibition was removed by 2-mercaptoethanol, indicating that a sulphhydryl group in the protein is concerned with the active centre of the enzyme.

In *Euglena* ornithine carbamoylphosphate transcarbamylase was located in cytosol and mitochondria. Mitochondrial enzyme had a role in synthesis of citrulline (data not shown). From these results it is assumed that accumulated arginine is metabolized in *Euglena* cells as shown in Fig. 1. Inhibition of deiminase by citrulline and ornithine indicates that these compounds serve as regulatory end products in the arginine dihydrolase pathway.

References

1. Hill, D.L. & Chambers, P. (1967): The biosynthesis of proline by *Tetrahymena pyriformis. Biochim. Biophys. Acta* **148**, 435–447.

2. Oda, Y., Nakano, Y. & Kitaoka, S. (1982): Utilization and toxicity of exogenous amino acids in *Euglena gracilis. J. Gen. Microbiol.* **128**, 853–858.

3. Park, B.-S., Hirotani, A., Nakano, Y. & Kitaoka, S. (1983): The physiological role and catabolism of arginine in *Euglena gracilis. Agric. Biol. Chem.* **47**, 2561–2567.

4. Park, B.-S., Hirotani, A., Nakano, Y. & Kitaoka, S.(1984): Purification and some properties of arginine deiminase in *Euglena gracilis* Z. *Agric. Biol. Chem.* **48**, 483–489.

5. Park, B.-S., Hirotani, A., Nakano, Y. & Kitaoka, S. (1985): Subcellular distribution and some properties of citrullinase in *Euglena gracilis* Z. *Agric. Biol. Chem.* **49**, 2205–2206.

6. Shigeoka, S., Nakano, Y. & Kitaoka, S. (1979): Some properties and subcellular localization of L-glono-γ-lactone dehydrogenase in *Euglena gracilis* Z. *Agric. Biol. Chem.* **43**, 2187–2188.

7. Tokunaga, M., Nakano, Y. & Kitaoka, S. (1979): Subcellular localization of the GABA-shunt enzymes in *Euglena gracilis. J. Protozool.* **26**, 471–473.

Guanidino Compounds in Biology and Medicine, eds. by P.P. De Deyn, B. Marescau, V. Stalon and I.A. Qureshi. ©1992 John Libbey & Company Ltd., pp. 39–45.

Chapter 6

Compartmentation and regulation of the enzymes of arginine biosynthesis in plant cells

P.D. SHARGOOL[1], J.C. JAIN[2] and S.K. BHADULA[1]

[1]*Department of Biochemistry, University of Saskatchewan, Saskatoon S7N OWO, Canada;* [2]*Department of Geological Sciences, University of Saskatchewan, Saskatoon S7N OWO, Canada*

Summary

Arginine is synthesized in plants from glutamate via a pathway that involves acetylornithine as an intermediate. Acetylation of glutamic acid is the first step in the pathway and the synthesis of acetylglutamate can be catalysed by two enzymes, namely N^2-acetyl-L-ornithine: L-glutamate N-acetyltransferase (AOGA) and acetyl-CoA: L-glutamate N-acetyltransferase (AGAT). The experiments utilizing an active site directed analogue of acetylornithine indicated that the reaction catalysed by AOGA plays a major role while the reaction catalysed by AGAT plays a minor role in the synthesis of acetylglutamate. Compartmentation studies using soybean cell cultures revealed that ornithine is synthesized in plastids while synthesis of arginine takes place in the cytoplasm. This suggests the involvement of transport systems during biosynthesis of arginine. Our results and also those obtained by other workers indicate that arginine plays a key role as a feedback regulator of the activity of certain enzymes of the pathway. Based on these results, a reasonably simple but coherent scheme for the synthesis of arginine in plant cells has been derived.

Introduction

The amino acid arginine which is an important constituent of proteins is also known to serve under certain circumstances as a store of nitrogen in plants[2]. In addition, arginine is utilized in plants for the biosynthesis of secondary products such as polyamines[10]. Arginine is synthesized in plants from glutamate via a pathway that involves a series of acetylated intermediates[6,12,21]. In this pathway glutamate is first converted to ornithine via a sequence of reactions as shown in Fig. 1. The enzyme N^2-acetyl-L-ornithine: L-glutamate N-acetyltransferase (AOGA) plays a key role in transferring an acetyl group from acetylornithine to glutamate thus conserving the acetyl group with the formation of acetylglutamate and ornithine. Arginine appears to be synthesized from ornithine via the reactions of the urea cycle, originally proposed by Krebs and Henseleit[9]. In order to develop an understanding of any biochemical pathway, we must appreciate first of all, the sequence of intermediates and enzymes involved. Secondly, we must understand the compartmentation of these enzymes and intermediates, and how they are regulated in metabolic fashion. The present paper reports the compartmentation and regulation of activity of the enzymes involved in the biosynthesis of arginine in plant cells and provides a simple comprehensive scheme for understanding these concepts.

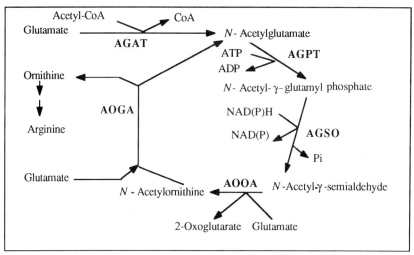

Fig. 1. Biosynthesis of ornithine and arginine in plants. AGSO = N-acetyl-L-glutamate-γ-semialdehyde: NADP⁺ oxidoreductase. Other abbreviations are defined in 'Enzyme assays'.

Materials and methods

Plant material

All cellular and enzymic fractions used in these studies were obtained from suspension cultures of soybean cells (*Glycine max* L. var. Mandarin). The media and growth conditions used for these cultures have been described previously[18].

Protoplast preparation

Protoplasts were prepared from soybean cells grown in B5 medium using the methods described earlier[4]. The crude protoplast preparation was resuspended in 5.0 ml of 20 per cent (w/v) sucrose solution. One ml of glucose was layered on top of the protoplast suspension and samples were centrifuged at $200 \times g$ for 5 min. The intact protoplast band formed at the interface of sucrose and glucose was collected, diluted 1: 8 (v/v) with 0.55 M sorbitol, and the protoplasts sedimented at 1000 $\times g$.

Subcellular fractionation

Purified protoplasts were resuspended in 10 ml of buffer A (0.3 M mannitol, 1 mM EDTA, 0.1 per cent BSA and 50 mM HEPES, pH 7.4). The protoplasts were lysed by passing through a 20 µm nylon net held on top of a 24–gauge needle, using a 10 ml disposable plastic syringe. A portion of this lysate was retained and the rest centrifuged at $3000 \times g$ for 20 min using a Sorvall SS–34 rotor. The 3000 $\times g$ pellet was resuspended in a small volume of buffer A and centrifuged at $200 \times g$ for 10 min. The pellet was discarded and the supernatant was centrifuged at $3000 \times g$ for 20 min to get a plastid enriched pellet, which was resuspended in a small volume of buffer B (1 mM phosphate buffer, pH 7.5, containing 1 mM β-mercaptoethanol and 1 mM EDTA)[8].

The first $3000 \times g$ supernatant was centrifuged twice at $12,000 \times g$ for 20 min in a Sorvall SS–34 rotor. The pellets were resuspended in buffer B, pooled together and used as the crude mitochondrial preparation, and the $12,000 \times g$ supernatants as soluble fraction.

Enzyme assays

N^2-acetyl-L-ornithine:L-glutamate N-acetyltransferase (AOGA) and acetyl-CoA:L-glutamate N-acetyltransferase (AGAT) were assayed according to the methods of Jain and Shargool[7]. The activity of ATP: N-acetyl-L-glutamyl 5-phosphotransferase (AGPT) was measured as described by McKay[11], and N^2-acetyl-L-ornithine:2-oxoglutarate aminotransferase (AOOA) according to the method of Albrecht and Vogel[1]. Argininosuccinate synthetase (ARSS) and argininosuccinate lyase (ARSL) were assayed as described by Shargool *et al.*[19] using a labelled substrate (L-[carbamyl[14]C]citrulline). For ornithine transcarbamoylase (OTC) and carbamoyl phosphate synthetase (CPS) assay, L-[1–[14]C]ornithine was used as substrate[19].

Assay of marker enzymes

Triose phosphate isomerase (TPI) was used as a marker enzyme for plastids[14]. The activity was measured by monitoring NADH oxidation at 340 nm as described by Gibbs and Turner[5]. Succinate dehydrogenase (SDH), a marker for mitochondria, and catalase, a marker for microbodies were assayed using the methods of Cooper and Beevers[3].

Effect of N^2-bromoacetyl-L-ornithine (NBAO) on ornithine biosynthesis

NBAO, which inhibits AOGA but not AGAT[17] was used to determine which of the two enzymes is involved in ornithine biosynthesis. This inhibitor (6 mM) was added to the culture medium and cells were harvested after 144 h of growth. The enzyme extracts were prepared using the methods of Jain and Shargool[7], and the activity of AOGA and AGAT were determined as described in the section 'Enzyme assays'. The level of ornithine was also measured as described by Shargool and Jain[17].

Results and discussion

Compartmentation studies

It should be noted that plastids, mitochondria and cytosol indicated in Fig. 2 (A and B) represent mainly $3000 \times g$ pellet, $12,000 \times g$ pellet and $12,000 \times g$ supernatant, respectively, obtained during cell fractionation procedures (see 'Materials and Methods'). The distribution of marker enzymes in these fractions showed that about 70 per cent of triosephosphate isomerase activity was associated with the plastidial fraction. The mitochondrial fraction was found to be rich in succinate dehydrogenase activity. The relative activity of catalase indicated that microbodies were found mainly in the cytosolic fraction. which showed 65 per cent of catalase, whereas a smaller fraction (27 per cent) was found in the mitochondria. The reason for the presence of catalase activity in the mitochondrial fraction could be due to the presence of contaminating microbodies, not removed by differential centrifugation procedures. The distribution of the enzymes of ornithine and arginine biosynthesis showed that more than 90 per cent of AGPT and AGAT activities were found in the cytosolic fraction. Similarly, approximately 70 per cent of the ARSL and ARSS activities were also recovered in this fraction (Fig. 2B). Thus it appears that these four enzymes are located in the cytoplasm. Taylor and Stewart[20] also detected the enzymes ARSS and ARSL in the cytoplasm of pea cotyledon cells. While over 80 per cent of the OTC and CPS activities were recovered in the plastidial fraction, none of these enzymes were detected in the mitochondrial or cytosolic fractions (Fig. 2A). Similarly, more than 55 per cent of the activities of the enzymes AOOA and AOGA were associated with plastids, and less than 40 per cent with the supernatant fraction. The distribution pattern indicates that these enzymes are located primarily in the plastidial fraction. The chloroplastic location of the enzymes OTC, CPS and AOGA has been shown in earlier studies[19,20]. The smaller amount of these enzymes found in the soluble fraction may be indicative of some plastid breakage and subsequent leakage of enzymes.

The relative distribution of the above mentioned enzymes of ornithine and arginine biosynthesis (Fig. 2A and B) clearly indicates that while acetylglutamate is synthesized in the plastids and also in the cytosol, acetyl-γ-glutamyl phosphate is synthesized in the cytosol and then transported to the plastids. Acetylornithine is synthesized in the plastids from acetyl-γ-glutamyl phosphate via acetyl-γ-semi-aldehyde. The presence of a major fraction of CPS and OTC in the plastids also indicates that citrulline is synthesized in the plastids and then transported to the cytosol where it is used for arginine synthesis (Fig. 3). This is further supported by the localization of ARSS and ARSL in the cytosol (Fig. 2B).

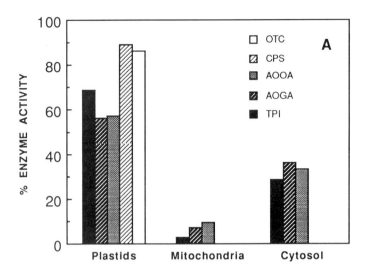

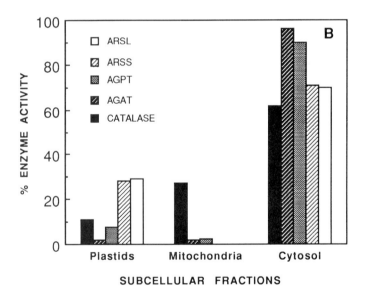

Fig. 2. Subcellular distribution of organelle marker enzymes and the enzymes involved in ornithine and arginine biosynthesis in soybean cell protoplasts. The activities are given as percentages of total units in whole protoplast lysate. Enzymatic abbreviations are defined in 'Enzyme assays'.

Regulation of enzyme activities

Glutamate is used as the precursor for the synthesis of ornithine and arginine. In arginine biosynthesis, glutamate is first converted to the non-protein amino acid, ornithine via a series of *N*-acetylated intermediates, acetylglutamate being the first in this series (Fig. 1). Since the synthesis of arginine and proline involves somewhat similar reactions, the use of *N*-acetylated intermediates in arginine biosynthesis seems to be a mechanism developed to avoid competition between arginine and ornithine biosynthesis in plants.

The enzymes AGAT and AOGA are both involved in the synthesis of *N*-acetylglutamate, an

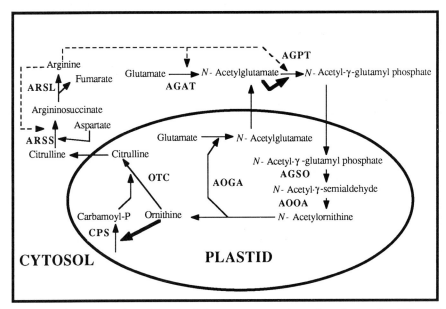

Fig. 3. Diagrammatic representation of the subcellular compartmentation and regulation of ornithine and arginine biosynthesis in plants. Enzymatic abbreviations are defined in 'Enzyme assays'. Solid lines = enzyme catalysed reaction, dotted lines = inhibition of the reaction by arginine, thick lines = activation of the reaction by acetylglutamate and ornithine.

intermediate during ornithine and arginine biosynthesis (Fig. 1). The experiment described here utilized N-bromoacetyl ornithine (NBAO), an active site directed reagent, to determine which of the enzymes predominates in the synthesis of L-ornithine in soybean cells grown in suspension culture. The activities of the enzymes AGAT and AOGA in cells grown in the presence or absence of NBAO are shown in Table 1. It is evident that the enzyme AOGA exhibited significantly lower activity when the cells were grown in the presence of NBAO. By contrast, NBAO did not appear to alter the activity of the enzyme AGAT. Interestingly, the concentration of free ornithine was also reduced, compared with control experiments (Table 1). This result indicates that enzyme AOGA plays a major role in the transfer of the acetyl group to glutamate to generate acetylglutamate, while AGAT plays an essentially anaplerotic role. Other work[11,13] using enzymes isolated from germinating pea cotyledons was concerned with the regulation of AGAT, AOGA and AGPT activities. Both AGAT and AGPT were found to be inhibited by arginine. However, the activity of AOGA was unaffected by arginine. The inhibition of AGPT activity by arginine was relieved by acetylglutamate which activated the enzyme. Thus AGPT appears to be a key regulatory point for metabolic control of the pathway and arginine plays a key role as a feedback regulator. O'Neil and Naylor[15] observed that the enzyme CPS was subject to feedback inhibition by certain nucleotides, while this inhibition was relieved by ornithine. ARSS activity was also found to be regulated by energy charge, with arginine acting as a modifier of this energy regulation[16].

Table 1. Effect of NBAO on N^2-acetyl-L-ornithine:L-glutamate N-acetyltransferase (AOGA), acetyl coenzyme A:L-glutamate N-acetyl-transferase (AGAT) and ornithine levels in soybean cells at 144 h of growth

Treatment	AOGA (units/mg protein)	AGAT (units/mg protein)	Ornithine (µmoles/g fr.wt)
B5	2.45 ± 0.03	1.30 ± 0.15	0.75 ± 0.06
B5 + NBAO	1.60 ± 0.20	1.40 ± 0.20	0.45 ± 0.05

Overall pathway

Figure 3 shows the compartmentation of the reactions of ornithine and arginine biosynthesis and their metabolic regulation. The synthesis of acetyl-γ-glutamyl phosphate seems to be one of the major regulatory points where arginine plays a key role as an inhibitor, and acetylglutamate as an activator. AGAT, which catalyses the synthesis of acetylglutamate from glutamate, and AGPT, which is involved in acetyl-γ-glutamyl phosphate synthesis, are both inhibited by arginine (Fig. 3). McKay[11] reported that 1 mM arginine inhibited 56 per cent of AGAT, whereas AGPT was completely inhibited by this concentration of arginine. Thus, AGPT is far more sensitive to arginine inhibition than is AGAT. The differential inhibition of these two enzymes by arginine probably allows AGAT to catalyse acetylglutamate synthesis while AGPT is inhibited. As acetylglutamate concentration rises, the inhibitory effect of arginine on AGPT is overcome and the cycle starts to generate more ornithine and arginine. Again, when arginine accumulates, it also inhibits argininosuccinate synthetase and acts as a feedback inhibitor. Another regulatory mechanism in this cycle is the activation of CPS reaction by ornithine to ensure a continuous supply of carbamoyl phosphate for citrulline synthesis which is required for the synthesis of arginine. Although the scheme proposed in Fig. 3 explains some of the aspects of ornithine and arginine biosynthesis and its metabolic regulation in plants, not much is known about the mechanisms involved in transport of intermediates across plastidial membranes. Thus, further studies on the transport mechanisms should yield valuable information to develop a better understanding of ornithine and arginine biosynthesis in plants.

Acknowledgements

This work was supported by an NSERC operating grant to PDS. The authors are thankful to Professor Ramji Khandelwal for helpful suggestions on the manuscript.

References

1.   Albrecht, A.M. & Vogel, H.J. (1964): Acetylornithine δ-transaminase. Partial purification and repression behaviour. *J. Biol. Chem.* **239**, 1872–1876.

2.   Beevers, L. (1976): In *Nitrogen metabolism in plants*, pp. 61. London: Edward Arnold.

3.   Cooper, T.G. & Beevers, H (1969): Mitochondria and glyoxysomes from castor bean endosperm. Enzyme constituents and catalytic capacity. *J. Biol. Chem.* **244**, 3507–3513.

4.   Gamborg, O.L. & Wetter, L.R. (1975): In *Plant tissue culture methods*, 1st edn, p. 17. Ottawa, Onatario: NRC Publ. National Research Council of Canada.

5.   Gibbs, M. & Turner, J.F. (1964): Enzymes of glycolysis. In *Modern methods of plant analysis*, Vol. 7, eds. H.F. Linskins, B.D. Sanwal & M.V. Tracey, p. 520. Berlin: Springer-Verlag.

6.   Goodwin, T.W. & Mercer, E.I. (1983): In *Introduction to plant biochemistry*, 2nd edn, Oxford: Pergamon Press.

7.   Jain, J.C. & Shargool, P.D. (1984): A simple assay system for enzymes involved in *N*-acetyl group transfer reactions: its use to study enzymes involved in ornithine biosynthesis in plants. *Anal. Biochem.* **138**, 25–29.

8.   Jain, J.C., Shargool, P.D. & Chung, S. (1987): Compartmentation studies on enzymes of ornithine biosynthesis in plant cells. *Plant Sci.* **51**, 17–20.

9.   Krebs, H.A. & Henseleit, K. (1932): Untersuchungen über die Harnstoffbildung im Tierkorper. *Hoppe-Seyler's Z. Physiol. Chem.* **210**, 33–66.

10.   Mazelis, M. (1980): Amino acid catabolism, In *The Biochemistry of plants. A comprehensive treatise*. Vol. 5, ed. B.J. Miflin, pp. 541–567. New York: Academic Press.

11.   McKay, G. (1980): Ornithine biosynthesis in higher plants. *PhD Thesis*. Canada: University of Saskatchewan.

12.   McKay, G. & Shargool, P.D. (1981): The biosynthesis of ornithine from glutamate in higher plant tissues. *Plant Sci. Lett.* **9**, 189–193.

13.   McKay, G. & Shargool, P.D. (1981): Purification and characterization of *N*-acetylglutamate 5-phosphotransferase from pea (*Pisum sativum*) cotyledons. *Biochem. J.* **195**, 71–81.

14.   Miflin, B.J. (1974): The location of nitrite reductase and other enzymes related to amino acid biosynthesis in the plastids of roots and leaves. *Plant Physiol.* **54**, 550–555.

15. O'Neil, T.D. & Naylor, A.W. (1968): Purine and pyrimidine nucleotide inhibition of carbamoyl phosphate synthetase from pea seedlings. *Biochem. Biophys. Res. Commun.* **31**, 322–327.

16. Shargool, P.D. (1973): The response of Soybean argininosuccinate synthetase to different energy charge values. *FEBS Lett.* **33**, 348.

17. Shargool, P.D. & Jain, J.C. (1985): Use of N^2-bromoacetyl- L-ornithine to study L-ornithine biosynthesis in soybean (*Glycine max* L.) cell culture. *Plant Physiol.* **78**, 795–798.

18. Shargool, P.D. & Jain, J.C. (1987): The responses of different soybean cell cultures to growth in media containing high levels of ammonia. *J. Plant Physiol.* **129**, 443–451.

19. Shargool, P.D., Steeves, T., Weaver, M. & Russell, M (1978): The localization within plant cells of enzymes involved in arginine biosynthesis. *Can. J. Biochem.* **56**, 273–279.

20. Taylor, A.A. & Stewart, G.R. (1981): Tissue and subcellular localization of enzymes of arginine metabolism in *Pisum sativum. Biochem. Biophys. Res. Commun.* **101**, 1281–1289.

21. Thompson, J.F. (1980): Arginine synthesis, proline synthesis, and related processes. In *The biochemistry of plants. A comprehensive treatise.* Vol. 5, ed. B.J. Miflin, pp. 375–402. New York: Academic Press.

Guanidino Compounds in Biology and Medicine, eds. by P.P. De Deyn, B. Marescau, V. Stalon and I.A. Qureshi. ©1992 John Libbey & Company Ltd., pp. 47–51.

Chapter 7

Possible function of canavanine during germination of alfalfa seeds (*Medicago sativa* L.)

J. MIERSCH

Martin-Luther-University Halle-Wittenberg, Institute of Biochemistry, Weinbergweg 16a, O-4050 Halle(S.), Germany

Summary

The structural analogue of arginine, canavanine (2-amino- guanidinooxybutyric acid), is found in 11 cultivars of alfalfa (*Medicago sativa* L.). Its concentration covers a range from 0.6 to 1.6 per cent in dry seeds. Canavanine represents more than 70 per cent of total soluble nitrogen in seeds and is stored in the cotyledons. During germination the guanidinooxy compound is translocated into hypocotyl and radicle. Previously canavanine was thought of as a nitrogen-storage compound catabolizing rapidly during seed germination. Contrary to this statement, in the early stage of seedling development (2nd day) the level of canavanine increased threefold. Therefore, canavanine must possess an additional function in nitrogen metabolism interpreted as an effective bypass for ammonia detoxification. In seedlings grown in the time range of 24 days, the guanidino compounds canavanine and arginine were metabolized rapidly whereas asparagine dominated. Furthermore, the toxic canavanine got into the environment of imbibed seeds or in the rhizosphere of young seedlings and increased in the milieu to concentrations of 3 to 57 µM. In biotests canavanine at 10 µM inhibited the radicle growth of cress (*Lepidium sativum* L.) and cabbage (*Brassica oleracea* L.). Therefore, canavanine could act as a potential allelochemical to sensitive plants. In experiments with alfalfa seedlings the inhibition of cabbage radicle growth is reduced but we cannot conclude unambiguously that this inhibition is due to canavanine only.

Introduction

Many higher plants synthesize non-protein amino acids as part of their rich array of secondary metabolites[2]. Some of these compounds are related structurally to the amino acids of common proteins. An excellent example is L-canavanine, 2-amino-4-guanidinooxybutyric acid, a structural analogue of L-arginine. According to Bleiler[3] and Rosenthal[17] canavanine acts as toxin against some microorganisms, plants and animals, especially insects.

The amino acid canavanine is stored in the seeds of many leguminous plants[11,17]. In general, during the imbibition and germination of seeds this nitrogen-rich compound is mobilized and degraded resulting in, e.g. the amino-oxy acid L-canaline and urea[2,4,14,22]. This paper describes the first experiments relating to the following points: (1) The distribution of canavanine in the developing seedlings of alfalfa, (2) the possible transport of canavanine in the root space, and (3) the inhibition of plant growth by canavanine.

Materials and methods

Plant material

Seeds of alfalfa (*Medicago sativa* L. cv. Verko) were obtained from VEG Pflanzenproduktion Langenstein, Germany, the other cultivars were received from Institut für Saat- und Pflanzgut Halle, Germany, and from Christov Institute of Fodder, Pleven (cv. Obnova 10), Bulgaria. The details of cultivation and extraction procedures were given by Miersch[13].

Biotests

The influence of canavanine was measured on the cell growth of tomato suspension culture (*Lycopersicon esculentum* Mill) as described by Grancharov[9]. The experiment was started by inoculation and application of canavanine at the concentrations of 10^{-3} to 10^{-6} M.

The inhibition of radicle growth of cabbage (*Brassica oleracea* L.) and cress (*Lepidium sativum* L.) was estimated during a period of 7 days at 20°C and 6000 Lux. For example, 40 seeds of cabbage (control) and 40 seeds of cabbage and 40 seeds of alfalfa were numbered onto a paper disk in Petri dishes of 9 cm in diameter, saturated with 5 ml water. After 7 days the radicle length of cabbage was measured. All data were analysed statistically according to Weber[24] by variance (F-test) and homogeneity (χ^2-test).

Canavanine analysis

Canavanine containing fractions derived from biological materials (seed extracts, extracts of seedlings, watery milieu of paper disk or sandy bed etc.) were purified by cation exchange chromatography[18]. This amino acid was quantified by colorimetric estimation according to Fearon[5] in respect of the results obtained by other groups[15,20]. Aliquots of extracts evaporated in vacuo were dissolved in buffer and analysed with an amino acid analyser (type AAA 881, Czechoslovakia) to compare the amount of canavanine and histidine (results not shown).

Results and discussion

The amount of canavanine measured was from 0.6 to 1.6 per cent in the seeds of alfalfa cultivars (Table 1). These values are in agreement with results derived from investigations of American cultivars by Natelson[15].

Table 1. Canavanine in the seeds of alfalfa

Cultivars	Collection year	Origin	Canavanine	
			µg per 10 seeds	% (dry mass)
Tokmashskaja	1984	USSR	104	0.6
Warminska	1986	PL	138	0.7
Szaponex	1986	H	151	0.7
Verko	1984	F	157	0.9
MEV	1984	FRG	167	0.9
Verko	1985	GDR	189	1.0
Vertus	1984	GDR	196	1.1
DuPuits	1986	F	204	1.1
Europe	1984	GB	220	1.2
Klesczewska	1986	PL	264	1.2
Obnova	1988	BG	354	1.6

Practically, all the canavanine was stored in the cotyledons (results not shown) whereas absent in the testa. Additionally, Fujihara[6] found by investigation of sword bean (*Canavalia gladiata* (Jaqu.) DC.) that the embryo is free of this amino acid. The seeds of jack bean (*Canavalia ensiformis* (L.) DC.)

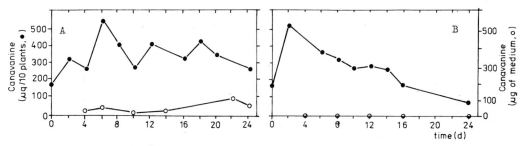

Fig. 1. Canavanine levels in seedlings and root exudations of alfalfa (Medicago sativa cv. Verko) cultured in nutrition solution (A) or water (B).

contained about 3.5 per cent canavanine but in the testa the guanidinooxy compound was only measured to 0.3 per cent.

During the imbibition of seeds and growth of seedlings the storage amino acid was metabolized, investigated e.g. in canavanine-containing leguminous plants *Medicago sativa* L.[14], *Canavalia ensiformis* (L.) DC.[10,19], *Colutea arborescens* L. and *Robinia pseudoacacia*[23]. During the development of *Canavalia ensiformis*[16] and *Medicago sativa* (Table 2) the toxic amino acid canavanine was localized in roots, leaves, and stems. Obviously, *Canavalia* is a canavanine-containing plant during the whole life cycle[6,17], but in other plants, e.g. *Medicago and Melilotus*[13] canavanine-free stages of distinct growth periods can be detected. Contrary to the results of Bell[2] and Nakatsu[14] the level of canavanine increased by 20 per cent during the imbibition of lucerne seeds and by 200 per cent during the first days of seedling growth (Fig. 1). Therefore, the storage compound should possess an additional function in the nitrogen metabolism, perhaps to be interpreted as an effective mechanism for detoxification of ammonia[7,21]. Besides canavanine synthesis, asparagine metabolism is found in alfalfa seedlings[1]. The amide asparagine is present for a long time in lucerne plants[12] whereas canavanine is catabolized completely.

Table 2. Distribution of canavanine in the seedlings of alfalfa (*Medicago sativa* cv. Verko)

Sample	1. Darkness		2. Light	
	Canavanine mg/g dry mass	%	Canavanine mg/g dry mass	%
Seedling	20.4	100	23.1	100
Root	3	15	3	13
Leaf	15.8	77	15.3	66
Stem	1.6	8	4.8	21

The seedlings were cultivated in darkness or in light (5500 Lux), harvested after 7 days and analysed as described in 'Materials and methods'

Furthermore, during the imbibition of seeds and growth of seedlings the storage compound was already mobilized, and got into the rhizosphere up to 1 per cent of total canavanine. This compound increased in the soil under experimental conditions to concentrations (Table 3) which inhibited the growth of receptor plants. The concentration of canavanine was measured and ranged from 3 to 57 µM respectively in the watery milieu of paper disk or sandy bed, with 500 seeds soaking for 24 h. Therefore, 500 seedlings cultivated under sterile conditions for 4 days, form a concentration of canavanine on the average of 8 to 26 µM in the watery-milieu of a sandy bed. Because 10 µM canavanine inhibited the growth of tomato cells to 35 per cent in 4 days, we assume an allelochemical influence of canavanine on other plants surrounding the imbibed as well as germinated seeds of alfalfa. Half-life periods of this compound in the soil must be estimated. But it is clear that no biological active saponin can penetrate the cell membranes of alfalfa roots[8]. In biotests the radicle growth of *Brassica* was inhibited, by co-growing alfalfa seedlings, to 65 per cent at 4 days[13]. It is not clear if canavanine alone caused the inhibition of plant growth because several basic toxins are present in lucerne plants.

Table 3. Canavanine in the environment of seeds and roots of alfalfa (*Medicago sativa* cv. Verko)

Variant	Volume ml	Number of seeds* or seedlings	Canavanine	
			µg	µM
Sand, non-sterile	30	500, 4 days	42	8
Sand, sterile (HgCl2)	30	500, 4 days	135	26
Sand, sterile (H2SO4)	30	500*, 1 day soaked	104	20
Sand, sterile (HgCl2)	50	500*, 1 day soaked	46	9
Sand, non-sterile	30	150, 4 days	16	3
Paper disk, non-sterile	10	500*, 1 day soaked	100	57

The experiments were carried out in plastic chambers (sandy beds) or Petri dishes (paper disk) under non-sterile or sterile conditions with the mentioned volumes water. The seeds were surface sterilized by using mercuric chloride or sulphuric acid, and washed 5 times with sterile water. After 1 day swelling or 4 days growth of seedlings the canavanine in the environment was estimated as described in 'Material and methods'.

Acknowledgement

I am grateful to Mrs M. Dübler for her skilful technical help.

References

1. Atkins, C.A., Pate, J.S. & Sharkey, P.J. (1975): Asparagine metabolism – key to the nitrogen nutrition of developing legume seeds. *Plant Physiol.* **56**, 807–812.

2. Bell, E.A. (1980): The non-protein amino acids of higher plants. *Endeavour, N.S.* **1**, 102–107.

3. Bleiler, J.A. & Rosenthal, G.A. (1988): Biochemical ecology of canavanine-eating seed predators. *Ecology* **69**, 427–433.

4. Downum, K.R., Rosenthal, G.A. & Cohen, W.S. (1983): L-Arginine and L-canavanine metabolism in jack bean, *Canavalia ensiformis* (L.) DC. and soybean, *Glycine max* (L.) Merr. *Plant Physiol.* **73**, 965–968.

5. Fearon, W.R. & Bell, E.A. (1955): Canavanine: detection and occurrence in *Colutea arborescens. Biochem. J.* **59**, 221.

6. Fujihara, S., Nakashima, T., Kurogochi, Y. & Yamaguchi, M. (1986): Distribution and metabolism of sym-homospermidine and canavaline in the sword bean (*Canavalia gladiata* cv. Shironata). *Plant Physiol.* **82**, 795–800.

7. Givan, C.V. (1979): Metabolic detoxification of ammonia in tissues of higher plants. *Phytochemistry* **18**, 375–382.

8. Gorski, P.M., Miersch, J. & Ploszynski, M. (1991): Production and biological activity of saponins and canavanille in alfalfa seedlings. *J. Chem. Ecol.* **17**, 1135–1143.

9. Grancharov, K., Krauβ, G.-J., Spassovska, N., Miersch, J., Maneva, L., Mladenova, J. & Golovinsky, E. (1985): Inhibitory effects of pyruvic acid semi- and thiosemicarbazone on the growth of bacteria, yeasts, experimental tumours and plant cells. *Pharmazie* **40**, 574–575.

10. Johnstone, J.M. (1956): Nitrogen metabolism in jack bean (*Canavalia ensiformis*). *Biochemistry* **64**, 21 P.

11. Lavin, M. (1986): The occurrence of canavanine in seeds of the tribe *Robiniae. Biochem. Syst. Ecol.* **14**, 71–74.

12. Mac Gregor, J.M., Taskovitsch, L. & Martin, W.P. (1961): Effect of phosphate and potash on amino acid content of alfalfa. *Agron. J.* **53**, 215–216.

13. Miersch, J., Jühlke, C. & Schlee, D. (1988): Zum Canavaninmetabolismus wahrend der Sämlingsentwicklung von Lucerne (*Medicago sativa* L. cv. Verko). *Wissensch. Beitrage der Martin Luther-Universität Halle-Wittenberg* **33** (S 65), 62–71.

14. Nakatsu, S., Haratake, S., Sakurai, J., Jo, N., Nishihara, J. & Hayashida, M. (1964): The change of the quantity of canavanine in leguminous plants in the process of germination, growth, and fructification. *Seikagaku (J. Jap. Biochem. Soc.)* **36**, 467–471.

15. Natelson, S. (1985): Canavanine in alfalfa (*Medicago sativa*). *Experientia* **41**, 257–259.

16. Rosenthal, G.A. (1972): Investigations of canavanine biochemistry in the jack bean plant, *Canavalia ensiformis* (L.) DC. II Canavanine biosynthesis in the developing plant. *Plant Physiol.* **50**, 328–331.

17. Rosenthal, G.A. (1977): The biological effects and mode of action of L-canavanine, a structural analogue of L-arginine. *Quart. Rev. Biol.* **52**, 155–178.

18. Rosenthal, G.A. (1977): Preparation and colorimetric analysis of L-canavanine. *Anal. Biochem.* **77**, 147–151.

19. Rosenthal, G.A. (1984): L-canavanine transport and utilization in developing jack bean, *Canavalia ensiformis* (L.) DC. (Leguminosae). *Plant Physiol.* **76**, 541–544.

20. Rosenthal, G.A. & Dahlman, D.L. (1982): A cautionary note on pentacyanoammoniumferrate use for determining L-canavanine occurrence in biological materials. *Experientia* **38**, 1034–1035.

21. Ta, T.C., Macdowall, F.D.H. & Faris, M.A. (1986): Excretion of nitrogen assimilates from N_2 fixed by nodulated roots of alfalfa (*Medicago sativa*). *Can. J. Bot.* **64**, 2063–2067.

22. Töpfer, R., Miersch, J. & Reinbothe, H. (1972): Untersuchungen zu Abbau von Canavanin in *Fabaceae*. *Biochem. Physiol. Pfl.* **161**, 231–242.

23. Tschiersch, B. (1959): Über Canavanin. *Flora* (Jena) **147**, 405–416.

24. Weber, E. (1986): *Grundriβ der biologischen Statistik*, 9. Aufl., Fischer-Verlag Jena.

Guanidino Compounds in Biology and Medicine, eds. by P.P. De Deyn, B. Marescau, V. Stalon and
I.A. Qureshi. ©1992 John Libbey & Company Ltd., pp. 53–60.

Chapter 8

Biologically active guanidine alkaloids from the Mediterranean sponge *Crambe crambe*

R.G.S. BERLINCK[1], J.C. BRAEKMAN[1]*, D. DALOZE[1], I. BRUNO[2], R. RICCIO[2], D. ROGEAU[3] and P. AMADE[3]

[1]*Laboratory of Bio-organic Chemistry, Faculty of Sciences, University of Brussels, 1050 Brussels, Belgium;*
[2]*Dipartimento di Chimica delle Sostanze Naturalli, Università di Napoli Federico II, Via D. Montesano 49, 80131
Naples, Italy;* [3]*Institut National de la Santé et de la Recherche Médicale; Unité 303, BP 3, La Darse, 06230
Villefranche-sur-Mer, France*

Summary

The isolation and structure determination of four new toxic guanidine alkaloids from the sponge *Crambe crambe* are
described. Crambine A, B and C$_1$ were isolated from *Crambe crambe* collected near Banyuls (France) while crambine
C$_2$, closely related to crambine C$_1$, was isolated from *Crambe crambe* collected off Favignana (Italy). A hypothetical
biosynthetic pathway of the crambines is proposed. Pure crambines A, B and C$_1$ were found to be ichthyotoxic for the
fish *Lebistes reticulatus*.

Introduction

Sponges are filter feeding, sessile invertebrates, present in almost all marine ecosystems. Since the early 1950s, sponge secondary metabolites have raised the interest of organic chemists. Today, nearly 1000 compounds have been isolated from about 300 different species of sponges. Many of these compounds present remarkable biological activities. Among them, acarnidine (I), isolated from the sponge *Acarnus erithacus*[8], is strongly antibacterial and cytotoxic.

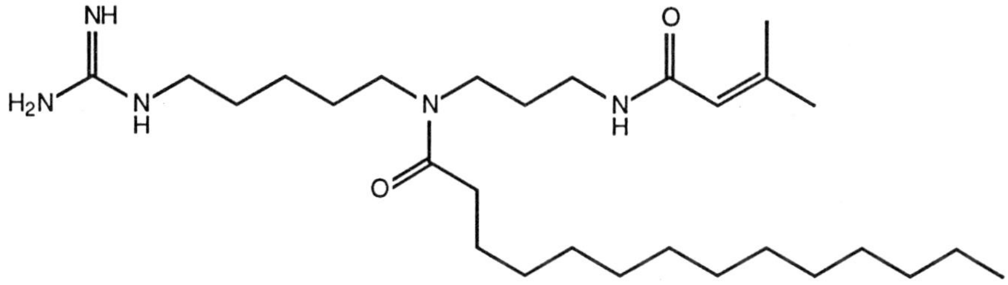

(I)

---

* *Principal author.*

Dysidenine (II) and isodysidenine (III) are two hexachlorinated metabolites isolated from the sponge *Dysidea herbacea*[13], that inhibit the iodide transport in dog's thyroid gland cells.

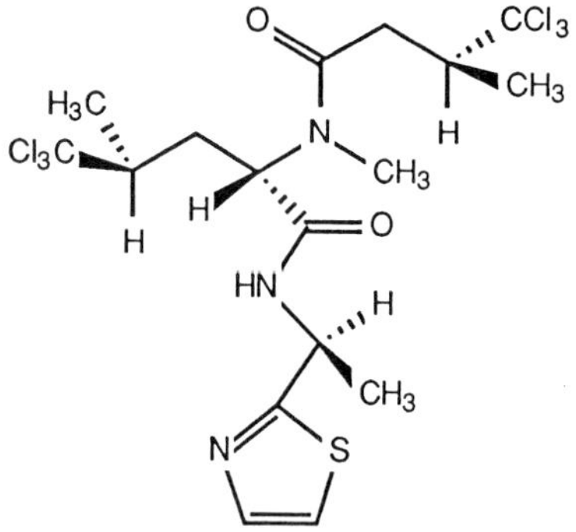

(II) 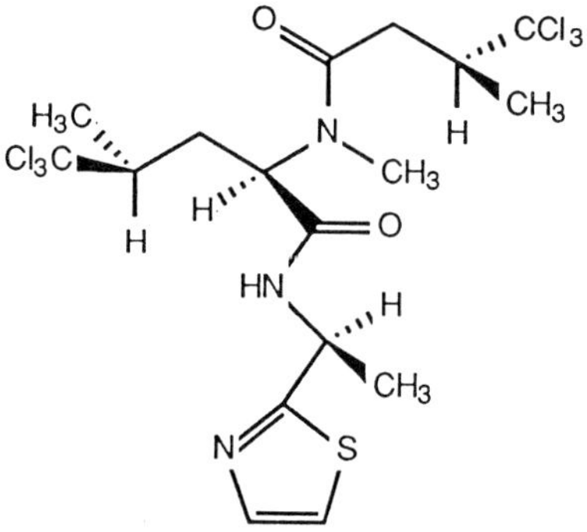   (III)

The dibromo-tyrosine derivative (IV), obtained from the Caribbean sponge *Aplysina fistularis*[9], is a structural and pharmacological hybrid of epinephrine and acetylcholine. It induced dual adrenergic activity in dogs, i.e. moderate increase in blood pressure, followed by a small short-lived decrease.

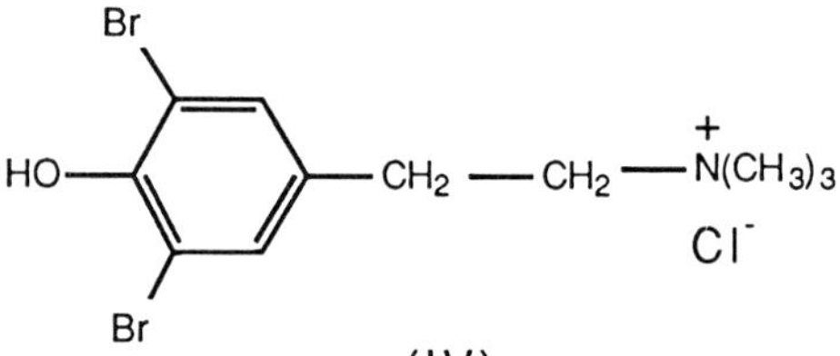

(IV)

Spongouridine (V) and spongothymidine (VI) are two nucleosides isolated from the sponge *Tethya crypta*. These compounds served as models for the synthesis of ARA-C (VII) and ARA-A (VIII), which are in clinical use, the former as an antitumour agent and the latter as an antiviral agent[7].

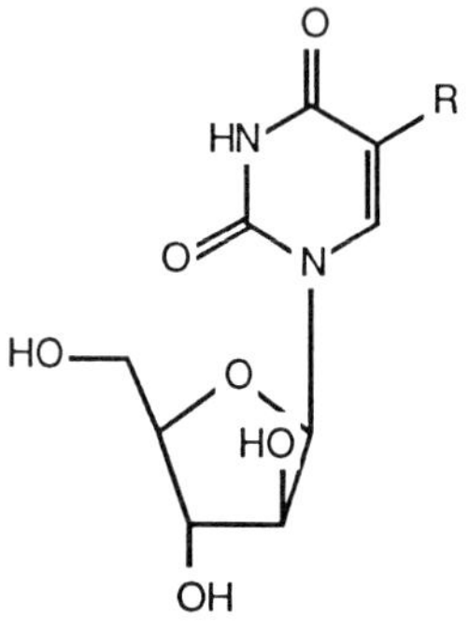

(V) R=H   (VII)   (VIII)
(VI) R=CH$_3$

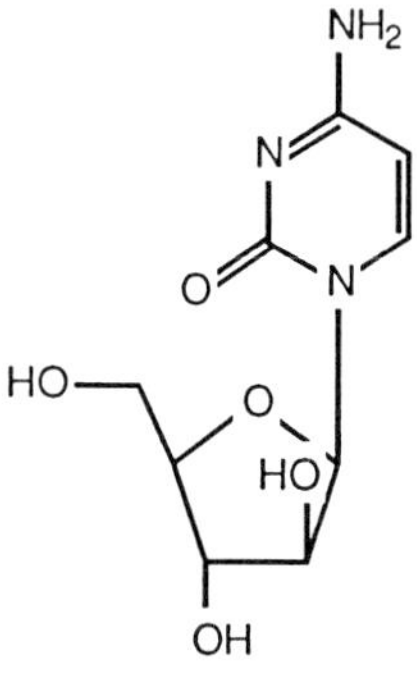

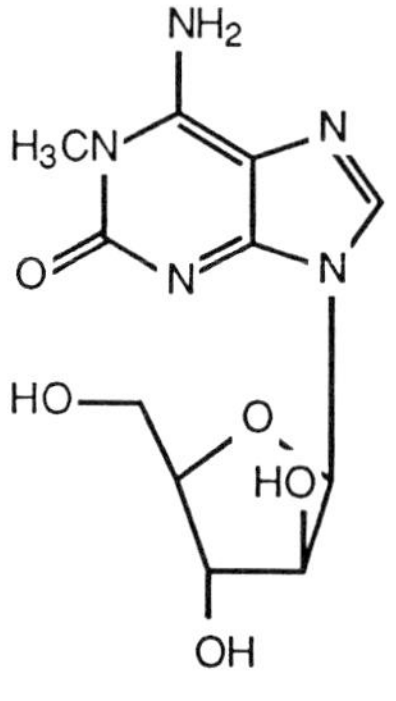

This short list, which could easily have been expanded, shows that sponges are a remarkable source of secondary metabolites with original structures and, in some cases, interesting pharmacological properties. This has stimulated our interest in the chemical study of marine sponges. We report here

54

on the structure of four new toxic compounds isolated from the Mediterranean sponge *Crambe crambe*.

It is a bright-red encrusting sponge, common in shallow waters of Mediterranean coasts. Preliminary investigations of organic extracts of *Crambe crambe* showed their high level of ichthyotoxicity, cytotoxicity[12] and inhibition of development of microorganisms[1,4]. Of 18 species of Mediterranean sponges tested, *Crambe crambe* was the only one that inhibited *Candida albicans*[4]. In another series of tests, *Crambe crambe* extracts were the most active (among 28 species of sponges tested) against 10 Gram-positive and Gram-negative terrestrial bacteria, 10 marine bacteria, 15 human pathogenic fungi and five phytopathogenic fungi[1].

Materials and methods

Specimens of *Crambe crambe* collected near Banyuls, France, were stored in methanol and exhaustively extracted with dichloromethane/methanol 1:1. After evaporation of the dichloromethane, the methanol/water suspension was extracted successively with hexane and tetrachloromethane. The methanol of the resulting aqueous phase was evaporated and the resulting water suspension extracted with n-butanol. This n-butanol extract (4.1 g) possessed most of the ichthyo- and cytotoxic activities.

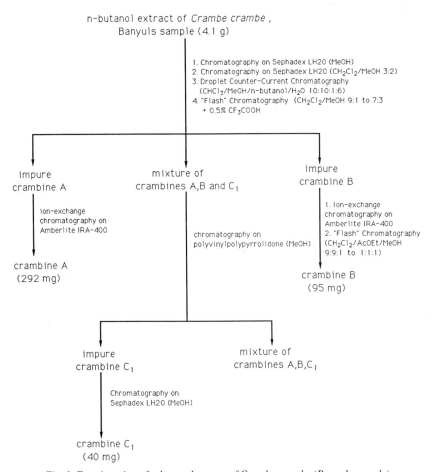

Fig. 1. Fractionation of n-butanol extract of Crambe crambe (Banyuls sample).

It was subjected to the fractionation procedure shown in Fig. 1. The separations were monitored by the ichthyotoxicity test against *Lebistes reticulatus*[4].

Results

Crambine A is a glassy solid $\{[\alpha]_D = +2°$ (MeOH, c = 0.7)$\}$. Its UV spectrum showed absorption λ_{max} 205 (10.600) and 288 nm (5.500) and its infra-red spectrum (NaCl film) at 3200–3120, 1694 and 1680–1650 cm^{-1}. High resolution fast atom bombardment mass spectrometry (HRFAB$^+$MS) of crambine A displays a quasi-molecular ion at m/z 449.3605 (calculated for $C_{24}H_{45}N_6O_2$: 449.3604) which corresponds to the molecular formula $C_{24}H_{44}N_6O_2$. This molecular formula was confirmed by the FAB$^-$MS which exhibits peaks at m/z 483 and 485 [3:1 ratio (M–H + HCl)$^-$] and 519, 521 and 523 [10:6:1 ratio, (M–H+2HCl)$^-$].

The ^{13}C-nuclear magnetic resonance (NMR) spectra of crambine A (broad band decoupling and distortionless enhancement by polarization transfer) confirm the presence of 24 carbon atoms in the major homologue. A quaternary carbon signal at 166.4 was assigned to the carbonyl of an ester group which, according to the IR and UV spectra should be conjugated. The presence of six nitrogen atoms together with the positive reaction to Sakaguchi reagent and quaternary carbon signals at δ 153.3 and 158.9 suggest that the compound contains two guanidine moieties. The two remaining sp^2 carbon signals at δ 153.0 and 103.6 may be assigned to a β-enaminoester moiety (Fig. 2).

The ^{1}H-NMR spectrum in dimethylsulphoxide (DMSO) indicates the presence of seven exchangeable protons, thus suggesting that crambine A was isolated as an hydrochloride after ion exchange chromatography. Analysis of its ^{1}H-NMR and correlated spectroscopy (COSY) ^{1}H-^{1}H spectra in

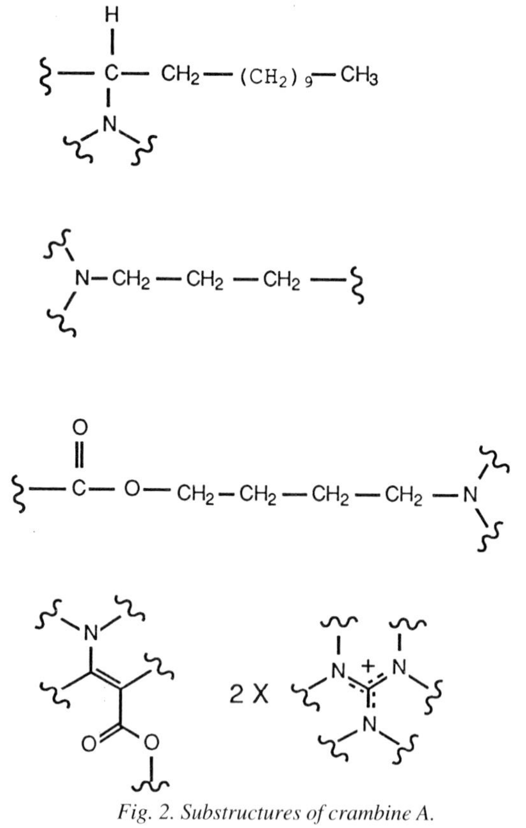

Fig. 2. Substructures of crambine A.

56

CD_3OD shows the presence of three separate spin systems: (a) a long alkyl chain with a terminal methyl group and a methine (δ 4.42) at the other end; (b) four methylenes in an open chain linked one side to an oxygen atom (H_2C-5: δ_H 4.22, δ_C 65.4) and to an NH group on the other side (H_2C-2: δ_H 3.24, δ_C 42.3). This methylene is coupled with the exchangeable proton at δ 7.93, as evidenced by decoupling experiments in DMSO; (c) three methylene groups involved in a ring system. These substructures are depicted in Fig. 2.

Heteronuclear multiple bond connectivity (HMBC) NMR experiments[2,11] permit us to connect all these substructures. Particularly relevant were the correlations observed between H-13 (δ 4.42) and the carbon atoms at δ 166.4 (C-6), 103.6 (C-7), 152.9 (C-8), 153.3 (C-12), 37.7 (C-14), 25.4 (C-15) and between H_2-14 and the signals at δ 103.6 (C-7), 51.3 (C-13), 25.4 (C-15) and 30.6 (C-16), allowing *inter alia* to make a distinction between two and three bond correlations with H-13. This led us to propose structure (IX) for crambine A (n = 9).

The spectral properties of crambine B indicate that its structure is closely related to that of crambine A. Nevertheless, crambine B does not present any UV absorption above 200 nm, and its IR spectrum shows, among others, a band at 1730 cm^{-1}, suggesting that the ester function is not conjugated. The

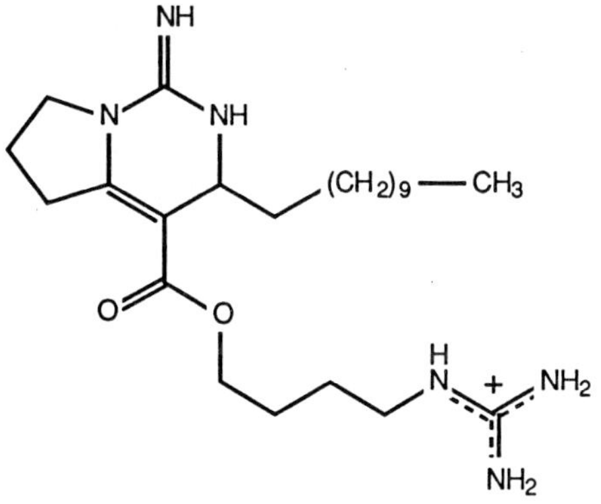

(IX): Crambine A

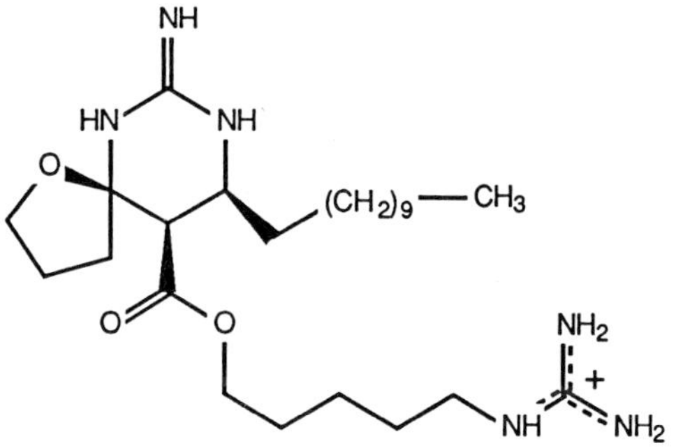

(X): Crambine B

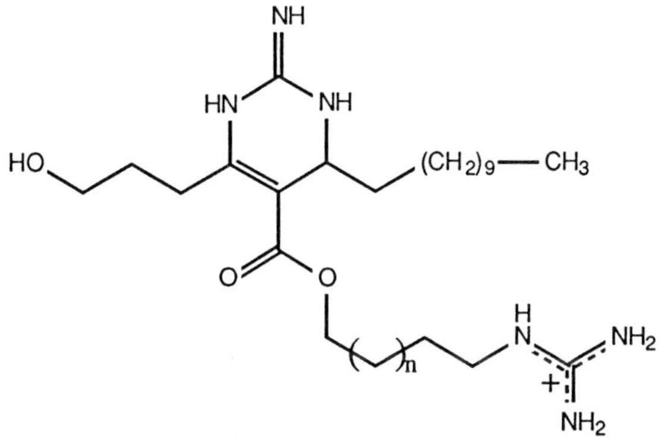

(XI): Crambine C_1, n=2
(XII): Crambine C_2, n=1

57

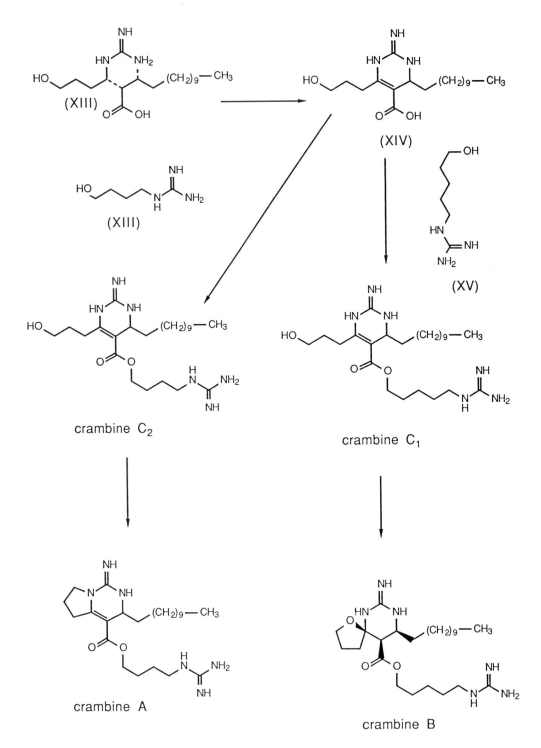

Fig. 3. Hypothetical biosynthetic pathway of the crambines.

FAB$^+$ mass spectrum displays a quasi-molecular ion at m/z 481. High resolution measurements (meas.: 481.3863; calc. for $C_{25}H_{49}N_6O_3$: 481.3866) point to the molecular formula $C_{25}H_{48}N_6O_3$, indicating that crambine B possesses one methylene group and one molecule of water more than crambine A. Crambine B has also three separate spin systems, and detailed analysis of its ^{1}H, ^{13}C, ^{1}H-^{1}H COSY, ^{1}H-^{13}C heteronuclear correlation (HETCOR) and HMBC NMR spectra led us to propose structure (X) for crambine B. Crambine B differs from crambine A by the length of the guanidinoalkyl chain linked to the ester group, and by the fact that the three methylene groups (H_2C-9 to H_2C-11) are now involved in a tetrahydrofurane ring. The relative configurations at C-8, C-7 and C-13 followed from nuclear Overhauser enhancement (nOe) experiments.

Crambine C_1 is an isomer of crambine B. Its quasi-molecular ion at m/z 481.390 [(M+H)$^+$, calculated for $C_{25}H_{49}N_6O_3$: 481.387] corresponds to the same molecular formula as crambine B, $C_{25}H_{48}N_6O_3$. Detailed analysis of its ^{1}H and ^{13}C NMR spectra also shows the presence of the three spin systems as crambines A and B, but the three methylene groups (C-9 to C-11) are now in an open chain, with a primary alcohol function at the end. Crambine C_1 possesses the same enaminoester chromophore as crambine A and the same guanidinoalkyl chain as crambine B. All these data led us to establish the structure of crambine C_1 as XI.

From specimens of *Crambe crambe* collected off Favignana (Italy), a fourth compound, crambine C_2, closely related to crambine C_1 could be isolated in minute amounts[3]. Analysis of its spectral properties clearly indicated that crambine C_2 (XII) differs from crambine C_1 only by the length of the guanidinoalkyl chain linked to the ester group. It has four methylene groups, as in crambine A.

Discussion

Biogenetically, the crambines could arise from the condensation of a fatty acid with alcohol (XIII) to give the intermediate (XIV). The alcohol (XIII) is a reduced form of γ-guanidinobutyric acid, a degradation product of arginine that is encountered in many marine invertebrates[6]. The esterification of the intermediate (XIV) by a second molecule of γ-guanidinobutanol (XIII) gives crambine C_2. Further cyclization to form the pyrrolidine ring, could explain the formation of crambine A. On the other hand, crambine B and C could arise from the esterification of (XIV) by δ-guanidinopentanol (XV), as depicted in Fig. 3. The fact that we have found two homologous guanidinoalkyl chains is

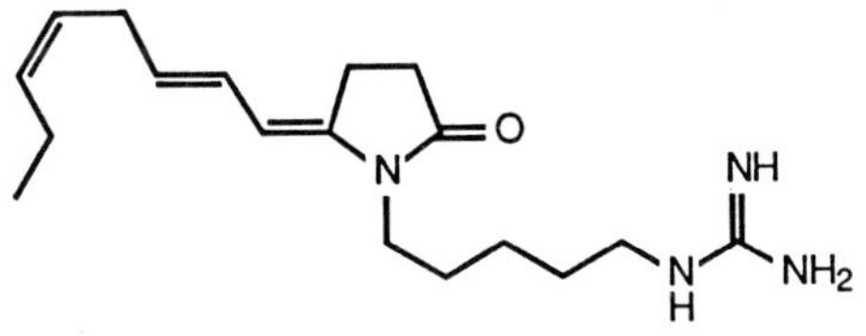

(XVI)

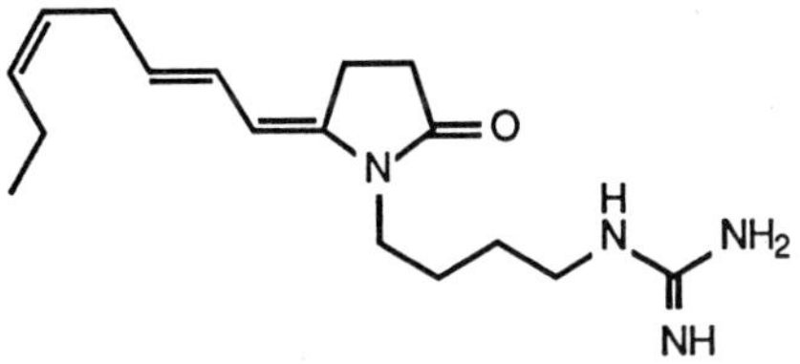

(XVII)

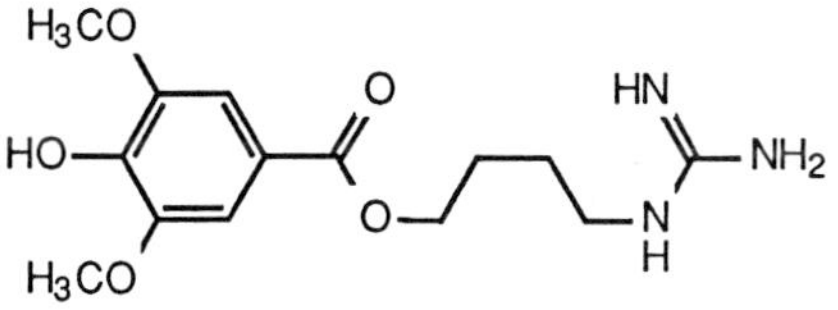

(XVIII)

59

not without precedent. Indeed, the pyrrolo lactam ring of the polyandrocarpidines A (XVI) and C (XVII) are alkylated by agmatine and homoagmatine, respectively[5]. The only naturally-occurring γ-guanidinobutanol derivative reported till now is the terrestrial plant secondary metabolite, leonurine (XVIII)[10].

Pure crambines A, B and C_1 are ichthyotoxic at the level of 2 ppm for the fish *Lebistes reticulatus*. Tests of cytotoxicity showed that crambine A is the most active of the series. All of them present antibacterial activities (E. Richelle and G. Van de Vyver, unpublished results).

Acknowledgements

We would like to thank Drs. G. Van de Vyver and N. Boury-Esnault for sponge collection and identification. Mr. C. Maerschalk for the NMR spectra at 600 MHz and Dr. M. Herin (Searle) for the FAB MS. One of us (RGSB) is indebted to the 'Conselho Nacional de Desenvolvimento Cientifico e Tecnologico' for financial support (proc. 200383/88.4) and to the 'Centro Pluridisciplinar de Pesquisas Quimicas, Biologicas e Agricolas' of UNICAMP (Campinas, Brazil). This work was supported by FRFC, Belgium (n° 2.4513.85 and 2.4554.87), by a NATO grant for collaboration research (Ref. D.210/86) and by CNR, Italy (n° 89.03744.03).

References

1. Amade, P., Charroin, C., Baby, C. & Vacelet, J. (1987): Antimicrobial activities of marine sponges from the Mediterranean sea. *Mar. Biol.* **94**, 271–275.

2. Bax, A. & Summers, M.F. (1986): ^1H and ^{13}C assignments from selectivity-enhanced detection of heteronuclear multiple-bond connectivity by 2D multiple quantum NMR. *J. Am. Chem. Soc.* **108**, 2093–2094.

3. Berlinck, R.G.S, Braekman, J.C., Daloze, D., Bruno, I., Riccio, R., Rogeau, D. & Amade, P. (1992): Crambines C_1 and C_2: two further toxic guanidine alkaloids the sponge *Crambe crambe*. *J. Nat. Prod.* **55**, 528–532.

4. Burkholder, P.R. & Ruetzler, K. (1969): Antimicrobial activity of some marine sponges. *Nature* **222**, 983–984.

5. Carté, B. & Faulkner, D.J. (1982): Revised structures for the polyandrocarpidines. *Tetrahedron Lett.* **23**, 3863–3866.

6. Chevolot, L. (1981): In *Marine natural products*, Vol. IV., ed. P.J. Scheuer, pp. 54. New York: Academic Press.

7. Ireland, C.M., Roll, D.M., Molinski, T.F., McKee, T.C., Zabrinskie, T.M. & Swersey, J.C. (1988): In *Biomedical importance of marine organisms*, n° 13, ed. D.G. Fautin, pp. 46. San Francisco: Memoirs of the California Academy of Sciences.

8. Munro, M.H.G., Luidbrand, R.T. & Blunt, J.T. (1987): In *Bioorganic marine chemistry* Vol. 1, ed. P.J. Scheuer, p. 100. Berlin: Springer-Verlag.

9. Munro, M.H.G., Luidbrand, R.T. & Blunt, J.T. (1987): In *Bioorganic marine chemistry* Vol. 1, ed. P.J. Scheuer, p. 146. Berlin: Springer-Verlag.

10. Reuter, G. & Diehl, H.J. (1971): Guanidin derivate in *Leonurus sibiricus* L. *Pharmazie* **26**, 777.

11. Summers, M.F., Marzilli, L.G. & Bax, A. (1986): Complete ^1H and ^{13}C assignments of coenzyme B_{12}, through the use of new two-dimensional NMR experiments. *J. Am. Chem. Soc.* **108**, 4285–4294.

12. Van de Vyver, G., Huysecom, J., Braekman, J.C. & Daloze, D. (1990): Screening and bioassays for toxic substances in sponges from western Mediterranean sea and north Brittany. *Vie Milieu* **40**, 285–292.

13. Van Sande, J., Deneubourg, F., Beauwens, R., Braekman, J.C., Daloze, D. & Dumont, D. (1990): Inhibition of iodide transport in thyroid cells by dysidenin, a marine toxin, and some of its analogs. *Mol. Pharm.* **37**, 583–589.

Section II

Metabolic, physiological and nutritional importance of arginine

Guanidino Compounds in Biology and Medicine, eds. by P.P. De Deyn, B. Marescau, V. Stalon and I.A. Qureshi. ©1992 John Libbey & Company Ltd., pp. 63–70.

Chapter 9

Arginine synthesis along the mammalian nephron

Olivier LEVILLAIN[1], Annette HUS-CITHAREL[2], François MOREL[2] and Lise BANKIR[1]

[1]*INSERM, Unité 90, Hôpital Necker, 75743 Paris Cedex 15, France;* [2]*Laboratoire de Physiologie Cellulaire, CNRS URA 219, Collège de France, 75231 Paris Cedex 05, France*

Summary

The importance of the kidney in producing arginine for whole body protein synthesis and metabolism is well recognized but the nephron segments responsible for this synthesis have not been identified. A micro-method was designed to measure the production of arginine in single pieces of the rat nephron isolated by microdissection from collagenase treated kidneys. Pieces of tubules were incubated with L-[ureido-^{14}C]-citrulline and aspartate in the incubation medium in a sealed glass chamber. Arginase and urease were added to the incubation medium so as to hydrolyse the newly formed arginine and to release $^{14}CO_2$. The $^{14}CO_2$ was trapped in a KOH droplet and subsequently counted. In the rat kidney, these experiments showed that arginine is synthesized only in the proximal tubule. Arginine synthesis decreases from the convoluted part (PCT) to the terminal straight part (PST): 122 ± 15 fmoles/min per mm tubular length in PCT, 71 ± 6 in cortical PST (CPST), and 41 ± 4 in PST of outer medulla (OSPST). In the mouse and the rabbit, similar results were obtained. The main site of arginine synthesis was the proximal tubule, but arginine was also synthesized in the thick ascending limb of the rabbit. Arginine synthesis was higher in PCT than in PST. Comparison of arginine synthesis in the PCT of the three species showed that the rabbit PCT has the lowest production of arginine: about three- to fourfold lower than in the rat or the mouse PCT. In these three species, arginine synthesis was higher in the first mm of the PCT than in its later part. Arginine synthesis was strongly dependent on precursor concentration in the medium, showing a 58 per cent reduction when citrulline concentration was decreased by half and an increase of 44 per cent or 136 per cent when citrulline concentration was doubled or quadrupled, respectively. This observation suggests that arginine formed in the convoluted part of the proximal tubule contributes to the maintenance of the whole body pool of arginine. Arginine synthesized in the PCT may also available, *in situ*, for different metabolic pathways, e.g. guanidino compound and nitric oxide synthesis.

Introduction

It is well known that arginine is synthesized, in large amounts, in the liver. Research of the localization of the enzymes involved in arginine synthesis was performed in several organs. The results showed that these enzymes were presents in the kidney and the brain. Arginine synthesis in the mammalian kidney was demonstrated at the beginning of this century. Kidney slices incubated with citrulline and amino acids produced arginine. When the amino acids were either glutamine or aspartate, arginine synthesis was significantly increased[1]. Then, Ratner *et al.* clearly showed that arginine synthesis in kidney homogenates required citrulline and aspartate to produce equivalent amounts of arginine and malic or fumaric acid[24]. The first step of arginine synthesis is achieved by the condensation of citrulline and aspartate. This enzymatic step, catalysed by argininosuccinate synthetase, ASS (EC 6.3.4.5), requires energy in the form of ATP. The intermediate product obtained (argininosuccinate) is split by argininosuccinate lyase, ASL (EC 4.3.2.1), to form arginine and fumaric acid[23]. Comparison of catalytic, physical and immunological properties of renal and hepatic ASS showed that the enzyme was similar in the two organs[3]. The liver and the kidney are both capable of synthesizing arginine. Citrulline availability is the limiting factor of arginine synthesis in both organs.

In the liver, the activity of ornithine carbamoyl transferase, OTC (EC 3.5.1.2), is very high. Thus, hepatocytes synthesize large amounts of citrulline but its conversion to arginine is reduced because of the low amount of arginine synthetase complex (ASS + ASL)[6,9,11,25]. The citrulline which is not converted into arginine is released in the bloodstream. In contrast, arginine formed in hepatocytes is rapidly hydrolysed into urea and ornithine because these cells exhibit a very high arginase activity (EC 3.5.3.1)[25]. The fate of arginine in the liver is to be hydrolysed as previously described. Because the liver extracts arginine from the blood but does not release it in the blood, the main source of arginine for the whole body is the kidney.

Nevertheless, the kidney is unable to synthetize citrulline and the main source of citrulline is the small intestine. Several metabolic steps are required to synthesize citrulline. The small intestine possesses the enzymes which convert glutamine into citrulline[32,34]. Glutamine is converted into δ-pyrroline-5-carboxylate and leads to ornithine[10,32]. Carbamoyl phosphate synthetase (EC 2.1.3.3) and OTC activities have been detected in the small intestine[32,33]. Thus, ornithine is converted into citrulline which is released into the bloodstream. Because of the low permeability to citrulline, hepatocytes do not remove citrulline from the blood and the citrulline synthesized by the small intestine is available for the kidney.

Featherston et al. clearly showed the importance of the kidney in arginine synthesis for the whole body[7]. Recently, measurements of arteriovenous differences through the kidney for arginine and citrulline showed that in the normal rat and in man, citrulline disappearance was of the same order of magnitude as arginine synthesis[5,30]. However, the previous studies did not reveal the exact sites of arginine synthesis in the mammalian kidney. The kidney is composed of numerous nephrons each of them comprising different successive segments characterized by different cell types. The purpose of this work was to identify the site(s) of arginine synthesis in the kidney. To achieve this study, arginine synthesis was measured with a micro-method in pieces of the different nephron segments obtained by microdissection.

Material and methods

Measurement of arginine production was based on the use of the radiolabelled precursor, L-ureido-^{14}C-citrulline. In the cells of microdissected tubules, labelled citrulline and cold aspartate, in the presence of tissue arginine synthetase (ASS + ASL), leads to L-guanidino-^{14}C-arginine. Arginine is released in the incubation medium, to which we previously added commercial arginase (EC 3.5.3.1) and urease (EC 3.5.1.5). Thus, arginine was split into ornithine and ^{14}C urea by exogeneous arginase. Then, ^{14}C urea was cleaved into $2NH_3$ and $^{14}CO_2$ by urease. CO_2 was trapped in potassium hydroxide and quantitated by liquid scintillation counting (Figs. 2 and 3).

The experiments were performed in adult male rats (Sprague Dawley), mice (OF1) and rabbits (New Zealand). Animals were anaesthetized with Nembutal®. The left kidney was prepared for microdissection as previously described[16,18]. The kidney was perfused via the abdominal aorta, first with cold NaCl 0.9 per cent solution to wash out the blood and reduce cellular functions, then with an appropriate collagenase solution (2.5 mg/ml for rat and mouse and 1 mg/ml for rabbit) in Hanks salts supplemented with energetic substrates[18]. The kidney was sliced and representative pyramids were incubated at 30 °C for a period of 20–30 min in a collagenase solution (microdissection of tubules from cortex and outer medulla) or for 60 min (for inner medulla).

Microdissection of the different tubules was performed at 0–4 °C under stereo-microscopic observation, using the usual anatomical and morphological criteria. Pieces of the following segments were obtained, Fig. 1: glomerulus (G); proximal convoluted tubule (PCT); straight part of the proximal

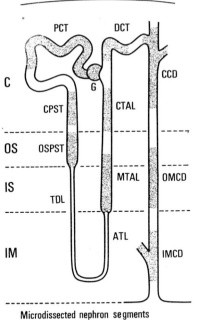

Microdissected nephron segments

Fig. 1. Localization of the tubular segments microdissected for the study of arginine synthesis along the rat nephron is shown by the shaded areas. G = glomerulus; PCT = proximal convoluted tubule; CPST and OSPST = straight part of the proximal tubule in the medullary rays of the cortex and in the outer stripe of the outer medulla, respectively; MTAL and CTAL = thick ascending limb of Henle's loop in the outer medulla and in the cortex respectively; DCT = distal convoluted tubule; CCD = cortical collecting duct; OMCD = medullary collecting duct in the outer medulla; IMCD = inner medullary collecting duct (reproduced from[18]).

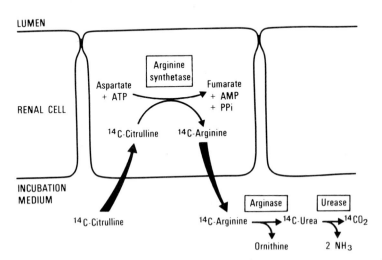

Fig. 2. Localization of the reactions involved in the production and measurement of arginine from ^{14}C-labelled citrulline. The amount of arginine produced during the incubation was evaluated from the counts of $^{14}CO_2$ recovered when arginase and urease were added to the incubation medium. Thick arrows indicate transport of citrulline and arginine. Thin arrows indicate reactions involved in the enzymatic synthesis and measurement of arginine.

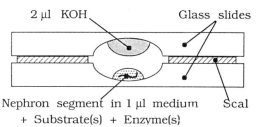

2 μl KOH   Glass slides

Nephron segment in 1 μl medium   Scal
+ Substrate(s) + Enzyme(s)

Fig. 3. Diagram of the incubation system. The nephron segments are in the incubation medium containing ^{14}C-labelled citrulline, arginase and urease (lower glass side). The newly formed ^{14}C arginine is split into ^{14}C urea, then urea is hydrolysed into 2 NH_3 and $^{14}CO_2$ which is continuously trapped in KOH (upper glass slide). The two glass slides are seal with vaseline.

tubule taken from the medullary rays of the cortex (CPST) and from the outer stripe of the outer medulla (OSPST); thick ascending limb from the outer medulla (MTAL) and from the cortex (CTAL); distal convoluted tubule (DCT); cortical collecting duct (CCD); medullary collecting duct from the outer medulla (OMCD); inner medullary collecting duct (IMCD) taken at variable depths of the inner medulla.

After microdissection, the glomeruli or tubules were transferred onto a siliconized and BSA-coated hollow glass slide in 0.5 μl incubation medium, tightly sealed with a cover glass slide (Fig. 3), and photographed for the subsequent determination of their length.

Incubation was started by adding to the 0.5 μl droplet containing the tubule, another 0.5 μl of incubation solution containing L-[ureido ^{14}C]-citrulline (≈230 Bq or 6.3 nCi per sample), arginase (230 U/ml) and urease (248 U/ml). Labelled citrulline was used without carrier to produce sufficient counts. Final citrulline concentration during incubation was 108 μM, a concentration in the range of plasma concentration *in vivo*. The samples were sealed by a cover glass slide containing a 2 μl droplet of an isoosmotic KOH solution. After 1 h incubation, in a water bath maintained at 37 °C, the cover slide was removed and the KOH, containing the $^{14}CO_2$ trapped during the incubation, was transferred into a counting vial. The amount of arginine formed during incubation of individual nephron segments was deduced from the amount of $^{14}CO_2$ counted in the KOH.

Results and discussion

The purpose of this study was to identify the sites of arginine synthesis in the mammalian nephron and to compare the results in different species: rat, mouse and rabbit. Unfortunately, thin descending and thin ascending limbs in the rat, and IMCD in the rabbit could not be obtained during microdissection. Our results showed that, in the rat, only the proximal tubule synthesizes arginine (Fig. 4). Neither glomerulus nor distal tubule (MTAL, CTAL, and DCT) and collecting duct (CCD, OMCD, and IMCD) synthetize arginine. Along the proximal tubule, arginine production is heterogeneous. The main site of arginine production is the convoluted part of the proximal tubule (125–200 fmoles/min.mm). The pars recta of the proximal tubule synthesizes more arginine in its cortical than in its medullary part, 72 ± 6 *vs* 41 ± 4 fmoles/min.mm, respectively.

Previous works have attempted to localize ASS and ASL in the kidney. Morris *et al.* demonstrated that mRNA of both enzymes were found exclusively in the rat cortex and colocalized with phosphoenolpyruvate carboxykinase, (EC 4.1.1.32) a gluconeogenic enzyme known to be restricted only to the proximal tubule[21]. The same group confirmed the proximal localization of these enzymes in the mouse kidney by an immunohistochemical technique[22]. In spite of the initial report of Szepesi[29] that 86 per cent of arginine synthetase activity was found in the medulla of the rat kidney, it seems reasonable to conclude from the two other reports[21,22] and from the present experiments that arginine synthesis takes place only in the proximal tubule.

We investigated whether the results obtained in the rat kidney were observed in other laboratory

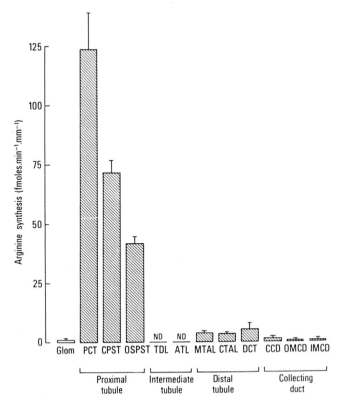

Fig. 4. Pattern of arginine synthesis along the rat nephron. For abbreviations, see Fig. 1. Mean ± SEM of 4–6 experiments (of two only for DCT). ND = not determined, (reproduced from[18]).

animals. The same experiments were performed in mice and rabbits, and the results clearly show that arginine is also synthesized mostly in the proximal tubule in these two species (Table 1). The rate of arginine production is about three- to fourfold higher in the mouse and the rat than in the rabbit PCT. In the three species, arginine synthesis is higher in the convoluted than in the straight part of the proximal tubule. However, arginine synthesis increases from the CPST to the OSPST in mouse and rabbit whereas it decreases in rat.

Table 1. Rate of arginine synthesis in individual nephron segments of the rat, mouse and rabbit kidney

		PCT	CPST	OSPST	MTAL	CTAL
Rat[a]	(6)	123 ± 15	72 ± 6	41.3 ± 3.6	3.8 ± 0.5	3.6 ± 0.6
Rat[b]	(4)	188 ± 9	61.1 ± 5.1	33.0 ± 6.0	–	–
Mouse	(6)	191 ± 28	21.2 ± 3.4	48.2 ± 7.9	1.1 ± 0.2	1.0 ± 0.1
Rabbit	(5)	57.0 ± 3.2	14.4 ± 4.0	22.7 ± 3.6	4.2 ± 0.7	2.3 ± 0.4

Values are means ± SEM in fmoles.min^{-1}.mm^{-1}. Number of animals is given in parentheses.
[a]First series of experiments, same data as that of Fig. 3 in [18].
[b]Second series of experiments, same data as that of Fig. 8 in [18].

We verified whether arginine synthesis was also heterogeneous along the proximal convoluted tubule. In rats, arginine synthesis was studied in two groups of proximal convoluted tubule: (1) PCT immediately adjacent to the glomerulus (early PCT: the first 0.6 mm of the PCT), (2) PCT taken at random but most probably more distant from the glomerulus (random PCT). The results clearly

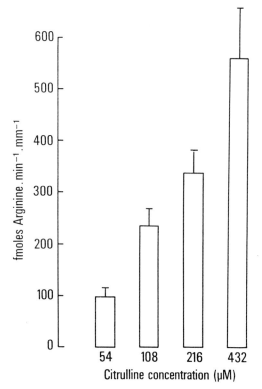

Fig. 5. Relationship between arginine synthesis and citrulline concentration in the medium. PCT segments of rat kidney were incubated for 60 min at 37 °C with citrulline concentrations of 54, 108, 216, or 432 μM. Means ± SEM of four rats (10 samples per rat for each citrulline concentration).

showed that arginine synthesis was higher in the early than the more distant part of the PCT; values are: 328 ± 35 vs 207 ± 22 fmoles Arg/min.mm, respectively, $P < 0.01$[18]. The pattern of arginine synthesis along the proximal convoluted tubule is similar to the pattern of reabsorption of filtered citrulline along the PCT. Citrulline transport across the epithelium is high in the first millimetre of the PCT[12,13,27] and, in normal physiological conditions, the luminal concentration of citrulline decreases along the PCT and probably falls to zero in the CPST and OSPST. Citrulline supply for these segments originates only from the basolateral side.

Along the mouse PCT, arginine synthesis decreased exponentially from the glomerulus to the end of the PCT. In the rabbit PCT, the profile of arginine synthesis was different from that of the rat and the mouse. A decrease of 50 per cent was observed between the first and the second millimetre of the PCT, then the production remained at the same level in the second and the third millimetre of the PCT.

Arginine production, in the thick ascending limb (both medullary and cortical) was 2–3 per cent of that observed in the PCT, in the rat and 0.6 per cent in the mouse. In the rabbit, MTAL and CTAL arginine production was 7 per cent and 4 per cent, respectively (Table 1). The functional significance of this very low arginine production in the straight distal tubule is not yet understood. We recently showed that arginase activity was present in the rabbit MTAL[19]. It is possible to assume that arginine is required in this segment to produce urea or ornithine.

Normal plasma level of citrulline is about 100 μM[26]. Several works showed that citrulline concentration may vary *in vivo*, especially during kidney diseases. Citrullinaemia increases three- to fourfold during chronic renal failure[4,28]. Influence of citrulline concentration on arginine synthesis was studied in three additional experiments. Four concentrations of citrulline ranging from 0.05 to 0.4 mM (i.e. spanning physio-pathological changes in the rat) were tested by varying the amount of radioactive citrulline added to the medium. Aspartate concentration in the incubation medium was also raised.

The relationship between arginine synthesis and precursor concentration was shown in Fig. 5. When citrulline concentration was 108 μM, as in other experiments, arginine production was 235 ± 32 fmoles per min and mm length. It rose to 337 ± 45 and 560 ± 97 (+ 44 per cent and + 136 per cent) for citrulline concentrations of 216 and 432 μM, respectively. Reducing citrulline concentration by half reduced arginine synthesis to 99 ± 15 fmoles per min and mm length (–58 per cent). These results show that arginine production is strongly dependent on citrulline concentration, *in vitro*. The same results were observed by Dhanakoti *et al.* in suspensions of rat cortical tubules. In addition, these authors increased plasma citrulline concentration, *in vivo*, and observed an almost parallel increase in renal arginine production[5]. Thus, *in vivo*, arginine synthesis is most probably regulated by the circulating concentration of citrulline.

The fate of arginine formed in the kidney probably depends on the metabolic pathways occurring in the different cell types. The cells of the PCT synthesize large amounts of arginine which are transported across the basolateral membrane and enters in the venous bloodstream. This arginine will be available for the different needs of the organs: protein synthesis, growth, intermediary metabolism. In the straight proximal tubule, the fate of arginine is probably different. We recently reported that 40 per cent and 60 per cent of the arginine produced by CPST and OSPST, respectively, were immediately hydrolysed into ornithine and urea[18]. This is explained by the presence of an arginase activity in these cells[17]. Consequently, the quantity of arginine released in the blood by the PST is probably negligible when compared to that released by the PCT.

Recently, arginine was shown to be the precursor of nitric oxide (NO) synthesis. Nitric oxide is an important mediator of cellular communication and is responsible for different physiological effects[20]. NO is a strong relaxing factor and modulates the vascular smooth muscle tone. Inhibition of NO formation leads to a very prompt increase in systemic blood pressure thus demonstrating the basal role of NO controlling peripheral resistance. Because arginine is produced in abundance in the kidney, it is possible to assume that NO formation could be more important to regulate haemodynamics in the kidney than in other organs and vascular beds. This is indeed suggested by the work of Walder *et al.*. They observed that inhibition of NO production by a low dose of N_{ω}-nitro-L-arginine methyl ester reduced renal cortical blood flow considerably without increasing blood pressure[31].

Arginine is also involved in the biosynthesis of other compounds such as glycocyamines. The kidney was shown to synthesize these molecules[2]. Glycocyamines or guanidino compounds originate from the condensation of the amidine group of arginine on glycine[8]. Among guanidino compounds, guanidinoacetate is of interest because its methylation leads to creatine biosynthesis[8,14,15]. Creatine may be phosphorylated into phosphoryl creatine, an energy-rich metabolite. Hydrolysis of this compound gives creatine and ATP. This is a possible storage of energy for the kidney in order to meet energy requirements during anoxia or critical situations. Up to date, the exact site of guanidino compound synthesis in the kidney remains unknown.

In conclusion, our results show that, in the rat, mouse and rabbit, arginine synthesis is much higher in the PCT than in PST and negligible in other segments of the nephron (Table 1). Taking into account the fact that PCT is much longer than the PST and that its early part shows the highest arginine synthetic rate, the PCT is responsible for most of the arginine released in the renal venous blood. Metabolic pathways of arginine and guanidino compounds, in the different organs, depend on the availability of citrulline and its transformation into arginine. Whole body arginine supply should thus be critically affected by impairment of PCT cell function by drugs or diseases.

Acknowledgement: O. Levillain received a scholarship from the Foundation de la Recherche Médicale.

References

1. Borsook, H. & Dubnoff, J.W. (1941): The conversion of citrulline to arginine in kidney. *J. Biol. Chem.* **140**, 717–738.

2. Borsook, H. & Dubnoff, J.W. (1941): The formation of glycocyamine in animal tissues. *J. Biol. Chem.* **138**, 389–403.

3. Bray, R.C. & Ratner, S. (1971): Argininosuccinase from bovine kidney: comparison of catalytic, physical, and chemical properties with the enzyme from bovine liver. *Arch. Biochem. Biophys.* **146**, 531–541.

4. Chan, W., Wang, M., Kopple, J.D. & Swendseid, M.E. (1974): Citrulline levels and urea cycle enzymes in uremic rats. *J. Nutr.* **104**, 678–683.

5. Dhanakoti, S.N., Brosnan, J.T., Herzberg, G.R. & Brosnan, M.E. (1990): Renal arginine synthesis: studies *in vitro* and *in vivo*. *Am. J. Physiol.* **259**, E437–E442.

6. Drotman, R.B. & Freedland, R.A. (1972): Citrulline metabolism in the perfused rat liver. *Am. J. Physiol.* **222**, 973–975.

7. Featherston, W.R., Rogers, Q.R. & Freedland, R.A. (1973): Relative importance of kidney and liver in synthesis of arginine by the rat. *Am. J. Physiol.* **224**, (1), 127–129.

8. Gerber, G.B., Koszalka, T.R., Gerber, G. & Altman, K.I. (1962): Biosynthesis of creatine by the kidney. *Nature* **196**, 286–287.

9. Gornall, A.G. & Hunter, A. (1943): The synthesis of urea in the liver, with special reference to citrulline as an intermediary in the ornithine cycle. *J. Biol. Chem.* **147**, 593–615.

10. Hartmann, F. & Plauth, M. (1989): Intestinal glutamine metabolism. *Metabolism* **38** (Suppl. 1), 18–24.

11. Hensgens, H.E.S., Verhoeven, A.J. & Meijer, A.J. (1980): The relationship between intramitochondrial N-acetylglutamate and activity of carbamyl-phosphate synthetase (ammonia). *Eur. J. Biochem.* **107**, 197–205.

12. Kettner, A. & Silbernagl, S. (1984): The renal fate of citrulline. Metabolism and specific transport. *Pflügers Arch.* **402**, R5.

13. Kettner, A. & Silbernagl, S. (1985): Renal handling of citrulline. In *Kidney metabolism and function*, eds. R. Dzurik, B. Lichardus & W. Guder, pp. 51–60. The Hague: Nijnoff press.

14. Koszalka, T.R. (1967): Extrahepatic creatine synthesis in the rat. *Arch. Biochem. Biophys.* **122**, 400–405.

15. Koszalka, T.R. & Bauman, A.N. (1966): Biosynthesis of creatine by the perfused rat kidney. *Nature* **212**, 691–693.

16. Le Bouffant, F., Hus-Citharel, A. & Morel, F. (1984): Metabolic CO_2 production by isolated single pieces of rat distal nephron segments. *Pflügers Arch.* **401**, 346–353.

17. Levillain, O., Hus-Citharel, A., Morel, F. & Bankir, L. (1989): Production of urea from arginine in pars recta and collecting duct of the rat kidney. *Renal Physiol. Biochem.* **12**, 302–312.

18. Levillain, O., Hus-Citharel, A., Morel, F. & Bankir, L. (1990): Localization of arginine synthesis along rat nephron. *Am. J. Physiol.* **259**, F916–F923.

19. Levillain, O., Hus-Citharel, A., Morel, F. & Bankir, L. (1990): Study of urea production along the mouse and rabbit nephron; comparison with the rat (abstract). *J. Am. Soc. Nephrol.* **703**, 21 P.

20. Moncada, S., Palmer, R.M.J. & Higgs, E.A. (1989): Biosynthesis of nitric oxide from L-arginine. *Biochem. Pharmacol.* **38**, 1709–1715.

21. Morris, S.M., Moncman, J.C.L., Holub, J.S. & Hod, Y. (1989): Nutritional and hormonal regulation of mRNA abundance for arginine biosynthetic enzymes in kidney. *Arch. Biochem. Biophys.* **273** (1), 230–237.

22. Morris, S.M., Sweeney, J.W.E., Kepka, J.D.M., O'Brien, W.E. & Avner, E.D. (1991): Localization of arginine biosynthetic enzymes in renal proximal tubules and abundance of mRNA during development. *Ped. Res.* **25** (2), 151–154.

23. Ratner, S. (1973): Enzymes of arginine and urea synthesis. *Adv. Enzymol.* **39**, 1–90.

24. Ratner, S. & Petrack, B. (1953): The mechanism of arginine synthesis from citrulline in kidney. *J. Biol. Chem.* **200**, 175–185.

25. Rogers, Q.R., Freedland, R.A. & Symmons, R.A. (1972): *In vivo* synthesis and utilization of arginine in the rat. *Am. J. Physiol.* **223**, 236–240.

26. Scharff, R. & Wool, I.G. (1964): Concentration of amino acids in rat muscle and plasma. *Nature* **202**, 603–604.

27. Silbernagl, S. (1988): The renal handling of amino acids and oligopeptides. *Physiol. Rev.* **68**, 911–1007.

28. Swendseid, M.E., Wang, M., Chan, W. & Kopple, J.D. (1975): The urea cycle in uremia. *Kidney Int.* **7**, S280–S284.

29. Szepesi, B., Avery, E.H. & Freedland, R.A. (1970): Role of kidney in gluconeogenesis and amino acid catabolism. *Am. J. Physiol.* **219** (6), 1627–1631.

30. Tizianello, A., DeFerrari, G., Garibotto, G., Gurreri, G. & Robaudo, C. (1980): Renal metabolism of amino acids and ammonia in subjects with normal renal function and in patients with chronic renal insufficiency. *J. Clin. Invest.* **65**, 1162–1173.

31. Walder, C.E., Thiemermann, C. & Vane, J.R. (1991): The involvement of endothelium-derived relaxing factor in the regulation of renal cortical blood flow in the rat. *Br. J. Pharmacol.* **102**, 967–973.

32. Windmueller, H.G. & Spaeth, A.E. (1974): Uptake and metabolism of plasma glutamine by the small intestine. *J. Biol. Chem.* **249**, 5070–5079.

33. Windmueller, H.G. & Spaeth, A.E. (1980): Respiratory fuels and nitrogen metabolism *in vivo* in small intestine of fed rats. *J. Biol. Chem.* **255**, 107–112.

34. Windmueller, H.G. & Spaeth, A.E. (1981): Source and fate of circulating citrulline. *Am. J. Physiol.* **241** (5), E473–E480.

Guanidino Compounds in Biology and Medicine, eds. by P.P. De Deyn, B. Marescau, V. Stalon and
I.A. Qureshi. ©1992 John Libbey & Company Ltd., pp. 71–75.

Chapter 10

The nitric oxide: cyclic GMP
pathway in the brain

J. GARTHWAITE and E. SOUTHAM

Department of Physiology, University of Liverpool, Brownlow Hill, PO Box 147, Liverpool, L69 3BX, UK

Summary

In the central nervous system, nitric oxide (NO) is generated from the amino acid, L-arginine. A major stimulus for NO formation is the Ca^{2+} influx that follows activation of the NMDA class of glutamate receptors on postsynaptic neurons although, in certain pathways, NO may also be formed in presynaptic elements. Unlike conventional second messengers, NO can cross membranes easily to act on neighbouring cells; indeed, this appears to be the principal way it operates in the brain. Among the potential targets are other neurons and their processes (including presynaptic terminals) and non-neuronal cells (astrocytes). At these targets sites, NO activates soluble guanylate cyclase and so raises cyclic GMP levels. Various roles of NO in normal and abnormal brain function are emerging.

Introduction

Glutamate is now regarded as the major excitatory transmitter in the mammalian central nervous system. When agonists active at glutamate receptors or when excitatory pathways in the brain are activated, an elevation in the levels of cyclic GMP has long been known to occur. We found strong evidence to suggest that the cells which responded with increased cGMP were different from those primarily stimulated by receptor agonists. This, of course, implied that there was an intercellular mediator[9] and we subsequently identified this messenger as 'endothelium-derived relaxing factor' (EDRF) which operates in blood vessels[10] and which is now recognized to be NO[16,18].

Synthesis of NO

NO in brain (and endothelial cells) is generated enzymatically from the amino acid, L-arginine, and L-citrulline is a co-product[13,16]. The formation of NO by intact cells is Ca^{2+}-dependent[10], a property which relates to the fact that the NO synthase enzyme has a strict requirement for Ca^{2+} in physiologically relevant concentrations (half-maximal stimulation at about 200 nM), together with calmodulin, for activity[3,13]. NADPH and oxygen are also needed. The cerebellar NO synthase has now been purified and subjected to molecular cloning[5]. It is apparently a 150 kDa monomer with binding sites for flavins (as well as for NADPH and calmodulin) and it expresses phosphorylation sites for cAMP- and Ca^{2+}/calmodulin-dependent protein kinases and protein kinase C making it a complex, but probably highly regulated, enzyme.

The elucidation of the enzymatic pathway for NO formation has led to the development of arginine derivatives that are active as competitive NO synthase inhibitors. Among those in widest use are L-N^G-methylarginine (L-MeArg) and L-N^G-nitroarginine (L-NOArg). We have shown that both these inhibitors are highly effective blockers of the cGMP elevating effects of glutamate receptor activation in incubated slices of cerebellum and hippocampus and that their inhibitory effects can be neutralized by additional arginine. Significant differences were noted in the potency of L-NOArg in the hippocampus and cerebellum and also in the cerebellum at different developmental stages, raising the possibility of enzyme heterogeneity [11].

Properties of NO

NO has several properties which set it apart from conventional messenger molecules. It is firstly, of course, a potent activator of soluble guanylate cyclase and this is believed to be its major mechanism of action physiologically. The 'NO receptor' is a haem moiety associated with the heterodimer enzyme protein. Secondly, NO is highly lipophilic which means that it can diffuse readily from the sites of its production, across membranes, and act on neighbouring cells. Indeed, this is the principal way in which it functions in blood vessels and, where studied, in the brain: in the former, NO (EDRF) is formed in the endothelial cells lining the vessel lumen and exerts its vasodilator effects by diffusing into the underlying smooth muscle and stimulating cGMP formation[16]. In the brain, NO is produced by certain populations of neurones and can act on other neurones or on astrocytes (see below). Thirdly, NO is an unstable free radical species: its physiological lifetime is not known but when perfused in oxygenated Krebs solution over a tissue, it decays 50 per cent in about 3 s. This inherent instability is thought to be what limits the sphere of influence of NO generated endogenously in intact tissues but it is still unknown whether it acts only in a very local manner or whether it is able to influence function over a wider, three-dimensional space. Finally, NO is potentially toxic. Certain cells (notably macrophages) have an inducible (and Ca^{2+}-independent) form of NO synthase which can generate large amount of NO for prolonged periods; the NO produced in this way contributes to the cytostatic and cytotoxic effects of the phagocytes by inhibiting key enzymes in invading organisms and tumour cells [14].

Stimuli for neuronal NO formation

Glutamate receptors

The discovery of the NO:cGMP pathway in the brain resulted from a study of second messengers associated with glutamate receptor activation in preparations from the developing rat cerebellum. In the immature tissue, glutamate-induced NO/cGMP formation appears to be entirely mediated through the NMDA (N-methyl-D-aspartate) subtype of receptor. In the adult, other receptor subtypes are able to evoke this response, including AMPA and metabotropic receptors, when exogenous agonists are added[21]. However, the main mediator of the effect of the natural neurotransmitter under more physiological situations *in vitro* and *in vivo*, remains the NMDA receptor[21,22]. The involvement of this receptor is of interest because, as is now well known, it has been implicated in the triggering of long-term changes in the efficacy of synaptic transmission and it has several special properties, including a Ca^{2+}-permeant ion channel. Presumably it is the Ca^{2+}-influx through NMDA channels that is responsible for activation of NO synthase.

Presynaptic action potentials

Recent studies have provided good evidence that NO is the 'transmitter' of several non-adrenergic, non-cholinergic autonomic nerves in the peripheral nervous system[11]. The NO synthase here is presumed to be located presynaptically and to be activated following action potential-dependent Ca^{2+}-influx. The same mechanism appears to operate in certain pathways in the brain: in the cerebellum, the climbing fibres, which provide a powerful excitatory input to Purkinje cells, also appear to release NO when stimulated[19,20].

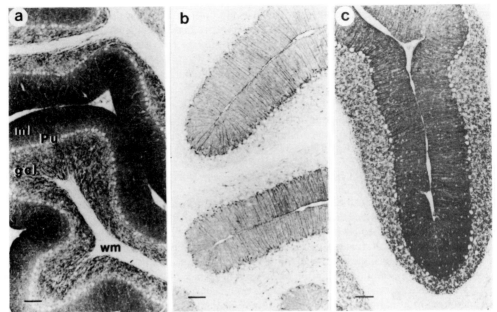

Fig. 1. Sources and targets of NO in adult rat cerebellum. NADPH-diaphorase (NO synthase) staining is shown in (a) and is seen to be prominent in the molecular layer (ml) and granule cell layer (gcl) but is very low in Purkinje cells (Pu) and in the white matter (wm). (b) cGMP immunostaining in incubated cerebellar slices under basal conditions is seen mainly in the molecular layer where the strongest labelling is of the Bergmann glial cells and their fibres. (c) When stimulated by the NO-donor, nitroprusside, cGMP immunostaining becomes very dense, both in the molecular layer and granule cell layer. In the latter the labelling is associated with several structures, including granule cells and presynaptic nerve terminals.

Anatomy of the NO:cGMP pathway

In order to understand how this pathway operates in the brain, it is essential to understand the anatomy and, accordingly, we have been actively pursuing the identification of the cells that produce NO and those that respond to it by increasing their cGMP content. The particular focus so far has been on the cerebellum, which is the area on which most of the previous work had been carried out. To identify NO synthase-containing neurones we have used a simple method, NADPH diaphorase histochemistry. This method has been known for many years to selectively stain neuronal populations but only recently has it become apparent that those populations are the ones that contain NO synthase[12]. To identify the targets, we have used a cGMP antiserum developed, and kindly given to us, by De Vente, Steinbusch *et al.*[7]. In the cerebellum the most abundant cells expressing NO synthase are the granule cells (Fig. 1a), in agreement with immunohistochemical findings using an antibody raised against purified NO synthase[4]. Most of the NMDA receptors in the cerebellum are also located on these neurones. There is also dense staining in the molecular layer where granule cell axons and climbing fibres synapse with Purkinje cell dendrites and where inhibitory interneurones (basket and stellate cells) are found.

cGMP levels in slices of cerebellum under resting conditions are low (Fig. 1b); this basal staining can be substantially reduced, or abolished, by NO synthase inhibitors or by incubation of the slices in Ca^{2+}-free media. When slices are exposed to a NO-donor, such as nitroprusside, the levels become greatly increased (Fig. 1c). The major loci include the excitatory mossy fibre nerve terminals (which release glutamate onto granule cells), terminals of inhibitory interneurons (basket cells) contacting Purkinje cells, Bergmann glia in the Purkinje and molecular layers, astrocytes in the granule cell layer and, perhaps surprisingly, granule cells. Granule cells thus emerge as neurons capable both of generating NO and of responding to it and yet they do not respond measurably to their own NO produced following NMDA receptor activation, probably because the Ca^{2+} which enters the neurons and activates NO synthase, simultaneously inhibits the effector enzyme, guanylate cyclase[13]. This may be an important device for promoting an inter- rather than intracellular action of NO and it is also likely to be significant in determining whether a potential target cell will respond to NO or not.

The above findings suggest a hypothesis in which NO is generated by granule cells in response to activation of their NMDA receptors by the mossy fibre transmitter (glutamate) and that the NO then diffuses out to influence the behaviour of (i) the afferent nerve terminals, possibly to affect neurotransmitter release, (ii) other granule cells and (iii) glial cells. High resolution techniques will be needed to understand the anatomy of the pathway in the molecular layer.

Possible functions of the NO:cGMP pathway in the central nervous system

Though concentrated in certain areas such as the cerebellum and olfactory bulb, NO synthase is found in discrete neuronal populations throughout the brain[4] and NO is becoming implicated in an increasing number of phenomena. It has been suggested to be necessary for a type of synaptic plasticity in the cerebellum known as long-term depression. In this, brief simultaneous activation of the climbing and parallel fibre input to Purkinje cells leads to a selective decrease in efficacy of the parallel fibre pathway[19]. In the hippocampus, inhibition of NO synthesis blocks the induction of another form of synaptic plasticity, long-term potentiation, which is induced as a result of NMDA receptor activation occurring during brief high frequency stimulation of afferent fibres[2]. Both these forms of plasticity are thought to represent cellular correlates of memory formation. NO synthase inhibitors have opiate-independent antinociceptive effects that seem to involve a central mechanism[17] and they also inhibit the blood-pressure lowering effect of glutamate in the nucleus of the solitary tract which, again, appears to involve NMDA receptors[8]. In the hypothalamus, NO may play a role in thermogenesis[1] and, in the cerebral cortex, it has been implicated in epileptogenesis[6,15]. The cellular mechanisms underlying these effects, however, are unknown and their elucidation will form an important future task.

Acknowledgements

The research in our laboratory is supported by the Medical Research Council (UK). We thank Dr. J. De Vente for supplying the cGMP antiserum.

References

1. Amir, S., De Blasio, E. & English, A.M. (1991): NG-monomethyl-L-arginine co-injection attenuates the thermogenic and hyperthermic effects of E2 prostaglandin microinjection into the anterior hypothalamic preoptic area in rats. *Brain Res.* **556**, 157–160.

2. Bohme, G.A., Bon, C., Stutzmann, J-M., Doble, A. & Blanchard, J-C. (1991): Possible involvement of nitric oxide in long-term potentiation. *Eur. J. Pharmacol.* **199**, 379–381.

3. Bredt, D.S. & Snyder, S.H. (1990): Isolation of nitric oxide synthetase, a calmodulin-requiring enzyme. *Proc. Natl. Acad. Sci. USA* **87**, 682–685.

4. Bredt, D.S., Hwang, P.M. & Snyder, S.H. (1990): Localization of nitric oxide synthase indicating a neural role for nitric oxide. *Nature* **347**, 768–770.

5. Bredt, D.S., Hwang, P.M., Glatt, C.E., Lowenstein, C., Reed, R.R. & Snyder, S.H. (1991): Cloned and expressed nitric oxide synthase structurally resembles cytochrome P-450 reductase. *Nature* **351**, 714–718.

6. De Sarro, G.B., Di Paola, E.D., De Sarro, A. & Vidal, M.J. (1991): Role of nitric oxide in the genesis of excitatory amino acid-induced seizures from the deep prepyriform cortex. *Fund. Clin. Pharmacol.* (in press).

7. De Vente, J., Bol, J.G.J.M., Berkelmans, H.S., Schipper, J. & Steinbusch, H.M.W. (1990): Immunocytochemistry of cGMP in the cerebellum of the immature, adult, and aged rat: the involvement of nitric oxide. A micropharmacological study. *Eur. J. Neurosci.* **2**, 845–862.

8. Di Paola, E.D., Vidal, M.J. & Nistico, G. (1991): L-glutamate evokes the release of an endothelium-derived relaxing factor-like substance from the rat nucleus tractus solitarius. *J. Cardiovasc. Pharmacol.* (Suppl. 3) **17**, S269–S272.

9. Garthwaite, J. & Garthwaite, G. (1987): Cellular origins of cyclic GMP responses to excitatory amino acid receptor agonists in rat cerebellum *in vitro. J. Neurochem.* **48**, 29–39.

10. Garthwaite, J., Charles, S.L. & Chess-Williams, R. (1988): Endothelium-derived relaxing factor release on activation of NMDA receptors suggests role as intercellular messenger in the brain. *Nature* **336**, 385–388.

11. Garthwaite, J. (1991): Glutamate, nitric oxide and cell–cell signalling in the nervous system. *Trends Neurosci.* **14**, 60–67.

12. Hope, B.T., Michael, G.J., Knigge, K.M. & Vincent, S.R. (1991): Neuronal NADPH diaphorase is a nitric oxide synthase. *Proc. Natl. Acad. Sci. USA* **88**, 2811–2814.

13. Knowles, R.G., Palacios, M., Palmer, R.M.J. & Moncada, S. (1989): Formation of nitric oxide from L-arginine in the central nervous system: a transduction mechanism for stimulation of the soluble guanylate cyclase. *Proc. Natl. Acad. Sci. USA* **89**, 5159–5162.

14. Marletta, M.A. (1989): Nitric oxide: biosynthesis and biological significance. *Trends Biochem. Sci.* **14**, 488–492.

15. Mollace, V., Bagetta, G. & Nistico, G. (1991): Evidence that L-arginine possesses proconvulsant effects mediated through nitric oxide. *Neuroreport* **2**, 269–272.

16. Moncada, S., Palmer, R.M.J. & Higgs, E.A. (1989): Biosynthesis of nitric oxide from L-arginine. A pathway for the regulation of cell function and communication. *Biochem. Pharmacol.* **38**, 1709–1715.

17. Moore, P.K., Oluyomi, A.O., Babbedge, R.C., Wallace, P. & Hart, S.L. (1991): L-NG-nitro arginine methyl ester exhibits antinociceptive activity in the mouse. *Br. J. Pharmacol.* **102**, 198–202.

18. Palmer, R.M.J., Ferrige, A.G. & Moncada, S. (1987): Nitric oxide release accounts for the biological activity of endothelium-derived relaxing factor. *Nature* **327**, 524–526.

19. Shibuki, K. & Okada, D. (1991): Endogenous nitric oxide release required for long-term synaptic depression in the cerebellum. *Nature* **349**, 326–328.

20. Southam, E. & Garthwaite, J. (1991): Climbing fibres as a source of nitric oxide in the cerebellum. *Eur. J. Neurosci.* **3**, 379–382.

21. Southam, E., East, S.J. & Garthwaite, J. (1991): Excitatory amino acid receptors coupled to the nitric oxide:cyclic GMP pathway in rat cerebellum during development. *J. Neurochem.* **56**, 2072–2081.

22. Wood, P.L., Richard, J.W., Pilapil, C. & Nair, N.P.V. (1982): Antagonists of excitatory amino acids and cyclic guanosine monophosphate in cerebellum. *Neuropharmacol.* **21**, 1235–1238.

Guanidino Compounds in Biology and Medicine, eds. by P.P. De Deyn, B. Marescau, V. Stalon and I.A. Qureshi. ©1992 John Libbey & Company Ltd., pp. 77–81.

Chapter 11

Regulation of tetrahydrobiopterin biosynthesis by cytokines. Significance for the conversion of arginine to citrulline and nitric oxide

Ernst R. WERNER, Gabriele WERNER-FELMAYER and Helmut WACHTER

Institute of Medical Chemistry and Biochemistry, University of Innsbruck, Fritz-Pregl-Str. 3, A-6020 Innsbruck, Austria

Summary

Tetrahydrobiopterin has been shown to be a cofactor of the conversion of arginine to critrulline and nitric oxide (NO) in NO synthase preparations from both cytokine-activated murine macrophages and porcine cerebellum. We investigated regulation of pteridine synthesis by cytokines in cultured cells of man and mouse *in vitro*, the impact of inhibition of pteridine synthesis on cytokine-induced NO formation, as well as pteridine synthesis in human diseases challenging the cell-mediated immunity. We find that cytokines (interferon-γ, tumour necrosis factor-α) stimulate up to 100-fold the activity of GTP-cyclohydrolase I, the first step in biosynthesis of tetrahydrobiopterin from guanosine triphosphate (GTP) in both murine and human cells. This leads to an up to 30-fold increase in intracellular tetrahydrobiopterin concentrations. In human cells, this is accompanied by the accumulation of neopterin derivatives due to a comparatively lower 6-pyruvoyl tetrahydropterin synthase activity. Neopterin leaks from the cells and can be determined in body fluids of humans. Neopterin is strongly increased in diseases with endogenous cytokine formation (e.g. parasitic and viral infection, sepsis, etc.). Our results suggest that cytokines strongly stimulate pteridine synthesis in order to provide increased amounts of cofactor required for the conversion of arginine to citrulline and nitric oxide.

Introduction

Pteridines are a class of compounds with a common pyrazino-pyrimidine ring structure. These are either embedded in a comparatively complex molecule in folic acid or in riboflavins and are then called 'conjugated' pteridines. Alternatively, the heterocyclic ring is substituted with a comparatively small side chain (most commonly three carbon atoms or less). These compounds are then called 'unconjugated' pteridines and comprise the compounds covered in this article, neopterin and biopterin. The biosynthesis of pteridines (reviewed by Nichol *et al.*[13]) starts from guanosine 5′ triphosphate (GTP), which is cleaved by GTP-cyclohydrolase I to form 7,8-dihydroneopterin triphosphate. This labile compound is then either cleaved by phosphatases to yield dihydroneopterin and neopterin derivatives, which are found elevated in body fluids of humans with diseases challenging the cell-mediated immunity (reviewed by Wachter *et al.*[18]). Alternatively, the enzymes 6-pyruvoyl-tetrahydropterin synthase and sepiapterin reductase convert 7,8-dihydroneopterin triphosphate to tetrahydrobiopterin, which can function as a cofactor in selected oxygen-utilizing reactions. The reactions which today are established to be pteridine dependent are phenylalanine 4-monooxygenase[8], tyrosine 3-monooxygenase[14], tryptophan 5-monooxygenase[3], alkylglycol-ether monooxygenase[16], and nitric oxide synthase[10,11,12,15]. The first four of these have been known to require pteridine cofactors for 20 years or more. The fifth, the conversion of arginine to citrulline and nitric oxide (NO), has been detected recently and provides the link of pteridines to the guanidino compounds, which are the topic of this book. The present contribution will summarize the knowledge of induction of pteridine biosynthesis by cytokines and discuss the impact of pteridines on the NO synthase reaction.

Materials and methods

For the analytical determination of pteridines in body fluids, in tissue culture material and in enzyme incubations we use reversed phase HPLC with fluorescence detection. Since the naturally occurring, labile di- and tetrahydroderivatives are non-fluorescent, these are oxidized to the stable fluorescent derivatives immediately upon sample collection[5]. In acidic medium, all species (di- and tetrahydro-biopterin) are converted to the fluorescent biopterin, whereas in basic medium tetrahydrobiopterin is cleaved to pterin and only dihydrobiopterin yields the fluorescent biopterin. Therefore, the amount of tetrahydrobiopterin, the biologically active species, can be determined from the difference of biopterin concentration after oxidation in acidic or alkaline media. For the HPLC analysis we use a conventional reversed phase HPLC system with fluorescence detection equipped with an AASP instrument (Varian, Palo Alto, CA, USA), which can insert solid-phase cartridges upstream of the HPLC column in a high pressure seal chamber. This provides a convenient set-up in which the acidic oxidation mixtures can be applied to solid phase cartridges (SCX, Varian), which extract and thereby enrich the pteridines. The cartridges are then automatically inserted into the HPLC system by the AASP instrument (see Werner et al.[19] for details).

For the special case of determination of neopterin in body fluids as an indicator of endogenous cytokine activity in humans in diseases, the determination of neopterin without any oxidation procedures has been proven useful for clinical studies[18]. Although the amount of the fluorescent neopterin is only a small part of neopterin derivatives present in the body fluids, it is sufficient to allow precise quantitation. In particular, in clinical settings, samples can not always be oxidized in due course after sample collection. In this case, the determination of only the stable, fluorescent neopterin gives results with better reproducibility, whereas the clinical significance of neopterin determination with and without oxidative procedures has turned out to be comparable[4].

Results and discussion

Dependence of NO synthase on tetrahydrobiopterin

Two types of NO synthase have been described. One constitutive, 'brain' type, which has been found in brain and endothelial cells and showed an activity dependence on Ca^{2+}/calmodulin[2]. The other type, the inducible 'macrophage' type has been investigated in activated macrophages. It was characterized to be independent of Ca^{2+}/calmodulin[6] and to require a low molecular weight factor of the cytosol for activity, which turned out to be tetrahydrobiopterin[10,15]. Based on the observation of enzyme activity in the absence of exogenously added tetrahydrobiopterin, the brain enzyme was claimed to be independent of the pteridine cofactor[1,9]. Mayer et al.[11] purified NO synthase from porcine cerebellum, added tetrahydrobiopterin and found a stimulation of the activity comparable to enzyme activity present in macrophages. We checked the porcine cerebellum NO synthase for tightly bound pteridines and found that the homogenous enzyme contains tetrahydrobiopterin[12]. This explains the activity in absence of added tetrahydrobiopterin and strongly suggests that the 'brain' type of NO synthase depends on tetrahydrobiopterin in a manner similar to the 'macrophage' type. Further support of the pteridine dependence of the NO synthase reaction comes from manipulation of NO synthase turnover in intact cells by drugs influencing pteridine metabolism[23], which will be detailed in Chapter 12.

Induction of pteridine synthesis by cytokines in cultured cells

Whereas a challenge of the cell-mediated immunity leads to massive increases in neopterin formation in humans (see below), no neopterin could be detected in body fluids or tissues of mice despite strong stimulation of the immune system. To investigate the action of cytokines on pteridine metabolism in more detail, we studied activities of pteridine biosynthetic enzymes in a panel of human and murine cells[20,21]. We found that in both human and murine cells, cytokines stimulate the activity of the key enzyme of pteridine synthesis, the GTP-cyclohydrolase I up to 100-fold, in a dose-and time-dependent manner. This leads to increase of intracellular concentrations of tetrahydrobiopterin. Depending on the activity of the second enzyme of the biosynthetic pathway, the 6-pyruvoyl tetrahydropterin synthase, this increase in tetrahydrobiopterin is accompanied by accumulation of varying concentrations of the first intermediate, 7,8-dihydroneopterin triphosphate. After cleavage of the phosphate groups by phosphatases, the neopterin derivatives leak from the cells and give rise to increased concentrations of neopterin in culture supernatants and body fluids. In murine cells, the activity of 6-pyruvoyl tetrahydropterin synthase is two orders of magnitude higher than in human cells so that the biosynthesis of pteridines proceeds to tetrahydrobiopterin without the accumulation of neopterin derivatives (Fig. 1).

Pteridine synthesis in human diseases

In the past 10 years, our group has studied in detail the excretion of neopterin in various clinical settings (reviewed by Wachter *et al.*[18]). Consistent with the finding of neopterin formation by interferon-γ activated macrophages *in vitro*[7], increase of neopterin is found in disease states with endogenous formation of cytokines. While interferon-γ is the most potent cytokine to stimulate pteridine synthesis, other cytokines like tumour necrosis factor α and interleukin-1 have been found

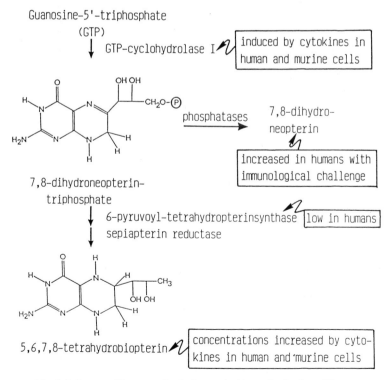

Fig. 1. Influence of immunostimulation on the biosynthesis of pteridines.

to be stimulatory as well and, more importantly, to potentiate the interferon-γ stimulus[17,22]. Thus, the neopterin amount excreted is a measure of the action of the cytokine network on pteridine biosynthesis. Increased endogenous synthesis of cytokines and hence increased neopterin excretion occurs in a variety of diseases, including viral infection, allograft rejection and autoimmune disorders, to mention just a few. In the light of the results outlined above showing that (i) formation of neopterin indicates increased tetrahydrobiopterin biosynthesis of the cells and (ii) that the NO synthase reaction is strongly stimulated by tetrahydrobiopterin, it becomes clear that enhanced turnover of NO synthase is to be expected in all clinical settings with increased neopterin excretion due to immunostimulation. This may lead to an understanding of the mechanism of clinical observations, e.g. of pronounced hypotension as a dose-limiting side effect of cytokine treatments of cancer patients.

Acknowledgements

Support by the Austrian Research Funds 'zur Förderung der wissenschaftlichen Forschung', project N° 8231, is gratefully acknowledged.

References

1. Bredt, D.S. & Snyder, S.H. (1990): Isolation of nitric oxide synthase, a calmodulin requiring enzyme. *Proc. Natl. Acad. Sci. USA* **87**, 682–685.

2. Bredt, D.S., Hwang, P.M., Glatt, C.E., Lowenstein, C., Reed, R.R. & Snyder, S.H. (1991): Cloned and expressed nitric oxide synthase structurally resembles cytochrome P-450 reductase. *Nature* **351**, 714–718.

3. Friedman, P.A., Kappelman, A.H. & Kaufman, S. (1972): Partial purification and characterization of tryptophan hydroxylase from rabbit hindbrain. *J. Biol. Chem.* **247**, 4165–4173.

4. Fuchs, D., Milstien, S., Krämer, A., Reibnegger, G., Werner, E.R., Goedert, J.J., Kaufman, S. & Wachter, H. (1989): Urinary neopterin concentrations *vs* total neopterins for clinical utility. *Clin. Chem.* **35**, 2305–2307.

5. Fukushima, T. & Nixon J.C. (1980): Analysis of reduced forms of biopterin in biological tissues and fluids. *Anal. Biochem.* **102**, 176–188.

6. Hauschildt, S., Lückhoff, A., Mülsch, A., Kohler, J., Bessler, W. & Busse, R. (1990): Induction and activity of NO synthase in bone-marrow derived macrophages are independent of Ca^{2+}. *Biochem. J.* **270**, 351–356.

7. Huber, C., Batchelor, R.J., Fuchs, D., Hausen, A., Lang, A., Niederwieser, D., Reibnegger, G., Swetly, P., Troppmair, J. & Wachter, H. (1984): Immune response associated production of neopterin. Release from macrophages primarily under control of interferon-γ. *J. Exp. Med.* **160**, 310–316.

8. Kaufmann, S. (1959): Studies on the mechanism of the enzymatic conversion of phenylalanine to tyrosine. *J. Biol. Chem.* **234**, 2677–2684.

9. Knowles, R.G., Palacio, M., Palmer, M.J. & Moncada, S. (1990): Kinetic characteristics of nitric oxide synthase from rat brain. *Biochem. J.* **269**, 207–210.

10. Kwon, N.S. Nathan. C.F. & Stuehr, D.J. (1989): Reduced biopterin as a cofactor in the generation of nitrogen oxides by murine macrophages. *J Biol. Chem.* **264**, 20496–20501.

11. Mayer, B., John, M. & Boehme, E. (1990): Purification of Ca^{2+}/calmodulin dependent nitric oxide synthase from porcine cerebellum. Cofactor role of tetrahydrobiopterin. *F.E.B.S. Lett.* **277**, 215–219.

12. Mayer, B., John, M., Heinzel, B., Werner, E.R., Wachter, H., Schulz, G. & Boehme, E. (1991): Brain nitric oxide synthase is a biopterin and flavin containing multifunctional oxido-reductase. *F.E.B.S. Lett.* **288**, 187–191.

13. Nichol, C.A., Smith, G.K. & Duch, D.S. (1985): Biosynthesis and metabolism of tetrahydrobiopterin and molybdopterin. *Ann. Rev. Biochem.* **54**, 729–764.

14. Shiman, R., Akino, M. & Kaufman, S. (1971): Solubilization and partial purification of tyrosine hydroxylase from bovine adrenal medulla. *J. Biol. Chem.* **246**, 1330–1340.

15. Tayeh, M.A. & Marletta, M.M. (1989): Macrophage oxidation of L-arginine to nitric oxide, nitrite and nitrate. Tetrahydrobiopterin is an intermediate. *J. Biol. Chem.* **264**, 19654–19658.

16. Tietz, A., Lindberg, M. & Kennedy, E.P. (1964): A new pteridine-requiring enzyme system for the oxidation of glyceryl ethers. *J. Biol. Chem.* **239**, 4081–4090.

17.   Troppmair, J., Nachbaur, K., Herold, M., Aulitzky, W., Tilg, H., Gastl, G., Bieling, P., Kotlan, B., Flener, R., Mull, B., Aulitzky, W.O., Rokos, H. & Huber, C. (1988): *In-vitro and in-vivo* studies on the induction of neopterin biosynthesis by cytokines, alloantigens and lipopolysaccharide. *Clin. Exp. Immunol.* **74**, 392–397.

18.   Wachter, H., Fuchs, D., Hausen, A., Reibnegger, G. & Werner, E.R. (1989): Neopterin as marker for activation of cellular immunity. Immunological basis and clinical application. *Adv. Clin. Chem.* **27**, 81–141.

19.   Werner, E.R., Fuchs, D., Hausen, A., Reibnegger, G. & Wachter, H. (1987): Simultaneous determination of neopterin and creatinine in serum with solid-phase extraction and on-line elution chromatography. *Clin. Chem.* **33**, 2028–2033.

20.   Werner, E.R., Werner-Felmayer, G., Fuchs, D., Hausen, A., Reibnegger, G., Yim, J.J., Pfleiderer, W. & Wachter, H. (1990): Tetrahydrobiopterin biosynthetic activities in human macrophages, fibroblasts, THP-1 and T 24-cells. GTP-cyclohydrolase I is stimulated by interferon-γ, and 6-pyruvol tetrahydropterin synthase and sepiapterin reductase are constitutively present. *J. Biol. Chem.* **265**, 3189–3192.

21.   Werner, E.R., Werner-Felmayer, G., Fuchs, D., Hausen, A., Reibnegger, G., Yim, J.J. & Wachter, H. (1991): Impact of tumour necrosis factor α and interferon-γ on tetrahydrobiopterin synthesis in murine fibroblasts and macrophages. *Biochem. J.* **280**, 709–714.

22.   Werner-Felmayer, G., Werner, E.R., Fuchs, D., Hausen, A., Reibnegger, G. & Wachter, H. (1989): Tumour necrosis factor α and lipopolysaccharide enhance interferon-induced tryptophan degradation and pteridine synthesis in human cells. *Biol. Chem. Hoppe Seyler* **370**, 1063–1069.

23.   Werner-Felmayer, G., Werner, E.R., Fuchs, D., Hausen, A., Reibnegger, G. & Wachter, H. (1990): Tetrahydrobiopterin-dependent formation of nitrite and nitrate by murine fibroblasts. *J. Exp. Med.* **172**, 1599–1607.

Guanidino Compounds in Biology and Medicine, eds. by P.P. De Deyn, B. Marescau, V. Stalon and I.A. Qureshi. ©1992 John Libbey & Company Ltd., pp. 83–88.

Chapter 12

Formation of nitrite and nitrate from L-arginine in murine fibroblasts

Gabriele WERNER-FELMAYER, Ernst R. WERNER and Helmut WACHTER

Institute of Medical Chemistry and Biochemistry, University of Innsbruck, Fritz-Pregl-Str. 3, A-6020 Innsbruck, Austria

Summary

Murine fibroblasts, isolated from ear dermis of adult Balb/c mice form nitrite (NO_2^-) and nitrate (NO_3^-) when treated with interferon-γ alone or in combination with tumour necrosis factor-α. Formation of nitrogen oxides can be inhibited by 2,4-diamino-6-hydroxy-pyrimidine, an inhibitor of tetrahydrobiopterin biosynthesis. Production of nitrogen oxides is restored by adding sepiapterin, which is converted intracellularly into tetrahydrobiopterin. This salvage pathway is inhibited by concomitant treatment with methotrexate. Besides L-arginine intact cells use L-citrulline, L-arginino-L-aspartate and L-argininosuccinate, but not L-ornithine for nitrite formation. L-glutamine stimulates nitrite formation from L-arginine or L-citrulline. Our results indicate that intracellular tetrahydropiopterin (BH_4) levels control cytokine-induced nitric oxide (NO) synthase in murine fibroblasts and suggest that a citrulline-arginine recycling, as described for endothelial cells, may operate in these cells.

Introduction

We observed previously that excretion of the pyrazinopyrimidine compound neopterin is increased in patients with diseases challenging the cellular immune system (for review see Wachter *et al.*[8]). Neopterin is formed in the course of tetrahydrobiopterin synthesis from guanosine triphosphate (GTP) and was found to be produced by macrophages upon stimulation with interferon-γ[2]. Further *in vitro* studies revealed that not only macrophages, but a variety of human cells can be induced to synthesize increased amounts of pteridines when treated with cytokines[9,12]. Cell types other than macrophages, however, produce tetrahydrobiopterin in excess over neopterin which can be explained by the relative activities of the first two enzymes of tetrahydrobiopterin synthesis, i.e. GTP-cyclohydrolase I and 6-pyruvoyl tetrahydropterin synthase[10].

In mice, no neopterin is detectable neither *in vivo* nor *in vitro*. Murine macrophages and macrophage lines already produce high levels of tetrahydrobiopterin but murine fibroblasts behave like their human counterparts: when untreated these cells have only a low tetrahydrobiopterin synthesizing capacity, which is stimulated several-fold by cytokines, especially by tumour necrosis factor-α or interleukin-1[11]. Thus, in both species cytokine-regulated pteridine synthesis can be observed.

In 1989 it was shown for cytosol preparations from cytokine-treated murine macrophages that tetrahydrobiopterin stimulates the formation of nitric oxide (NO) and citrulline from L-arginine by NO synthase[3,7], (for details concerning cytokine-induced pteridine synthesis and nitric oxide formation *see* Chapter 11).

In order to demonstrate a functional role for tetrahydrobiopterin in this reaction in intact cells we studied murine fibroblasts, which have both cytokine-inducible tetrahydrobiopterin synthesis[11] and a macrophage-like, cytokine-inducible NO synthase[13]. For further characterization of NO synthase from murine fibroblasts we tested a number of arginine-related compounds for their potential to substitute for L-arginine in the formation of nitrogen oxides as well as the effect of L-glutamine on nitrite formation.

Materials and methods

Murine fibroblasts were isolated from ear dermis and cultured in RPMI, containing 2 mM L-glutamine, 100 U/ml penicillin, 0.1 mg/ml streptomycin, and 10 per cent heat-inactivated foetal calf serum. For experiments, confluent monolayers, grown in 24- or 96-well plates, were treated with 50 U/ml of murine recombinant interferon-γ (Holland Biotechnology, Leiden, The Netherlands) in combination with 500 U/ml of murine recombinant tumour necrosis factor-α (a kind gift of Dr. G.R. Adolf, Bender Co., Vienna, Austria).

Nitrate was reduced enzymatically and nitrite was determined in supernatants by the Griess reaction.

Tetrahydrobiopterin synthesis was manipulated by treating cells with 5 mM 2,4-diamino-6-hydroxy-pyrimidine (DAHP) (Sigma, Munich, Germany), 100 μM sepiapterin (Dr. B. Schircks, Jona, Switzerland) and 10 μM methotrexate, which was applied in combination with 40 μM thymine and 100 μM inosine.

Intracellular biopterin levels were determined by high performance liquid chromatography after oxidation at alkaline or acidic pH, as detailed in Werner-Felmayer *et al.*[13].

For testing which compounds can substitute for arginine in the formation of nitrite, cells were cultured in arginine-free culture medium (MEM/Earle Select Amine Kit, Gibco Laboratories, Grand Island, New York), containing 1 mM of L-citrulline, L-arginino-L-aspartate, L-argininosuccinate or L-ornithine (all from Serva, Heidelberg, Germany). In some experiments, arginine- and glutamine-free medium was used and supplemented with either 1 mM L-arginine or 1 mM L-citrulline.

Results and discussion

The strategy for demonstrating the functional link between tetrahydrobiopterin synthesis and formation of nitrogen oxides is outlined in Fig. 1. Using an inhibitor of GTP-cyclohydrolase I, i.e. 2,4-diamino-6-hydroxy-pyrimidine, *de novo* synthesis of tetrahydrobiopterin was inhibited. Reconstitution of the cofactor was achieved by parallel treatment of cells with sepiapterin. This pteridine is converted into tetrahydrobiopterin in the intact cell via a 'salvage pathway', employing sepiapterin reductase and dihydrofolate reductase[4]. Methotrexate, an inhibitor of dihydrofolate reductase was used to inhibit this salvage pathway.

The results of these experiments are shown in Fig. 2. Untreated control cells contain little tetrahydrobiopterin and do not release nitrite or nitrate into supernatants. Treatment with 50 U/ml interferon-γ in combination with 500 U/ml tumour necrosis factor-α leads to an about threefold increase of intracellular tetrahydrobiopterin levels, and formation of nitrite and nitrate is induced. When sepiapterin (100 μM) is applied to the cells, the intracellular tetrahydrobiopterin concentration is strongly increased and also nitrogen oxides formation is further stimulated, suggesting that cytokine-induced NO-synthase is not saturated with the cofactor. Inhibition of GTP-cyclohydrolase I via 2,4-diamino-6-hydroxy-pyrimidine inhibits tetrahydrobiopterin formation as well as synthesis of nitrite and nitrate by about 50 per cent. Considering that both metabolic pathways are under control of cytokines, involving *de novo* synthesis of the key enzymes, this 50 per cent inhibition is quite remarkable. Nitrite/nitrate formation is restored when 2,4-diamino-6-hydroxy-pyrimidine is applied in combination with sepiapterin. Methotrexate, which itself has no effect on nitrogen oxide formation, counteracts the beneficial effect of sepiapterin on 2,4-diamino-6-hydroxy-pyrimidine treated cells. Due to inhibition of dihydrofolate reductase by methotrexate dihydrobiopterin but not the tetrahydro-deri-

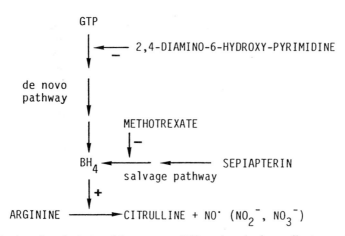

Fig. 1. Schematic view of manipulation of the turnover of NO-synthase by drugs affecting pteridine metabolism.

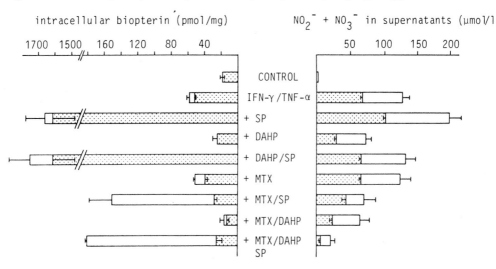

Fig. 2. Influence of drugs on intracellular biopterin levels and nitrogen oxides formation in cultured murine fibroblasts.
On the left hand side, open bars show total biopterin, dotted bars tetrahydrobiopterin concentrations. On the right hand side open bars show the sum of nitrite plus nitrate, dotted bars show the nitrite concentrations. Abbreviations: IFN = interferon; TNF = tumour necrosis factor; SP = sepiapterin; DAHP = 2,4-diamino-6-hydroxypyrimidine; MTX = methotrexate.

vative accumulates in the cells. These data demonstrate that intracellular tetrahydrobiopterin concentrations control the amount of nitrogen oxides formed. In parallel we could show that the cytotoxic effect of interferon-γ/tumour necrosis factor-α treatment is linked to formation of nitrogen oxides and that this cytokine effect can be influenced by manipulating tetrahydrobiopterin synthesis[13].

In order to further characterize nitrite formation in murine fibroblasts we treated cells with interferon-γ/tumour necrosis factor-α in arginine-free media containing L-citrulline, L-arginino-L-aspartate, L-argininosuccinate or L-ornithine (Fig. 3). Cells could not metabolize L-ornithine, but used the other tested compounds for nitrite production. L-citrulline and L-arginino-L-aspartate led to formation of equal amounts of nitrite when compared to L-arginine, whereas L-argininosuccinate was not metabolized as efficiently which may be explained by transport phenomena. These results are in accord-

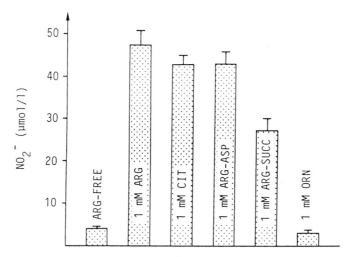

Fig. 3. Dependence of nitrite formation by cytokine-treated murine fibroblasts on the supply of L-arginine or related compounds in the culture medium.

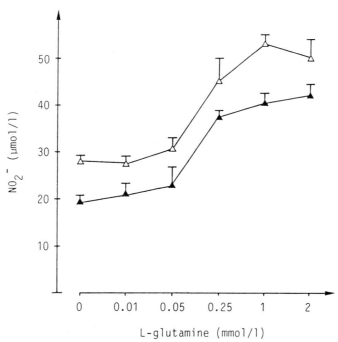

Fig. 4. Dependence of nitrite formation by cytokine-treated murine fibroblasts on the concentration of L-glutamine in the culture medium.
Open triangles: culture medium with L-arginine; filled triangles: culture medium with L-citrulline instead of L-arginine.

ance with observations previously made with bovine aortic endothelial cells where recycling of L-citrulline to L-arginine and participation of this process in nitric oxide formation could be demonstrated[1].

Furthermore, it was demonstrated for rabbit aorta that L-glutamine inhibits release of nitric oxide[6] and in bovine endothelial cells it could be demonstrated that this inhibition may occur due to inhibition of citrulline-arginine recycling by L-glutamine[5]. Since L-glutamine at 2 mM is a standard medium supplement we were interested to see whether also in murine fibroblasts the formation of nitric oxide is influenced by L-glutamine. As shown in Fig. 4, nitrite formation from L-arginine as well as from L-citrulline was enhanced by L-glutamine in a dose-dependent way. This was not due to improved cell-growth in presence of L-glutamine, since cell viability of cultures kept glutamine-free during cytokine-treatment was 96 ± 7 per cent (mean of eight cultures $\pm$ SD) from controls cultivated in medium containing 2 mM of L-glutamine. One possibility to explain this stimulatory effect of L-glutamine is that L-glutamine may donate the nitrogen incorporated into citrulline in the arginine-citrulline recycling process. However, the basis for the contrasting effects of L-glutamine in murine fibroblasts as compared to endothelial cells remains to be clarified.

Acknowledgements

Support by the Austrian Research Funds 'Zur Förderung der wissenschaftlichen Forschung', P 8231, is gratefully acknowledged.

References

1.   Hecker, M., Sessa, W.C., Harris, H.J., Anggard, E.E. & Vane J.R. (1990): The metabolism of L-arginine and its significance for the biosynthesis of endothelium derived relaxing factor. Cultured endothelial cells recycle L-citrulline to L-arginine. *Proc. Natl. Acad. Sci. USA* **87**, 8612–8616.

2.   Huber, C., Batchelor, R.J. Fuchs, D., Hausen, A., Lang, A., Niederwieser, D., Reibnegger, G., Swetly, P., Troppmair, J. & Wachter, H. (1984): Immune response associated production of neopterin. Release for macrophages primarily under control of interferon-γ. *J. Exp. Med.* **160**, 310–316.

3.   Kwon, N.S., Nathan, C.F. & Stuehr, D.J. (1989): Reduced biopterin as a cofactor in the generation of nitrogen oxides by murine macrophages. *J. Biol. Chem.* **264**, 20496–20501.

4.   Nichol, C.A., Smith, G.K. & Duch, D.S. (1985): Biosynthesis and metabolism of tetrahydrobiopterin and molybdopterin. *Ann. Rev. Biochem.* **54**, 729–764.

5.   Sessa, W.C., Hecker, M., Mitchell, J.A. & Vane, J.R. (1990): The metabolism of L-arginine and its significance for the biosynthesis of endothelium derived relaxing factor: L-glutamine inhibits the generation of L-arginine by cultured endothelial cells. *Proc. Natl. Acad. Sci. USA* **87**, 8607–8611.

6.   Swierkosz, A., Mitchell, J.A., Sessa, W.C., Hecker, M. & Vane, J.R. (1990): L-glutamine inhibits the release of endothelium derived relaxing factor from the rabbit aorta. *Biochem. Biophys. Res. Commun.* **172**, 143–148.

7.   Tayeh, M.A. & Marletta, M.M. (1989): Macrophage oxidation of L-arginine to nitric oxide, nitrite and nitrate. Tetrahydrobiopterin is an intermediate. *J. Biol. Chem.* **264**, 19654–19658.

8.   Wachter, H., Fuchs, D., Hausen, A., Reibnegger, G. & Werner, E.R. (1989): Neopterin as marker for activation of cellular immunity. Immunological basis and clinical application. *Adv. Clin. Chem.* **27**, 81–141.

9.   Werner, E.R., Werner-Felmayer, G., Fuchs, D., Hausen, A., Reibnegger, G. & Wachter, H. (1989): Parallel induction of tetrahydrobiopterin synthesis and indoleamine 2,3-dioxygenase activity in human cells and cell lines by interferon-γ. *Biochem. J.* **262**, 861–866.

10. Werner, E.R., Werner-Felmayer, G., Fuchs, D., Hausen, A., Reibnegger, G., Yim, J.J., Pfleiderer, W. & Wachter, H. (1990): Tetrahydrobiopterin biosynthetic activities in human macrophages, fibroblasts, THP-1 and T 24-cells. GTP-cyclohydrolase I is stimulated by interferon-γ, and 6-pyruvoyl tetrahydropterin synthase and sepiapterin reductase are constitutively present. *J. Biol. Chem.* **265**, 3189–3192.

11. Werner, E.R., Werner-Felmayer, G., Fuchs, D., Hausen, A., Reibnegger, G., Yim, J.J. & Wachter, H. (1991): Impact of tumour necrosis factor α and interferon-γ on tetrahydrobiopterin synthesis in murine fibroblasts and macrophages. *Biochem. J.* **280**, 709–714.

12. Werner-Felmayer, G., Werner, E.R., Fuchs, D., Hausen, A., Reibnegger, G. & Wachter, H. (1990): Neopterin formation and tryptophan degradation by a human myelomonocytic cell cline (THP-1) upon cytokine treatment. *Cancer Res.* **50**, 2863–2867.

13. Werner-Felmayer, G., Werner, E.R., Fuchs, D., Hausen, A., Reibnegger, G. & Wachter, H. (1990): Tetrahydrobiopterin-dependent formation of nitrite and nitrate by murine fibroblasts. *J. Exp. Med.* **172**, 1599–1607.

Guanidino Compounds in Biology and Medicine, eds. by P.P. De Deyn, B. Marescau, V. Stalon and
I.A. Qureshi. ©1992 John Libbey & Company Ltd., pp. 89–96.

Chapter 13

The arginine-nitric oxide pathway mediates non-adrenergic non-cholinergic neurotransmission in gastrointestinal tissue

G.E. BOECKXSTAENS, P.A. PELCKMANS, H. BULT, J.G. De MAN,
A.G. HERMAN and Y.M. Van MAERCKE

*Divisions of Gastroenterology and Pharmacology, Faculty of Medicine, University of Antwerp (UIA), B-2610
Antwerpen-Wilrijk, Belgium*

Summary

The role of nitric oxide (NO) was investigated in the inhibitory non-adrenergic non-cholinergic (NANC) innervation
of the rat gastric fundus and the canine lower oesophageal sphincter and ileocolonic junction. Exogenous administration
of NO induced concentration-dependent and tetrodotoxin-resistant relaxations, which mimicked those in response to
short periods (10 s) of electrical stimulation. N^G-nitro-L-arginine (L-NNA) and N^G-monomethyl-L-arginine (L-
NMMA), stereospecific inhibitors of the NO biosynthesis, inhibited the relaxations induced by electrical stimulation,
an effect that was prevented by L-arginine but not by D-arginine. Furthermore, haemoglobin abolished the NO-induced
responses and reduced those to NANC nerve stimulation.

Using a superfusion bioassay, we demonstrated the release of a vasorelaxant factor which pharmacologically behaves
like NO and is synthesized from L-arginine in the rat gastric fundus and ileocolonic junction. The release was inhibited
by L-NNA and by tetrodotoxin, a blocker of the neuronal conductance.

In conclusion, our results demonstrate that NO or a NO-releasing substance is formed from L-arginine upon NANC
nerve stimulation and suggest that NO is an inhibitory NANC neurotransmitter in the gut.

Introduction

The inhibitory innervation of the gastrointestinal tract is mainly provided by intramural neurons which release a neurotransmitter different from acetylcholine and noradrenaline. Hence, these nerves are called non-adrenergic non-cholinergic or NANC nerves. They are concerned with the reflex opening of the lower oesophageal and the internal anal sphincter, the receptive relaxation of the stomach and the descending inhibition during intestinal peristalsis[9]. Also in pathophysiological conditions such as achalasia, nausea, vomiting, gastric dysrhythmia, paralytic ileus and Hirschsprung's disease, these nerves are believed to play an important role[1].

The identity of the inhibitory NANC neurotransmitter in the gastrointestinal tract however is still debated. According to the purinergic theory[8] adenosine 5′-triphosphate (ATP) or a related purine nucleotide has been proposed as putative NANC neurotransmitter, whereas the peptidergic theory[15] suggests vasoactive intestinal polypeptide (VIP) as mediator of the NANC inhibition. However, neither of these substances fulfils all the criteria of a neurotransmitter[13]. Recently, a role as neurotransmitter in the brain[12,18] and the rat anococcygeus muscle[14] has been ascribed to nitric oxide (NO), a labile gas which accounts for the biological activity of the endothelium-derived relaxing factor or EDRF[21]. Therefore, we have investigated the possible role of NO, which is synthesized from its precursor L-arginine[22], in the NANC neurotransmission of the rat gastric fundus[2], and the canine ileocolonic junction[3–5,7] and lower oesophageal sphincter[10].

Materials and methods

Experiments in organ baths

Wistar rats of both sexes (150–250 g) were reserpinized (5 mg/kg, ip) 24 h before sacrificing; they were fasted but had free access to water. Longitudinal muscle strips of the gastric fundus were prepared as described[2].

Mongrel dogs of either sex (body weight 10–30 kg) were anaesthetized with sodium pentobarbitone (30 mg/kg, iv) and a laparotomy was performed. The lower oesophageal sphincter, the ileocolonic junction and the terminal ileum were resected and after removal of the mucosa, circular muscle strips were prepared as described[10,23].

The muscle strips were mounted in organ baths (25 ml) filled with a modified Krebs–Ringer solution (mM: NaCl 118.3, KCl 4.7, $MgSO_4$ 1.2, KH_2PO_4 1.2, $CaCl_2$ 2.5, $NaHCO_3$ 25, CaEDTA 0.026 and glucose 11.1). The solution was maintained at 37 °C and aerated with a mixture of 95 per cent O_2 and 5 per cent CO_2. The muscle strips were pulled through two platinum ring electrodes and connected to a strain gauge transducer for continuous recording of isometric tension. Electrical impulses (rectangular waves, 9 V, 1–2 ms) were provided by a GRASS stimulator and a direct current amplifier.

All experiments were performed in the presence of atropine (3×10^{-7}–10^{-6} M) and during a contraction induced by 5-hydroxytryptamine (10^{-7} M in the rat gastric fundus; 3×10^{-6} M in the canine lower oesophageal sphincter) or noradrenaline (3×10^{-5} M in the canine ileocolonic junction), to better evaluate the inhibitory effects. The contractions are expressed as g contraction, whereas the relaxations are expressed as per cent decrease of the initial contraction. The negative logarithm of the concentration of agonist (pD_2) that produced a half-maximal response was calculated by linear regression analysis.

Superfusion bioassay cascade

The gastric fundus, folded as a strip, or a muscle strip of the canine ileocolonic junction, from which the mucosa and submucosa had been removed, were mounted in a perfusion chamber tube with a volume of about 0.5 ml and was perfused with a modified Krebs–Ringer solution (3 ml/min) containing superoxide dismutase (SOD, 20 U/ml), L-arginine (5×10^{-5} M) and guanethidine (3×10^{-6} M). The solution was aerated with a mixture of 75 per cent N_2, 20 per cent O_2 and 5 per cent CO_2. This perfusion chamber contained two platinum ring electrodes, through which the strip was pulled[5].

The effluent then superfused rabbit aortic rings (3 mm wide) denuded of their endothelium by gentle rubbing. These rings were arranged either in a cascade with a delay of 2–3 s between the two rings, or in parallel. The tissues were contracted by an infusion of noradrenaline (10^{-7} M) and their isometric tension was recorded. A bolus of acetylcholine (3×10^{-9} mol) was injected directly onto the aortic rings to verify the absence of the endothelium and the sensitivity of the bioassay tissues was standardized by a bolus injection of nitroglycerine (3×10^{-10} mol). Subsequently, atropine (3×10^{-7} M) was introduced into the superfusate[2,5,7].

Results

Experiments in organ baths

In the presence of atropine (3×10^{-7}–10^{-6} M) and guanethidine (3×10^{-6} M), NO (10^{-7}–3×10^{-5} M) induced transient relaxations in the rat gastric fundus and the canine ileocolonic junction and lower

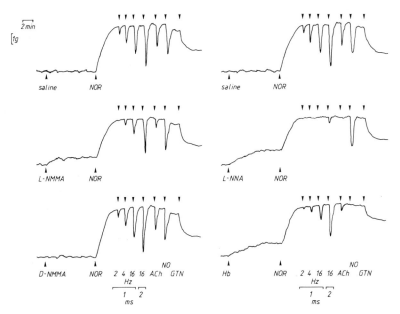

Fig. 1. Isometric tension recordings of two circular muscle strips of the canine ileocolonic junction showing the effect of L-NMMA (3×10^{-4} M), D-NMMA (3×10^{-4} M), L-NNA (10^{-4} M), and haemoglobin (3×10^{-5} M) on the relaxations to electrical stimulation (2–16 Hz, 1–2 ms), acetylcholine (ACh, 3×10^{-5} M), NO (10^{-5} M) and nitroglycerine (GTN, 10^{-6} M). All experiments were performed during a noradrenaline (NOR, 3×10^{-5} M)-induced contraction and in the presence of 3×10^{-7} M atropine. Similar results were obtained from five additional experiments. The tracings shown on the left and the right panels were obtained from separate muscle strips. Reprinted from reference[4] with permission.

oesophageal sphincter (Figs. 1 and 2). The pD_2 values were 6.11 ± 0.11 (n = 6) in the rat gastric fundus, 5.45 ± 0.09 (n = 5) in the canine ileocolonic junction and 5.40 ± 0.14 (n = 5) in the canine lower oesophageal sphincter. These inhibitory effects were resistant to blockade of the nerve conductance by tetrodotoxin (10^{-6} M) and mimicked the relaxations induced by short periods (10 s) of electrical stimulation (0.25–16 Hz, 0.5–2 ms). The relaxations induced by electrical impulses were abolished by tetrodotoxin (10^{-6} M) in all tissues studied.

Inhibition of the L-arginine:NO pathway by N^G-nitro-L-arginine (L-NNA, 3×10^{-5}–10^{-4} M) and N^G-monomethyl-L-arginine (L-NMMA, 10^{-4}–3×10^{-4} M) concentration-dependently raised the basal tension in the rat gastric fundus and the canine ileocolonic junction, but not in the canine lower oesophageal sphincter (Figs. 1 and 2). Furthermore, inhibition of the NO biosynthesis reduced the nerve-induced relaxations in all three tissues, but not those in response to NO, nitroglycerin or isoprenaline (Figs. 1 and 2). The enantiomer of L-NMMA, D-NMMA (10^{-4}–3×10^{-4} M) had no effect (Fig. 1). The inhibitory effect of L-NNA and L-NMMA was prevented by the substrate of the NO biosynthesis, L-arginine (2–5×10^{-3} M), but not by D-arginine (2–5×10^{-3} M). Finally, haemoglobin (10^{-5}–3×10^{-5} M), known to avidly bind NO[16], reduced the nerve-induced relaxations and abolished the NO-induced responses in the canine ileocolonic junction and lower oesophageal sphincter (Figs. 1 and 2).

Superfusion bioassay cascade

In the presence of atropine (3×10^{-7}–10^{-6} M), guanethidine (3×10^{-6} M), L-arginine (5×10^{-5} M) and superoxide dismutase (SOD, 20 units/ml), the rat gastric fundus and the canine ileocolonic junction released a vasorelaxant factor upon electrical stimulation (Fig. 3) (0.5–32 Hz, 1–2 ms, pulse

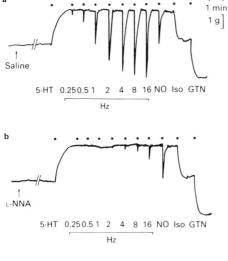

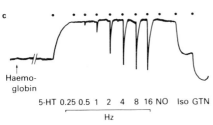

Fig. 2. Isometric tension recordings of canine lower oesophageal sphincter circular muscle strips showing (a) the control and the effects of (b) L-NNA (10^{-4} M) and (c) haemoglobin (3×10^{-5} M) on the relaxations induced by electrical stimulation (0.25–16 Hz, 0.5 ms), NO (3×10^{-6} M), isoprenaline (Iso, 10^{-4} M) and nitroglycerine (GTN, 10^{-5} M). All experiments were performed during a 5-hydroxytryptamine (5-HT, 3×10^{-6} M)-induced contraction and in the presence of atropine (10^{-6} M). Similar results were obtained from five additional experiments. Tracing-breaks represent periods of tissue equilibration. Modified from reference[10] with permission.

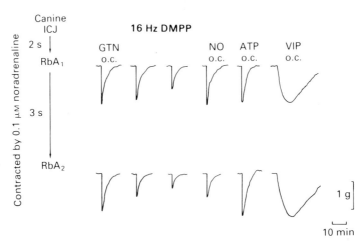

Fig. 3. Relaxation of rabbit aortic (RbA) rings arranged in a cascade by a transferable factor released by the canine ileocolonic junction (ICJ) in response to electrical impulses (16 Hz, 2 ms) and the nicotinic receptor agonist DMPP (3×10^{-5} M) and by injection of NO (10^{-10} mol), ATP (10^{-6} mol) and VIP (5×10^{-10} mol) onto the cascade (o.c.). Breaks in the tracings represent periods of tissue equilibration. The experiments were performed in the presence of atropine (3×10^{-7} M), guanethidine (3×10^{-6} M), L-arginine (5×10^{-5} M) and SOD (20 units/ml). The rings of rabbit aorta were denuded of endothelium and contracted submaximally by a continuous infusion of noradrenaline (10^{-7} M). The sensitivity of the bioassay was standardised by injection of nitroglycerine (GTN, 10^{-10} mol). Similar results were obtained in four other experiments. Reprinted from reference[5] with permission.

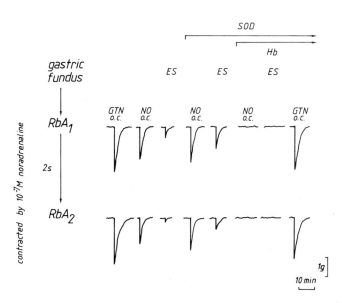

Fig. 4. Relaxation of rabbit aortic (RbA) rings arranged in a cascade by injection of NO (10^{-10} mol) over the cascade (o.c.) and by a transferable factor released by the rat gastric fundus in response to electrical stimulation (ES, 4 Hz, 1 ms). The relaxations induced by NO and the factor were increased by SOD (20 units/ml), but abolished by infusion of haemoglobin (Hb, 10^{-7} M). Breaks in the tracing represent periods of tissue equilibration. The experiments were performed in the presence of L-arginine (5×10^{-5} M) and atropine (10^{-6} M). The rings of rabbit aorta were denuded of endothelium and contracted submaximally by a continuous infusion of noradrenaline (10^{-7} M). The sensitivity of the bioassay was standardized by injection of nitroglycerine (GTN, 10^{-10} mol). Similar results were obtained from 3–5 other experiments. Reprinted from reference[2] with permission.

trains lasting 20 s) in a frequency-dependent manner. In the canine ileocolonic junction, the nicotinic agonist 1,1-dimethyl-4-piperazinium (DMPP, 3×10^{-5} M) also induced the release of this factor (Fig. 3). The relaxations of the rabbit aortic rings induced by the factor had the same morphology as those induced by nitroglycerin (10^{-10} mol), NO (10^{-10} mol) or ATP (10^{-6} mol) directly injected onto the cascade. Injection of VIP (5×10^{-10} mol) over the bioassay tissues on the other hand resulted in a slowly developing and long-lasting relaxation (Fig. 3). The biological activity of the factor declined to the same extent (44 ± 6 per cent, n = 7) as NO (42 ± 5 per cent, n = 4) during passage down the cascade, whereas that of ATP and VIP remained unchanged (Fig. 3). The relaxation of the aortic rings induced by the factor released by the rat gastric fundus or the canine ileocolonic junction was significantly augmented by SOD (Fig. 4) and L-arginine. On the other hand, these relaxations were inhibited by haemoglobin (10^{-7} M) (Fig. 4), L-NNA (10^{-4} M) and tetrodotoxin (10^{-6} M). The relaxations induced by nitroglycerin, ATP or VIP were not affected by SOD or haemoglobin.

Discussion

Endothelium-derived relaxing factor or EDRF is now thought to be nitric oxide[21] and/or S-nitroso-cysteine which spontaneously decomposes with the liberation of NO[20]. NO is enzymatically synthesized from its precursor L-arginine[22], a pathway which has been demonstrated in several cell types underlying a variety of biological actions[19]. A role for the L-arginine:NO pathway in neurotransmission has been reported in the rat cerebellum[11,12] and forebrain[18] and the rat anococcygeus muscle[14]. Our experiments have provided evidence extending the role of the L-arginine:NO pathway to the gut and demonstrated that NO or a NO-releasing molecule plays an important role in the inhibitory NANC neurotransmission of the gut[2-5,7,10].

93

One of the first requirements of any putative neurotransmitter is that it should mimic the response to nerve stimulation. In the rat gastric fundus, the canine lower oesophageal sphincter and ileocolonic junction, exogenous administration of NO indeed induced tetrodotoxin-resistant relaxations that were similar to those obtained by short periods of NANC-nerve stimulation[2–4,10]. In addition, nitroglycerine, from which NO is liberated intracellularly as active substance, relaxed the smooth muscle strips[2–4,10]. These relaxations were sustained as a result of the continuous generation of NO, whereas NO gas injected as aqueous solution rapidly disappears from the bathing solution resulting in a transient relaxation. Thus, NO is able to mimic the NANC relaxations by a direct action on the smooth muscle and hence fulfils the first criterion to be considered as putative NANC neurotransmitter.

A second criterion is that drugs which block or potentiate the response to the exogenous putative neurotransmitter should have parallel effects on the response to nerve stimulation. Haemoglobin, known to avidly bind NO[17], indeed had similar effects on the NO-induced relaxations as on the NANC nerve-mediated responses[3,4,10]. The inhibitory effect of haemoglobin on the nerve-induced relaxations was not as pronounced as the total blockade of the response to exogenous NO. However, considering the molecular size of haemoglobin, only a small portion may have reached the neuromuscular junction to be effective, explaining this discrepancy. Next, inhibition of NO biosynthesis reduced the NANC nerve-mediated relaxations in all three tissues[2–5,10]. This effect was stereospecific and competitive, as it was reversed by the substrate L-arginine but not by D-arginine, and since the enantiomer of L-NMMA, D-NMMA, was without effect. These results thus demonstrate that NO also fulfils the second criterion and thus provide further evidence for NO or a NO-releasing substance as inhibitory NANC neurotransmitter in the gut.

In both the rat gastric fundus and the canine ileocolonic junction[2–4] but not the canine lower oesophageal sphincter[10], haemoglobin and/or inhibition of the L-arginine:NO pathway raised the basal tension. These findings might indicate the presence of an inhibitory innervation resulting from a constant background release of NO. Alternatively, the rise in tension might result from an action directly on the smooth muscle.

A third fundamental criterion for NO to be considered as putative neurotransmitter is the demonstration of its release on nerve stimulation. Using a superfusion bioassay cascade, we were able to demonstrate the release of an unstable vasorelaxant factor with characteristics of NO in response to NANC nerve stimulation in the canine ileocolonic junction[5,7] and rat gastric fundus[2]: the release of this factor was tetrodotoxin sensitive, indicating that neuronal structures are activated rather than other cell populations such as endothelial or mononuclear cells or microorganisms. Recently, evidence has been reported that the enzyme which catalyses NO formation, NO synthase, is present in the intestine and is most highly concentrated in neuronal axons[6]. In addition, it has also been demonstrated that intact neuronal tissue can release NO upon stimulation[11]. Although we cannot exclude the possibility that an as yet unidentified neurotransmitter induced the release of the factor from some non-neuronal cell, e.g. the smooth muscle, the latter data favour our view that NO or a NO-releasing substance is the neurotransmitter itself and is released by the NANC nerves[2,5,7].

The nature of the factor released by the rat gastric fundus and the canine ileocolonic junction in response to electrical stimulation is indeed non-adrenergic non-cholinergic since both atropine and guanethidine were present. In addition, the aortic rings failed to relax to acetylcholine or noradrenaline[2,5,7]. As illustrated by the decrease of the biological activity down the cascade, the factor released by the ileocolonic junction is unstable, a feature that is not compatible with peptides like VIP, amines, vasodilator prostanoids or ATP, but rather with NO or a NO-releasing molecule[2,5,7]. Indeed, relaxations induced by NO declined to the same extent as those in response to the transferable factor. In addition, the biological activity of both NO and the factor[2,5,7], but not VIP and ATP[2,5] was increased by superoxide dismutase, an enzyme that prolongs the half-life of NO through suppression of its inactivation by superoxide anion[16]. Furthermore, L-arginine, the precursor of NO[22], increased the release of the factor whereas inhibitors of the NO biosynthesis decreased the release of the factor, demonstrating the presence of the L-arginine:NO pathway in the inhibitory NANC nerves of the gut[2,5,7]. Finally, haemoglobin abolished the biological activity of the factor, but not that of ATP and VIP. These results indicate that NO, but not ATP or VIP, accounts for the vasorelaxant effect of the transferable factor[2,5,7], suggesting NO or a NO-releasing substance as inhibitory NANC neurotransmitter.

In conclusion, we were able to demonstrate that NO mimics and mediates the NANC relaxations induced by short periods of electrical stimulation in the rat gastric fundus[2] and the canine lower oesophageal sphincter[10] and ileocolonic junction[3,4]. Furthermore, using a superfusion bioassay, we demonstrated the release of an unstable vasorelaxant factor in response to NANC nerve stimulation in the rat gastric fundus[2] and canine ileocolonic junction[5,7]. This factor behaves like NO and is synthesized from L-arginine, extending the role of the L-arginine:NO pathway to the NANC inhibitory neurotransmission of the gut. Based on these results, we have proposed NO or a NO-releasing substance as inhibitory NANC neurotransmitter in the gut[2–5,7,10]. In contrast to classical neurotransmitters, the instability of NO makes classical inactivation by uptake or enzyme systems unnecessary. Furthermore, since NO is a highly lipophilic substance, it will diffuse easily through the postjunctional cellular membrane without binding to an extracellular receptor. Inside the cell, the haem group of guanylate cyclase will bind NO and thus acts as intracellular receptor. This results in activation of guanylate cyclase and induces the synthesis of cyclic GMP resulting in smooth muscle relaxation. Therefore, if future experiments confirm the role of NO in NANC neurotransmission, it will be an odd and primitive neurotransmitter which is also reflected by its chemical composition.

Acknowledgements

This work was supported by the Belgian Fund for Medical Research (Grant 3.0014.90). GEB is a Research Assistant of the National Fund for Scientific Research Belgium (NFWO). The authors gratefully acknowledge F.H. Jordaens for the technical assistance and Mrs L. Van de Noort for typing the manuscript.

References

1. Abrahamsson, H. (1986): Non-adrenergic non-cholinergic nervous control of gastrointestinal motility patterns. *Arch. Int. Pharmacodyn. Ther.* **280**, 50–61.

2. Boeckxstaens, G.E., Pelckmans, P.A., Bogers, J.J., Bult, H., De Man, J.G., Oosterbosch, L., Herman, A.G. & Van Maercke, Y.M. (1991): Release of nitric oxide upon stimulation of non-adrenergic non-cholinergic nerves in the rat gastric fundus. *J. Pharmacol. Exp. Ther.* **256**, 441–447.

3. Boeckxstaens, G.E., Pelckmans, P.A., Bult, H., De Man, J.G., Herman, A.G. & Van Maercke, Y.M. (1991): Evidence for nitric oxide as mediator of non-adrenergic non-cholinergic relaxations induced by ATP and GABA in the canine gut. *Br. J. Pharmacol.* **102**, 434–438.

4. Boeckxstaens, G.E., Pelckmans, P.A., Bult, H., De Man, J.G., Herman, A.G. & Van Maercke, Y.M. (1990): Non-adrenergic non-cholinergic relaxation mediated by nitric oxide in the canine ileocolonic junction. *Eur. J. Pharmacol.* **190**, 239–246.

5. Boeckxstaens, G.E., Pelckmans, P.A., Ruytjens, I.F., Bult, H., De Man, J.G., Herman, A.G. & Van Maercke, Y.M. (1991): Bioassay of nitric oxide released upon stimulation of non-adrenergic non-cholinergic nerves in the canine ileocolonic junction. *Br. J. Pharmacol.* **103**, 1085–1091.

6. Bredt, D.S., Hwang, P.M. & Syder, S.H. (1990): Localization of nitric oxide synthase indicating a neural role for nitric oxide. *Nature* **347**, 768–770.

7. Bult, H., Boeckxstaens, G.E., Pelckmans, P.A., Jordaens, F.H., Van Maercke, Y.M. & Herman, A.G. (1990): Nitric oxide as an inhibitory non-adrenergic non-cholinergic neurotransmitter. *Nature* **345**, 346–347.

8. Burnstock, G. (1972): Purinergic nerves. *Pharmacol. Rev.* **24**, 5O9–581.

9. Burnstock, G. (1981): Review lecture. Neurotransmitters and trophic factors in the autonomic nervous system. *J. Physiol. Lond.* **313**, 1–35.

10. De Man, J.G., Pelckmans, P.A., Boeckxstaens, G.E., Bult, H., Oosterbosch, L., Herman, A.G. & Van Maercke, Y.M. (1991): The role of nitric oxide in inhibitory non-adrenergic, non-cholinergic neurotransmission in the canine lower oesophageal sphincter. *Br. J. Pharmacol.* **103**, 1092–1096.

11. Dickie, B.G.M., Lewis, M.J. & Davies, J.A. (1990): Potassium-stimulated release of nitric oxide from cerebellar slices. *Br. J. Pharmacol.* **101**, 8–9.

12. Garthwaite, J., Charles, S.L. & Chess Williams, R. (1988): Endothelium-derived relaxing factor release on activation of NMDA receptors suggests role as intercellular messenger in the brain. *Nature* **336**, 385–388.

13. Gillespie, J.S. (1982): Non-adrenergic non-cholinergic inhibitory control of gastrointestinal motility. In *Motility of the digestive tract*, ed. M. Wienbeck, pp. 51–66. New York: Raven Press.

14. Gillespie, J.S., Liu, X.R. & Martin, W. (1989): The effects of L-arginine and N^G-monomethyl L-arginine on the response of the rat anococcygeus muscle to NANC nerve stimulation. *Br. J. Pharmacol.* **98**, 1080–1082.

15. Goyal, R.K., Rattan, S. & Said, S.I. (1980): VIP as a possible neurotransmitter of non-cholinergic non-adrenergic inhibitory neurones. *Nature* **288**, 378–380.

16. Gryglewski, R.J., Palmer, R.M. & Moncada, S. (1986): Superoxide anion is involved in the breakdown of endothelium-derived vascular relaxing factor. *Nature* **320**, 454–456.

17. Kelm, M., Feelisch, M., Spahr, R., Piper, H.M., Noack, E. & Schrader, J. (1988): Quantitative and kinetic characterization of nitric oxide and EDRF released from cultured endothelial cells. *Biochem. Biophys. Res. Commun.* **154**, 236–244.

18. Knowles, R.G., Palacios, M., Palmer, R.M. & Moncada, S. (1989): Formation of nitric oxide from L-arginine in the central nervous system: a transduction mechanism for stimulation of the soluble guanylate cyclase. *Proc. Natl. Acad. Sci. USA* **86**, 5159–5162.

19. Moncada, S., Palmer, R.M. & Higgs, E.A. (1989): Biosynthesis of nitric oxide from L-arginine. A pathway for the regulation of cell function and communication. *Biochem. Pharmacol.* **38**, 1709–1715.

20. Myers, P.R., Minor, R.L. Jr, Guerra, R. Jr, Bates, J.N. & Harrison, D.G. (1990): Vasorelaxant properties of the endothelium-derived relaxing factor more closely resemble S-nitrosocysteine than nitric oxide. *Nature* **345**, 161–163.

21. Palmer, R.M., Ferrige, A.G. & Moncada, S. (1987): Nitric oxide release accounts for the biological activity of endothelium-derived relaxing factor. *Nature* **327**, 524–526.

22. Palmer, R.M., Rees, D.D., Ashton, D.S. & Moncada, S. (1988): L-arginine is the physiological precursor for the formation of nitric oxide in endothelium-dependent relaxation. *Biochem. Biophys. Res. Commun.* **153**, 1251–1256.

23. Pelckmans, P.A., Boeckxstaens, G.E., Van Maercke, Y.M., Herman, A.G. & Verbeuren, T.J. (1989): Acetylcholine is an indirect inhibitory transmitter in the canine ileocolonic junction. *Eur. J. Pharmacol.* **170**, 235–242.

Guanidino Compounds in Biology and Medicine, eds. by P.P. De Deyn, B. Marescau, V. Stalon and I.A. Qureshi. ©1992 John Libbey & Company Ltd., pp. 97–101.

Chapter 14

Corticosteroids prevent the induction of nitric oxide-synthase activity in different pulmonary cells

P.G. JORENS, F.J. Van OVERVELD, H. BULT[1], P.A. VERMEIRE and A.G. HERMAN[1]

Divisions of Respiratory Medicine and [1]Pharmacology, University of Antwerp (UIA), Universiteitsplein 1, B-2610 Wilrijk, Belgium

Summary

Rat alveolar macrophages and rat lung fibroblasts secreted nitrite (NO_2^-) upon stimulation with recombinant interferon-γ. This production was dependent on the presence of L-arginine in the incubation medium. Dexamethasone inhibited the induction of NO_2^- production in both cells. Cortexolone, a partial agonist of the glucorticoid receptor, and RU 38 486, a pure antagonist, were able to modulate and partially inhibit this suppressive effect. These data indicate that glucocorticoids are capable of inhibiting the induction of nitric oxide synthase activity via a receptor-mediated mechanism in the two lung cell types investigated.

Introduction

It is known that many cell types, including macrophages[4] and endothelial cells[10], produce nitric oxide (NO) which is rapidly decomposed into nitrite (NO_2^-) and nitrate (NO_3^-). This pathway is L-arginine (L-arg) dependent with citrulline as a by-product[4]. NO is supposed to contribute to the cytotoxicity of activated macrophages[3].

Recently, it was discovered that a calcium-independent nitric oxide synthase is induced in the lungs of rats treated with endotoxin (lipopolysaccharide, LPS)[6] and that anti-inflammatory glucocorticoids are able to inhibit this induction[7]. In view of our recent observation that alveolar and pleural macrophages are capable of producing L-arginine derived nitrogen oxides[5], this prompted us to investigate if glucocorticoids were able to inhibit nitric oxide synthase in rat alveolar macrophages as well as in another lung cell, fibroblasts, and to elucidate the mechanism involved.

Materials and methods

Culture of rat alveolar macrophages

Specific pathogen-free male Wistar rats (weight 200–250 g; Proefdierencentrum Leuven, Belgium) were killed and alveolar macrophages were harvested according to standard procedures[11]. Briefly, the recovered and washed cells were suspended in Dulbecco's modified Eagle's medium without phenol red (DMEM) supplemented with 5 per cent foetal calf serum (FCS) and penicillin/streptomycin. The macrophage populations were enriched by adherence to plastic in 24-well sterile dishes (Nunc, Roskilde, Denmark) with 0.25×10^6 macrophages added to each well. After 90 min, the nonadherent cells were removed by washing with prewarmed DMEM with penicillin/streptomycin. Then, 1050 µl of medium with or without the stimuli or corticosteroids dissolved in 50 µl medium were added to the wells: the cells were then incubated at 37 °C and 5 per cent CO_2 during the period indicated. The adherent cells contained more than 95 per cent macrophages as assessed by May Grunwald–Giemsa and non-specific esterase staining. Cell viability was always greater than 95 per cent for all wells and in all incubation circumstances (trypan blue exclusion).

Culture of rat lung fibroblasts

A standard explant procedure was used to cultivate rat pulmonary fibroblasts[1]. Small lung tissue fragments (± 1 cm^3) were anchored on the surface of tissue culture flasks (Nunc, Roskilde, Denmark) and cultured in DMEM supplemented with 2 mM L-glutamine, non-essential amino acids, 10 per cent FCS, 100 pg/ml streptomycin and 100 U/ml penicillin. Once confluent, the fibroblasts were trypsinized and seeded at 5×10^4 cells/well (24-well plastic dishes, Nunc, Roskilde, Denmark). Second-passage confluent rat lung fibroblasts ($3.4 \times 10^5 \pm 0.2$ cells/well, as determined by a Coulter counter after trypsinization) were then stimulated with rIFN-γ with or without addition of corticosteroids. The wells contained more than 95 per cent fibroblasts as assessed by morphological criteria and positive vimentine immune-staining.

Nitrite assay

Nitrite (NO_2^-) was determined on cell-free supernatant by a spectrophotometric assay based on the Griess reaction[5,13]. This technique determines NO_2^- and not NO_3^-. HCl (90 µl 6.5 M) and 90 µl 37.5 mM sulphanilic acid were added to 900 µl supernatant, followed by 90 µl 12.5 mM N-(naphthyl)-ethylene diamine HCl 30 min later. After 30 min the absorbance was measured at 540 nm. The NO_2^- content was expressed as nmol/10^6 macrophages or nmol/ confluent monolayer of rat lung fibroblasts. The NO_2^- content of DMEM or DMEM supplemented with the corticosteroids was always less than 0.5 nmol/ml (= detection limit).

Materials and drugs

Dulbecco's modified Eagle's medium without phenol red (DMEM) and heat-inactivated foetal calf serum (FCS), L-glutamine and non-essential amino acids were purchased from Gibco Ltd., Paisley, UK. Cortexolone (4-pregnene-17, 21-diol-3, 20-dione), dexamethasone, *Escherichia coli* lipopolysaccharide serotype O111:B4 (LPS), hydrocortisone, N-(1-naphthyl)-ethylene diamine hydrochloride, penicillin, progesterone, streptomycin and sulphanilic acid came from Sigma Chemical Co. (St. Louis, MO, USA). Rat recombinant interferon-γ (rIFN-γ) was a kind gift from P.H. Van der Meide (TNO, Rijswijk, The Netherlands). Roussel UCLAF (Romainville, France) supplied RU 38 486 (17β-hydroxy-11β-(4-dimethylamino-phenol) 17α-(prop-1-ynyl) estra-4,9-diene-3-one).

Statistical analysis

The data are expressed as mean $\pm$ SEM for six cell lines from different rats.

Results

rIFN-γ induced NO_2^- release

Rat alveolar macrophages incubated over a period of 48 h released considerable amounts of NO_2^- when stimulated with 500 ng/ml LPS or 50 U/ml rIFN-γ (Table 1). The biosynthesis required the presence of L-arg in the medium: culture in L-arg deficient medium resulted in a complete loss of the observed NO_2^- production. Longer incubation periods were not used since this resulted in considerable decline in cell viability (< 75 per cent) as assessed by trypan blue exclusion. Rat lung fibroblasts, passaged and stimulated after forming a monolayer, released 30.2 ± 1.6 nmol NO_2^-/72 h/10^6 lung fibroblasts when treated with 50 U/ml rIFN-γ, but they did not produce any NO_2^- after stimulation with LPS. Again, this secretion was dependent on the presence of L-arg in the medium.

Table 1. Secretion of NO_2^- (nmol/10^6 cells) after stimulation of alveolar macrophages with LPS or IFN-γ for 48 h (mean ± SEM, n = 6)

	Medium with L-arg	Medium without L-arg
Control	1.0 ± 0.3	0.8 ± 0.4
LPS (500 ng/ml)	21.5 ± 4.5	0.7 ± 0.3
IFN-γ (50 U/ml)	63.0 ± 9.1	0.7 ± 0.3

Inhibition of NO_2^- release by dexamethasone

Co-incubation of alveolar macrophages with dexamethasone (10^{-9} to 10^{-5} M) or hydrocortisone (10^{-9} to 10^{-5} M) caused a concentration dependent inhibition of the NO_2^- formation induced by LPS (500 ng/ml) or rIFN-γ (50 U/ml). Figure 1 shows the results for the incubations with rIFN-γ and dexamethasone. Similar results were obtained for rat lung fibroblasts: both hydrocortisone (10^{-7}–10^{-4} M) and dexamethasone (10^{-7}–10^{-4} M) inhibited the rIFN-γ induced NO_2^- production. Figure 2 represents the data obtained for dexamethasone. In both cell types, progesterone (10^{-9}–10^{-3} M) failed to inhibit this secretion.

Modulation of the inhibition by dexamethasone

Addition of equimolar concentrations of cortexolone to the incubations with IFN-γ and dexamethasone blocked partially the inhibitory action of dexamethasone on NO_2^- secretion by rat alveolar macrophages (Fig. 1). Comparable results were seen in rat lung fibroblasts (Fig. 2).

Moreover, the steroid analogue RU 38 486, added in equimolar concentrations to the incubations with dexamethasone, almost completely reversed the effect of the glucocorticoid, both in rat alveolar macrophages and rat lung fibroblasts stimulated with rIFN-γ (Figs. 1 and 2).

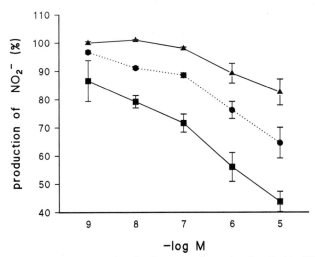

Fig. 1. Inhibition of the production of nitrite by alveolar macrophages incubated with rIFN-γ (50 U/ml) by dexamethasone (■), expressed as per cent of the incubation with rIFN-γ alone. The inhibitory effect was partially reversed by the addition of equimolar concentrations of cortexolone (●) or RU 38 486 (▲) to incubations with dexamethasone. The results are expressed as mean ± SEM for six experiments.

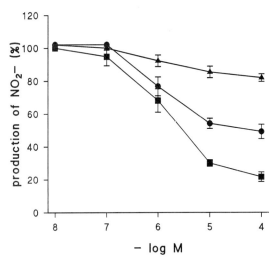

Fig. 2. Inhibition of the production of nitrite by rat lung fibroblasts incubated with rIFN-γ (50 U/ml) in the presence of dexamethasone (■). Addition of equimolar concentrations of cortexolone (●) or RU 38 486 (▲) to incubations with dexamethasone partially reversed the inhibitory effect.
Data are presented as per cent of the incubation with rIFN-γ alone, and given as mean ± SEM, n = 6.

Discussion

Our findings show that alveolar macrophages and lung fibroblasts can produce NO_2^- after stimulation with the appropriate stimuli (LPS and rIFN-γ for macrophages, rIFN-γ for lung fibroblasts). This secretion was dependent on the presence of L-arg in the medium. Previously, we confirmed that NO was an intermediate in the NO_2^- biosynthesis by rat alveolar macrophages, and that the induction of the NO synthase activity required *de novo* protein synthesis[5]. The observed amounts of NO_2^- produced by the rat alveolar macrophages and lung fibroblasts were in agreement with the amounts reported for cytokine-induced NO_2^- production in murine peritoneal macrophages[3,4,15] and murine skin fibroblasts[17] respectively.

The main finding is the fact that glucocorticoids were able to block the induction of NO synthase by cytokines in both alveolar macrophages and lung fibroblasts. Moreover, the underlying mechanism of this inhibition was suggested to be receptor-mediated, since the addition of equimolar concentrations of cortexolone, a partial agonist of the glucocorticoid receptor, partially blocked the observed inhibitory action of dexamethasone. These findings in both pulmonary cell types confirm previous reports of the inhibitory effect of glucocorticoids on the induction of nitric oxide synthase in the macrophage cell line J774[2].

Furthermore, RU 38 486 almost completely abolished the observed inhibitory effect of dexamethasone in both lung cell types. RU 38 486 appears to be a potent progestin and glucocorticoid antagonist. Its effect is caused by its high affinity for the glucocorticoid receptor in different cells. Although it exhibits no agonistic effect in certain experimental conditions, both *in vivo* and *in vitro*[8,12,16], RU 38 486 in concentrations of 10^{-9} to 10^{-4} M exerted a potent antagonism on the dexamethasone induced inhibition of NO_2^- in both lung cell types.

These findings suggest that glucocorticoids could influence lung defence mechanisms by inhibiting the induction of L-arg dependent production of nitric oxide in different pulmonary cells. Indeed, the nitric oxide synthase pathway is believed to play an important role in the cytotoxic effector mechanisms of immune-stimulated murine macrophages[3,4]. It remains to be determined if glucocorticoids administered *in vivo* could influence local immune reactions in the (rat) lung by inhibiting nitric oxide synthase induction in different lung cells, including alveolar macrophages, the resident phagocyte of the alveolar space[14]. This could add further proof to the assumption that the inhibitory action of glucocorticoids on the induction of nitric oxide synthase activity could contribute to the observed anti-inflammatory effect of glucocorticoids *in vivo*[9].

Acknowledgements

The authors wish to thank Mrs Greta Van de Vyver and Mrs Denise Andries for typing the manuscript, and wish to express their gratitude to Dr. P.H. Van der Meide (TNO, Rijswijk, The Netherlands) for the gift of rIFN-γ and to Mrs M. Garnier (Roussel-UCLAF, Romainville, France) for the gift of RU 38 486.

References

1. Absher, M. (1989): Fibroblasts. In *Lung cell biology*, ed. D. Massaro, pp. 401–439. New York: Marcel Dekker Inc.

2. Di Rosa, M., Radomski, M., Carnuccio, R.R. & Moncada, S. (1990): Glucocorticoids inhibit the induction of nitric oxide synthase in macrophages. *Biochem. Biophys. Res. Commun.* **172**, 1042–1048.

3. Drapier, J.C., Wietzerbin, J. & Hibbs, J.B. Jr. (1988): Interferon-γ and tumor necrosis factor induce the L-arginine-dependent cytotoxic effector mechanism in murine macrophages. *Eur. J. Immunol.* **18**, 1587–1592.

4. Hibbs, J.B. Jr., Taintor, R.R. & Vavrin, Z. (1987): Macrophage cytotoxicity: role for L-arginine deiminase and imino nitrogen oxidation to nitrite. *Science* **235**, 473–476.

5. Jorens, P.G., Van Overveld, F.J., Bult, H., Vermeire, P.A. & Herman, A.G. (1991): L-Arginine-dependent production of nitrogen oxides by rat pulmonary macrophages. *Eur. J. Pharmacol.* **200**, 205–209.

6. Knowles, R.G., Merret, M., Salter, M. & Moncada, S. (1990): Differential induction of brain, lung and liver nitric oxide synthase by endotoxin in the rat. *Biochem. J.* **270**, 883–886.

7. Knowles, R.G., Salter, M., Brooks, S.L. & Moncada, S. (1990): Anti-inflammatory glucocorticoids inhibit the induction by endotoxin of nitric oxide synthase in the lung, liver and aorta of the rat. *Biochem. Biophys. Res. Commun.* **172**, 1246–1248.

8. Moguilensky, M. & Philibert, D. (1984): RU 38 486: potent antiglucocorticoid activity correlated with strong binding to the cytosolic glucocorticoid receptor followed by an impaired activation. *J. Steroid Biochem.* **20**, 271–276.

9. Moncada, S. & Palmer, R.M.J. (1991): Inhibition of the induction of nitric oxide synthase by glucocorticoids: yet another explanation for their anti-inflammatory effects? *TIPS* **12**, 130–131.

10. Palmer, R.M.J., Ferrige, A.G. & Moncada, S. (1987): Nitric oxide release accounts for the biological activity of endothelium-derived relaxing factor. *Nature* **327**, 524–526.

11. Pauwels, R.A., Kips, J.C., Peleman, R.A. & Van der Straeten, M.E. (1990): The effect of endotoxin inhalation on airway responsiveness and cellular influx in rats. *Am. Rev. Respir. Dis.* **141**, 540–545.

12. Philibert, D. (1984): RU 38 486: an original multifaceted antihormone *in vivo*. In *Adrenal steroid antagonism*, ed. M.K. Agarwal, pp. 77–100. Berlin: Walter de Gruyter & Co.

13. Schmidt, H.H.H.M., Nau, H., Wittfoht, W., Gerlach, J., Prescher, K.E., Klein, M.M., Niroomand, F. & Böhme, E. (1988): Arginine is a physiological precursor of endothelium-derived nitric oxide. *Eur. J. Pharmacol.* **154**, 213–216.

14. Sibille, Y. & Reynolds, H.J. (1990): Macrophages and polymorphonuclear neutrophils in lung defense and injury. *Am. Rev. Respir. Dis.* **41**, 471–501.

15. Stuehr, D.J., Gross, S.S., Sakuma, I., Levi, R. & Nathan, C.F. (1989): Activated murine macrophages secrete a metabolite of arginine with the bioactivity of endothelium-derived relaxing factor and the chemical reactivity of nitric ozide. *J. Exp. Med.* **169**, 1011–1020.

16. Van Voorhis, B.J., Anderson, D.J. & Hill, J.A. (1989): The effects of RU 38486 on immune function and steroid-induced immunosuppression *in vitro*. *J. Clin. Endocrin. Metabol.* **69**, 1195–1199.

17. Werner-Felmayer, G., Werner, E.R., Fuchs, D., Hausen, A., Reibnegger, G. & Wachter, H. (1990): Tetrahydrobiopterin-dependent formation of nitrite and nitrate in murine fibroblasts. *J. Exp. Med.* **172**, 1599–1607.

Guanidino Compounds in Biology and Medicine, eds. by P.P. De Deyn, B. Marescau, V. Stalon and I.A. Qureshi. ©1992 John Libbey & Company Ltd., pp. 103–108.

Chapter 15

Effect of guanidino compounds on endothelium-dependent relaxation and superoxide production

George THOMAS, Kurt F. HEIM and Peter W. RAMWELL

Department of Physiology and Biophysics, Georgetown University Medical Center, 3900 Reservoir Road NW, Washington DC 20007, USA

Summary

Guanidine containing compounds such as Nα-benzoyl L-arginine ethyl ester (BAEE) and other substituted arginines elicit relaxation in the rat aorta preparation and inhibit human platelet aggregation. L-arginine has no vasodilatory effect except in millimolar concentrations and this effect is neither endothelium-dependent nor stereospecific. The endothelium-dependent relaxing factor (EDRF) is inhibited by superoxide anion. The vasodilatory effect of BAEE is enhanced by superoxide dismutase (SOD) and attenuated by several superoxide generating systems. Incubation of rat aortic rings with BAEE attenuates superoxide anion concentration. This effect of BAEE is antagonized by the inhibitor of EDRF generation, N^G-monomethyl L-arginine (L-NMMA). In contrast, L-arginine stimulates superoxide production. These results indicate that substituted guanidines such as BAEE generate an EDRF-like agent, whereas L-arginine has little or no effect. This lends support to the notion that EDRF may be derived from a derivative of L-arginine and not L-arginine *per se*.

Introduction

Most of the peptides which elicit endothelium-dependent relaxation possess the amino acid L-arginine in their sequence[13]. However, L-arginine is a poor vasodilator except in high millimolar concentrations and the relaxation is not endothelium dependent[12,13]. In contrast, several modified guanidine containing compounds such as Nα-benzoyl L-arginine ethyl ester (BAEE) are potent vasodilators in the rat aorta, mesentery, heart and kidney[1,11,14,15]. These effects of BAEE are antagonized by the EDRF inhibitors, haemoglobin and superoxide anion[1,11,14,15]. The scavenger of superoxide anion, superoxide dismutase (SOD) enhances these effects[1,11,14,15]. These *in vitro* studies indicate that guanidine-containing compounds such as BAEE generate an EDRF-like agent. In this report we give further evidence to support this hypothesis and show that under the same conditions, the amino acid L-arginine does not serve as a precursor of EDRF.

Materials and methods

Vascular activity of guanidino compounds in the rat aorta

The bio-assays using isolated rat aortic rings were performed as suggested previously[13]. The non-specific nature of arginine-induced relaxation was investigated by adding L- and D-arginine (6.5 $\times$ 10^{-2} M) to the rat aorta preparation. From some preparations the endothelium was removed by rubbing a metal spatula through the lumen of the artery for 30 s and the endothelium removal was confirmed as suggested previously[2].

In order to study whether guanidines other than L-arginine elicit vascular relaxation the dipeptide L-arginine–L-arginine was added cumulatively to the rat aorta preparation and the relaxation response was calculated. Methylene blue is an inhibitor of soluble guanylate cyclase. The relaxation response elicited by di-arginine was studied by adding methylene blue (3×10^{-5} M) to the bio-assay bath and comparing with that in the absence of methylene blue.

The EDRF inhibitor, L-NMMA, inhibits the endothelium-dependent relaxation[8]. In order to study the specificity of the vascular relaxation elicited by compounds such as BAEE, rat aortic rings were relaxed with BAEE in a dose-dependent manner. This relaxation-response was then compared with that obtained in the presence of L-NMMA.

In vivo effect of guanidino compounds

The *in vivo* effects of the guanidino compounds were studied by measuring the change in mean arterial pressure (MAP) in pentobarbital-anaesthetized (65 mg/kg, i.p.) Sprague-Dawley rats. The various guanidines were administered through a cannula in the jugular vein in a dose-dependent manner. Methylene blue, an inhibitor of soluble guanylate cyclase, attenuates endothelium-dependent relaxation. This suggests that EDRF activates soluble guanylate cyclase. In order to study whether the *in vivo* effects of guanidino compounds such as BAEE are also mediated by the activation of soluble guanylate cyclase, the change in MAP was measured by the co-administration of methylene blue (100 µg, i.v.).

Effect of substituted guanidino compounds on human platelet aggregation

Human platelet rich plasma (PRP) was collected as suggested previously[10]. Platelet aggregation was initiated by the addition of collagen (2 µg/ml). The various guanidino compounds were added 1 min prior to the addition of collagen. To study the interaction between endothelial cells and platelets, endothelial cells were harvested from freshly isolated porcine aorta and treated with indomethacin as suggested previously[10]. The platelet aggregation was studied in the presence of endothelial cells before and after the addition of the guanidino compounds.

Effect of guanidino compounds on superoxide production

Endothelium-dependent relaxation is inhibited by the superoxide anion[3,9]. Compounds such as sodium nitroprusside which generate nitric oxide, scavenge the superoxide anion[5]. In this study, the effect of various guanidino compounds on superoxide production was investigated. The experimental conditions for the detection of superoxide anion were the same as suggested previously[5,6].

Results

Effect of L- and D-arginine on vascular relaxation

L-arginine has no effect on vascular relaxation except in millimolar concentrations[13]. Our results show that at extremely large concentrations (6.5×10^{-2} M), L-arginine elicited a significant relaxation of the rat aorta (Fig. 1) irrespective of the presence of endothelium. In addition, the relaxation was not stereo-specific since, at high concentrations, D-arginine also elicited the same degree of relaxation (Fig. 1).

Effect of di-arginine on vascular relaxation

In contrast to the monomeric form of L-arginine, the L-arginine dimer was a potent vasodilator in the rat aorta (Fig. 2). Similar results were obtained with other arginine-containing dipeptides (Thomas and Ramwell, unpublished observation). The relaxation elicited by di-arginine was significantly

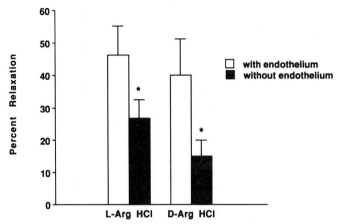

Fig. 1. Vascular relaxation elicited by L- and D-arginine HCl (6.5×10^{-2} M) in the presence and absence of the endothelium.

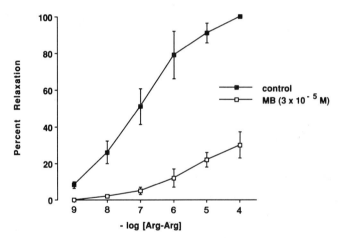

Fig. 2. Vascular relaxation mediated by the dipeptide L-arg–L-arg and its inhibition by methylene blue (MB).

attenuated by pre-treatment with methylene blue (Fig. 2). This shows that in the rat aorta preparation, modified arginine-containing compounds are superior vasodilators to L-arginine.

In vivo **effects of guanidines**

Figure 3 shows that administration of BAEE dose-dependently decreased the mean arterial pressure. Comparable concentrations of L-arginine had no effect. In Table 1 we show the relative potencies of several guanidino compounds on MAP. It is clear that BAEE is the most potent hypotensive agent followed by L-arginine ethyl ester (AEE). The substituted guanidine, benzoyl glycine L-arginine (BGA) which has an additional glycine moiety, is a relatively poor hypotensive agent when compared to BAEE. Similar to L-arginine, unsubstituted guanidine also has no effect on the mean arterial pressure. The hypotensive effect elicited by BAEE is antagonized by the co-administration of methylene blue or L-NMMA (Thomas and Ramwell, unpublished observation). These *in vivo* studies demonstrate that properly substituted guanidino compounds are superior to free L-arginine as candidates for EDRF generation.

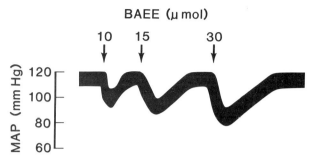

Fig. 3. Change in the mean arterial pressure (MAP) induced by the administration of BAEE (i.v.) in the anaesthetized rat.

Table 1. Hypotensive effect elicited by the various guanidines each at a dose of 50 µmol (i.v.) in the anaesthetized rat

Compound	Change in mean arterial pressure (MAP) mmHg
L-arginine HCl	no change
Benzoyl glycine arginine (BGA)	8 ± 2
Benzoyl arginine ethyl ester (BAEE)	55 ± 4
Arginine ethyl ester (AEE)	26 ± 7

Table 2. Effect of guanidines on platelet aggregation in the presence (E^+) and absence (E^-) of endothelial cells (50,000/ml)

Compound	EC_{50} (M) for inhibition of aggregation	
	E^+	E^-
Guanidine	no effect	no effect
L-arginine HCl	$> 1 \times 10^{-2}$	$> 1 \times 10^{-2}$
Arginine ethyl ester (AEE)	0.9×10^{-5}	5.3×10^{-3}
Benzoyl arginine ethyl ester (BAEE)	1.2×10^{-6}	4.8×10^{-4}

Table 3. Effect of guanidines (3 mM) on the production of superoxide anion induced by alloxan (0.6 mM) in the rabbit aorta

Compound	Superoxide formation, nmol/cm^2/min
Basal	1.2 ± 0.35
Alloxan	7.5 ± 1.8
Alloxan + guanidine	10.1 ± 2.6
Alloxan + L-arginine	19.1 ± 3.2
Alloxan + BAEE	6.2 ± 2.6

Effect of guanidines on human platelet aggregation

In Table 2 we show the effect of various guanidines on platelet aggregation. It is clear that both L-arginine and unsubstituted guanidine had no effect on platelet aggregation in the presence or absence of endothelial cells.

Effect of L-arginine and BAEE on superoxide production

Alloxan generates superoxide in the rat and rabbit aorta. Table 3 compares the effect of various guanidines on superoxide production induced by alloxan. Compared to all the guanidines BAEE was

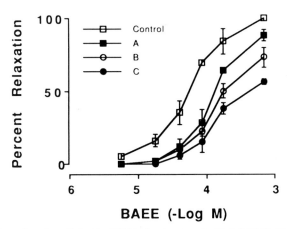

Fig.4. Inhibition of vascular relaxation elicited by BAEE after treatment with L-NMMA. Concentrations of L-NMMA were (A) 10 μM, (B) 50 μM and (C) 100 μM.

the most potent inhibitor of superoxide production. Paradoxically, L-arginine stimulated superoxide formation. The effect of BAEE was specifically due to the production of EDRF, since L-NMMA antagonized its effect.

Interaction of EDRF inhibitor L-NMMA with substituted guanidines

Endothelium-dependent relaxation is inhibited by L-NMMA[8]. This inhibition is reversed by treatment with excess L-arginine[8]. Hence it is concluded that L-NMMA is a specific inhibitor of EDRF formation from L-arginine. Our data show that pre-treatment with L-NMMA dose-dependently attenuated the relaxation elicited by BAEE (Fig. 4). This indicates that in the rat aorta BAEE generates an EDRF-like agent and that L-NMMA antagonizes not only L-arginine but other arginine compounds as well.

Discussion

Our results lend support to the hypothesis that EDRF is generated from the metabolism of an arginine derivative and not L-arginine *per se*. It has been reported by many investigators that the lack of response by externally added L-arginine is due to the presence of a saturating amount of L-arginine[4,7]. However, we recently discovered that rat aorta contains the same amount of L-arginine whether endothelium is present or not[17]. In addition, there is no change in the concentration of L-arginine during endothelium-dependent relaxation and HPLC analyses indicate that there was no formation of L-citrulline. In contrast, we identified the formation of a compound which gives a positive colour reaction characteristic for the ureido group of citrulline[17]. This clearly demonstrates that during endothelium-dependent relaxation some guanidino compounds other than L-arginine are utilized for EDRF synthesis.

Interaction of guanidino compounds with L-NMMA indicates that compounds such as L-homo-arginine antagonize the vasoconstrictor effects of L-NMMA even though these guanidines by themselves have no vascular activities[17]. This suggests that properly substituted guanidino compounds will antagonize L-NMMA even though they themselves are not substrates for EDRF generation. Thus the reported antagonism between L-arginine and L-NMMA in freshly isolated tissues may be due to the structural similarities between L-arginine and L-NMMA and not due to the generation of an EDRF-like material from L-arginine. However, it is possible that L-arginine may generate an EDRF-like agent under conditions where the endogenous guanidines are decreased.

The effect of EDRF is antagonized by superoxide anion[3,9]. Thus superoxide may have a significant role in the regulation of vascular tone. The diabetogenic compound alloxan is a potent inhibitor of endothelium-dependent relaxation[16]. This effect of alloxan is mediated by the formation of superoxide anion. Our results indicate that BAEE scavenges superoxide anion when incubated in the presence of rabbit aorta. Similar results are obtained with rat aorta also. This effect of BAEE is antagonized by the EDRF inhibitor, L-NMMA. In contrast to these, in freshly isolated tissues L-arginine stimulates superoxide anion. This is further evidence that substituted arginine compounds generate EDRF and the precursor of EDRF may be an arginine derivative and not L-arginine *per se*.

The above hypothesis is further supported by the results obtained from platelet aggregation and from *in vivo* blood pressure measurements. All these studies demonstrate that properly substituted guanidino compounds such as BAEE generate an EDRF-like agent and L-arginine has no effect either alone or in the presence of endothelial cells.

Acknowledgements

This work was supported in part by NIH grants HL-36802 and HL-40069.

References

1. Farhat, M., Ramwell. P.W. & Thomas, G. (1990) Endothelium mediated effect of N-substituted arginine in the isolated perfused rat kidney. *J. Pharmacol. Exp. Ther.* **255**, 473–477.

2. Furchgott, R.F.(1983): Role of endothelium in the vascular responses of smooth muscle. *Circ. Res.* **53**, 557–573.

3. Gryglewski, R.J., Palmer, R.M.J. & Moncada, S. (1986): Superoxide anion is involved in the breakdown of endothelium-derived vascular relaxing factor. *Nature* **320**, 454–456.

4. Hecker, M., Mitchell, J.A., Harris, H.J., Katsura, M., Thiemermann, C. & Vane, J.R. (1990): Endothelial cells metabolize N^G-monomethyl L-arginine to L-citrulline and subsequently to L-arginine. *Biochem. Biophys. Res. Commun.* **167**, 1037–1043.

5. Heim, K.F., Thomas, G. & Ramwell, P.W. (1991): Effect of substituted arginine compounds on superoxide production in the rabbit aorta. *J. Pharmacol. Exp. Ther.* **257**, 1130–1135.

6. Heim, K.F., Thomas, G. & Ramwell, P.W. (1991): Superoxide production in the isolated rabbit aorta and the effect of alloxan, indomethacin and nitrovasodilators. *J. Pharmacol Exp. Ther.* **256**, 537–541.

7. Palmer, R.M.J., Ashton, D.S. & Moncada, S. (1988): Vascular endothelial cells synthesize nitric oxide from L-arginine. *Nature* **333**, 664–666.

8. Rees, D.D., Palmer, R.M.J., Hodson, H.F. & Moncada, S. (1989): A specific inhibitor of nitric oxide formation from L-arginine attenuates endothelium dependent relaxation. *Br. J. Pharmacol.* **96**, 418–424.

9. Rubanyi, G.M. & Vanhoutte, P.M. (1986): Superoxide anions and hyperoxia inactivate endothelium-derived relaxing factor. *Am. J. Physiol.* **250**, H822–H827.

10. Thomas, G., Farhat, M. & Ramwell, P.W. (1990): Effect of L-arginine and substituted arginine compounds on platelet aggregation: role of endothelium. *Thromb. Res.* **60**, 425–429.

11. Thomas, G., Farhat, M. & Ramwell, P.W. (1990): Effect of Nα-benzoyl L-arginine ethyl ester on coronary perfusion pressure in isolated guinea pig heart. *Eur. J. Pharmacol.* **178**, 251–254.

12. Thomas, G., Hecker, M. & Ramwell, P.W. (1989): Vascular activity of polycations and basic amino acids: L-arginine does not specifically elicit endothelium dependent relaxation. *Biochem. Biophys. Res. Commun.* **158**, 177–180.

13. Thomas, G. , Mostaghim, R. & Ramwell, P.W. (1986): Endothelium dependent vascular relaxation by arginine containing polypeptides. *Biochem. Biophys. Res. Commun.* **141**, 446–451.

14. Thomas, G. & Ramwell, P.W. (1988): Peptidyl arginine deiminase and endothelium dependent relaxation. *Eur. J. Pharmacol.* **153**, 147–148.

15. Thomas, G. & Ramwell, P.W. (1988): Vasodilatory properties of mono L-arginine containing compounds. *Biochem. Biophys. Res. Commun.* **154**, 332–338.

16. Thomas, G. & Ramwell, P.W. (1989): Streptozotocin: a nitric oxide carrying molecule and its effect on vasodilation. *Eur. J. Pharmacol.* **161**, 279–280.

17. Thomas, G. & Ramwell, P.W. (1991): Effects of guanidino compounds on the EDRF inhibitor N^G-monomethyl L-arginine. *J. Pharmacol. Exp. Ther.* (in press).

Guanidino Compounds in Biology and Medicine, eds. by P.P. De Deyn, B. Marescau, V. Stalon and
I.A. Qureshi. ©1992 John Libbey & Company Ltd., pp. 109–116.

Chapter 16

The biophysical events involved in the stimulation of insulin release by arginine

Jean-Claude HENQUIN

Unité de Diabétologie et Nutrition, Université de Louvain, UCL 54.74, B-1200 Brussels, Belgium;
Physiologisches Institut, University of Saarland, D-6650 Homburg/Saar, Germany

Summary

Arginine, which is only poorly metabolized in pancreatic islet cells, depolarizes the B-cell membrane, stimulates Ca^{2+} influx through voltage-dependent Ca^{2+} channels, increases cytosolic Ca^{2+} and triggers insulin release. This depolarization, unlike that produced by glucose and leucine which are well metabolized, is not secondary to a decrease in K^+ permeability of the plasma membrane. It is also not due to an increase in Na^+ permeability. Citrulline, which contains a non-ionized ureido group instead of the guanidino group, and guanidinoacetic acid, which lacks an α-amino group are practically inactive. Guanidine has effects compatible with a weak inhibition of K^+ channels. In contrast, lysine, which has no guanidino group but is also a dibasic amino acid, mimics all effects of arginine. It is concluded that the ability of arginine to depolarize the B-cell membrane, hence to increase insulin release, is due to its electrogenic transport (in a positively charged form) in the cell.

Introduction

The rate of insulin release from pancreatic B-cells is primarily controlled by variations in the concentrations of circulating nutrients. Among these, glucose is by far the most important regulator of B-cell function. The basic mechanisms by which the sugar induces insulin release are schematized in Fig. 1. The metabolism of glucose in B-cells generates signals which close ATP-sensitive K^+ channels in the plasma membrane. The decrease in K^+ permeability that results from this closure leads to depolarization of the membrane with subsequent activation of voltage-dependent Ca^{2+} channels. This permits Ca^{2+} influx, and the resulting rise in cytoplasmic Ca^{2+} then activates an effector system eventually responsible for exocytosis of insulin granules[21].

Besides glucose, amino acids are also important regulators of insulin release. They can be subdivided into two broad categories. Amino acids like leucine, which are well metabolized in B-cells[19,29], trigger a similar sequence of events to that produced by glucose[3,22]. Amino acids like arginine or alanine, which are only poorly metabolized in B-cells[19,28], exert their effects by other mechanisms. Alanine is a weak insulin secretagogue, the effects of which are ascribed to the depolarization of the B-cell membrane that it produces because of its cotransport with sodium[9,13,18,24,32].

The mode of action of arginine: historical perspective

The ability of arginine to increase plasma insulin levels in man was recognized more than 25 years

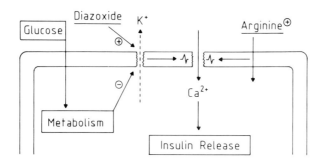

Fig. 1. Schematic representation of the mechanisms by which glucose and arginine induce insulin release, and by which diazoxide inhibits insulin release.

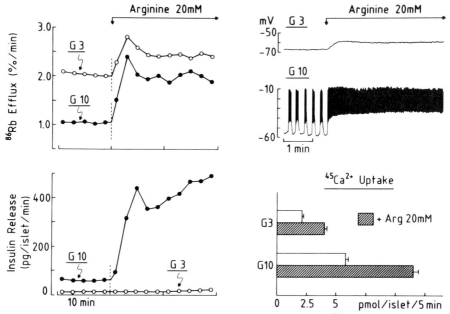

Fig. 2. Characteristics of the effects of L-arginine on mouse pancreatic B-cells. [86]Rb efflux (which estimates K^+ efflux) and insulin release were measured simultaneously with islets perifused with a medium containing 3 or 10 mM glucose (G). The membrane potential of B-cells was measured with intracellular microelectrodes. Ca^{2+} influx in B-cells was estimated by the 5-min uptake of [45]$CaCl_2$ in isolated islets. Results are shown by representative experiments, except for [45]Ca uptake where values are means ± SEM for 15 batches of islets.

ago[16]. It was then rapidly suspected, from both in vivo[14,15] and in vitro studies[6,17,27], that different mechanisms were involved in the stimulation of insulin release by arginine and by glucose or leucine. One intriguing observation was that the insulinotropic action of arginine was rather resistant to diazoxide[6,14], but the significance of this finding could not be fully appreciated because it was not yet known that diazoxide inhibits insulin release by opening ATP-sensitive K^+ channels in the B-cell membrane[23,33] (Fig. 1).

The idea of testing whether the guanidino group of arginine is important for the effects of the amino acid on B-cells probably stemmed from the observations that guanidine itself, monoguanidine compounds (e.g. galegine) or diguanidine compounds (e.g. synthalin A) possess hypoglycaemic properties[11]. It was first reported that guanidine and guanidinoacetic acid were able to increase insulin release from the perfused rat pancreas, and this finding prompted the suggestion that the arginine molecule had to be converted into such products to stimulate B-cells[2]. Intravenous injection of guanidine to rats increased plasma insulin levels, but this effect was inhibited by atropine[4]. With isolated islets, guanidine proved to be ineffective[1] or stimulatory[31] on insulin release. It should also be noted that the biguanides, that are used in the treatment of non-insulin-dependent diabetes, do not stimulate insulin secretion[5].

Characteristics of the effects of arginine

To define the biophysical events involved in the stimulation of insulin release by arginine, the membrane potential of mouse B-cells was recorded with intracellular microelectrodes, K^+ efflux from mouse islet cells was estimated with ^{86}Rb as tracer for K, and Ca^{2+} influx was measured with $^{45}CaCl_2$.

In the presence of a low concentration of glucose (3 mM), the membrane potential of B-cells is stable, between −60 and −70 mV, the rate of ^{86}Rb efflux is high, and insulin release is not stimulated (Fig. 2). Addition of 20 mM arginine to the medium caused a rapid and sustained depolarization by almost 10 mV but did not induce electrical activity. This depolarization was accompanied by an acceleration of ^{86}Rb efflux and a stimulation of ^{45}Ca influx, but insulin release was hardly affected.

In the presence of 10 mM glucose the rate of ^{86}Rb efflux is lower, B-cells are depolarized and exhibit typical rhythmic electrical activity, ^{45}Ca influx is increased, and insulin release is stimulated (Fig. 2). Arginine now caused a large and persistent depolarization with continuous electrical activity, markedly accelerated ^{86}Rb efflux and ^{45}Ca influx, and strongly increased insulin release.

The observation that the depolarization brought about by arginine is consistently associated with an acceleration of ^{86}Rb efflux prompted the conclusion[22] that, unlike glucose and leucine, arginine does not decrease the K^+ permeability of the B-cell membrane. This conclusion was subsequently confirmed by reports showing that the depolarizing action of arginine in low glucose persists in the presence of diazoxide[26] and that arginine does not affect ATP-sensitive K^+ channels that are closed by glucose[3].

Ionic requirements of the effects of arginine

To investigate the mechanisms by which arginine could exert its effects on the membrane potential, experiments were performed in the absence of extracellular Ca and Na. Figure 3 (left panels) illustrates the rapid and large increases in ^{86}Rb efflux and insulin release that are produced by 10 mM arginine in a control medium containing 10 mM glucose. It also shows that these changes are reversible on removal of arginine. When the medium was devoid of Ca, the control rate of ^{86}Rb efflux was not significantly affected, but glucose-induced insulin release was abolished. Arginine again accelerated ^{86}Rb efflux, but failed to induce insulin release. Because we are unable to record the membrane potential of B-cells with intracellular microelectrodes in the absence of Ca, the demonstration that the depolarizing effect of arginine is Ca-independent cannot be obtained directly. This is, however, inferred from the increase in ^{86}Rb efflux that arginine causes in the Ca-free medium.

In the absence of Na and presence of Ca, the rates of ^{86}Rb efflux and insulin release are high (Fig. 3, right panels) partly because of a large influx of Ca^{2+}, probably mediated by the Na^+/Ca^{2+} exchanger[12]. Under these conditions, arginine still accelerated ^{86}Rb efflux and increased insulin release. It also depolarized the B-cell membrane (not shown). Omission of Ca from the Na-free medium lowered ^{86}Rb efflux and suppressed insulin release. It also prevented arginine from increasing release, but not from accelerating ^{86}Rb efflux.

These results obtained with mouse islets entirely confirm those we previously obtained with rat islets[10]. From these experiments, we may conclude that the depolarizing action of arginine is not primarily due to an increase in Ca^{2+} or Na^+ permeability of the B-cell membrane. That the stimulation of Ca^{2+} influx brought about by arginine results from the activation of voltage-dependent Ca^{2+} channels was directly demonstrated by experiments using blockers of these channels[20]. However,

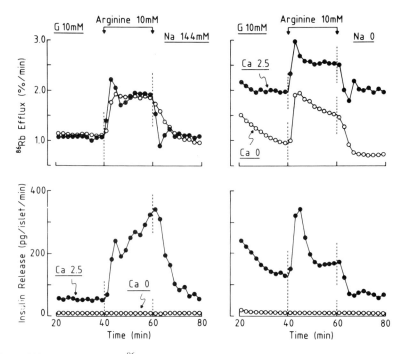

Fig. 3. Effects of 10 mM L-arginine on ^{86}Rb efflux (upper panels) and insulin release (lower panels) from mouse islets perifused with a medium containing 10 mM glucose (G). The experiments were performed in the presence (left panels) or absence (right panels) of Na in the medium. In the latter solutions, choline salts were used as substitutes for Na salts, and 1 μM atropine was present throughout. The solutions contained 2.5 mM Ca (●) or were devoid of Ca (O). Results are shown by representative experiments.

others have suggested that arginine also promotes Ca^{2+} influx through voltage-insensitive Ca^{2+} channels[25].

Molecular requirements of the effects of arginine

The next experiments were performed to determine whether the guanidino group present in the arginine molecule is involved in the biophysical and secretory effects of the amino acid. To this end, we compared the effects of a series of related compounds (Fig. 4) on ^{86}Rb efflux, insulin release and B-cell membrane potential.

Citrulline, which differs from arginine by the substitution of a non-ionized ureido group for the guanidino group did not significantly change ^{86}Rb efflux but marginally increased insulin release from mouse islets perifused with a medium containing 10 mM glucose (Fig. 5). In a previous study performed with rat islets, citrulline was found not to affect ^{86}Rb efflux and insulin release in the presence of 7 mM glucose[10]. Citrulline also caused a slight increase in the frequency of the periods with electrical activity recorded in mouse B-cells stimulated by 10 mM glucose (Fig. 6B). This effect is reminiscent of, but much smaller than, that produced by alanine[24], and could reflect the entry of citrulline in B-cells with Na+.

Guanidine itself slightly increased ^{86}Rb efflux and insulin release (Fig. 5), and augmented glucose-induced electrical activity in B-cells (Fig. 6C). In spite of these apparent analogies between the B-cell responses to guanidine and arginine, the hypothesis that the effects of arginine are simply due to its guanidino group is untenable for several reasons. First, guanidine was a much weaker stimulator of B-cells than arginine. Second, omission of extracellular Ca abolished the effects of guanidine on

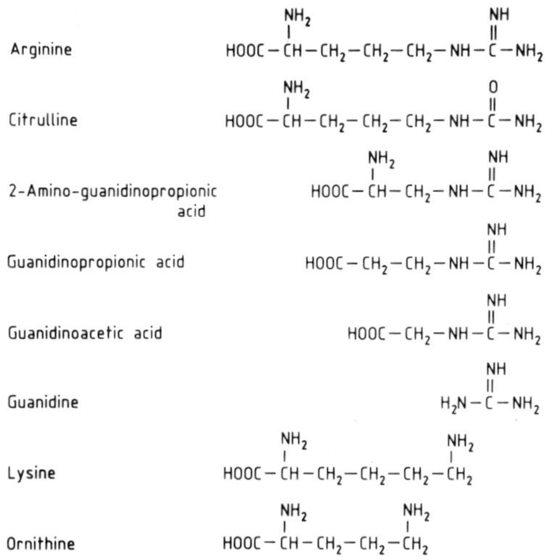

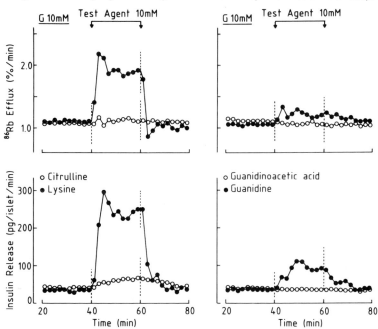

Fig. 4. Structure of arginine and of several related compounds tested in this study.

Fig. 5. Effects of lysine, citrulline, guanidine and guanidinoacetic acid on ^{86}Rb efflux and insulin release from mouse islets perifused with a medium containing 10 mM glucose (G), 2.5 mM CaCl$_2$ and 144 mM NaCl. Each substance was tested at a concentration of 10 mM. Results are shown by representative experiments.

both ^{86}Rb efflux and insulin release (not shown), whereas the effect of arginine on ^{86}Rb efflux persisted (Fig. 3). Third, guanidine increased the duration of the periods with electrical activity and decreased that of the intervals (Fig. 6C), an effect that is not seen with arginine even when it is tested at low concentrations. Fourth, and most importantly, guanidine decreased ^{86}Rb efflux from mouse islets perifused with a medium containing 3 mM glucose (not shown). In a previous study, guanidine produced a small, insignificant decrease in ^{86}Rb efflux from rat islets perifused with a glucose-free

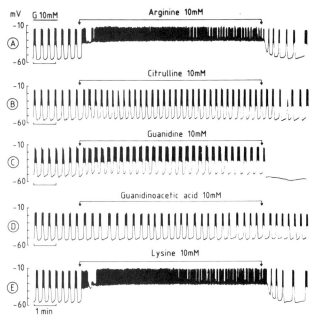

Fig. 6. Effects of arginine, citrulline, guanidine, guanidinoacetic acid and lysine on the membrane potential of mouse B-cells stimulated by 10 mM glucose (G). All recordings were obtained in the same cell, in the order C, B, E, A, D. In C, normal electrical activity reappeared 3 min after removal of guanidine.

medium[10]. All these changes are compatible with the hypothesis that guanidine weakly inhibits ATP-sensitive K$^+$ channels in B-cells. Experiments with the patch-clamp technique would be necessary to determine whether this effect is direct or indirect.

In contrast to guanidine, guanidinoacetic acid was totally ineffective on ^{86}Rb efflux and insulin release (Fig. 5), and on B-cell membrane potential (Fig. 6D). In previous studies, performed with rat islets, guanidinoacetic acid and guanidinopropionic acid were found to be ineffective on ^{86}Rb efflux and insulin release, whereas 2-amino-guanidinopropionic acid (Fig. 4) mimicked the effects of arginine[10].

It is thus clear that while omission of the guanidino group from arginine makes the compound inactive, the mere presence of the guanidino group in the molecule is not sufficient to make it active. On the other hand, lysine, which does not contain a guanidino group (Fig. 4), produced effects almost indistinguishable from those of arginine on ^{86}Rb efflux and insulin release (Fig. 5) and on B-cell membrane potential (Fig. 6E). This is also the case for other cationic amino acids like ornithine[8,10].

Conclusions

Experiments similar to those presented above have led us to suggest, several years ago[10,22], that arginine depolarizes the B-cell membrane because its entry in the cell, through the transport system of cationic amino acids, is electrogenic. The same mechanism may underlie the biophysical effects of all dibasic amino acids which bear a net positive charge at physiological pH. The depolarization then activates voltage-dependent Ca^{2+} channels, leading to Ca^{2+} influx and insulin release (Fig. 1).

This model is entirely compatible with the glucose dependence of the effects of arginine. Although arginine uptake by islet cells is not increased by glucose[8,18], the current, that the entry of the amino acid causes, produces a larger depolarization when the electrical resistance of the membrane is high. This happens when ATP-sensitive K$^+$ channels are closed by either an increase in glucose concentration or a pharmacological agent like tolbutamide[33]. Tolbutamide was indeed found to augment the depolarizing and secretory effects of arginine in low glucose[26]. On the other hand, opening of these channels with diazoxide is without effect on the B-cell response to arginine in low glucose but decreases it in high glucose, by counteracting the increase in membrane electrical resistance that the sugar normally causes.

The emphasis put on the biophysical effects of arginine in B-cells does not imply that they entirely account for the changes in insulin release brought about by the amino acid. Intracellular events may also be involved. For instance, the metabolism of glucose may increase the efficacy of the initiating signal (Ca^{2+} influx) triggered by arginine. On the other hand, arginine itself might modulate the secretory response by acting on yet unidentified intracellular targets[7,30].

References

1. Alberti, K.G.M.M. & Whalley, M.D. (1973): Short chain analogues of arginine: potent stimulators of insulin secretion. In *Proceedings of the 8th Congress of the International Diabetes Federation*, pp. 21 (abstr.). Amsterdam: Excerpta Medica.

2. Alsever, R.N., Georg, R.H. & Sussman, K.E. (1970): Stimulation of insulin secretion by guanidinoacetic acid and other guanidine derivatives. *Endocrinology* **86**, 332–336.

3. Ashcroft, F.M., Ashcroft, S.J.H. & Harrison, D.E. (1987): Effects of 2-ketoisocaproate on insulin release and single potassium channel activity in dispersed rat pancreatic β-cells. *J. Physiol.* **385**, 517–529.

4. Aynsley-Green, A. & Alberti, K.G.M.M. (1974): *In vivo* stimulation of insulin secretion by guanidine derivatives in the rat. *Horm. Metab. Res.* **6**, 115–120.

5. Bailey, C.J. (1988): Metformin revisited, its actions and indications for use. *Diabetic Med.* **5**, 315–320.

6. Basabe, J.C., Lopez, N.L., Viktora, J.K. & Wolff, F.W. (1971): Insulin secretion studied in the perfused rat pancreas. I. Effect of tolbutamide, leucine and arginine: their interaction with diazoxide and relation to glucose. *Diabetes* **20**, 449–456.

7. Bjaaland, T. & Howell, S.L. (1989): Stimulation of insulin secretion from electrically permeabilised islets of Langerhans by L-arginine. *Diabetologia* **32**, 467A.

8. Blachier, F., Mourtada, A., Sener, A. & Malaisse, W.J. (1989): Stimulus-secretion coupling of arginine-induced insulin release. Uptake of metabolized and nonmetabolized cationic amino acids by pancreatic islets. *Endocrinology* **124**, 134–141.

9. Charles, S. & Henquin, J.C. (1983): Distinct effects of various amino acids on $^{45}Ca^{2+}$ fluxes in rat pancreatic islets. *Biochem. J.* **214**, 899–907.

10. Charles, S., Tamagawa, T. & Henquin, J.C. (1982): A single mechanism for the stimulation of insulin release and $^{86}Rb^+$ efflux from rat islets by cationic amino acids. *Biochem. J.* **208**, 301–308.

11. Davidoff, F. (1973): Guanidine derivatives in medicine. *N. Engl. J. Med.* **289**, 141–146.

12. De Miguel, R., Tamagawa, T., Schmeer, W., Nenquin, M. & Henquin, J.C. (1988): Effects of acute sodium omission on insulin release, ionic flux and membrane potential in mouse pancreatic B-cells. *Biochim. Biophys. Acta* **969**, 198–207.

13. Dunne, M.J., Yule, D.I., Gallacher, D.V. & Petersen, O.H. (1990): Effects of alanine on insulin-secreting cells: patch-clamp and single cell intracellular Ca^{2+} measurements. *Biochim. Biophys. Acta* **1055**, 157–164.

14. Fajans, S.S., Floyd, J.C., Knopf, R.F., Guntsche, E.M., Rull, J.A., Thiffault, C.A. & Conn, J.W. (1967): A difference in mechanism by which leucine and other amino acids induce insulin release. *J. Clin. Endocrinol. Metab.* **27**, 1600–1606.

15. Fajans, S.S., Floyd, J.C., Knopf, R.F., Pek, S., Weissman, P. & Conn, J.W. (1972): Amino acids and insulin release *in vivo*. *Isr. J. Med. Sci.* **8**, 233–243.

16. Floyd, J.C., Fajans, S.S., Conn, J.W., Knopf, R.F. & Rull, J. (1966): Stimulation of insulin secretion by amino acids. *J. Clin. Invest.* **45**, 1487–1502.

17. Gerich, J.E., Charles, M.A. & Grodsky, G.M. (1974): Characterization of the effects of arginine and glucose on glucagon and insulin release from the perfused rat pancreas. *J. Clin. Invest.* **54**, 833–841.

18. Hellman, B., Sehlin, J. & Täljedal, I.B. (1971): Uptake of alanine, arginine, and leucine by mammalian pancreatic β-cells. *Endocrinology* **89**, 1432–1439.

19. Hellman, B., Sehlin, J. & Täljedal, I.B. (1971): Effects of glucose and other modifiers of insulin release on the oxidative metabolism of amino acids in microdissected pancreatic islets. *Biochem. J.* **123**, 513–521.

20. Henquin, J.C., Charles, S., Nenquin, M., Mathot, F. & Tamagawa, T. (1982): Diazoxide and D600 inhibition of insulin release. Distinct mechanisms explain the specificity for different stimuli. *Diabetes* **31** 776–783.

21. Henquin, J.C., Debuyser, A., Drews, G. & Plant, T.D. (1991): Regulation of K^+ permeability and membrane potential of insulin-secreting cells. In *Nutrient regulation of insulin secretion*, ed. P.R. Flatt, pp. 173–191. London: Portland Press.

22. Henquin, J.C. & Meissner, H.P. (1981): Effects of amino acids on membrane potential and $^{86}Rb^+$ fluxes in pancreatic β-cells. *Am. J. Physiol.* **240**, E245–E252.

23. Henquin, J.C. & Meissner, H.P. (1982): Opposite effects of tolbutamide and diazoxide on $^{86}Rb^+$ fluxes and membrane potential in pancreatic B-cells. *Biochem. Pharmacol.* **31**, 1407–1415.

24. Henquin, J.C. & Meissner, H.P. (1986): Cyclic adenosine monophosphate differently affects the response of mouse pancreatic β-cells to various amino acids. *J. Physiol.* **381**, 77–93.

25. Herchuelz, A., Lebrun, P., Boschero, A.C. & Malaisse W.J. (1984): Mechanism of arginine-stimulated Ca^{2+} influx into pancreatic B-cell. *Am J. Physiol.* **246**, E38–E43.

26. Hermans, M.P., Schmeer, W. & Henquin, J.C. (1987): The permissive effect of glucose, tolbutamide and high K^+ on arginine stimulation of insulin release in isolated mouse islets. *Diabetologia* **30**, 659–665.

27. Levin, J.R., Grodsky, G.M., Hagura, R., Smith, D.F. & Forsham, P.H. (1972): Relationship between arginine and glucose in the induction of insulin secretion from the isolated, perfused rat pancreas. *Endocrinology* **90**, 624–631.

28. Malaisse, W.J., Blachier, F., Mourtada, A., Camara, J., Albor, A., Valverde, I. & Sener, A. (1989): Stimulus-secretion coupling of arginine-induced insulin release. Metabolism of L-arginine and L-ornithine in pancreatic islets. *Biochim. Biophys. Acta* **1013**, 133–143.

29. Malaisse, W.J., Hutton, J.C., Carpinelli, A.R., Herchuelz, A. & Sener, A. (1980): The stimulus-secretion coupling of amino acid-induced insulin release. Metabolism and cationic effects of leucine. *Diabetes* **29**, 431–437.

30. Malaisse, W.J., Plasman, P.O., Blachier, F., Herchuelz, A. & Sener, A. (1991): Stimulus-secretion coupling of arginine-induced insulin release. Significance of changes in extracellular and intracellular pH. *Cell Biochem. Funct.* **9**, 1–7.

31. Marco, J., Calle, C., Hedo, J.A. & Villanueva, M.L. (1976): Glucagon-releasing activity of guanidine compounds in mouse pancreatic islets. *FEBS Lett.* **64**, 52–54.

32. Prentki, M. & Renold, A.E. (1983): Neutral amino acid transport in isolated rat pancreatic islets. *J. Biol. Chem.* **258**, 14239–14244.

33. Trube, G., Rorsman, P. & Ohno-Shosaku, T. (1986): Opposite effects of tolbutamide and diazoxide on the ATP-dependent K^+ channel in mouse pancreatic β-cells. *Pflugers Arch.* **407**, 493–499.

Guanidino Compounds in Biology and Medicine, eds. by P.P. De Deyn, B. Marescau, V. Stalon and
I.A. Qureshi. ©1992 John Libbey & Company Ltd., pp. 117–122.

Chapter 17

Metabolism and secretory effects of L-arginine and other cationic amino acids in normal and tumoral islet cells

Willy J. MALAISSE

*Laboratory of Experimental Medicine, Erasmus School of Medicine, Brussels Free University, 808 Route de Lennik,
B-1070 Brussels, Belgium*

Summary

Several cationic amino acids including L-arginine, L-ornithine, L-lysine and L-histidine, as well as L-homoarginine and 4-amino-1-guanylpiperidine-4-carboxylic acid augment ^{86}Rb and ^{45}Ca outflow from prelabelled pancreatic islets and stimulate insulin release. The results obtained with the non-metabolized analogues suggest that the mere cellular accumulation of these positively charged amino acids might be sufficient to cause cell depolarization and subsequent gating of voltage-sensitive Ca^{2+} channels. This view was questioned, however, taking into consideration the pK of the side chain on some of these amino acids. Yet, alternative hypotheses such as the fuel function of the amino acids in islet cells, the *de novo* generation of polyamines, the availability of a substrate for transglutaminase, the biosynthesis of nitric oxide, a change in either intracellular pH or cyclic AMP content all also met with severe limitations. This is not meant to deny that L-arginine, L-ornithine, L-lysine and L-histidine are indeed efficiently metabolized in islet cells. In this respect, the metabolism of L-arginine differed significantly in normal and tumoral islet cells, with, in the latter case, lower arginase but much higher ornithine decarboxylase activity. In conclusion, it is acknowledged that the mechanism by which L-arginine, one of the most commonly used insulinotropic agents, stimulates insulin release from the pancreatic B-cell remains a largely unsettled issue.

Introduction

L-arginine is a potent insulinotropic agent, at least when the pancreatic B-cell is exposed to a permissive concentration of D-glucose. This amino acid is widely used therefore, both *in vitro* and *in vivo*, as a stimulus for insulin release. Yet, the intimate mode of action of L-arginine in the pancreatic B-cell remains an unsettled matter. The present report briefly reviews recent experimental work dealing with the metabolic, ionic and secretory response of pancreatic islet cells to L-arginine and other cationic amino acids.

The electrostatic hypothesis

A current view ascribes the insulinotropic action of L-arginine to the mere accumulation of this positively charged amino acid in the pancreatic B-cell, leading to depolarization of the plasma membrane and, hence, to the gating of voltage-sensitive Ca^{2+} channels[8]. This hypothesis should not be ruled out. It meets, however, with certain limitations. Thus, even when islets are already exposed

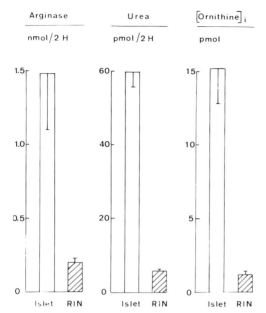

Fig. 1. Comparison between normal (open columns) and tumoral (shaded columns) islet cells in terms of arginase maximal activity in cell homogenates (left) and urea production (middle) and L-ornithine content (right) in intact cells exposed to 0.5 mM L-arginine. All results are expressed per µg protein.

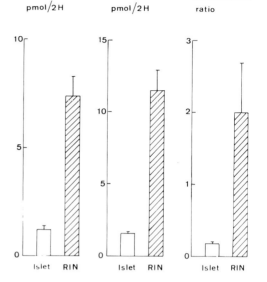

Fig. 2. Comparison between normal (open columns) and tumoral (shaded columns) islet cells in terms of ornithine decarboxylase maximal activity in cell homogenates (left), the generation of di- and polyamines in intact cells exposed to 0.5 mM L-ornithine (middle) and the ratio between the generation of the latter amines and oxidation of succinate derived from the cationic amino acid in intact cells exposed to 0.5 mM L-arginine (right). All results are expressed per µg protein, except the ratio between metabolic flows which is expressed in absolute terms.

to a high concentration of K^+ (50 mM), L-arginine further enhances insulin release[4]. Moreover, in the case of insulin release evoked by L-histidine, the amount of the amino acid present in the islet cells with a positively charged side chain was estimated to be below the threshold value required for stimulation of insulin secretion by fully ionized cationic amino acids, such as L-arginine[21]. Last, L-arginine and other cationic amino acids also stimulate secretion in cells thought to be electrically inexcitable[3,19,24].

With these considerations in mind, advantage was taken of the specific features of the cationic response of islet cells to L-arginine, especially the stimulation of K^+ efflux, to explore the possible role of the cell K^+ content in the phenomenon of B-cell memory[9].

Metabolism of L-arginine in islet cells

Several arguments suggest that L-arginine does not act, to any significant extent, as a nutrient secretagogue in the pancreatic B-cells. Thus, at variance with true nutrient secretagogues, L-arginine fails to decrease ^{86}Rb outflow from prelabelled islets perifused in the absence of exogenous nutrient[12], fails to provoke a phosphate flush in islets prelabelled with [^{32}P]orthophosphate[7], and fails to cause the hydrolysis of inositol-containing phospholipids[2].

This is not meant to deny that L-arginine is efficiently metabolized in islet cells[15]. Because of its

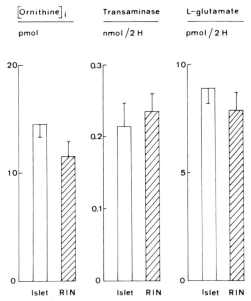

Fig. 3. Comparison between normal (open columns) and tumoural (shaded columns) islet cells in terms of the L-ornithine content of cells exposed to 0.5 mM L-ornithine (left), the maximal activity of ornithine-glutamate transaminase in cell homogenates (middle) and the generation of L-glutamate in intact cells exposed to 0.5 mM L-ornithine (right). All results are expressed per µg protein.

possible relevance to the stimulation of insulin release, e.g. through the *de novo* generation of polyamines, the metabolism of L-arginine was extensively investigated in both normal and tumoral islet cells[15,16].

There were obvious similarities but also striking differences between normal and tumoral islet cells in terms of cationic amino acid metabolism. Three sets of comparisons deserve to be underlined. First, the maximal velocity of arginase, as well as either the production of ^{14}C-urea or the steady-state content in ^{14}C-labelled L-ornithine in intact cells exposed to L-[U-^{14}C]arginine (0.5 mM), were all about one order of magnitude lower in tumoral than normal islet cells, all results being expressed per µg protein (Fig. 1). Second, the maximal activity of ornithine decarboxylase was much higher in RINm5F cell than islet homogenates (Fig. 2) and this coincided with a generation rate of di- and polyamines about one order of magnitude higher in tumoral than normal islet cells exposed to exogenous L-ornithine (0.5 mM). Likewise, the ratio between the rate of amine oxidation and that of succinate oxidation was about 10 times higher in RINm5F cells than in normal pancreatic islets exposed to L-arginine (0.5 mM). At this point, it should be underlined that the cell content in ^{14}C-labelled L-ornithine is comparable in normal and tumoral cells exposed to exogenous L-ornithine (0.5 mM). Third, the activity of ornithine-glutamate transaminase is roughly identical in pancreatic islet and RINm5F cell homogenates, and this coincides with a comparable rate of ^{14}C-labelled L-glutamate generation by intact cells exposed to L-[^{14}C]ornithine (Fig. 3).

These results, especially those relative to the generation of polyamines, might be relevant to the growth of tumoral islet cells[6].

Alternative hypotheses for the insulinotropic action of L-arginine

Several alternative hypotheses for the insulinotropic action of L-arginine were also scrutinized. The idea that L-arginine might, by causing activation of adenylate cyclase[11], lead to an increased production of cyclic AMP is not supported by the results of a recent study dealing with the effects of L-arginine, L-homoarginine, L-ornithine, L-lysine and L-histidine upon cyclic AMP production by pancreatic islets[18]. Likewise, since the secretory response to L-arginine or L-homoarginine is modulated by primary changes in either extracellular or intracellular pH[17,22], a possible effect of cationic amino acids upon the intracellular pH of islet cells was taken into consideration. Yet, L-arginine, L-homoarginine, L-ornithine and L-lysine, all tested at a 10 mM concentration, failed to affect the intracellular pH of islet cells prelabelled with the tetraacetoxymethyl ester of 2,7-biscarboxy-ethyl-(5,6)carboxyfluorescein[17]. A direct or indirect provision by the cationic amino acids of a substrate for the reaction catalysed by the Ca^{2+}-sensitive transglutaminase cannot be dogmatically ruled out, but is not supported by the findings recorded in the presence of L-histidine[21]. Except in the case of L-lysine[22], no evidence was found to suggest that cationic amino acids stimulate insulin release in a manner comparable to that seen with true nutrient secretagogues (see above). Several findings also argue against the view that cationic amino acids act upon insulin release by causing the *de novo* synthesis of putrescine, spermidine and spermine. Thus, the secretory response to L-arginine or L-ornithine is little affected in situations associated with either a marked increase[25] or decrease[14] in islet ornithine decarbamylase activity. The secretory response of islet cells to L-arginine also differs from that evoked by agmatine, which is generated from L-arginine in bacteria but not in mammalian cells[23].

Generation of nitric oxide

In the most recent work, the possible role of NO generation in the secretory response to L-arginine was explored[5]. Since the synthesis of NO is coupled to the generation of L-citrulline and displays NADPH-dependency, advantage was taken of the absence of ornithine transcarbamylase activity in islet cells[20] to monitor the NADPH-dependent direct generation of L-citrulline from L-[2,3,4,5-^{3}H]-arginine in islet homogenates prepared by sonication in a Tris-HCl buffer (100 mM, pH 7.2). The reaction was conducted over 30 min incubation at 37 °C after mixing the islet homogenate (30 µl, corresponding to 300–500 islets) with a solution of L-[2,3,4,5-^{3}H]arginine prepared in the same Tris-HCl buffer. The reaction was halted by addition of 10 µl perchloric acid (10 per cent, v/v). The separation of L-arginine, L-ornithine and L-citrulline (measured together with L-glutamate and 2-ketoglutarate) was achieved by reversed-phase high-performance liquid chromatography, as described elsewhere[1]. All readings were corrected for both the blank value found in the absence of islet homogenate and the recovery of corresponding standards.

The data summarized in Table 1 indicate that, in sonicated pancreatic islets incubated in the presence of L-[2,3,4,5-^{3}H]arginine (2.4 µM), the incorporation of NADPH (0.15 and 1.5 mM) in the assay medium caused a concentration-related increase in the generation of tritiated L-citrulline. In these experiments, the production of tritiated L-ornithine largely exceeded that of tritiated L-citrulline and failed to be significantly affected by the presence of NADPH. It was commensurate with the activity of L-arginase in islet homogenates, provided that allowance was made for the low concentration of L-arginine used in the present experiments, relative to the K_m of arginase[15]. The islet homogenates failed to catalyse the generation of labelled L-citrulline from L-[U-^{14}C]ornithine (2.4 µM), whether in the absence (n = 4) or presence (n = 3) of NADPH (data not shown).

Table 1. Generation of tritiated L-citrulline and L-ornithine from L-[2,3,4,5-^{3}H]arginine (2.4 µM) in islet homogenates. Mean values (± SEM) are shown together with the number of individual observations (in parentheses)

NADPH (mM)	L-citrulline (fmol/30 min per islet)	L-ornithine (fmol/30 min per islet)
Nil	2.49 ± 0.25 (9)	83.2 ± 6.6 (9)
0.15	4.35 ± 0.47 (6)	74.0 ± 3.6 (6)
1.50	6.24 ± 0.07 (3)	77.3 ± 0.9 (3)

In considering these results, it should be kept in mind that they do not rule out the participation of contaminating vascular endothelial cells and macrophages to the direct conversion of L-arginine to L-citrulline and NO.

Nevertheless, the finding that pancreatic islets are apparently equipped to catalyse the synthesis of NO from L-arginine led us, in a last series of experiments, to assess the effect of nitroprusside upon glucose-stimulated insulin release. As shown in Table 2, nitroprusside (1.0 mM) inhibited ($P < 0.001$) over 30 min incubation the release of insulin evoked by either 8.3 or 16.7 mM D-glucose. Over 90 min incubation at the high concentration of the hexose, as little as 0.1 mM nitroprusside already caused a severe decrease in insulin output.

Table 2. Effect of nitroprusside upon glucose-stimulated insulin release. Secretory rates (μU/islet per min) were measured over either 30 min (†) or 90 min (*) incubation. Mean values ($\pm$ SEM) are shown together with the number of individual observations (in parentheses)

Nitroprusside (mM)	D-glucose (mM)		
	8.3	16.7	16.7
Nil	$1.14 \pm 0.12\,(10)^{\dagger}$	$1.86 \pm 0.12\,(10)^{\dagger}$	$1.67 \pm 0.16\,(9)^{*}$
0.1			$0.61 \pm 0.05\,(9)^{*}$
0.3			$0.56 \pm 0.06\,(9)^{*}$
1.0	$0.39 \pm 0.05\,(10)^{\dagger}$	$0.83 \pm 0.05\,(10)^{\dagger}$	$0.46 \pm 0.03\,(9)^{*}$

It is currently thought that the effect of nitroprusside upon intact cells, acellular systems or purified enzymes is mediated through the generation of NO. Although an alternative mechanism cannot be ruled out, the data relative to the effect of nitroprusside upon glucose-stimulated insulin release, which extend a prior observation[13], do not suggest that NO exerts, in the presence of D-glucose, a positive insulinotropic action similar to that of L-arginine. A further argument against the view that the secretory response to L-arginine is mediated by the generation of NO resides in the fact that L-ornithine, which tightly reproduces the ionic and secretory effects of L-arginine in islet cells, does not generate NO, being deprived of the guanidino group. Moreover, NO and nitroprusside activate guanylate cyclase, but little evidence is available to suggest that such an activation markedly affects insulin secretion[10].

In summary, the present data indicate that islet cells are equipped to generate L-citrulline and presumably NO from L-arginine, but do not suggest that such a process participates in the stimulation of insulin release by the latter cationic amino acid.

Concluding remark

In concluding this far-from-exhaustive review on the metabolism and secretory effects of L-arginine and other cationic amino acids in normal and tumoral islet cells, there is little risk to propose that the mode of action of these amino acids in the pancreatic B-cell remains a field open for further investigations. Nevertheless, any explanation for their insulinotropic action should account for the modulation of their secretory potential by factors such as the ambient concentration of D-glucose or intracellular pH.

Acknowledgements

The experimental work mentioned in this review was supported by grants from the Belgian Foundation for Scientific Medical Research. The authors are grateful to C. Demesmaeker for secretarial help.

References

1. Blachier, F., Darcy-Vrillon, B., Sener, A., Duée, P.-H. & Malaisse, W.J. (1991): Arginine metabolism in rat enterocytes. *Biochim. Biophys. Acta* **1092**, 304–310.

2. Blachier, F., Leclercq-Meyer, V., Marchand, J., Woussen-Colle, M.-C., Mathias, P.C.F., Sener, A. & Malaisse, W.J. (1989): Stimulus-secretion coupling of arginine-induced insulin release. Functional response of islets to L-arginine and L-ornithine. *Biochim. Biophys. Acta* **1013**, 144–151.

3. Blachier, F., Mourtada, A., Gomis, R., Sener, A. & Malaisse, W.J. (1991): Metabolic and secretory response of parotid cells to cationic amino acids. Uptake and catabolism of L-arginine and L-ornithine. *Biochim. Biophys. Acta* **1091**, 151–157.

4. Blachier, F., Mourtada, A., Sener, A. & Malaisse, W.J. (1989): Stimulus-secretion coupling of arginine-induced insulin release. Uptake of metabolized and nonmetabolized cationic amino acids by pancreatic islets. *Endocrinology* **124**, 134–141.

5. Blachier, F., Rasschaert, J., Bodur, H., Sener, A. & Malaisse, W.J. (1992): NADPH-dependent generation of L-citrulline from L-arginine in pancreatic islet homogenates. *Med. Sci. Res.*, **20**, 23–24.

6.   Camara, J., Valverde, I. & Malaisse, W.J. (1990): Depletion of polyamines and inhibition of growth by difluoromethylornithine in malignant pancreatic islet cells. *Med. Sci. Res.* **18**, 61–62.

7.   Carpinelli, A.R. & Malaisse, W.J. (1980): The stimulus-secretion coupling of glucose-induced insulin release. XLIV. A possible link between glucose metabolism and phosphate flush. *Diabetologia* **19**, 458–464.

8.   Charles, S., Tamagawa, T. & Henquin, J.C. (1982): A single mechanism for the stimulation of insulin release and $^{86}Rb^+$ efflux from rat islets by cationic amino acids. *Biochem. J.* **208**, 301–308.

9.   Fichaux, F. & Malaisse, W.J. (1990): Possible role of intracellular K^+ in the B-cell memory to insulin secretagogues. *Diabetes, Nutr. Metab.* **3**, 9–16.

10.   Gagerman, E. & Hellman, B. (1977): Islet contents of cyclic 3',5'-guanosine monophosphate under conditions which affect the cyclic 3',5'-adenosine monophosphate. *Acta Endocrinol.* **86**, 344–354.

11.   Garcia-Morales, P., Dufrane, S.P., Sener, A., Valverde, I. & Malaisse, W.J. (1984): Inhibitory effect of clonidine upon adenylate cyclase activity, cyclic AMP production, and insulin release in rat pancreatic islets. *Biosci. Rep.***4**, 511–521.

12.   Herchuelz, A., Lebrun, P., Boschero, A.C. & Malaisse, W.J. (1984): Mechanism of arginine-stimulated Ca^{2+} influx into pancreatic B-cell. *Am. J. Physiol.* **246**, E38–E43.

13.   Laychock, S.G. (1981): Evidence for guanosine 3,5'-monophosphate as a putative mediator of insulin secretion from isolated rat islets. *Endocrinology* **108**, 1197–1205.

14.   Leclercq-Meyer, V., Marchand, J. & Malaisee, W.J. (1990): Stimulus-secretion coupling of arginine-induced insulin release. Resistance of arginine- and ornithine-stimulated glucagon and insulin release to D,L-α-difluoromethylornithine. *Biochem. Pharmacol.* **39**, 537–547.

15.   Malaisse, W.J., Blachier, F., Mourtada, A., Camara, J., Albor, A., Valverde, I. & Sener, A. (1989): Stimulus-secretion coupling of arginine-induced insulin release. Metabolism of L-arginine and L-ornithine in pancreatic islets. *Biochim. Biophys. Acta* **1013**, 133–143.

16.   Malaisse, W.J., Blachier, F., Mourtada, A., Camara, J., Albor, A., Valverde, I. & Sener, A. (1989): Stimulus-secretion coupling of arginine-induced insulin release: metabolism of L-arginine and L-ornithine in tumoral islet cells. *Mol. Cell. Endocrinol.* **67**, 81–91.

17.   Malaisse, W.J., Plasman, P.O., Blachier, F., Herchuelz, A. & Sener, A. (1991): Stimulus-secretion coupling of arginine-induced insulin release: significance of changes in extracellular and intracellular pH. *Cell Biochem. Function* **9**, 1–7.

18.   Malaisse, W.J., Sener, A., Poloczek, P. & Winand, J. (1990): Arginine-induced insulin release: effect of cationic amino acids on cyclic AMP production by pancreatic islets. *Med. Sci. Res.* **18**, 711–712.

19.   Mourtada, A., Blachier, F., Plasman, P.O., Sener, A. & Malaisse, W.J. (1991): Stimulation of ^{45}Ca uptake and amylase release by cationic amino acids in parotid cells. *J. Biol. Buccale* **19**, 119–124.

20.   Sener, A., Blachier, F. & Malaisse, W.J. (1988): Production of urea but absence of urea cycle in pancreatic islet cells. *Med. Sci. Res.* **16**, 483-484.

21.   Sener, A., Blachier, F., Rasschaert, J. & Malaisse, W.J. (1990): Stimulus-secretion coupling of arginine-induced insulin release: comparison with histidine-induced insulin release. *Endocrinology* **127**, 107–113.

22.   Sener, A., Blachier, F., Rasschaert, J., Mourtada, A., Malaisse-Lagae, F. & Malaisse, W.J. (1989): Stimulus-secretion coupling of arginine-induced insulin release: comparison with lysine-induced insulin secretion. *Endocrinology* **124**, 2558–2567.

23.   Sener, A., Lebrun, P., Blachier, F. & Malaisse, W.J. (1989): Stimulus-secretion coupling of arginine-induced insulin release. Insulinotropic action of agmatine. *Biochem. Pharmacol.* **38**, 327–330.

24.   Sener, A., Mourtada, A., Blachier, F. & Malaisse, W.J. (1990): Metabolic and secretory responses of parotid cells to cationic amino acids. Oxidation of the amino acids and interference with the oxidation of D-glucose or endogenous nutrients. *Biochimie* **72**, 685–688.

25.   Sener, A., Owen, A., Malaisse-Lagae, F. & Malaisse, W.J. (1989): Stimulus-secretion coupling of arginine-induced insulin release. Effect of 2-aminoisobutyric acid upon ornithine decarboxylase activity and insulin secretion. *Res. Commun. Chem. Pathol. Pharmacol.* **65**, 65–80.

Guanidino Compounds in Biology and Medicine, eds. by P.P. De Deyn, B. Marescau, V. Stalon and I.A. Qureshi. ©1992 John Libbey & Company Ltd., pp. 123–131.

Chapter 18

· The response of arginine differs with species and associated nutrients

Willard J. VISEK

University of Illinois, College of Medicine, 190 Medical Sciences Building, 506 South Mathews Avenue, Urbana, Illinois 61801, USA

Summary

The three urea cycle amino acids, arginine, ornithine and citrulline are synthesized in the liver and tissues of ureotelic species but commonly consumed proteins are virtually free of ornithine and citrulline. This places practical importance upon the supply of arginine in the diet. Arginine, utilized in urea, protein and creatine synthesis, is required for optimal growth of cats, rats, guinea-pigs, dogs, rabbits, swine and mink. Birds also require dietary arginine. The balance of evidence shows that adult rats, human adults and pregnant swine meet their arginine needs for maintenance from endogenous synthesis. However, nearly mature cats die with severe hyperammonaemia within 1–3 h after eating a meal devoid of arginine and human beings with metabolic impairments may require additional arginine in their nutritional regimen to avoid hyperammonaemia. Ureotelic species excrete elevated amounts of orotic acid when the arginine content of their diet is insufficient. Dietary arginine intake modulates biochemical, hormonal and immunological processes and has been reported to modulate tumour growth. The evidence indicates that species like cats that are especially susceptible to arginine deficiency lack enzymatic capacity in their intestinal mucosa for synthesizing ornithine which serves as a precursor for arginine synthesis in other tissues. The author has calculated in agreement with others that arginine intake in usual human Western diets may be insufficient for the needs of creatine synthesis when total protein intake is below 50 g per day. Because the needs for arginine depend upon species and conditions of use, the term dispensable (non-essential) and essential (indispensable) may have special meaning and require definition when dietary arginine requirements are being considered.

Introduction

Magendie showed that nitrogen in the body is derived from nitrogen in the food[11]. Immediately after the turn of this century Kossel and Kutscher and Emil Fischer described methods for isolating amino acids and Hopkins and Cole and Ehrlich were identifying specific amino acids[20]. Another important early contribution to the knowledge of protein nutrition was that of Osborne and Mendel[17] showing that certain proteins causing nutritional failure could be made satisfactory with supplemental amino acids and that protein metabolism depends upon the biochemical behaviour of amino acids. Their investigations highlighted the need for determining which amino acids are indispensable and which are dispensable in the diet without interfering with normal physiological processes. Rose, a student of Mendel was the leader in developing these concepts with his associates at the University of Illinois[20]. Classification of arginine given special consideration by Rose remains complex and sometimes uncertain. Readers desiring more extensive information about arginine than can be discussed here are directed to the following citations[2,23–27].

The discovery and identification of threonine by Rose and associates in 1935–1936 completed the list of about 20 amino acids in plant and animal proteins of dietary importance[20]. This made it possible to prepare diets containing mixtures of highly purified amino acids as the only sources of dietary N and to determine the quantitative requirements for each. Using this approach they found that 10 amino acids were required by weanling rats. These were: valine, leucine, isoleucine, methionine, threonine, lysine, phenylalanine, tryptophan, histidine and arginine. Excluding any one except arginine diminished appetite which was followed by weight loss, nutritive failure and eventual death. Exclusion of arginine decreased weight gain but allowed the animals to live. This was attributed to endogenous synthesis which was adequate for survival but not for optimum growth. From their studies Rose and associates defined indispensable amino acids as those that cannot be synthesized by the species in question from materials ordinarily available to cells of the body at a rate commensurate with needs for optimum growth[5]. Rose[20] summarized their conclusions as follows: (1) 'by that definition arginine must be classified as indispensable, although it alone of its group may be excluded from food without occasioning loss of weight'; (2) classification of arginine as dispensable or indispensable for rats is a matter of definition; (3) because arginine can be synthesized endogenously only in limited quantities differentiates it from other nonessential amino acids which appear to be synthesized in sufficient quantities to meet fully the requirements of growing rats; and (4) evidence obtained for one species may not apply to another.

Determination of the amino acid requirements of rats was a starting point for determining the amino acid requirements of healthy young men[20]. These showed that of the 10 amino acids indispensable for rats eight were required by mature human males and that histidine and arginine were the exceptions. Based upon more recent evidence histidine is now considered indispensable for human infants and adults[16]. Claims that arginine is not required for human infants appear to rest upon data from three subjects and the arginine needs for growth of normal or premature infants are not known[25]. Lethargy and hyperammonaemia observed in human infants maintained by total parenteral nutrition has been reported preventable by supplementation of the infusate with arginine. This suggests that arginine synthesis may be inadequate to meet the needs for ammonia detoxification by the urea cycle when intravenous solutions include small amounts of arginine or when liver function is compromised[8]. Studies in man and lower animals continue to expand the biochemical roles of arginine, ornithine and citrulline. Although these amino acids are synthesized in the liver and other tissues of ureotelic species, ornithine and citrulline are virtually absent in dietary proteins. Thus, maintaining urea cycle capacity for detoxifying ammonia, providing for creatine synthesis and meeting other needs for urea cycle amino acids make dietary arginine supply critical under practical conditions.

Distinction between dispensable and indispensable amino acids

The distinction between dispensable and indispensable amino acids is strictly dietary. All of the known 20–22 amino acids are components of animal tissues whether or not they are dietary essentials. The distinction, based upon the inability of the body to synthesize their carbon skeletons, has little meaning metabolically except that one group is synthesized by mammals and the other is not. The distinction also has uncertain value for animals with a functional rumen because the microflora synthesize and degrade amino acids and make it possible for ruminants to grow when they are fed no preformed amino acids[3]. All of the indispensable amino acids other than lysine and threonine can be replaced by their α keto analogues at efficiencies which vary with the amino acid and the animal species. Thus lysine and threonine are the only amino acids which are truly indispensable. The dispensable amino acids are also needed in the diet for optimal growth[25]. Chicks, like rats[22,25], fed all of their dietary N in indispensable amino acid form have significantly depressed growth rates compared to those fed diets containing sources of nonessential nitrogen (Table 1). Nutrients that are dispensable for normal animals may be essential when there is metabolic impairment or a physiological state that imposes a need which exceeds synthesis by the target organism[4]. The terminology regarding dispensability and indispensability remains cumbersome and sometimes confusing. When applied to amino acid nutrition the terms dispensable (non-essential) and indispensable (essential) need to be defined. This is particularly true for arginine which imposes considerations specific for the species and the conditions of use.

Table 1. Weight gain of weanling rats fed indispensable and dispensable amino acids in different ratios for 17 days with different percentages of total dietary nitrogen[a]

Indispensable	Dispensable	Weight gain		
(%)	(%)	1.2% N	1.8% N	2.4 %N
100	–	33	37	21
80	20	52	69	57
50	50	64	71	67

[a]Modified from Stucki and Harper[22].

Dietary requirements for arginine

Dietary arginine is required for optimal growth of cats, rats, guinea pigs, dogs, rabbits, swine and mink. The balance of evidence shows that adult rats, human adults and pregnant and non-pregnant swine meet their arginine needs for maintenance and some nitrogen storage from endogenous synthesis. Early reports indicated that arginine is not required by adult dogs but recent studies suggest that they, like other carnivores, have a requirement for arginine[1,25]. Birds have a definite need for arginine[1]. The finding that cats die within a few hours with severe ammonia intoxication after a meal lacking arginine was unexpected and places arginine in a unique category that may be shared to varying degrees by ornithine and citrulline. The withholding of no other nutrient is known to cause the drastic consequences seen with the omission of arginine from the diet of cats and young ferrets[6]. Animals conditioned to a low protein diet show depressed feed intake and perhaps weight loss within a few hours after eating a meal deficient in one of the amino acids considered essential by classic criteria. However, the acute life-threatening responses seen in susceptible species with arginine-free diets appear not to have been observed with other nutrient deficiencies. Death from deficiencies of other generally recognized essential nutrients requires days or weeks. However, near mature cats fed an arginine free-diet develop severe hyperammonaemia and die within 2–3 h after eating a meal devoid of arginine[14,19]. Cats show hyperglycaemia also seen with experimental or congenital hyper-ammonaemia in other species. Similar signs were observed in simple-stomached species or ruminants given toxic doses of ammonium salts, urease, urea or amino acids. The severity of intoxication depends upon previous dietary history. Plasma amino acid profiles in the arginine deficient cats were similar to those in dogs and human subjects during hepatic encephalopathy[24,25].

Immature or mature dogs given an arginine-free meal by stomach tube show tremors, vomiting, profuse salivation and hyperglycaemia. Dogs also show profound citric and orotic aciduria characteristic of arginine deficiency in other animals but noncarnivorous species show minimal signs of intoxication[24,25]. Dogs fed *ad libitum* appear to adjust their intake of arginine-free diets to avoid severe consequences. When fed an arginine-free diet their total *ad libitum* feed intake was severely depressed and urinary orotic acid excretion in the urine was elevated 50–100 times. With diets containing 0.28 per cent dietary arginine mature dogs showed no significant changes in ammonia, urea or orotate excretion[25]. As in studies of Rose and Rice dogs in more recent studies did not show differences in N balance whether or not the dogs were fed arginine. It is important to recognize that the basal diet may influence the response to arginine. More recent studies of arginine requirements have been based upon formulations that more closely meet optimal growth requirements for all nutrients and the animals have grown more rapidly. When Rose and associates conducted their investigations their control rats gained 1.29 and 1.65 g per day compared to 1.03 and 1.35 g per day for their arginine-deficient rats. Milner and co-workers observed gains of 3.6 and 1.03 g per day in their arginine-fed and arginine-deficient animals, respectively[25]. Higher rates of growth have been obtained in studies with purified diets[1]. Ornithine and citrulline can replace arginine for correcting the orotic aciduria but only citrulline has been shown to replace arginine for growth in rats (Table 2)[13,25].

Table 2. Blood NH_3—N, blood urea, weight change, feed consumption, urinary citric acid and urinary orotic acid summarized for weanling rats fed an amino acid diet devoid of the indicated amino acid[a]

Treatment	4-day feed intake (g)	3-day weight change (g)	Peripheral blood		Urine	
			NH_3—N (µg/ml)	Urea (mg/100 ml)	Citrate (µg/24 h)	Orotate (µg/24 h)
Control	37	+12	2	8	262	46
-Arg	29	+7	2	29	5739	3982
-Lys	29	+3	2	14	1213	58
-Try	22	0	2	14	140	105
-Phe	21	-3	2	16	154	27
-Val	18	-3	2	16	184	30
-Thr	17	-9	3	58	206	33
-His	14	-10	2	53	221	31
-Met	15	-7	2	62	232	36
-Iso	14	-11	2	64	167	27
-Leu	15	-10	2	61	164	32
-Arg, Pro, Glu, Asp,Asn	27	+6	3	34	8690	2840
-Arg + Gly	28	+4	–	46	4368	2753
-Arg + Cit	43	+13	–	28	2708	64
-Arg + Orn	32	+5	–	54	261	49

[a]Arginine was replaced on an equal nitrogenous basis by glycine unless otherwise stated[25].

Arginine, protein repletion and hormonal responses

Intravenous administration of arginine has been widely used to study hormone secretion in humans. Injections of arginine also prevent intoxication by ammonium salts or amino acids. Hyperammonaemic rats are refractory to exogenous insulin, their endogenous plasma insulin concentrations are elevated and their secretion of insulin in response to injected arginine is depressed[9,15,25]. Mature rats depleted of 40 per cent of their body weight and then repleted with diets containing 0, 0.25, 0.5, 0.75 and 1.0 per cent arginine showed weight gains significantly enhanced by arginine (Table 3).

Table 3. Weight gain, feed intake, urinary orotate and liver lipids of mature rats repleted with different dietary intakes of arginine in an L-amino acid diet[a]

Measure	Dietary arginine (%)					
	0	0.25	0.5	0.75	1.0	1.5
Weight gain, g/15 d	42	85	106	94	94	124
Feed intake, g/15 d	301	290	319	236	311	264
Gain/feed ratio	0.14	0.30	0.33	0.40	0.31	0.47
Urinary orotate, mg/15 d	56	34	25	10	0.8	2.2
Liver weight, g/100 g body wt	4.3	4.7	3.8	3.5	3.7	2.2
Liver lipids, mg/100 mg tissue	11	12	6	4	4	
Plasma albumin,[2] g/100 ml	2.3			2.5		2.4
Testes, g	2.5	2.6		3.7		3.2

[a]From Kari et al.[9] and Mulloy et al.[15]: where possible, means of data from both studies have been calculated; [b]Average 1.9 g/100 ml at end of depletion[25].

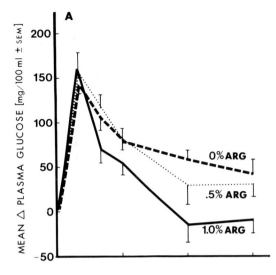

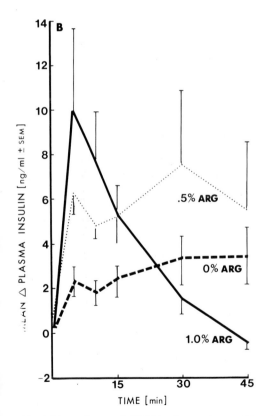

Fig. 1. Mean changes (± SEM) in plasma glucose (A) and insulin (B) during an intravenous glucose tolerance test in rats repleted with diets varying in arginine content. From Mulloy et al. (1982)[15].

Accompanying the highly significant differences in body mass were higher food intakes as arginine content of the diet was raised from 0 to 0.5 per cent. After an intravenous challenge with glucose the plasma insulin concentrations of the group fed 1 per cent arginine reached a peak in 5 min and declined steadily to near basal levels at 45 min. Plasma insulin concentrations for animals fed 0 or 0.5 per cent arginine failed to respond normally and remained above basal concentrations throughout the 45 min of measurement. The differences in plasma insulin were concurrent with differences in glucose disappearance and indicated a refractoriness to endogenous insulin when there was a dietary deficiency of arginine. The coefficient of glucose disappearance and the ratio of insulin to glucose was decidedly greater when 1 per cent arginine was fed (Fig. 3). It is well known that pharmacological doses of injected arginine cause insulin release. Figure 1 shows that dietary arginine at physiological concentrations influences the secretion of insulin in response to a glucose challenge in rats[9,15] and Fig. 2 and 3 show differences in the coefficient of glucose disappearance and the insulinogenic index. Plasma glucagon concentrations were also influenced by arginine intake in these studies.

Arginine and other amino acids are rapidly degraded by the rumenal microflora while lysine, an arginine antagonist, is among the most resistant to microbial degradation[3]. It is unknown if such degradation creates a functional deficiency of arginine and depresses detoxification of ammonia. Ruminants and other species enhance their urinary orotic acid as their protein intake is increased suggesting that high protein intake may cause disparity between the load of ammonia presented for detoxification and urea synthesizing capacity[25,27].

Arginine, recovery from trauma and immunocompetence

Fisher and associates supplemented high casein diets with arginine and glycine and increased nitrogen retention significantly after experimental trauma in rats[18,21]. Similar observations were reported by others[2]. The biochemical, hormonal, immunological and tumour modulating effects of arginine are reviewed by Barbul[2].

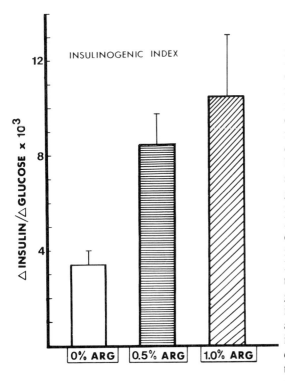

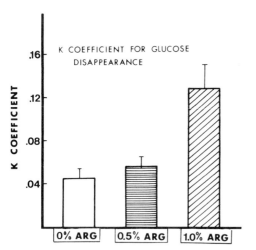

Fig. 2. Mean values (± SEM) of the k coefficients for
glucose disappearance and of the insulinogenic index
▲insulin
▲glucose
after intravenous glucose load. From Mulloy et al.
(1982)[15].

Orotic acid excretion

Abnormally high urinary excretion of orotic acid
has been documented during urea cycle enzyme
deficiency, therapy with allopurinol or 5-azau-
ridine, pregnancy, feeding of diets high in lysine
in relation to arginine, diets deficient in urea
cycle amino acids, or ammonia intoxication
from excessive doses of ammonium salts, urease
or urea[24–27]. Urinary orotic acid excretion varies
between species but the quantity excreted within
species is inversely related to the supply of arginine in the diet. Orotic acid in the urine of rats
is elevated within 24 h after their ingestion of an
L-amino acid diet lacking arginine, ornithine and
citrulline. The quantity excreted rises as N intake
increases at a constant intake of arginine.
Marked increases in urinary orotic acid excretion have been observed in growing mice,
guinea-pigs, hamsters, rabbits and dogs. Growing pigs and kittens show relatively smaller elevations[25,27]. Growing rats show more orotic acid
in their urine than mature rats during arginine
deficiency[13,24,25]. Liver slices from rats adapted
to low arginine diets and incubated with ammonia synthesize more orotate than controls but
extra hepatic tissues show insignificant changes
characteristic of tissues lacking urea cycle enzymes including the liver of birds[24]. Arginine
supplementation of diets of pregnant rats reduces
orotate excretion and increases the birth weight
and weaning weight of the young[24,25]. Enhanced
orotic acid excretion by women during the third
trimester of pregnancy has been reported but it
is unknown whether that is related to arginine
intake[24,25]. Orotic acid, first discovered in cows,
milk, where it occurs in significant quantities,
does not appear to be associated with ammonia
detoxification and its normal concentrations in
milk have not been shown to pose a health ha-
zard[7]. However orotic acid feeding and arginine
deficient diets have been used to promote experimentally induced liver tumours in rats[10]. The
fatty liver in rats which is seen with arginine
deficient diets and orotic acid feeding appears to
be a species-specific response[7]. The rationale for
using orotic acid excretion as an indicator of urea
cycle capacity and ammonia intoxication has
been reviewed[24].

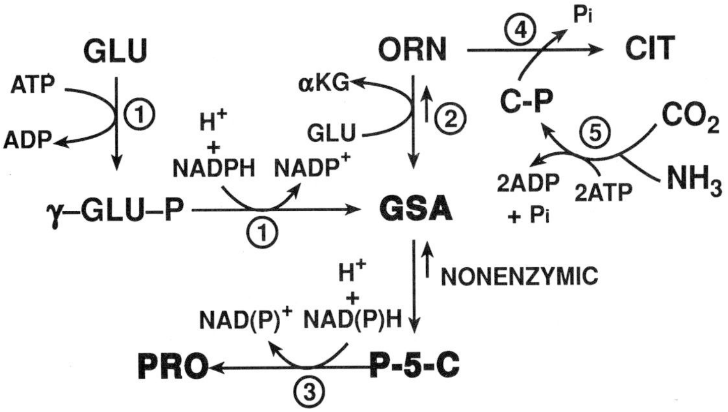

Fig. 3. Pathway of citrulline (CIT), ornithine (ORN), and proline (PRO) formation from glutamate (GLU) in the intestinal mucosa. Glutamate-γ-semialdehyde (GSA), pyrroline-5-carboxylate (P-5-C), carbamoyl phosphate (C~P), γ-glutamyl phosphate (γ-GLU~P). (1) P-5-C synthase; (2) ornithine aminotransferase; (3) P-5-C synthase; (4) ornithine carbomoyltransferase; (5) carbamoyl phosphate synthase. Diagram and legend taken from Rogers & Phang[19].

Intestinal pyrroline-5-carboxylate synthase activity and arginine needs

Observations of severe ammonia intoxication during arginine deficiency in cats were a forerunner of enzyme studies to explore the capacity of cats to synthesize ornithine in their small intestinal mucosa[19]. Other investigators had shown that the small intestinal mucosa is the only site for synthesis of pyrroline-5-carboxylate (P-5-C), ornithine and citrulline from glutamate in rats (Fig. 3). These studies revealed that P-C-5 synthetase activity per g of small intestinal mucosa of cats is 18 per cent as high as in rats and only 5 per cent as high per kg body weight. P-5-C synthetase catalyses the conversion of glutamate to glutamic-γ-semialdehyde which is converted to ornithine and spontaneously and non-enzymatically to P-5-C. Morris and Rogers have provided other evidence that the activity of enzymes for conversion of ornithine to citrulline may also be lower than in other species because the plasma concentrations of citrulline are normally lower in cats than in rats, dogs or human beings. This indicates that the severe hyperammonaemia found in cats and other carnivores fed an arginine-deficient diet is due to the low capacity for endogenous synthesis of ornithine. It is also likely that malnutrition, and chemotherapy of malignant disease, which can cause thinning and protein depletion of the intestinal mucosa may lower endogenous synthesis of ornithine via this pathway.

Arginine and creatine synthesis

Walser has called attention to the need for guanidino groups which must be provided by arginine for the synthesis of creatine[28]. Consequently the molar quantities of creatinine, an anhydride of creatine excreted in the urine, closely approximate the millimoles of arginine used for creatine synthesis. Walser has estimated that a 70 kg person eating 50 g of protein consumes about 0.2 mmol of arginine per kg body weight per day. According to his calculations a child gaining 15 ml of extracellular water per day would require an additional 45 mmol/kg of arginine to provide for the increase in the muscle creatine pool. I have calculated that the usual diet in the USA which supplies an average of 100 g of protein/day supplies about 5400 mg of arginine or 0.31 mmol/g of protein (Table 4). The recommended daily allowance (RDA) for protein is 0.8 g/kg of body weight which in the USA diet supplies 0.25 mmol of arginine/kg body weight. An average 18- to 29-year-old 70 kg male excretes an average of 23.6 mg or 0.21 mmol of creatinine per kg body weight per day. This translates to 0.21 mmol of arginine/kg body weight used for synthesizing creatine from an equimolar quantity of arginine. Thus the estimate of Walser based upon creatinine excretion and the estimate from survey data described here agree with the statement that 50 g of protein in the usual American diet barely supply the guanidino groups required for creatine synthesis in adult human males[27].

Table 4. Average per capita intake of protein and arginine in the United States by source

Source	Estimated intake	
	Protein (g/d)	Arginine (mg/d)
Meat	30.3	2051
Poultry and fish	12.1	680
Dairy products	22.4	720
Eggs	6.3	405
Cereals	18.4	762
Other[a]	10.5	799
Total	100.0	5397

[a]The supply of arginine by other foods was calculated by assuming that the percentage contribution to this category was as follows: legumes, 35; nuts and seeds, 35; potatoes and sweet potatoes, 20; fruits, 5; and green and yellow vegetables, 5[27].

Optimizing dietary amino acid ratios

The ratios of dietary amino acids to achieve optimal nutrition of lower animals and man have been studied extensively. A number of the studies have been based upon diets with crystalline amino acids as the only source of N. Such purified diets provide important qualitative data but fail to adequately simulate conditions encountered in practical circumstances with diets composed of nonpurified or natural ingredients. To overcome such shortcomings data from a vast array of studies have been used in calculating dietary amino acid ratios most appropriate for a particular purpose. Baker & Czarnecki-Maulden[1] have calculated such ratios for growth of dogs, pigs, cats, and chicks using lysine as the reference amino acid. The percentages of particular amino acids required for optimal growth were divided by the optimal percentage of lysine and the quotient was multiplied by 100. As expected the ratios of arginine to lysine for cats and dogs were relatively high compared to pigs. This was also true for chicks which lack an urea cycle in the liver and no capacity for arginine synthesis in the kidneys. The calculations also revealed unusual ratios which may be explained on the basis of physiological mechanisms or represent anomalies which will require more investigation. Similar calculations for human beings which Dr. Vernon Young and associates of the Massachusetts Institute of Technology have shared with the author indicate that the amino acid ratios for preschool children compare well with tissue composition data and the needs for adults when the needs for methionine are used as a reference. The critical factor in such calculations are reliable data for the requirements of the reference amino acid. Since actual requirements for amino acids are not known for some physiological and many clinical conditions this approach deserves further evaluation for guiding research and formulating appropriate mixtures of amino acids for circumstances where data are limited or non-existent.

References

1. Baker, D.H. & Czarnecki-Maulden, G.L. (1991): Comparative nutrition of cats and dogs. *Ann. Rev. Nutr.* **11**, 239–363.

2. Barbul, A. (1986): Arginine biochemistry, physiology, and therapeutic implications. *J. Parenter. Enteral. Nutr.* **10**, 227–238.

3. Chalupa, W. (1976): Degradation of amino acids by the mixed rumen microbial population. *J. An. Sci.* **43**, 828–834.

4. Chipponi, J.K., Bleier, J.C., Santi, M.C. & Rudman, D. (1983): Deficiencies of essential and conditionally essential nutrients. *Am. J. Clin. Nutr.* **35**, 1112–1116.

5. Cox, G.J. & Rose, W.C. (1926): The availability of synthetic imidazoles in supplementing diets deficient in histidine. *J. Biol. Chem.* **68**, 781–799.

6. Deshmukh, D. & Shope, T.C. (1983): Arginine requirement and ammonia toxicity in ferrets. *J. Nutr.* **113**, 1664–1667.

7. Durschlag, R. & Robinson (1980): Species specificity in the metabolic consequences of orotic acid. *J. Nutr.* **110**, 822–828.

8. Heird, W.C., Nicholson, J.F., Driscoll, J.R. Jr., Schullinger J.N. & Winters, R.W. (1972): Hyperammonemia resulting from intravenous alimentation using a mixture of synthetic L-amino acids. A preliminary report. *J. Pediatr.* **81**, 162–165.

9. Kari, F.W., Ulman, E.A., Mulloy, A.L. & Visek, W.J. (1981): Arginine requirement of mature protein-malnourished rats for maximal rate of repletion. *J. Nutr.* **111**, 1489–1493.

10. Laurier, C., Tatematsu, M., Rao, P.M., Rajalakshmi, S. & Sarma, D.S.R. (1984): Promotion by orotic acid of liver carcinogenesis in rats initiated by 1,2-dimethylhydrasine. *Cancer Res.* **44**, 2186–2191.

11. Magendie F. (1816): Sur les propriétés nutritives des substances qui ne contiennent pas d'azote. *Ann. Chim. Phys.*, **1st Ser 3**, 66–77.

12. Milner, J.A. & Visek, W.J. (1973): Orotic aciduria and arginine deficiency. *Nature* **245**, 211–213.

13. Milner, J.A. & Visek, W.J. (1975): Urinary metabolites characteristic of urea-cycle amino acid deficiency. *Metabolism* **24**, 643–651.

14. Morris, J.G., Rogers, Q.R., Winterrowd, D.L. & Kamikawa, E.M. (1979): The utilization of ornithine and citrulline by the growing kitten. *J. Nutr.* **109**, 724–729.

15. Mulloy, W.L., Kari, F.W. & Visek, W.J. (1982): Dietary arginine, insulin secretion, glucose tolerance and liver lipids during repletion of protein-depleted rats. *Horm. Metab. Res.* **14**, 471–475.

16. National Research Council (1989): *Recommended Dietary Allowances.* Washington, DC: National Academy Press.

17. Osborne, T.B. & Mendel, L.B. (1914): Amino acids in nutrition and growth. *J. Biol. Chem.* **17**, 325–349.

18. Pui, Y.M.L. & Fisher, H. (1979): Factorial supplementation with arginine and glycine on nitrogen retention and body weight gain in the traumatized rat. *J. Nutr.* **109**, 240–246.

19. Rogers, Q.R. & Phang, J.M. (1985): Deficiency of pyrroline-5-carboxylate synthase in the intestinal mucosa of the cat. *J. Nutr.* **115**, 146–150.

20. Rose, W.C. (1957): The amino acid requirements of adult man. *Nutr. Abst. Rev.* **27**, 631–647.

21. Sitren, H.S. & Fisher, H. (1977): Nitrogen retention in rats fed on diets enriched with arginine and glycine. Improved N retention after trauma. *Br. J. Nutr.* **37**, 195–208.

22. Stucki, W.P. & Harper, A.E. (1962): Effects of altering indispensable to dispensable amino acids in diets for rats. *J. Nutr.* **78**, 278–286.

23. Swanson, D.R. (1990): Somatomedin C and arginine: implicit connections between mutually isolated literatures. *Perspect. Biol. Med.* **33**, 157–186.

24. Visek, W.J. (1979): Ammonia metabolism, urea cycle capacity and their biochemical assessment. *Nutr. Rev.* **37**, 273–282.

25. Visek, W.J. (1984): An update of concepts of essential amino acids. *Ann. Rev.* **4**, 137–155.

26. Visek, W.J. (1985): Arginine and disease states. *J. Nutr.* **115**, 532–541.

27. Visek, W.J. (1986): Arginine needs physiological state and usual diets. *J. Nutr.* **116**, 36–46.

28. Walser, M. (1983): Urea cycle disorders and other hereditary hyperammonemic syndromes. In *The metabolic basis of inherited disease*, eds. J.B. Stanbury, J.B. Wyngaarden, D.S. Fredrickson, J.L. Goldstein & M.S. Brown, pp. 402-438, New York: McGraw-Hill Book Co.

Guanidino Compounds in Biology and Medicine, eds. by P.P. De Deyn, B. Marescau, V. Stalon and
I.A. Qureshi. ©1992 John Libbey & Company Ltd., pp. 133–138.

Chapter 19

Guanidino compound metabolism in arginine-free diet induced hyperammonaemia

Devendra R. DESHMUKH, Kathleen MEERT, Ashok P. SARNAIK, Bart MARESCAU[1]
and Peter P. De Deyn[1]

Department of Pediatrics, Children's Hospital of Michigan, Wayne State University, Detroit, Michigan, USA;
[1]*Department of Medicine-UIA, Laboratory of Neurochemistry-BBF, University of Antwerp, 2610 Antwerp, Belgium*

Summary

The concentration of plasma and urinary guanidino compounds is altered in many pathological conditions, especially those involving the urea cycle. Our objective was to investigate the relationship between guanidino compounds and hyperammonaemia in young ferrets. Acute ammonia toxicity was induced in 2-month-old ferrets by feeding a single meal of an arginine-free diet. Plasma and kidney arginine was decreased whereas guanidinosuccinic acid was increased in ferrets fed arginine-free diet. Hepatic creatine and kidney and brain guanidinoacetic acid were significantly decreased in this group. These results indicate that hyperammonaemia produced decreased methylation activity in the liver and transamidination activity in kidney. Increased guanidinosuccinate levels coupled with decreased hepatic creatine synthesis may play a role in the pathophysiology of arginine-free diet-induced hyperammonaemia in young ferrets.

Introduction

Arginine is an important intermediate of the urea cycle, the main function of which is to detoxify ammonia by converting it into urea. Although the complete urea cycle occurs only in the liver, the metabolism of urea cycle intermediates and guanidino compounds occurs in many tissues. For example, citrulline is taken up by the kidney and converted into arginine[18]. Arginine formed within the kidney is either metabolized into ornithine and guanidinoacetic by the action of arginine:glycine transamidinase or released into the blood and utilized by other tissues[17]. Guanidinoacetic acid may be either excreted into urine or taken up by the liver where it is converted into creatine by the action of guanidinoacetate methyltransferase. Creatine is metabolized in the skeletal muscle into creatinine and then excreted in the urine.

The concentration of guanidino compounds is altered in many pathological conditions such as renal failure[2,8] and hyperargininaemia[11]. The objective of our study was to investigate the relationship between acute hyperammonaemia and metabolism of guanidino compounds. Ferrets were selected for this study because they develop severe hyperammonaemia immediately after ingesting a small amount of an arginine-free diet[6].

Materials and methods

Two-month-old, male sable-coated ferrets, vaccinated for canine distemper, were purchased from

Marshall Research Laboratory (North Rose, New York). They were kept in groups of three, in cages with grid flooring, in an isolated room with controlled light and temperature (22–23 °C, 12 h light/dark cycle). Water and stock diet (Cat Chow, Ralston Purina Co., St Louis, MO) were provided *ad libitum* unless otherwise indicated. The composition of an arginine-containing diet which consists of free amino acids, carbohydrates, commercial vitamins and salt mixtures, was the same as reported earlier[6,15]. An arginine-free diet was prepared by substituting alanine for an isonitrogenous amount of arginine.

Ferrets were fasted overnight and fed, *ad libitum*, either arginine-containing diet or arginine-free diet. Three h later, they were lightly anaesthetized with ether and 3 ml of blood was drawn by cardiac puncture. Blood was collected into chilled heparinized tubes (Vacutainer) and centrifuged at 8000 × *g* for 15 min in a refrigerated centrifuge. A portion of the plasma was stored at –20 °C for guanidino compound analysis. Animals were sacrificed under deep ether anaesthesia and liver and kidney were removed. In addition, brain samples were removed from five animals in each group. Tissues were kept in ice during dissections and subsequently stored at –20 °C.

Portions of liver, kidney and brain samples were homogenized in chilled perchloric acid (1 M). Homogenates were centrifuged at 10,000×*g* for 15 min at 0–2 °C. The supernatants were neutralized to pH 7.0 with 1 N sodium hydroxide. Ammonia in the supernatants and plasma was assayed by the glutamate dehydrogenase reaction[14]. The concentration of urea was determined as described by Ceriotti[1].

Other portions of tissue samples were homogenized in 3 ml water, deproteinized with 1 ml of 30 per cent trichloracetic acid and then centrifuged at 100,000 × *g* for 30 min at 4 °C. After decanting the supernatant, the pellet was vortex-mixed with 1 ml of 10 per cent tricholoracetic acid and centrifuged at 100,000 × *g* for 30 min. The first and second supernatants were pooled and analysed for guanidino compounds using a Biotronic LC 5001 amino acid analyser, adapted for guanidino compound determination. The guanidino compounds were separated over a cation-exchange column using sodium citrate buffers and detected by the fluorescence ninhydrin method[11,12].

Student's *t*-test was used to calculate statistical difference. the *P* values < 0.05 were considered statistically significant.

Table 1. Effect of arginine-free diet on plasma guanidino compounds in young ferrets

Guanidino compound (nmoles/l)	Control n = 9	Arginine-free diet n = 12
α-N-Acetylarginine	< DL	< DL
Ammonia	57 ± 7	2890 ± 361*
Arginine	145 ± 21	36 ± 5*
Argininic acid	0.12 ± 0.015	0.05 ± 0.01*
Creatine	17.3 ± 3	12.3 ± 2
Creatinine	13.5 ± 0.8	16.4 ± 2.1
Guanidine	0.11 ± 0.02	0.12 ± 0.03
Guanidinoacetic acid	2.29 ± 0.35	1.24 ± 0.18*
γ-Guanidinobutyric acid	0.11 ± 0.05	0.42 ± 0.38
β-Guanidinopropionic acid	< DL	< DL
Guanidinosuccinic acid	0.42 ± 0.03	1.11 ± 0.13*
α-Keto-δ-guanidinovaleric acid	3.64 ± 1.1	3.21 ± 2.4
Homoarginine	0.1 ± 0.02	0.13 ± 0.02
Methylguanidine	< DL	< DL
Urea (mmol/l)	14.1 ± 0.6	8.3 ± 0.3*

Control = Arginine-containing diet, < DL = below detection limit. Values are mean ± SEM. *Significantly different (*P* < 0.05) from the control group.

Results

Ferrets fed arginine-free diet developed typical signs of hyperammonaemia such as seizures and coma within 2 h. At 3 h following arginine-free diet, plasma ammonia was significantly elevated. Plasma urea levels were significantly lower than those in the arginine-containing diet group. Plasma guanidinoacetic acid, argininic acid and arginine were significantly decreased, whereas guanidino-succinic acid was significantly elevated following arginine-free diet (Table 1).

Arginine-free diet caused a significant decrease in the hepatic concentrations of arginine, guanidino-acetic acid, β-guanidinopropionic acid and creatine (Table 2). A significant decrease in the renal concentration of arginine, guanidinoacetic acid, argininic acid and β-guanidinopropionic acid was observed following arginine-free diet (Table 3). Guanidinosuccinic acid was significantly increased in the kidneys of young ferrets fed arginine-free diet compared to those fed arginine-containing diet.

Table 2. Effect of arginine-free diet on hepatic guanidino compounds in young ferrets

Guanidino compound (μmoles/g)	Control n = 10	Arginine-free diet n = 12
α-N-Acetylarginine	< DL	< DL
Arginine	22 ± 5	6.5 ± 0.7*
Argininic acid	< DL	< DL
Creatine	146 ± 7	104 ± 13*
Creatinine	< DL	2.4 ± 2.4
Guanidine	1.6 ± 0.3	1.1 ± 0.3
Guanidinoacetic acid	8.4 ± 1.9	2.2 ± 0.2*
γ-Guanidinobutyric acid	< DL	< DL
β-Guanidinopropionic acid	1.56 ± 0.14	0.94 ± 0.11*
Guanidinosuccinic acid	8.9 ± 0.9	8.7 ± 1.1
α-Keto-δ-guanidinovaleric acid	0.59 ± 0.30	0.22 ± 0.2
Homoarginine	< DL	0.07 ± 0.07
Methylguanidine	< DL	< DL

Control = Arginine-containing diet, < DL = Below detection limit. Values are mean ± SEM. *Significantly different ($P < 0.05$) from the control group.

Table 3. Effect of arginine-free diet on renal guanidino compounds in young ferrets

Guanidino compound (nmoles/g)	Control n = 10	Arginine-free diet n = 12
α-N-Acetylarginine	< DL	< DL
Arginine	428 ± 60	225 ± 27*
Argininic acid	3.3 ± 0.4	0.53 ± 0.2*
Creatine	541 ± 37	491 ± 39
Creatinine	60 ± 6	75 ± 6
Guanidine	0.51 ± 0.22	1.45 ± 0.55
Guanidinoacetic acid	190 ± 14	51 ± 8*
γ-Guanidinobutyric acid	0.24 ± 0.12	< DL
β-Guanidinopropionic acid	1.89 ± 0.29	1.13 ± 0.12*
Guanidinosuccinic acid	1.6 ± 0.16	4.43 ± 0.61*
α-Keto-δ-guanidinovaleric acid	< DL	< DL
Homoarginine	0.92 ± 0.4	0.91 ± 0.2
Methylguanidine	< DL	< DL

Control = Arginine-containing diet, < DL = below detection limit. Values are mean ± SEM. *Significantly different ($P < 0.05$) from the control group.

Table 4. Effect of arginine-free diet on brain guanidino compounds in young ferrets

Guanidino compound (nmoles/g)	Control n = 5	Arginine-free diet n = 5
α-N-Acetylarginine	0.33 ± 0.33	0.57 ± 0.57
Arginine	215 ± 26	168 ± 24
Argininic acid	< DL	< DL
Creatine	7514 ± 452	7781 + 306
Creatinine	136 ± 5	181 ± 10*
Guanidine	2.41 ± 1.2	0.15 ± 0.15
Guanidinoacetic acid	9.9 ± 0.9	4.52 ± 0.2*
γ-Guanidinobutyric acid	3.71 ± 0.26	3.98 ± 1.34
β-Guanidinopropionic acid	0.23 ± 0.09	0.13 ± 0.08
Guanidinosuccinic acid	< DL	< DL
α-Keto-δ-guanidinovaleric acid	< DL	< DL
Homoarginine	< DL	0.29 ± 0.29
Methylguanidine	0.23 ± 0.07	0.22 ± 0.11

Control = Arginine-containing diet, < DL = below detection limit. Values are mean ± SEM. *Significantly different ($P < 0.05$) from the control group.

Although the brain arginine concentration was lower in ferrets fed arginine-free diet than those fed arginine-containing diet, the difference was not statistically significant. Arginine-free diet caused a significant decrease in the level of guanidinoacetic acid in brain (Table 4).

Discussion

The guanidino compound values reported in this study are from the samples obtained 3 h after feeding the specified diets. A period of 3 h after feeding was selected becaused plasma ammonia levels and the degree of sickness were maximum at this time.

Arginine is transaminated into α-keto-δ-guanidinovaleric acid which in turn is converted argininic acid by the hydrogenation reaction. Arginine-free diet caused a significant decrease in plasma argininic acid, although α-keto-δ-guanidinovaleric acid was not altered (Table 1). The decrease in argininic acid may be due to lower levels of arginine in ferrets fed arginine-free diet. Ferrets fed control diet had higher plasma levels of α-keto-δ-guanidinovaleric acid than those seen in other species[13], indicating that in ferrets, more arginine is converted into α-keto-δ-guanidinovaleric acid. Hydrogenation of α-keto-δ-guanidinovaleric acid into argininic acid may be decreased in ferrets because plasma argininic acid was not high as compared to that in other species[13]. Previous studies indicate that α-keto-δ-guanidinovaleric acid is epileptogenic in rabbits[5]. α-Keto-δ-guanidinovaleric acid levels were either low or below the detection limit in the kidney and brains of young ferrets (Tables 3 and 4).

A single meal of arginine-free diet caused a significant decrease in arginine concentration in the plasma, kidney, and liver of young ferrets (Tables 1, 2 and 3). Arginine deficiency also caused a significant decrease in guanidinoacetate in the plasma, liver, kidney and brain of young ferrets. Arginine is the only amino acid that provides the amidino group for the synthesis of creatine, the major source of high energy phosphate in muscle. The low levels of creatine in young ferrets are consistent with our earlier observation of low activity of arginine:glycine transamidinase[7], an enzyme involved in the synthesis of creatine. Creatine deficiency may contribute to the pathophysiology of hyperammonaemia observed in this animal model.

Arginine deficiency caused a significant increase in plasma and kidney guanidinosuccinic acid (Tables 1 and 2). The metabolism of guanidinosuccinate has not been completely elucidated. Cohen et al.[2] suggested that guanidinosuccinic acid formation represents an alternate route of ammonia detoxication and proposed two possible mechanisms for its biosynthesis. The first mechanism is based on the repression of arginine:glycine transamidinase as well as on the appearance or activation of a new enzyme, arginine:asparate transamidinase. This scheme depends upon the postulate of a transamidinase which uses aspartate rather than glycine as the amidine acceptor. Although our earlier studies demonstrated a decrease in activity of arginine:glycine transamidinase in kidneys of ferrets when compared to other species, no significant difference was previously detected in arginine:glycine transmidinase activity among ferrets fed different diets[7]. However, the decrease in guanidinoacetate which accompanies the elevated guanidinosuccinic acid in this study suggests this hypothesis as a possible mechanism.

The second mechanism of guanidinosuccinic acid formation proposed by Cohen is based upon its structural similarity to the known urea cycle intermediate, argininosuccinic acid. In the urea cycle, argininosuccinic acid is reduced by argininosuccinase to form arginine and fumaric acid. Repression of argininosuccinase could result in cleavage of argininosuccinic acid at a different site resulting in formation of carbamyl aspartate which could be converted to guanidinosuccinic acid. Argininosuccinase activity has also been previously studied in the ferret model[7]. Repression of argininosuccinase was demonstrated in the kidney of young ferrets compared to adults, although this difference was not significant.

Natelson & Sherwin[16] proposed a guanidine cycle, a pathway closely related to the urea cycle which provides a mechanism for the reutilization of urea nitrogen. According to this scheme, urea is recycled in the liver by oxidation and hydrolysis to carbamate and hydroxylamine. Both of these enter the guanidine cycle, ultimately generating creatine for muscle activity but resulting in formation of guanidinosuccinic acid as a by-product. Via this mechanism, urea serves as a precursor for guanidinosuccinic acid formation. However, young, arginine deficient ferrets had lower plasma urea concentration (Table 1), making this an unlikely route for guanidinosuccinic acid synthesis in our model.

Guanidinosuccinic acid is a known toxin which is elevated in various biological fluids in uraemia[4,9]. It stimulates the central nervous system as well as it depresses the action of the peripheral nervous system and spinal polysynaptic arcs[10]. Recently, guanidinosuccinic acid has been shown to inhibit responses to the inhibitory neurotransmitters γ-aminobutyric acid and glycine on mouse neurons in cell culture[3]. Our data show an increase in guanidinosuccinic acid in plasma and kidney which suggest that guanidinosuccinic acid may play an important role in hyperammonaemia and related disorders.

Acknowledgements

This work was supported by the Evergreen Endowment Fund (KM) of Children's Hospital of Michigan, by the Born-Bunge Foundation, the Universitaire Instelling Antwerpen, the United Fund of Belgium and the NFWO grants N° 3.0044.92 and NDE 58.

References

1. Ceriotti, G. (1971): Ultramicrodetermination of plasma urea by reaction with diacetylmonoxime-antipyrene without deproteinization. *Clin. Chem.* **17**, 400–402.

2. Cohen, B.D., Stein, I.M. & Bonas, J.E. (1968): Guanidinosuccinic aciduria in uremia, a possible alternate pathway for urea synthesis. *Am. J. Med.* **45**, 63–68.

3. De Deyn, P.P. & MacDonald, R.L. (1990): Guanidino compounds that are increased in uremia inhibit GABA and glycine responses on mouse neurons in cell culture. *Ann. Neurol.* **28**, 627–633.

4. De Deyn, P.P., Marescau, B., Cuykens, J.J., Van Gorp, L., Lowenthal, A. & De Potter, W.P. (1987): Guanidino compounds in the serum and cerebrospinal fluid of non-dialyzed patients with renal insufficiency. *Clin. Chim. Acta* **167**, 81–88.

5. De Deyn, P.P., Marescau, B. & Macdonald, R.L. (1989): Effects of α-keto-δ-guanidinovaleric acid on inhibitory amino acid responses on mouse neurons in cell culture. *Brain Res.* **449**, 54–60.

6. Deshmukh, D.R. & Shope, T.C. (1983): Arginine requirement and ammonia toxicity in ferrets. *J. Nutr.* **13**, 1664–1667.

7.   Deshmukh, D.R. & Rusk, C.D. (1989): Effects of arginine-free diet on urea cycle enzymes in young and adult ferrets. *Enzyme* **41**, 168–174.

8.   Giovannetti, S., Balestri, P.L. & Barsotti, G. (1973): Methylguanidine in uremia. *Arch. Intern. Med.* **131**, 709–713.

9.   Kishore, B.K., Kallay, Z. & Tulkens, P.M. (1989): Clinico-biochemical aspects of guanidine compounds in uraemic toxicity. *Int. Urol. Nephrol.* **21**, 223–232.

10.   Kishore, B. K. (1983): Some observations on the *in vivo* and *in vitro* effects of guanidinosuccinic acid on the nervous system of the laboratory animals. *Acta Med. Biol.* **31**, 79–86.

11.   Marescau, B., Qureshi, I.A., De Deyn, P.P., Letarte, J., Ryba, R. & Lowenthal, A. (1985): Guanidino compounds in plasma, urine, and cerebrospinal fluid of hyperargininemic patients during therapy. *Clin. Chim. Acta* **146**, 21–27.

12.   Marescau, B., De Deyn, P.P., Van Gorp, L. & Lowenthal, A. (1986): Purification procedure for some urinary guanidino compounds. *J. Chromat.* **337**, 334–338.

13.   Marescau, B., De Deyn, P.P., Wiechert, P., Van Gorp, L. & Lowenthal, A. (1986): Comparative study of guanidino compounds in serum and brain of mouse, rat, rabbit and man. *J. Neurochem.* **46**, 717–720.

14.   Mondzack, A., Ehrlich, G.E. & Sigmiller, J.E. (1965): An enzymatic determination of ammonia in biological fluids. *J. Lab. Clin. Med.* **66**, 526–531.

15.   Morris, J.G. & Rogers, Q.R. (1978): Arginine: an essential amino acid for cat. *J. Nutr.* **108**, 1944–1953.

16.   Natelson, S. & Sherwin, J.W. (1979): Proposed mechanism for urea nitrogen reutilization: relationship between urea and proposed guanidine cycle. *Clin. Chem.* **25**, 1343–1344.

17.   Shindo, S. & Mori, A. (1985): Metabolism of L-[amidino-^{15}N]- arginine of guanidino compounds. In *Guanidines*, eds. A. Mori, B.D. Cohen & A. Lowenthal, pp. 71–81, New York: Plenum Press.

18.   Windmueller, H.G. & Spaeth, A.E. (1981): Source and fate of circulating citrulline. *Am. J. Physiol.* **241**, E473–E480.

138

Guanidino Compounds in Biology and Medicine, eds. by P.P. De Deyn, B. Marescau, V. Stalon and
I.A. Qureshi. ©1992 John Libbey & Company Ltd., pp. 139–144.

Chapter 20

Guanidino compound metabolism in hyperammonaemic rats

Kathleen L. MEERT, Devendra R. DESHMUKH, Bart MARESCAU[1],
Peter P. De DEYN[1] and Ashok P. SARNAIK

*Department of Pediatrics, Children's Hospital of Michigan, Wayne State University, 3901 Beaubien Boulevard,
Detroit, MI 48201, USA; [1]Department of Medicine-UIA, Laboratory of Neurochemistry-BBF, University of
Antwerp, Universiteitsplein 1, 2610 Antwerp, Belgium*

Summary

The effect of hyperammonaemia induced by an ammonium acetate injection on the levels of guanidino compounds in
plasma, liver, kidney and brain of rats was investigated. Blood and tissues were removed 1 h following intraperitoneal
injection of 6 mmol/kg ammonium acetate or saline (control). Guanidino compounds were analysed by high performance
liquid chromatography. Plasma and kidney levels of guanidinosuccinic acid were significantly elevated in rats
challenged with ammonium acetate. Brain α-N-acetylarginine levels were also significantly higher in rats injected with
ammonium acetate as compared to controls. Our results suggest that guanidinosuccinic acid and α-N-acetylarginine
may play an important role in hyperammonaemia.

Introduction

Hyperammonaemia leads to an increase in arginine catabolism and urea formation when an intact urea cycle exists. Guanidino compounds, catabolites of arginine, may also be altered during hyperammonaemia. Guanidino compounds are present in many tissues including liver, kidney and brain[14]. Several guanidino compounds are known toxins. Our objective was to investigate the effect of hyperammonaemia induced by an ammonium acetate challenge on the metabolism of guanidino compounds in rats[8].

Materials and methods

Adult male albino rats (Sprague-Dawley) weighing 200–250 g were purchased from Charles River Laboratories (Wilmington, DE) and maintained on rodent chow (Ralston-Purina Co., St Louis, MO). All animal experimentation was performed with Institutional Animal Investigation Committee Approval.

Hyperammonaemia was induced by an intraperitoneal (ip) injection of ammonium acetate (0.8 M, 6 mmol/kg). Control rats received an equal volume of ip saline (injectable form, pH 7.0). Fifty min after ammonium acetate or saline injection, rats were anaesthetized with an intramuscular injection of xylazine/ketamine (13 mg/kg and 87 mg/kg, respectively). One hour after ammonium acetate or saline injection, blood was collected by cardiac puncture into chilled heparinized tubes and centrifuged at $5000 \times g$ for 15 min. Liver, kidney and brain were quickly removed and all samples were stored at $-20\,°C$.

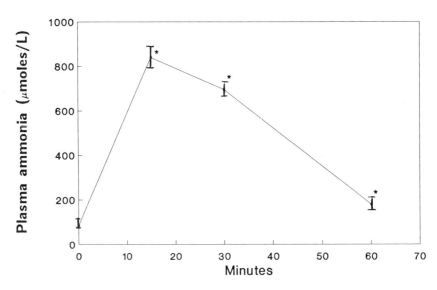

*Fig. 1. Plasma ammonia levels in rats injected with ammonium acetate. Rats were injected (ip) with 6 mmol/kg of ammonium acetate. Blood was collected in heparinized tubes from the tail vein. Plasma ammonia was assayed as described in the text. Results are means ± SEM of five animals. * = Significantly different (P < 0.05) from controls.*

Plasma ammonia was assayed immediately after collection by the glutamate dehydrogenase reaction[9]. Ammonia free water was used as a reference blank and for preparing all reagents. Plasma urea was determined by the method of Ceriotti[2].

Liver, kidney and brain samples were homogenized in 2 ml water, deproteinized with 1 ml of 30 per cent trichloroacetic acid and centrifuged at $100,000 \times g$ for 30 min. The supernatant was collected and the pellet was vortex-mixed with 1 ml of 10 per cent trichloroacetic acid, centrifuged at 100,000 $\times g$ for 30 min and the two supernatants were pooled. The guanidino compounds were separated over a cation exchange column using sodium citrate buffer and detected by the fluorescence ninhydrin method[6,7] using a Biotronic LC 6001 amino acid analyser.

Student's t-test was used to calculate statistical differences. The P values < 0.05 were considered statistically significant.

Results

A preliminary experiment was performed to determine the time course of plasma ammonia following an intraperitoneal injection of ammonium acetate (6 mmol/kg). Blood was drawn from the tail vein at 15, 30 and 60 min after ammonium acetate injection. Peak plasma ammonia levels were reached at 15 min, followed by a steady decline (Fig. 1).

Within 15 min of ammonium acetate injection, rats exhibited typical signs of hyperammonaemia ranging from lethargy to coma, followed by gradual resolution of symptoms over the next 45 min. Plasma ammonia and urea levels were significantly elevated in these rats at 60 min following ammonium acetate challenge (Table 1).

Rats receiving ammonium acetate had lower plasma arginine and higher creatine levels than controls, however the differences were not statistically significant (Table 1). Plasma levels of guanidinosuccinic acid and creatinine were significantly higher in hyperammonaemic animals as compared to controls.

Hepatic creatine was significantly increased following ammonium acetate challenge (Table 2). Renal concentrations of arginine, guanidinosuccinic acid, guanidinoacetic acid and creatinine were significantly elevated in the ammonium acetate group (Table 3). Brain α-N-acetylarginine was significantly higher in rats injected with ammonium acetate (Table 4).

Table 1. Effect of ammonium acetate on guanidino compounds in rat plasma

Guanidino compound (μmol/l)	Control	Ammonium acetate
α-N-Acetylarginine	< DL	< DL
Arginine	193 ± 16	162 ± 8
Argininic acid	< DL	< DL
Ammonia	85 ± 26	191 ± 25*
Creatine	162 ± 14	222 ± 26
Creatinine	17.7 ± 1.45	22.1 ± 1.24*
Guanidine	0.36 ± 0.02	0.38 ± 0.04
Guanidinoacetic acid	5.79 ± 0.75	5.09 ± 0.43
γ-Guanidinobutyric acid	0.45 ± 0.09	0.66 ± 0.14
β-Guanidinopropionic acid	0.04 ± 0.01	0.04 ± 0.01
Guanidinosuccinic acid	0.05 ± 0.01	0.07 ± 0.01*
α-Keto-δ-guanidinovaleric acid	0.24 ± 0.03	0.27 ± 0.03
Homoarginine	2.38 ± 0.31	2.06 ± 0.15
Methylguanidine	0.18 ± 0.02	0.13 ± 0.01
Urea	5.13 ± 0.26	7.70 ± 0.40*

Mean ± SEM; n = 10 in each group; * = significantly different ($P < 0.05$) from controls; < DL = below detection limit.

Table 2. Effect of ammonium acetate on guanidino compounds in rat liver

Guanidino compound (nmol/g)	Control	Ammonium acetate
α-N-Acetylarginine	< DL	< DL
Arginine	11.4 ± 0.7	11.9 ± 0.6
Argininic acid	< DL	< DL
Creatine	311 ± 25	499 ± 80*
Creatinine	2.98 ± 2.98	10.0 ± 6.5
Guanidine	1.79 ± 0.61	1.72 ± 0.57
Guanidinoacetic acid	9.00 ± 1.38	8.92 ± 2.12
γ-Guanidinobutyric acid	73.5 ± 6.5	75.4 ± 11.8
β-Guanidinopropionic acid	23.3 ± 1.1	22.5 ± 2.1
Guanidinosuccinic acid	2.91 ± 0.25	3.48 ± 0.26
α-Keto-δ-guanidinovaleric acid	< DL	< DL
Homoarginine	5.76 ± 0.38	6.41 ± 0.40
Methylguanidine	0.61 ± 0.19	0.56 ± 0.17

Mean ± SEM; n = 10 in each group; * = significantly different ($P < 0.05$) from controls; < DL = below detection limit.

Table 3. Effect of ammonium acetate on guanidino compounds in rat kidney

Guanidino compound (nmol/g)	Control	Ammonium acetate
α-N-Acetylarginine	< DL	< DL
Arginine	724 ± 68	1112 ± 165*
Argininic acid	< DL	< DL
Creatine	931 ± 157	1486 ± 253
Creatinine	120 ± 12	207 ± 35*
Guanidine	2.88 ± 0.80	4.25 ± 1.16
Guanidinoacetic acid	226 ± 21	290 ± 19*
γ-Guanidinobutyric acid	10.1 ± 1.1	13.3 ± 1.8
β-Guanidinopropionic acid	4.12 ± 0.40	4.45 ± 0.44
Guanidinosuccinic acid	0.52 ± 0.07	0.94 ± 0.09*
α-Keto-δ-guanidinovaleric acid	< DL	< DL
Homoarginine	5.37 ± 0.56	6.12 ± 0.47
Methylguanidine	1.26 ± 0.18	1.67 ± 0.34

Mean ± SEM; n = 10 in each group; * = significantly different ($P < 0.05$) from controls; < DL = below detection limit.

Table 4. Effect of ammonium acetate on guanidino compounds in rat brain

Guanidino compound (nmol/g)	Control	Ammonium acetate
α-N-Acetylarginine	2.36 ± 0.26	3.03 ± 0.16*
Arginine	206 ± 15	242 ± 12
Argininic acid	0.99 ± 0.04	1.04 ± 0.04
Creatine	8075 ± 367	8684 ± 175
Creatinine	109 ± 3	120 ± 8
Guanidine	0.95 ± 0.35	0.69 ± 0.52
Guanidinoacetic acid	2.99 ± 0.16	2.82 ± 0.19
γ-Guanidinobutyric acid	8.45 ± 0.18	8.71 ± 0.72
β-Guanidinopropionic acid	0.36 ± 0.06	0.45 ± 0.04
Guanidinosuccinic acid	< DL	< DL
α-Keto-δ-guanidinovaleric acid	< DL	< DL
Homoarginine	1.88 ± 0.19	2.18 ± 0.16
Methylguanidine	0.23 ± 0.08	0.06 ± 0.06I

Mean ± SEM; n = 10 in each group; * = significantly different ($P < 0.05$) from controls; < DL = below detection limit.

Discussion

Since the catabolism of arginine occurs in several tissues, we determined the levels of arginine and other guanidino compounds in the plasma, liver, kidney and brain of control and ammonium acetate treated rats. In order to allow a time period for the conversion of ammonia into urea by the urea cycle and for formation of guanidino compounds, tissues were removed 1 h after ammonium acetate challenge. The preliminary experiments indicated that animals remained hyperammonaemic during this period.

Although arginine is synthesized in large amounts by the liver, the high activity of hepatic arginase normally prevents release of arginine into the blood. Among extrahepatic tissues, the kidney has the greatest capacity to convert citrulline into arginine. The citrulline required for this process is

142

synthesized in the small intestine, released in the blood and taken up by the kidney[18]. Arginase, present in liver, kidney and brain, converts arginine into ornithine and urea. Urea is then excreted by the kidney eliminating excess nitrogen.

Arginine may also undergo transamidination in the kidney by the enzyme arginine:glycine transa-midinase to form guanidinoacetic acid[16]. We observed that renal guanidinoacetic acid was significantly elevated in hyperammonaemic animals indicating that arginine was converted into guanidinoacetic acid. Increased guanidinoacetic acid would lead to increased hepatic creatine because guanidinoacetic acid is taken up by the liver where it is methylated to form creatine. Creatine is metabolized by skeletal muscle to creatinine which is transported by the bloodstream back to the kidney and excreted in the urine. Hepatic creatine and plasma and renal creatinine levels were significantly elevated in rats injected with ammonium acetate.

Guanidinosuccinic acid was first isolated in the urine of uraemic patients[11]. It has been identified as a uraemic toxin contributing to the bleeding dyscrasia, increased haemolysis and immunological disturbances of patients with renal failure[3–5,15]. The synthesis of guanidinosuccinic acid from urea has been described in isolated rat hepatocytes[1] as well as in an intact rat model[12]. Urea would be oxidized and hydrolysed in the liver to form carbamate and hydroxylamine. These substrates would enter into the guanidine cycle[12,13], a proposed series of reactions analogous to the urea cycle in which homoserine rather than arginine would supply a structure for carrying nitrogen. The main suggested role of the guanidine cycle is to reutilize urea nitrogen for the formation of creatine. Guanidinosuccinic acid would be formed as a by-product of this pathway and excreted in the urine. We observed that ammonium acetate induced hyperammonaemia not only leads to elevated urea but also to elevated plasma and kidney guanidinosuccinic acid. Metabolism of urea via the guanidine cycle may be the mechanism of increased guanidinosuccinic acid formation in ammonium acetate induced hyperam-monaemia.

Brain α-N-acetylarginine was significantly higher in rats injected with ammonium acetate. The probable mechanism of α-N-acetylarginine formation is via the acetylation of arginine. α-N-Acetyl-arginine has been shown to be experimentally epileptogenic[10]. Elevated levels of α-N-acetylarginine have been demonstrated in the serum and brain of audiogenic sensitive rats during both the preconvulsive and tonic phase of cerebral seizures[17]. α-N-acetylarginine may contribute to the neurotoxicity associated with hyperammonaemia.

Elevated guanidinosuccinic acid and α-N-acetylarginine may represent alternate routes of nitrogen metabolism following ammonium acetate challenge. With their known toxic effects, both compounds may contribute to the pathophysiology of hyperammonaemia.

Acknowledgements

This work was supported by the Evergreen Endowment Fund (KM) of Children's Hospital of Michigan, by the Born-Bunge Foundation, the Universitaire Instelling Antwerpen, the United Fund of Belgium and the NFWO grants N° 3.0044.92 and NDE58.

References

1. Aoyagi, K., Ohba, S., Marita, M. & Tojo, S. (1982): Biosynthesis of guanidinosuccinic acid in isolated rat hepatocytes II. Inhibition of its synthesis by urea cycle members and D-L norvaline. *Jpn. J. Nephrol.* **24**, 1137–1146.

2. Ceriotti, G. (1971): Ultramicrodetermination of plasma urea by reaction with diacetyl monoxime-antipyrene without deproteinization. *Clin. Chem.* **17**, 400–402.

3. De Deyn, P.P., Marescau, B., Cuykans, J.J., Van Gorp, L., Lowenthal, A. & De Potter, W.P. (1987): Guanidino compounds in serum and cerebrospinal fluid of non-dialyzed patients with renal insufficiency. *Clin. Chim. Acta* **167**, 81–88.

4. Giovannetti, S., Cioni, L., Balestri, P.L. & Biagni, M. (1968): Evidence that guanidines and some related compounds cause hemolysis in chronic uremia. *Clin. Sci.* **34**, 141–148.

5. Horowitz, J.I., Stein, I.M., Cohen, B.D. & White, J.G. (1970): Further studies on the platelet inhibitory effect of guanidinosuccinic acid and its role in uremic bleeding. *Am. J. Med.* **49**, 336–345.

6. Marescau, B., Qureshi, I.A., De Deyn, P.P., Letarte, J., Ryba, R. & Lowenthal, A. (1985): Guanidino compounds in plasma, urine and cerebrospinal fluid of hyperargininemic patients during therapy. *Clin. Chim. Acta* **146**, 21–27.

7. Marescau, B., De Deyn, P.P., Van Gorp, L. & Lowenthal, A. (1986): Purification procedure for some urinary guanidino compounds. *J. Chromatogr.* **377**, 334–338.

8. Meert, K.L., Deshmukh, D.R., Marescau, B., De Deyn, P.P. & Sarnaik, A.P. (1991): Effect of ammonium acetate induced hyperammonemia on metabolism of guanidino compounds. *Biochem. Med. Metabol. Biol.* (in press).

9. Mondzack, A., Ehrlich, G.E. & Sigmiller, J.E. (1965): An enzymatic determination of ammonia in biological fluids. *J. Lab. Clin. Med.* **66**, 526–531.

10. Mori, A. & Ohkusu, H. (1971): Isolation of α-N-acetyl-L- arginine from calf brain in convulsive seizures induced by this substance. *Adv. Neurol. Sci.* **15**, 304–306.

11. Natelson, S., Stein, I. & Bonas, J.E. (1964): Improvements in the method of separation of guanidino organic acids by column chromatography. Isolation and identification of guanidinosuccinic acid from human urine. *Microchem. J.* **8**, 371–382.

12. Natelson, S. (1984): Metabolic relationship between urea and guanidino compounds as studied by automated fluorimetry of guanidino compounds in urine. *Clin. Chem.* **30**, 252–258.

13. Natelson, S. & Sherwin, J.E. (1979): Proposed mechanism of urea nitrogen reutilization: relationship between urea and proposed guanidine cycles. *Clin. Chem.* **25**, 1343–1344.

14. Robin, Y. & Marescau, B. (1985): Natural guanidino compounds. In *Guanidines*, eds. A. Mori, B.D. Cohen & A. Lowenthal. pp. 383–438, New York: Plenum Press.

15. Shainkin-Kesterbaum, R., Winikoff, Y., Dvilansky, A., Vhaimotiz, C. & Nathan, I. (1987): Effect of guanidino-propionic acid on lymphocyte proliferation. *Nephron* **44**, 295–298.

16. Van Pilsum, J.F., Taylor, D., Zaikis, B. & McCormick, P. (1970): Simplified assay for transamidinase from rat kidney homogenates. *Anal. Biochem.* **35**, 277–286.

17. Wiechert, P., Marescau, B., De Deyn, P.P. & Lowenthal, A. (1987): Guanidino compounds in serum and brain of audiogenic sensitive rats during the preconvulsive running phase of cerebral seizures. *Neurosciences* **13**, 35–39.

18. Windmuller, H.G. (1982): Glutamate utilization by the small intestine. In *Advances in enzymology*, Vol. 53, ed. A. Meister, pp. 201–237. New York: John Wiley and Sons.

Section III

Biochemistry – metabolism and physiology of creatine–creatinine and phosphocreatine biosynthesis pathways

Guanidino Compounds in Biology and Medicine, eds. by P.P. De Deyn, B. Marescau, V. Stalon and I.A. Qureshi. ©1992 John Libbey & Company Ltd., pp. 147–151.

Chapter 21

The antagonistic action of creatine and growth hormone on the expression of the gene for rat kidney L-arginine:glycine amidinotransferase

John F. VAN PILSUM, Denise M. McGUIRE and Cheryl A. MILLER

Department of Biochemistry, University of Minnesota, The Medical School, 4-225 Millard Hall, 435 Delaware Street SE, Minneapolis, Minnesota 55455; The Department of Biology, St Cloud State University, St Cloud, Minnesota, 56301, USA

Summary

Transamidinase activities and relative mRNA levels were determined in kidneys from hypophysectomized rats fed a creatine-free or a creatine-supplemented diet, both groups of which were maintained with and without injections of recombinant human growth hormone. The kidney transamidinase activities and mRNA levels after growth hormone administration were: (1) with rats fed a creatine-free diet, six and four times greater respectively, than the non-injected animals; (2) with rats fed a creatine-supplemented diet, both three times greater than the non-injected animals. The corresponding values after feeding creatine were: (1) with non-injected animals, 50 and 30 per cent respectively of the values from the rats fed a creatine-free diet; (2) with the hormone injected animals, similar to the values of the rats fed a creatine-free diet without hormone injections. All of the alterations in enzyme activities and mRNA levels were statistically significant whereas no statistically significant differences were found in the enzyme activity/mRNA ratios of any experimental group. The induction of transamidinase activities by growth hormone in the hypophysectomized rat (with or without creatine in the diet) and the repression of transamidinase activities by creatine in the diet (with and without the presence of growth hormone) were concluded to be at the pretranslational level. The repression of the expression of the gene for transamidinase by creatine in the diet was completely negated by the administration of growth hormone. The induction of the enzyme activities and mRNA levels by growth hormone in the creatine-fed hypophysectomized rats was only to the extent which these values had been repressed by creatine in the diet. Creatine and growth hormone have been concluded to have an antagonistic action on the expression of the gene for rat kidney transamidinase in the hypophysectomized rat.

Introduction

Rat kidney L-arginine:glycine amidinotransferase (transamidinase) activities are altered greatly in a variety of dietary and hormonal states. For example, Walker[14] and Fitch *et al.*[3] reported that rats fed complete diets supplemented with creatine had ~20 per cent of the kidney transamidinase activities found in rats fed the unsupplemented diets. Kidneys from hypophysectomized[10] or thyroidectomized[11] rats had ~20 per cent of the transamidinase activities found in kidneys from intact rats. Thyroidectomized rats given injections of thyroxine or hypophysectomized rats given injections of bovine growth hormone had kidney transamidinase activities similar to those found in intact rats[10,11]. Evidence has been obtained, using a translational assay for transamidinase mRNA, that the repression of enzyme activities by creatine in the diet occurs at the pretranslational level[6]. Alterations in the amount of enzyme protein have been reported to account for the low transamidinase activities of the hypophysectomized rat and the induction of the enzyme activities by growth hormone[7]. The transamidinase mRNA levels of kidneys from hypophysectomized rats maintained with and without injections of growth hormone was not determined in this report[7]. Since creatine and growth hormone have been shown to have diametrically different actions on rat kidney transamidinase activities, it seemed possible that their effect on the expression of the transamidinase gene might be interrelated. The effect of creatine in the diet and/or growth hormone injections on the transamidinase activities and mRNA levels of kidneys from hypophysectomized rats has been determined and is the subject of this report. We have interpreted the results to the effect that a relationship does indeed exist between the opposing actions of creatine and growth hormone on the expression of the rat kidney transamidinase gene.

Materials and methods

Sixty male weanling hypophysectomized rats were purchased from Harlan Sprague Dawley, Inc. (PO Box 4220, Madison, Wisconsin, 53711). The animals were shipped and arrived in Minneapolis on the day after surgery. The animals were housed individually in a room at ~80 °F and had access to 1/4th of an orange, a 5 per cent glucose solution in tap water and a complete purified diet[12], *ad libitum*. On the day after their arrival all animals were given an intraperitoneal injection of 0.25 ml of a suspension of 1.0 mg/ml of cortisol in 0.9 per cent NaCl solution. Five days after their arrival the drinking water was changed to tap water. Ten days after their arrival, ~40 rats that had gained no more than 0.4 g/day were used for the experiment. One half of the rats continued to be fed the purified complete diet and the other half of the rats were allowed to consume the complete purified diet supplemented with 1.0 g creatine/100 g diet. One half of the rats, fed either diet, were given subcutaneous injections in the dorsal neck region of 0.2 mg/100 g body wt of recombinant human growth hormone (in 0.9 per cent NaCl) daily for 15 days. The recombinant human growth hormone was a generous gift from Genentech, Inc. (460 Point San Bruno Boulevard, South San Francisco, CA 94080, USA).

The rats were killed by decapitation, their kidneys removed and one-half of one kidney used in the preparation of the kidney homogenate for transamidinase assays and total protein determinations. The remaining half of the kidney and the entire other kidney was frozen in liquid nitrogen to be used in the determination of transamidinase mRNA. Pools of kidneys from two rats were made for the mRNA determinations. Twenty per cent homogenates in ice-cold distilled water of the individual halves of the kidneys were made in a Potter-Elvehjem homogenizer and the homogenates were stored at –70 °F for 1–2 days prior to doing the transamidinase enzyme assays. The transamidinase assay was by a method reported previously[13], and the total protein of the kidney homogenates was determined by the method of Lowry *et al.*[5]. The isolation of the total mRNA was according to the procedure of Chirgwin *et al.*[1]. Transamidinase mRNA was determined by dot blot assay using a ^{32}P-labelled transamidinase cDNA according to the procedure of Sim *et al.*[8]. A β-actin cDNA was used to determine the amount of total RNA spotted in each sample. The quantitation of hybridization of the mRNA was performed with an image analysis system.

The specific cDNA for rat kidney transamidinase was isolated by the following procedure. An expression library of rat kidney mRNAs was prepared in lambda gt11. The library was screened with an anti-transamidinase antibody and seven positive colonies were identified. The identity of the cDNA from these positive colonies as the cDNA for transamidinase was confirmed based on the following evidence. First, the expressed protein in these colonies reacted with a monoclonal antibody prepared against transamidinase[4]. Second, the isolated cDNA was transcribed and translated *in vitro*. The protein product was immunoprecipitable with anti-transamidinase antibodies and had a size of 57 kD.

This was in good agreement with the size of the translated transamidinase product measured with endogenous mRNA from the rat kidney[6]. Third, the sequence of this identified cDNA was in excellent agreement with the amino terminal sequence of the purified transamidinase protein. Fourth, the sequence of the transamidinase cDNA had two strong areas of homology with a bacterial amidino-transferase enzyme from *Streptomyces griseus*[2,9,15]. This homology most likely indicates conservation of the functional protein areas involved in amidinotransferase activity.

Results and discussion

The transamidinase activities and the mRNA levels of kidneys from the hypophysectomized rats are shown in Table 1. The rats fed a creatine-free diet and given injections of human growth hormone (Group II) had six and four times greater kidney transamidinase activities and mRNA levels, respectively, than rats fed the creatine-free diet who received no growth hormone (Group I). Thus, induction of rat kidney transamidinase activities by growth hormone in hypophysectomized rats was concluded to be at the pretranslational level. Rats fed a creatine-supplemented diet and given no hormone injections (Group III) had ~50 and 30 per cent of the transamidinase activities and mRNA levels, respectively, of the rats fed a creatine-free diet and given no hormone injections (Group I). Rats fed a creatine-supplemented diet and given injections of growth hormone (Group IV) had ~25 per cent of the transamidinase activities and mRNA levels of the rats fed the creatine-free diet and given growth hormone injections (Group II). In other words, a repression of transamidinase activities and its mRNA by creatine in the diet was found in hypophysectomized rats devoid of growth hormone and in hypophysectomized rats injected with growth hormone. We have found previously that intact rats fed creatine supplemented diets had ~30 per cent of the kidney transamidinase activities and mRNA levels found in rats fed the creatine-free diet[6]. Therefore, creatine in the diet repressed the expression of the transamidinase gene in the presence of endogenous or exogenous growth hormone. It is considered of interest that creatine in the diet repressed the gene expression of transamidinase in rats that had mRNA levels of only ~30 per cent of that found in hypophysectomized rats injected with growth hormone. Further, the extent or degree of the repression of the transamidinase gene by creatine in the diet of hypophysectomized rats was similar with and without growth hormone administration and also similar to that reported in intact rats[6].

Table 1. The effects of growth hormone adminstration and/or dietary creatine on transamidinase activities and mRNA levels of kidneys from hypophysectomized rats

	Group		Activities		mRNA	
	(Cr)	(GH)	Mean ± SD	(n)	Mean ± SD	(n)
I	0	0	81 ± 29	(8)	45 ± 9	(4)
II	0	GH	473 ± 107	(12)	164 ± 38	(4)
III	Cr	0	41 ± 16	(10)	13 ± 2	(4)
IV	Cr	GH	114 ± 24	(11)	42 ± 9	(3)

P values I *vs* IV > 0.01: I *vs* II or III; II *vs* III or IV; III *vs* IV < 0.01.
The transamidinase activities are expressed in μmol ornithine formed per g kidney protein. The transamidinase mRNA levels are expressed as the relative ratios of the amounts of mRNA/total RNA in arbitary units.

The repression of transamidinase activities and its corresponding mRNA by creatine in the diet of hypophysectomized rats was completely negated by the administration of growth hormone (Table 1). The transamidinase activities and mRNA levels of the kidneys from the rats fed a creatine-supplemented diet and given growth hormone injections (Group IV) were similar to the values found in the kidneys of rats fed a creatine-free diet and given no growth hormone injections (Group I). The corresponding values of the kidneys from the rats fed the creatine-containing diet and given growth hormone injections were significantly greater than the values found in kidneys from rats fed the creatine-supplemented diet and that received no growth hormone injections (Group III). In other words, the induction of the enzyme activities and mRNA levels by growth hormone in the creatine-fed hypophysectomized rats was only to the extent to which the enzyme activities and mRNA levels had been repressed by creatine in the diet. We believe that the above results are logical in view of the data we have obtained previously with growth hormone administration to intact rats. Growth hormone administration had no effect on the kidney transamidinase activities of intact rats fed complete diets with or without added creatine [10]. Therefore, one would not expect the hypophysectomized rats that received both creatine and growth hormone to have enzyme activities and mRNA levels similar to the hypophysectomized rats that received growth hormone and that were fed a creatine-free diet. No statistically significant differences were found in the transamidinase activities/mRNA ratios of any experimental group of rats (Table 2). Therefore, the repression of transamidinase activities by creatine (both with and without growth hormone administration) and the induction of the activities by growth hormone (both with and without creatine in the diet) appears to be at the pretranslational level with no evidence of post- translational modification of the enzyme.

Table 2. The comparative effects of growth hormone and/or creatine on transamidinase activities/mRNA ratios of kidneys from hypopohysectomized rats

R_1 R_1' (Cr, GH)/(Cr, GH)	R_2 Mean ± SE
Groups	
0/Cr	0.60 ± 0.30
0/GH	0.54 ± 0.28
0/Cr, GH	0.62 ± 0.31
Cr/Cr, GH	1.02 ± 0.38
GH/Cr	1.12 ± 0.43
GH/Cr, GH	1.14 ± 0.45

$$R_2 = \frac{R_1 \text{ activity/mRNA (experimental group)}}{R'_1 \text{ activity/mRNA (experimental group')}};$$

All R_2 values were not statistically significantly different from 1.0. Therefore, there was no statistically significant difference between any R_1 and R_1'.

We have concluded that creatine in the diet and exogenous growth hormone have an antagonistic action on the expression of the rat kidney transamidinase gene of the hypophysectomized rat. We have been unable to detect any antagonistic action of growth hormone and creatine on kidney transamidinase activities of intact rats with the doses of creatine and growth hormone that were used with the hypophysectomized rats. This is believed to be the reason that the antagonism we found in our experiments was restricted to the extent to which the creatine repressed the expression of the gene in the hypophysectomized rat. Hypophysectomized rats were used in our experiment because we have been unable to induce transamidinase activities by growth hormone administration to intact rats. The antagonism of creatine and growth hormone on rat kidney transamidinase gene expression in hypophysectomized rats may be at one or more of several levels: membrane or cellular transport of creatine; membrane receptor sites for growth hormone; production, interaction and/or localization of nuclear regulatory factors. At present we are investigating the role of growth hormone on creatine transport into rat kidney proximal tubule cells in an attempt to elucidate the mechanism of the antagonism of growth hormone and creatine on rat kidney transamidinase gene expression in the hypophysectomized rat.

References

1. Chirgwin, J.M., Przybyla, A.E., McDonald, R.J. & Rutter, W.J. (1979): Isolation of biologically active ribonucleic acid from sources enriched in ribonuclease. *Biochemistry* **18**, 5297–5299.

2.   Distler, J., Ebert, A., Mansouri, K., Pissowotzki, K., Stockmann, M. & Piepersberg, W. (1987): Gene cluster for streptomycin biosynthesis in *Streptomyces griseus*: nucleotide sequence of three genes and analysis of transcriptional activity. *Nucl. Acids Res.* **15**, 8041–8056.

3.   Fitch, C.D., Hsu, C. & Dinning, J.S. (1960): Some factors affecting kidney transamidinase activity in rats. *J. Biol. Chem.* **235**, 2362–2364.

4.   Gross, M.D., McGuire, D.M. & Van Pilsum, J.F. (1985): The production and characterization of two monoclonal antibodies to rat kidney L-arginine:glycine amidinotransferase. *Hybridoma* **4**, 257–269.

5.   Lowry, O.H., Rosebrough, N.J., Farr, A.L. & Randall, R.J. (1951): Protein measurement with the Folin phenol reagent. *J. Biol. Chem.* **193**, 265–275.

6.   McGuire, D.M., Gross, M.D., Van Pilsum, J.F. & Towle, H.C. (1984): Repression of rat kidney L-arginine: glycine amidinotransferase synthesis by creatine at a pretranslational level. *J. Biol. Chem.* **259**, 12034–12038.

7.   McGuire, D.M., Tormanen, C.D., Segal, I.S. & Van Pilsum, J.F. (1980): The effect of growth hormone and thyroxine on the amount of L-arginine:glycine amidinotransferase in kidneys of hypophysectomized rats. *J. Biol. Chem.* **255**, 1152–1159.

8.   Sim, G.K., Kafatos, F.C., Jones, C.W., Koehler, M.D., Efstratiadis, A. & Maniatis, T. (1979): Use of a cDNA library for studies on evolution and developmental expression of the chorion multigene families. *Cell* **18**, 1303–1316.

9.   Tohyama, H., Okami, Y. & Umezawa, H. (1987): Nucleotide sequence of the streptomycinphosphotransferase and amidinotransferase genes from *Streptomyces griseus*. *Nucl. Acids Res.* **15**, 1819–1833.

10.   Ungar, F. & Van Pilsum, J.F. (1966): Hormonal regulation of rat kidney transamidinase; effect of growth hormone in the hypophysectomized rat. *Endocrinology* **78**, 1238–1247.

11.   Van Pilsum, J.F., Carlson, M., Boen, J.R., Taylor, D. & Zakis, B. (1970): A bioassay for thyroxine based on rat kidney transamidinase activities. *Endocrinology* **87**, 1237–1244.

12.   Van Pilsum, J.F., Taylor, D. & Boen, J.R. (1967): Evidence that creatine may be one factor in the low transamidinase activities of kidneys from protein-depleted rats. *J. Nutr.* **91**, 383–390.

13.   Van Pilsum, J.F., Taylor, D., Zakis, B. & McCormick, P. (1970): Simplified assay for transamidinase activities of rat kidney homogenates. *Anal. Biochem.* **35**, 277–286.

14.   Walker, J.B. (1960): Metabolic control of creatine biosynthesis: effect of dietary creatine. *J. Biol. Chem.* **235**, 2357–2361.

15.   Walker, J.B. (1975): Pathways of biosynthesis of the guanidinated inositol moieties of streptomycin and bluensomycin. In *Methods in enzymology*, 43, eds. J.H Hash, ed-in-chief S.P Colowick and N.O. Kaplan, pp. 431-470. New York, San Francisco, London: Academic Press.

Guanidino Compounds in Biology and Medicine, eds. by P.P. De Deyn, B. Marescau, V. Stalon and
I.A. Qureshi. ©1992 John Libbey & Company Ltd., pp. 153–157.

Chapter 22

Heterogeneity of transamidinase activity and creatine content along the rat nephron

Michio TAKEDA[1], Shuei NAKAYAMA[1], Yasuhiko TOMINO[1], Kyu Yong JUNG[2],
Hitoshi ENDOU[2] and Hikaru KOIDE[1]

[1]*Division of Nephrology, Department of Medicine, Juntendo University School of Medicine, 2-1-1 Hongo,
Bunkyo-ku, Tokyo 113 Japan;* [2]*Department of Pharmacology, Faculty of Medicine, University of Tokyo, 7-3-1
Hongo, Bunkyo-ku, Tokyo 113, Japan*

Summary

Guanidinoacetic acid (GAA) is synthesized by transamidinase mainly in the kidney, and transported to the liver where
it is methylated to yield creatine. To clarify the intranephron distribution of transamidinase activity and creatine content,
we microdissected individual nephron segments from the rat kidney. GAA synthesized from glycine and arginine or
canavanine was separated by HPLC and determined fluorometrically. Creatine was measured by an ultramicromethod
that we have established for microdissected nephron segments. Results obtained were as follows: (1) Transamidinase
activity was detected only in the first (S1) and the second portion (S2) of the proximal tubule. (2) Creatine was detected
in all nephron segments tested. However, the comparatively large amount of creatine was contained in the glomerulus,
cortical thick ascending limb of Henle's loop, and the distal tubule of the rat kidney. These results indicate that GAA
synthesis is a specific function of S1 and S2 of the proximal tubule, and creatine content is different from segment to
segment of the nephron.

Introduction

Guanidinoacetic acid (GAA) is synthesized from arginine and glycine by glycine-amidinotrans-
ferase (transamidinase) mainly in the kidney[3] (Fig. 1) . A part of GAA synthesized in the
kidney is excreted into the urine[11] and the remainder is postulated to be transported to the liver
where it is methylated to yield creatine by guanidinoacetate-methyltransferase (GAA-MT)[4] (Fig. 1).

Transamidinase is considered to be a rate-limiting enzyme for creatine synthesis[15]. Recently immunochemical studies revealed that the localization of transamidinase immunoreactivity was localized
in the proximal tubules of the kidney[7]. However intrarenal localization of this enzyme has not been
quantified.

Although Baker *et al.*[1], Borsook *et al.*[2] and Gerber *et al.*[5] have studied creatine synthesis in the kidney,
they could not recognize this activity by using homogenate[2,14]. We also could not detect creatine
synthetic activity using isolated nephrons and renal homogenate obtained from rats (unpublished
data).

In this paper, we report on the intranephron distribution of transamidinase activity and creatine content
using the procedure of microdissection.

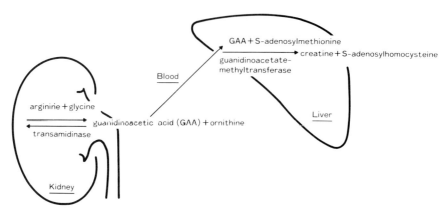

Fig. 1. Metabolic scheme of guanidinoacetic acid and creatine biosynthesis.

Materials and methods

Preparation of nephron segments

The isolation of nephron segments were carried out from the kidney of male Sprague-Dawley rats by the method of Morel *et al.*[8] with a slight modification.

Assay of GAA synthesized in individual nephron segments

Individual nephron segments were incubated in 500 µl of modified Hank's solution containing 2 mM arginine or 2 mM canavanine plus 2 mM glycine with shaking in a 1.5 ml microtube at 37 °C for 60 min. After incubation, the reaction was stopped by adding 10 per cent trichloroacetic acid (TCA). The TCA extracts were filtered and then pH was adjusted to 2.2. GAA in aliquots of the neutralized extracts was determined using HPLC and the fluorescence detector[16].

Assay of creatine in individual nephron segments

Creatine was assayed according to the method of Tanzer *et al.*[12] with a modification (Fig. 2). This reaction was coupled with the oxidation of NADH which resulted in a decrease of fluorescence intensity. We incubated the individual nephron segments in 60 µl of Tris-glycine (TG) buffer containing 20 mM ATP and 1600 U/ml creatine kinase (CK) at 37 °C for 30 min. After the incubation, we added 1 ml of phosphate buffer saline (PBS) containing 0.65 mM phosphoenolpyruvate (PEP), 16.65 U/ml lactate dehydrogenase (LDH), 7.5 U/ml pyruvate kinase (PK) and 15 µM NADH to T-G buffer and further incubated at 37 °C for 30 min. After the second incubation, we measured the fluorescence intensity at 365 nm for excitation and 495 nm for emission using a fluorophotometer (F-2000, Hitachi Co., Ltd, Tokyo, Japan).

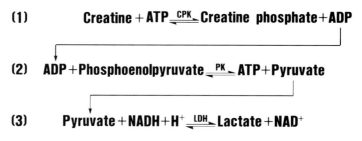

Fig. 2. Reaction scheme for enzymatic determination of creatine.

Statistics

The amount of GAA and creatine were expressed as means ± SEM. Statistical analysis was performed by the unpaired Student test (GAA) and by analysis of variance (ANOVA) (creatine content).

Results

Intranephron distribution of transamidinase activity in rats

As shown in Fig. 3, transamidinase activity was distributed only in the first (S1) and the second (S2) portions of the proximal tubule using arginine plus glycine as substrates. S1 portion showed a significantly higher activity than S2 portion ($P < 0.001$). The same result was obtained when canavanine plus glycine were used as substrates (data not shown). This suggests that no substrate specificity was recognized between arginine and canavanine.

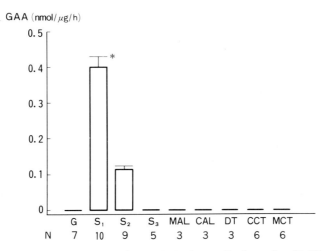

*Fig. 3. Intranephron distribution of transamidinase activity in rats. G: glomerulus; S1–S3: the first to the third portion of proximal tubule; MAL and CAL: medullary and cortical thick ascending limbs of Henle's loop; DT: distal tubule; CCT and MCT: cortical and medullary collecting ducts. Values are means ± SEM from indicated numbers of samples. *P < 0.001.*

Establishment of an ultramicromethod for determining pmol quantity of creatine

Figure 4 shows that a decrease of fluorescence was proportional to creatine content. We could determine pmol quantity of creatine from this calibration line.

Intranephron distribution of creatine content in rats

As shown in Fig. 5, all of nephron segments tested contained creatine. Creatine content (pmol) calculated per µg tissue protein was the highest in the glomerulus, cortical thick ascending limb of Henle's loop, and distal tubule among the nephron segments.

Discussion

Transamidinase activity has been reported to be located mainly in the kidney and pancreas in various mammals[13]. Since intrarenal localization of transamidinase protein has been reported to be in proximal tubules of the kidney by immunofluorescence technique[7], it is highly possible that the GAA synthesis is also limited to proximal tubules. However, transamidinase activity has not been deter-

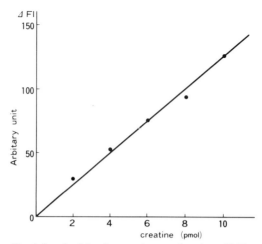

Fig. 4. Standard line for creatine as a function of DFI.

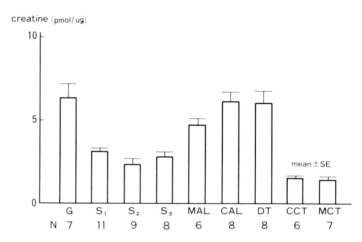

Fig. 5. Intranephron distribution of creatine content in rats. Values are means ± SEM from indicated numbers of samples. There is a significant difference between each nephron segment (P < 0.001).

mined quantitatively in the nephron segment, and substrate specificity about transamidinase in the nephron segment has not been investigated.

We microdissected individual nephron segments and used them as enzyme sources to provide direct evidence on GAA synthesis site within the kidney. We demonstrated that GAA synthesis was limited to S1 and S2 (S1 > S2), which was consistent with the previous immunofluorescence findings reported by McGuire et al.[7]. Our findings not only imply the biological significance, but also clarify the significance of GAA as a urinary indicator of gentamicin nephropathy which we have previously reported[9,10].

The amount of creatine content in the kidney has been reported to be small[6]. Therefore, we established an ultramicromethod for determining pmol quantities of creatine to clarify the intranephron distribution of creatine content in rats. Our results showed that the creatine content is different from segment to segment and two possible reasons will be postulated for this finding. One is the difference of creatine synthesis in each nephron segment, and the other is the difference of creatine uptake. We incubated microdissected proximal convoluted tubule, sieved glomeruli, and kidney homogenate with various substrates to detect creatine synthesis. However, we could not detect any creatine synthetic activity (unpublished data), which is inconsistent with the previous reports[1,2,5]. We speculate that mesangial cells of the glomerulus may take up creatine as an energetic source for their smooth muscle-like contraction. Distal tubule and cortical thick ascending limb of Henle's loop also contain large amounts of creatine. The physiological significance of creatine in these segments should be established by future studies. At present, we postulate that differences of creatine content in each nephron segment may be due to the difference in creatine uptake.

References

1. Baker, Z. & Miller, B.F. (1940): Studies on the metabolism of creatine and creatinine. III. Formation of creatine by isolated rat tissues. *J. Biol. Chem.* **132**, 233–237.

2. Borsook, H. & Dubnoff, J.W. (1940): Creatine formation in liver and in kidney. *J. Biol. Chem.* **134**, 635–659.

3. Borsook, H. & Dobnoff, J.W. (1941): The formation of glycocyamine in animal tissues. *J. Biol. Chem.* **138**, 389–403.

4. Gerber, G.B., Gerber, G., Koszalka, T.R. & Miller, L.L. (1962): The rate of creatine synthesis in the isolated, perfused rat liver. *J. Biol. Chem.* **237**, 2246–2250.

5. Gerber, G.B., Koszalka, T.R., Gerber, G. & Altman, K.I. (1962): Biosynthesis of creatine by the kidney. *Nature* **196**, 286–287.

6. Hamaguchi, C., Nozima, M., Nakata, T., Hatanaka, Y. & Ueda, K. (1981): Enzymatic studies on the mechanism of creatinuria in rats with experimental thyrotoxic myopathy and dystrophic mice. *Med. J. Osaka. Univ.* **31**, 75–81.

7. McGuire, D.M., Gross, M.D., Elde, R.P. & Van Pilsum, J.F. (1986): Localization of L-arginine-glycine amidinotransferase protein in rat tissues by immunofluorescence microscopy. *J. Histochem. Cytochem.* **34**, 429–435.

8. Morel, F., Chabarades, D. & Imbert, M. (1976): Functional segmentation of the rabbit distal tubule by microdetermination of hormone-dependent adenylate cyclase activity. *Kidney Int.* **9**, 264–277.

9. Nakayama, S., Junen, M., Kiyatake, I. & Koide, H. (1990): Urinary guanidinoacetic acid excretion as an indicator of gentamicin nephrotoxicity in rats. In *Guanidines II*, eds. A. Mori, B.D. Cohen & H. Koide, pp. 303–311. New York: Plenum Press.

10. Nakayama, S., Kiyatake, I., Shirokane, Y. & Koide, H. (1990): Effect of antibiotic administration of urinary guanidinoacetic acid excretion in renal disease. In *Guanidine II*, eds. A. Mori, B.D. Cohen & A. Lowenthal, pp. 313. New York: Plenum.

11. Sasaki, M., Takahara, K. & Natelson, S. (1973): Urinary guanidinoacetate/guanidinosuccinate ratio: an indicator of kidney dysfunction. *Clin. Chem.* **19**, 315–321.

12. Tanzer, M.L. & Gilvarg, C. (1959): Creatine and creatine kinase measurement. *J. Biol. Chem.* **234**, 3201–3204.

13. Van Pilsum, J.F., Olsen, B., Taylor, D., Rozycki, T. & Pierce, J.C. (1963): Transamidinase activities, *in vitro*, of tissues from various mammals and rats fed protein-free, creatine -supplemented and normal diets. *Arch. Biochem. Biophys.* **100**, 520–534.

14. Van Pilsum, J.F., Stephens, G.C. & Taylor, D. (1972): Distribution of creatine, guanidinoacetate and the enzymes for their biosynthesis in the animal kingdom. *Biochem. J.* **126**, 325–345.

15. Walker, J.B. (1960): Metabolic control of creatine biosynthesis, I. Effect of dietary creatine. *J. Biol. Chem.* **235**, 2357–2361.

16. Yamamoto, Y., Manji, T., Saito, A., Maeda, K. & Ohta, K. (1979): Ion-exchange chromatographic separation and fluorometric detection of guanidino compounds in physiologic fluids. *J. Chromatogr.* **162**, 327–340.

Guanidino Compounds in Biology and Medicine, eds. by P.P. De Deyn, B. Marescau, V. Stalon and
I.A. Qureshi. ©1992 John Libbey & Company Ltd., pp. 159–164.

Chapter 23

Structure-function relationships of guanidinoacetate methyltransferase as revealed by chemical modification and site-directed mutagenesis of cysteine residues

Motoji FUJIOKA, Yoshimi TAKATA, Kiyoshi KONISHI, Tomoharu GOMI and
Hirofumi OGAWA

*Department of Biochemistry, Toyama Medical and Pharmaceutical University, Faculty of Medicine, Sugitani,
Toyama 930-01, Japan*

Summary

Cys-15 of rat guanidinoacetate methyltransferase is highly reactive toward 5,5′-dithio-bis (2-nitrobenzoate) (DTNB), iodoacetate and glutathione disulphide. Incubation of the enzyme with an equimolar concentration of DTNB results in an almost quantitative disulphide bond formation between Cys-15 and Cys-90. Replacement of Cys-90 with Ala by site-directed mutagenesis results in a catalytically active enzyme having kinetic constants very similar to those of wild-type. Treatment of the mutant with one equivalent of DTNB leads to disulphide cross-linking between Cys-15 and Cys-219. These results suggest that Cys-15 lies exposed on the protein surface and is juxtaposed to Cys-90, and Cys-219 in the three-dimensional structure. Whereas the cross-linked enzymes are only ~10 per cent active, they have K_m values for substrates not greatly different from those of the unmodified enzymes. Guanidinoacetate methyltransferase having the Cys-15-glutathione mixed disulphide is devoid of activity, but binds *S*-adenosylmethionine. The data presented here suggest that the region around Cys-15 is not involved in substrate binding but the flexibility of the region is important for the expression of catalytic activity.

Introduction

Guanidinoacetate methyltransferase (GAMT), first described by Cantoni & Vignos[1] in 1954, catalyses the transfer of the methyl group of *S*-adenosyl-L-methionine (AdoMet) to guanidinoacetate yielding creatine. This enzyme is found in the liver of all vertebrates[14] and, at least in man, is believed to play a major role in the conversion of AdoMet to *S*-adenosyl-L-homocysteine (AdoHcy)[7]. Homocysteine formed by hydrolysis of AdoHcy by AdoHcy hydrolase is utilized for the synthesis of cysteine or recycled to methionine by accepting the methyl group from N^5-methyltetrahydropteroylglutamates. Since the folate coenzymes exist mainly as the N^5-methyl derivatives in the cell, the latter reaction is crucial for regeneration of tetrahydropteroylglutamates and eventually for the biosynthetic reactions utilizing C_1 compounds. Thus GAMT, in addition to providing creatine, appears to be the key enzyme in the metabolism of sulphur amino acids and C_1 compounds.

GAMT has been purified to homogeneity from the liver of rat[8] and pig[5]. The enzymes from both sources are monomeric proteins with a relatively small size. The cDNA for rat GAMT has been cloned, sequenced, and expressed in *Escherichia coli*[10]. The rat GAMT gene has also been isolated and characterized[9]. The cDNA-derived primary structure indicates that the enzyme consists of 235 amino acid residues, is relatively rich in aromatic residues, and possesses five half-cystines. The catalytically active enzyme has no disulphide bond. The X-ray crystallographic structure of the enzyme has not been determined, but it is suggested, based on the results of limited proteolysis with various proteases[3] and chemical modification with acetic anhydride in the absence and presence of AdoMet[12], that the N-terminal region up to around residue 25 is exposed to solvent, and that the AdoMet-binding site is in a hydrophobic interior of the protein molecule.

In this report we present some interesting aspects on the structure and function of rat GAMT that have emerged from the studies of its cysteine residues.

Materials and methods

Enzyme

GAMT used in this study was a recombinant enzyme produced in *E. coli* JM109 transformed with the plasmid pUCGAT9-1 that contained the coding region of the cDNA for rat liver GAMT[10]. The recombinant rat GAMT (rGAMT) lacks the N-terminal acyl group present in the liver enzyme, but, except for this difference, both enzymes show the same catalytic and physicochemical properties so far examined. Conversion of Cys-90 into Ala was carried out by oligonucleotide-directed mutagenesis of plasmid pUCGAT9–1 using a synthetic 23-mer oligonucleotide, 3′-TAATAACTT*C*GGTTGC-TACCCCA-5′, that introduced mutation into residue 90[13].

Enzyme assay

The GAMT activity was determined spectrophotometrically by a coupled assay with AdoHcy hydrolase and adenosine deaminase as described[2].

Chemical modification reactions

Reaction of rGAMT or Cys-90→Ala mutant (C90A) with 5,5′-dithio-bis(2-nitrobenzoate) (DTNB) was carried out in 0.1 M Tris-HCl buffer (pH 8.0), containing 1 mM EDTA at 30 °C. The concentrations of 2-nitro-5-mercaptobenzoate (TNB) anion released from DTNB were calculated using a value $\varepsilon_{412} = 1.415 \times 10^4 \, M^{-1} \, cm^{-1}$. Incubation of rGAMT with glutathione disulphide (GSSG) and of the GSSG-inactivated rGAMT with reduced glutathione (GSH) was carried out in 50 mM Tris-HCl buffer (pH 7.5), containing 1 mM EDTA.

Labelling of cysteine residues engaged in disulphides

The enzyme having a protein disulphide or mixed disulphide was first treated in 4 M guanidine hydrochloride with 5–6 mM *N*-ethylmaleimide (NEM) at pH 6.8 to block unmodified thiol groups. The NEM-modified enzyme was then reduced with dithiothreitol and the regenerated thiols were labelled with [^{14}C]iodoacetate.

Chymotryptic digestion and isolation of peptides

The radiolabelled sample was digested with α-chymotrypsin in 0.1 M NH$_4$HCO$_3$. The chymotryptic peptides were separated by HPLC on a reverse phase column TSK ODS 120T (Tosoh, Tokyo, Japan), using an acetonitrile gradient in 0.05 per cent trifluoroacetic acid[2]. The effluent was monitored by absorbance at 220 nm, and collected in 1.6 ml fractions. Aliquots from each fraction were determined for radioactivity, and the radioactive fractions were purified by rechromatography on the same column using a different solvent system[2].

Amino acid analysis and sequence determination

Amino acid analysis was performed based on reverse phase separation of phenylthiocarbamoyl derivatives[4]. Sequence determination was carried out by automated Edman degradation on an Applied Biosystems 470A sequenator.

Determination of total thiol groups

Free thiol groups of GAMT before and after various modification reactions were determined by the reaction with DTNB under denaturing conditions as described by Riddles *et al.*[11].

Results and discussion

Reaction of rGAMT with iodoacetate

Incubation of rGAMT with excess iodoacetate at pH 8.0 and 30 °C resulted in a time-dependent loss of activity. No activity was detected after prolonged incubation. Reverse phase HPLC of chymotryptic peptides derived from the [^{14}C]iodoacetate-inactivated enzyme separated four radioactive peptides which were identified as peptides containing Cys-15, Cys-90, Cys-207, and Cys-219, respectively. Thus, of the five cysteine residues of the enzyme, only Cys-168 was refractory to modification by iodoacetate. Kinetics of radioactivity incorporation from [^{14}C]iodoacetate indicated that Cys-15 and Cys-90 were the most reactive of the four residues.

Reaction of rGAMT with DTNB

When rGAMT was incubated with excess DTNB at pH 8.0, there was a rapid release of TNB anion. A total of 4.07 mol of TNB/mol of enzyme was released within a few min. The absorption spectrum of the modified enzyme showed a peak at ~330 nm in addition to an absorption maximum at 280 nm, indicating the formation of protein-TNB mixed disulphide. However, using a value of $\varepsilon_{330} = 7.5 \times 10^3$ M^{-1} cm^{-1} for the molar absorptivity of protein-TNB, the amount of TNB fixed to the enzyme was calculated to be 2.6 mol per mol of enzyme. This suggests that a fraction of the TNB mixed disulphide formed undergoes reaction with a neighbouring thiolate to form protein disulphide. To examine the point in more detail, we incubated rGAMT with limited, graded amounts of DTNB, and the reaction was allowed to go to completion.

Table 1. Reaction of rGAMT and C90A with graded amounts of DTNB

Enzyme	DTNB added mol/mol of enzyme	TNB released mol/mol of enzyme	Activity remaining %	Cys residues linked
rGAMT	1.0	1.90	16	Cys-15–Cys-90
	1.5	2.62	9	
	2.0	3.05	0	
	3.0	3.92		
C90A	1.0	1.88	12	Cys-15–Cys-219
	1.5	2.43	3	
	2.0	2.82	0	
	3.0	2.99		

rGAMT and C90A were treated successively with 1.0, 0.5, 0.5 and 1.0 equivalent of DTNB. After each addition, reaction was allowed to go to completion (monitored by the absorbance change at 412 nm). Residual activity is the percentage of activity obtained by comparison with the value before the addition of DTNB. Adapted from Fujioka *et al.*[2] with permission of *Biochemistry*, Copyright (1988) American Chemical Society, and from Takata *et al.*[13] with permission of *Biochem. J.*, Copyright (1991) The Biochemical Society.

As Table 1 shows, the addition of one equivalent of DTNB caused the appearance of roughly two equivalents of TNB, with a large loss of activity. Thereafter, a 1:1 stoichiometry was obtained between the amount of DTNB added and the amount of TNB produced. The spectrum of the enzyme treated with an equimolar concentration of DTNB showed a very weak absorption at ~330 nm, corresponding to only ~0.05 mol of bound TNB/mol of enzyme. To determine the residues oxidized, rGAMT modified with one equivalent of DTNB was treated with NEM under denaturing conditions to block unchanged thiol groups, and then with dithiothreitol and [^{14}C]iodoacetate, as described under 'Materials and methods'. Analysis of chymotryptic peptides derived therefrom revealed that roughly equal amounts of radioactivity were associated only with Cys-15 and Cys-90, indicating that a disulphide was formed specifically between Cys-15 and Cys-90. Thus it may be concluded that either of these cysteines is very reactive toward DTNB and the resulting mixed disulphide is susceptible to disulphide interchange. The reaction of DTNB was not influenced by AdoMet or sinefungin.

Reaction of mutant enzyme C90A with DTNB

Whereas Cys-15 and Cys-90 appear to reside outside of the AdoMet-binding site as seen by the lack of effect of AdoMet or sinefungin in the reaction with DTNB, there is a large loss of enzyme activity when these residues are cross-linked by a disulphide bond (Table 1). This suggests that either one or both of the cysteines are located in a region important for catalytic functioning of the enzyme. To better understand the role of individual residues, we attempted to replace each of Cys-15 and Cys-90 with Ala or Ser by site-directed mutagenesis. However, only Cys-90→Ala mutant (C90A) could be obtained in good yield. C90A was catalytically active and had kinetic constants very similar to those of rGAMT. Also, the UV absorption, CD, and fluorescence spectra were indistinguishable (Table 2). Thus Cys-90 has neither a catalytic nor a structural role, and the mutant enzyme appears to fold into a similar tertiary structure.

Table 2. Comparative properties of rGAMT and C90A

	rGAMT	C90A
Molecular weight	26,100	26,100
No. of SH groups (mol/mol of enzyme)	5	4
$[\theta]_{222}$(deg.cm^2.dmol^{-1})	2.05	2.10
A $^{0.1\%}_{280}$	2.2	2.3
Kinetic constants (pH 8.0, 30 °C)		
$\quad$ K_m (AdoMet) (μM)	2.7	4.5
$\quad$ K_m (GAA) (μM)	22.7	24.6
$\quad$ k_{cat} (min^{-1})	4.9	4.5

GAA, guanidinoacetate. Kinetic constants are from Takata *et al.*[13] with permission of *Biochem. J.*, Copyright (1991) The Biochemical Society.

Despite the absence of Cys-90, C90A again exhibited an anomalous behaviour in the reaction with excess DTNB; the amount of TNB released exceeded the amount of TNB bound. Titration with limited amounts of DTNB showed that a disulphide bond was formed upon addition of one equivalent of DTNB (Table 1). The residues engaged in the disulphide were Cys-15 and Cys-219 in this case. Since no significant structural alteration appears to accompany the mutation as described above, it is likely that Cys-15, Cys-90, and Cys-219 of GAMT occur close together in the three-dimensional structure. An almost exclusive formation of the Cys-15–Cys-90 cross-link observed when rGAMT is treated with DTNB may be due to the fact that Cys-90, compared with Cys-219, is a better nucleophile and/or positioned more favourably for the reaction with the TNB-modified Cys-15. It would also be possible that replacement of the charged cysteine with an uncharged alanine results in a slight alteration in the local structure, thereby enabling the mutant to form the Cys-15–Cys-219 cross-link more readily. In the reaction of rGAMT with DTNB it was not possible to determine which of Cys-15 and Cys-90 was the site of attack of DTNB. The finding that a disulphide is formed between Cys-15 and Cys-219 in C90A strongly suggests that Cys-15 is the residue that reacts with DTNB.

Reaction of rGAMT with GSSG

rGAMT was inactivated by incubation with GSSG and the inactivated enzyme was reactivated with GSH. Both reactions followed pseudo-first order kinetics, with apparent first order rate constants proportional to the reagent concentrations. Second order rate constants of 20.8 and 11.1 M^{-1} min^{-1} were obtained for inactivation and reactivation reactions, respectively. The rate of inactivation was not affected in the presence of AdoMet, sinefungin, or guanidinoacetate, but almost complete protection against inactivation was observed with sinefungin (0.1 mM) plus guanidinoacetate (1.0 mM). GAMT binds substrates in an obligatory order of AdoMet followed by guanidinoacetate, and sinefungin is a competitive inhibitor with respect to AdoMet.

When incubated in a buffer containing both GSSG and GSH, GAMT exhibited the same final activity regardless of whether the native or GSSG-inactivated enzyme was used, indicating an equilibrium between the active and inactive enzymes. The reaction may be written as:

$$E - SH + GSSG \overset{k_1}{\underset{k_2}{\rightleftharpoons}} E - SSG + GSH \tag{1}$$

or

$$E\begin{smallmatrix} SH \\ \diagup \\ \diagdown \\ SH \end{smallmatrix} + GSSG \rightleftharpoons E\begin{smallmatrix} S \\ \diagup \\ | \\ \diagdown \\ S \end{smallmatrix} + 2GSH. \tag{2}$$

For reaction 1,

$$K_{eq} = \frac{[E - SSG][GSH]}{[E - SH][GSSG]} \tag{3}$$

which is rearranged to

$$\frac{[E - SH]}{[E_t]} = \frac{[GSH]/[GSSG]}{K_{eq} + [GSH]/[GSSG]} \tag{4}$$

where E_t is the total concentration of the enzyme. Therefore, the plot of fractional activities against [GSH]/[GSSG] is a hyperbola which is dependent on the ratio but not on the absolute concentrations of GSH and GSSG. For Reaction 2, the [GSH] terms in Eqs. 3 and 4 must be squared, and the plot is not a hyperbola and the [GSH]/[GSSG] value required for half-maximal activity shifts with [GSH]. Figure 1 shows the experiment performed to distinguish the two alternatives. Plots of equilibrium

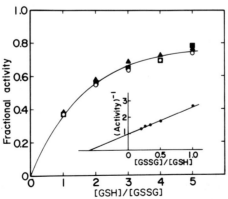

Fig. 1. GAMT activity as a function of [GSH]/[GSSG] ratio. rGAMT (19.1 μM) was incubated at 30 °C in 50 mM Tris-HCl buffer (pH 7.5)/1 mM EDTA, containing 5 (O), 10 (▲), and 20 mM (■) GSH and various amounts of GSSG to give the indicated ratios. The activities obtained after 60 min incubation were taken as equilibrium activities. The curve was drawn according to Eq. 4 with $K_{eq} = 1.69$. Reprinted from Konishi & Fujioka[6] with permission of Arch. Biochem. Biophys., Copyright (1991) Academic Press, Inc.

activities obtained at various concentrations of GSH and GSSG yielded a hyperbola that was independent of the absolute concentration of the reagents. A K_{eq} value of 1.69 was found in this experiment which compared well with the value calculated as the ratio of second order rate constants for inactivation and reactivation ($k_1/k_2 = 20.8$ M^{-1} min^{-1}/11.1 M^{-1} min^{-1} = 1.87).

Determination of the number of free thiol groups as a function of inactivation showed that the loss of a single cysteine residue is responsible for complete inactivation. By peptide analysis, Cys-15 was identified as the residue that formed the glutathione mixed disulphide (see 'Materials and methods').

The catalytically inactive GSSG-modified rGAMT binds AdoMet with a dissociation constant of 17.0 μM. While this value is similar to that found with the unmodified enzyme (10 μM)[2], no appreciable binding of guanidinoacetate was observed in the presence and absence of sinefungin. (The rGAMT–sinefungin complex binds guanidinoacetate with a K_d of 70.1 μM.)

The results presented here and the previous results of limited proteolysis[3] indicate that the N-terminal segment up to residue 25 is on the protein surface and flexible. While the region around Cys-15 appears to be apart from the active site region, this portion of the enzyme undergoes a substantial structural alteration upon formation of the *E*-AdoMet-guanidinoacetate ternary complex. Since disulphide cross-linking with or introduction of glutathione to Cys-15 leads to loss of activity, it may be suggested that flexibility of this segment is necessary for adopting the catalytically active conformation.

References

1. Cantoni, G. L. & Vignos, P.J., Jr. (1954): Enzymatic mechanism of creatine synthesis. *J. Biol. Chem.* **209**, 647–659.

2. Fujioka, M., Konishi, K. & Takata, Y. (1988): Recombinant rat liver guanidinoacetate methyltransferase: reactivity and function of sulfhydryl groups. *Biochemistry* **27**, 7658–7664.

3. Fujioka, M., Takata, Y. & Gomi, T. (1991): Recombinant rat guanidinoacetate methyltransferase: structure and function of the NH2-terminal region as deduced by limited proteolysis. *Arch. Biochem. Biophys.* **285**, 181–186.

4. Gomi, T., Ogawa, H. & Fujioka, M. (1986): S-Adenosylhomocysteinase from rat liver. Amino acid sequences of the peptides containing active site cysteine residues modified by treatment with 5′-*p*-fluorosulfonyl-benzoyladenosine. *J. Biol. Chem.* **261**, 13422–13425.

5. Im, Y. S., Chiang, P. K. & Cantoni, G. L. (1979): Guanidoacetate methyltransferase. Purification and molecular properties. *J. Biol. Chem.* **254**, 11047–11050.

6. Konishi, K. & Fujioka, M. (1991): Reversible inactivation of recombinant rat liver guanidinoacetate methyltransferase by glutathione disulfide. *Arch. Biochem. Biophys.* **289**, 90–96.

7. Mudd, S.H. & Poole, J.R. (1975): Labile methyl balances for normal humans on various dietary regimens. *Metabolism* **24**, 721–735.

8. Ogawa, H., Ishiguro, Y. & Fujioka, M. (1983): Guanidoacetate methyltransferase from rat liver: purification, properties, and evidence for the involvement of sulfhydryl groups for activity. *Arch. Biochem. Biophys.* **226**, 265–275.

9. Ogawa, H. & Fujioka, M. (1988): Nucleotide sequence of the rat guanidinoacetate methyltransferase gene. *Nucleic Acids Res.* **16**, 8715–8716.

10. Ogawa, H., Date, T., Gomi, T., Konishi, K., Pitot, H.C., Cantoni, G. L. & Fujioka, M. (1988): Molecular cloning, sequence analysis, and expression in *Escherichia coli* of the cDNA for guanidinoacetate methyltransferase from rat liver. *Proc. Natl. Acad. Sci. USA* **85**, 694–698.

11. Riddles, P.W., Blakeley, R.L. & Zerner, B. (1983): Reassessment of Ellman's reagent. *Methods Enzymol.* **91**, 49–60.

12. Takata, Y. & Fujioka, M. (1990): Recombinant rat guanidinoacetate methyltransferase: study of the structure by trace labeling lysine residues with acetic anhydride. *Int. J. Biochem.* **22**, 1333–1339.

13. Takata, Y., Date, T. & Fujioka, M. (1991): Rat liver guanidinoacetate methyltransferase. Proximity of cysteine residues at positions 15, 90, and 219 as revealed by site-directed mutagenesis and chemical modification. *Biochem. J.* **277**, 399–406.

14. Van Pilsum, J.F., Stephens, G.C. & Taylor, D. (1972): Distribution of creatine, guanidinoacetate and the enzymes for their biosynthesis in the animal kingdom. Implication for phylogeny. *Biochem. J.* **126**, 325–345.

Guanidino Compounds in Biology and Medicine, eds. by P.P. De Deyn, B. Marescau, V. Stalon and I.A. Qureshi. ©1992 John Libbey & Company Ltd., pp. 165–174.

Chapter 24

Mitochondrial contact sites: a dynamic compartment for creatine kinase activity

W. JACOB, W. BIERMANS and A. BAKKER

University of Antwerp (UIA), Dept of Medicine, Universiteitsplein 1, B-2610 Antwerp-Wilrijk, Belgium

Summary

Since their discovery mitochondrial contact sites were thought to play a role in the transport of metabolites from the cytoplasm to the mitochondrion and vice versa. Creatine phosphate is an energy-transporting molecule in the so-called creatine/creatinephosphate shuttle in cells with an intense energy metabolism, e.g. heart and brain. In the shuttle isozymes of creatine kinase play an important role through their specificity and location. Especially, mitochondrial creatine kinase plays a key role through its functional coupling with adenine nucleotide translocase. It was shown by an enzyme-cytochemical method that mitochondrial creatine kinase was exclusively active in the contact sites in a number of tissues. The tissue-specific presence of mitochondrial creatine kinase could be confirmed by agarose gel electrophoresis.

As shown earlier in isolated mitochondria, contact site formation depends on the energy metabolism. By morphometric analysis, we demonstrated that creatine kinase active contact sites (*in vivo* and *in vitro*) are correlated with the induced energy state. Recently it became clear that mitochondrial creatine kinase is present as two oligomers. It is possible that an equilibrium exists between the membrane-bound and the free oligomers in the intermembrane space. This equilibrium depends on the actual metabolic state and extent of contact sites. How the contact sites are modulated still remains an open question, but presumably calcium and the mitochondrial matrix granules play an important role.

Mitochondrial contact sites and creatine kinase

The occurrence of contact sites between IMM and OMM was first suggested on the basis of thin section electron micrographs of liver mitochondria[22]. It is obvious that membrane lipids must be actively involved in the actual formation of intermembrane contacts. This process assumes the solution of two biophysical problems. First, prior to fusion, the two membranes must move into close contact, an energetically unfavourable event due to the strong intermembrane repulsion force. Here divalent cation-negative phospholipid interactions are of special interest. Indeed, addition of Ca^{2+} to acidic phospholipids, especially cardiolipin, results in Ca-binding concomitant with head group dehydratation and decreased electrostatic repulsion[18].

Secondly, at some stage the lipids will have to locally adopt a (transiently) non-bilayer configuration. In the actual fusion event, one can distinguish different stages, which do not always occur: the 'adhesion', the 'joining' and the 'fission' of fusing membranes, each characterized by corresponding features during freeze-fracture experiments (for a review: see[57]). Although most of these results came from work on model membranes[57], analogous data were obtained from extensive freeze-fracture experiments on isolated mitochondria[24,56].

Knoll[33] measured the length of the fracture plane edge per standard area and compared it between isolated mitochondria of different functional states. They observed a marked increase of this length parameter in phosphorylating (state 3) mitochondria compared to energized (state 4), freshly isolated (state 1) or uncoupled mitochondria.

Furthermore, the degree of membrane contact formation upon transition into state 3 appeared to be influenced by metabolites, such as fatty acids[32] and by hormones, as observed in intact hepatocytes[47] and heart muscle[26]. In this last case a close correlation was found between the metabolic state of myocardium *in vivo*, influenced by catecholaminergic stimulation or inhibition with amytal, and the mitochondrial intermembrane distance and the appearance of contact sites (Table 1).

Table 1. Effect of stimulation, inhibition and uncoupling of energy metabolism on different parameters of contact site formation in myocardium

	EC	ATPase$_{mit}$	Ca^{2+}	NMG	S$_{Vmit}$	IMD
Control	0.856 ± 0.031	–	$\pm$	+	0.46 ± 0.03	13.1 ± 1.7
Simulated	0.775 ± 0.031	–	++	–	0.68 ± 0.03	11.7 ± 1.9
Inhibited	0.877 ± 0.025	ND	–	+++	0.33 ± 0.02	15.8 ± 1.7
Uncoupled	0.778 ± 0.038	+	ND	+++	0.11 ± 0.02	ND

EC = Energy charge; S$_{Vmit}$ = surface density of CK active contact sites (cm^{-1}); IMD = intermembrane distance in nanometers; Ca^{2+} = cytochemically localized Ca in mitochondria; NMG = number of native matrix granules; ND = not determined. Data represent mean (confidence limits 95 per cent).

From all these data it clearly follows that the coupling of phosphorylation to the electrochemical potential μ_{H+} might be correlated to the frequency of membrane contacts. Brdiczka[12] has shown that hexokinase bound to the contacts may serve to create a micro-environment facilitating a direct exchange of ATP and ADP between the enzyme and the compartment of oxidative phosphorylation. Such a functional coupling could increase the effective concentration of metabolites near the target enzyme, e.g. hexokinase, and therefore maintain high activity rates.

The preferential binding of an exogenous kinase, like hexokinase, to the contact sites of liver and brain mitochondria logically leads to the supposition that these structures might also be of fundamental importance for functional coupling of other kinases, e.g. creatine kinase, to the inner mitochondrial compartment.

The physiologically active isozymes of CK are formed by dimerization of the subunits to a muscle-type CK-MM, a brain-type CK-BB and a heterodimer-type CK-MB isoform (for a review: see[19]). A fourth isozyme of CK, mitochondrial CK (CK$_m$) was described by Jacobs *et al.*[28]

A dynamic role for the CK isozymes in heart energy metabolism has recently emerged. Indeed, it has been shown that a Cr/CrP system, together with CK isozymes localized both at sites of ATP production (mitochondria) and at sites of consumption can apparently act as an intracellular shuttle for the transfer of CrP as the high energy phosphate compound (Saks *et al.*[49] and for a review: see[7]).

According to Kammermeier[31] the Cr/CrP shuttle is a prerequisite for the high thermodynamic efficiency of excitable tissues, since a high level of free energy change of ATP is necessarily accompanied by low cytosolic ADP concentrations and consequently a low rate of ADP diffusion.

The efficiency of the shuttle itself is greatly enhanced by the binding of CK to cellular organelles, e.g. myofibrils, plasmalemma, sarcoplasmic reticulum and mitochondria. Hence the shuttle implies an effective functional compartmentation of CK isozymes and nucleotides, i.e. ADP and ATP.

In heart mitochondria CK$_m$ and the adenine nucleotide translocase (ANT) are functionally coupled[20,30,48]. Both enzymes CK$_m$ and ANT share the same phospholipid domain. Indeed cardiolipin has been found to be the membrane receptor for CK$_m$ [15,41,51], and the same phospholipid is connected with ANT[8].

Ultrastructural localization of CK$_m$ activity *in vivo* and *in vitro*

CK activity was cytochemically localized according to a modified method[9] of Baba[5]. The specificity of the cytochemical reaction was ascertained by extensive controls.

As illustrated by the results in Fig. 1 and Fig. 2, mitochondrial CK activity was exclusively localized in membrane contacts, where IMM and OMM were in close apposition. This phenomenon was not only demonstrated in myocardial and skeletal muscle (Fig. 1a,b), but also visualized in all other tissues where one could expect the presence of active CK$_m$, e.g. brain (Fig. 1c), kidney (Fig. 1d) and retina photoreceptor cells (not shown).

These data were further corroborated by the ultrastructural localization of CK$_m$ activity in isolated mitochondria. After adjustment of the *in vivo* cytochemical procedure, e.g. using short fixation and incubation times, CK$_m$ activity was clearly demonstrated in contact sites between IMM and OMM in isolated heart and brain mitochondria (Fig. 2a, b).

This relation between CK$_m$ activity and mitochondrial contact sites amply confirmed data from recent biochemical research. Indeed, CK$_m$ activity was found to be associated with an isolated membrane fraction enriched in contact sites from brain and kidney[1,34].

Moreover, Schlegel[54] localized CK$_m$ by immunogold labelling in mitochondria from chicken retina photoreceptor cells and visualized antigenic sites clustered at membrane contacts.

This can be explained by the presence of high amounts of cardiolipin in these contact sites[3]. Indeed, CK$_m$ is not only bound by strong interaction to a cardiolipin-containing domain of the IMM[15,41,51], but the presence of high amounts of cardiolipin may be a prerequisite for the enzyme to show activity.

As already stated by Sandermann[50], membrane-bound enzymes must in general be surrounded by membrane lipids in order to be active. Furthermore, the activity of membrane-bound enzymes is substantially enhanced in the presence of phospholipids, e.g. cardiolipin, forming non-bilayer structures[36].

The mitochondrial energetic state *in vivo* and *in vitro* and the extent of CK-active contact sites are correlated

In heart muscle 30 per cent of the CK activity is performed by the mitochondrial isozyme CK$_m$ [28,49]. Already in 1973 Jacobus[30] showed that CrP synthesis in heart mitochondria is stimulated by oxidative phosphorylation. Subsequent studies[2,17,44] proposed that the interaction between CK$_m$ and oxidative phosphorylation is mediated by ATP/ADP ratios in the extra-mitochondrial compartment. Moreover, as shown by our cytochemical data, CK$_m$ was exclusively visualized in contact sites between IMM and OMM. In this way the functional necessity of the OMM was emphasized, as already suggested by biochemical investigation[14,21].

To search for a possible link between the above-mentioned biochemical data and the morphological phenomena, i.e. the appearance of CK-active contact sites, an *in vivo* model for myocardium was developed.

Anaesthetized Wistar rats were divided into four groups, which received different treatments by intravenous injection of the adjusted agent (Table 1). Deviations from the regular metabolic state were induced by stimulation, inhibition or uncoupling of cardiac energy production, using noradrenaline, amytal or DNP. CK$_m$ activity was ultrastructurally localized in myocardium to a different extent as described before[11]. The morphological results were put on a firm quantitative base by a morphometric analysis of the surface density S$_v$ of CK-active contact sites. In a series of parallel experiments the metabolic state, defining the animal model, was investigated.

Therefore the energy charge (EC)[4] was ascertained by analysis of nucleotides (ATP, ADP, AMP) in lyophilized heart biopsies[59]. No statistically significant difference could be found between the EC

In recent thorough biochemical and structural investigations of CK_m from chicken brain and heart mitochondria[53,55] and bovine heart mitochondria[38] evidence was presented that native CK_m forms highly ordered octameric molecules (the m2 type) by association of four CK_m dimers (the m1 type). According to data from recent literature, the two isoforms show a distinct difference in kinetics and binding capacity to the IMM of heart mitochondria[38,40,45,46,52].

According to Lipskaya[37] CK_m not only can exist under physiological conditions as membrane-bound but also in a free state in the intermembrane space. In order to investigate the association behaviour of both CK_m isoforms with mitochondria, a series of experiments was performed in which the release of CK_m from heart mitochondria, incubated in different media, was studied.

Studying the release of CK_m from mitochondria under different pH conditions revealed a strong pH dependence, which was also found by alternative methods in earlier studies[13,29]. However, by combining incubation with electrophoretic separation, the association behaviour of the two isoforms could be differentiated. Clearly it is only the dimeric CK_{m1} type which is released at the high pH values (pH = 7.4 and 7.1), while at the low pH values (pH = 6.6 and 6.1) no CK_{m1} could be found in the supernatant and it is only the octameric CK_{m2} type which is retained in the mitochondrial pellet.

Obviously this corroborates the well-known preferential binding capacity of the octameric CK_{m2} to the IMM[52], but more important it also illustrates a possible octamer-dimer interconversion caused by changing pH.

This pH effect on the behaviour of the CK isoforms suggests that variations in the metabolic state of the mitochondria which change the pH in the intermembrane space may regulate the association and interconversion of the two CK_m isoforms.

The exact nature of this pH effect is at present unknown. pH-induced electrostatic interactions may be involved in the association of CK_{m2} to the IMM[13] or in the formation of both CK_{m1} and CK_{m2} oligomers[37,40]. There also might be a pH effect on ultrastructural changes in the intermembrane space, e.g. the formation of membrane contacts by promoting conversion of lamellar to H-II phase in the phospholipid bilayers[57].

By holding the pH at 7.4, but changing the incubation conditions, a very different picture emerged. After incubation of the mitochondria in a respiratory medium, thus with Mg^{2+} and ADP, the octameric CK_{m2} was the only CK_m isoform visualized in the supernatant (Fig. 3). This is not astonishing since in vitro studies with the isolated enzyme already showed the octamer as the preferred form in the presence of Mg^{2+} and ADP[40].

More intriguing are the following points.

(1) The appearance of the octameric CK_{m2} in itself in the supernatant. This may be due to desorption from the IMM or the existence of a CK octamer, which is not bound to membranes[39] in the intermembrane space during oxidative phosphorylation.

(2) The strong similarity in the CK_m releasing pattern after incubation in isolation medium with DNP or adriamycin (ADM)(Fig. 3.), which both have a known effect on contact site formation[33,42]. In these cases the dimeric CK_{m1} was the only form released in the supernatant, a phenomenon which was accompanied by a simultaneous decrease in octameric CK_{m2} in the mitochondrial pellet.

The octameric CK_{m2} is likely to be the active form at the contact sites[35]. In these membrane contacts, CK_m is thought to interact with the OMM pore protein, the IMM adenine nucleotide translocase and with lipids of IMM and OMM, like for instance cardiolipin, a major phospholipid of the IMM.

Beyond the contacts, a complex dynamic interplay between CK_m oligomers in the intermembrane space is probably modulated by nucleotides, pH[58] and by inhibitors, such as DNP or ADM.

Which parameters are modulating the contact sites to form a dynamic microcompartment for the efficient use of creatine kinase?

Contact sites create functional microcompartments in which a number of kinases, e.g. HK and CK_m, localized at the mitochondrial periphery, gain preferred access to ATP, generated during oxidative

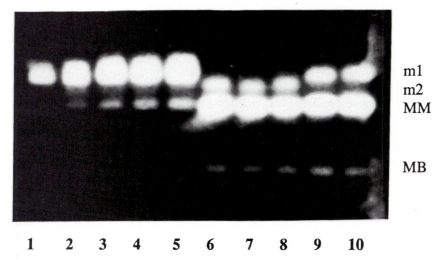

ml
m2
MM

MB

1　2　3　4　5　6　7　8　9　10

Fig. 3. CK isozyme pattern after electrophoresis of supernatants (1–5) and extracts of mitochondrial pellets (6–10) after incubation of mitochondria in respiration medium (1 and 6), in respiration medium with 1.8 mM amytal (2 and 7), in isolation medium (3 and 8), in isolation medium with 0.1 mM DNP (4 and 9), in isolation medium with 1 mM ADM (5 and 10).

phosphorylation[1,9,34]. The ultimate question concerning contact sites is: which physiological parameters are regulating their dynamic formation *in vivo*?

In our study a close correlation between the energetic state of myocardium, CK_m activity and mitochondrial contact sites was demonstrated. Hence, there has to be a morphological part in the process, leading from signal, i.e. hormonal stimulation, to reaction, i.e. enhanced energy production, namely the simultaneous formation of membrane contacts.

The presence of CK_m activity, exclusively associated with contact sites, illustrates the importance of the OMM for the regulation of CK_m activity, as was previously postulated[7,14,21]. However, at the moment essentially nothing is known about the physiological factors *in vivo*, that govern the formation of these membrane contacts.

Nevertheless, there are significant clues for a prime role of Ca^{2+}. First, physiological concentrations of Ca^{2+} stimulate membrane fusion by shielding negative surface charges between IMM and OMM, and by inducing H-II-phase transition in cardiolipin-containing membranes[56].

Secondly, all agents that disturb Ca^{2+} transport or Ca^{2+} metabolism as such, also inhibit either H-II-phase transition in cardiolipin-liposomes or contact site formation itself, e.g. ADM[42], ruthenium red, a specific IMM Ca^{2+} transport inhibitor[16], EGTA[12], DNP[9,33], amytal[25,26]. Thirdly, isolated contact site fractions from brain mitochondria show a preferential and high Ca^{2+}-binding capacity[34].

Of course, other factors might be involved in contact site formation, e.g. pH and μH^+ generated by the activity of the respiratory chain and/or matrix granules. These ubiquitous intramitochondrial structures are considered to be possible precursors of the IMM[6] and they show a dynamic behaviour that seems to be correlated with the energetic state of the mitochondrion, the presence of membrane contacts and the intramitochondrial Ca^{2+} concentration[25,26,27]. Major components of the matrix granules are calcium[43] and phospholipids, probably cardiolipin.

In inhibiting situations matrix granules are numerous and large (70–100 nm), mitochondrial calcium is not cytochemically measurable, the distance between IMM and OMM is large and contact sites are rarely seen (Table 1). Upon stimulation of the metabolism the granules move towards the periphery where they loose their electron density and are subsequently incorporated into the IMM, mitochondrial calcium concentration becomes detectable, the intermembrane distance is small and contact sites increase. After termination of the stimulation the granules reappear gradually[23] and the other parameters become normal again.

Interpreting these results, it can be stated that, by the fusion of the NMGs with the IMM, calcium is liberated, creating an environment favouring membrane approach and fusion. Incorporation of phospholipids together with enhanced calcium concentration promote lipid phase transitions and non-bilayer lipid structures, leading to increased transport possibilities, by which the mitochondrion can cope with sudden and demanding energy changes.

Acknowledgements: This study was supported by Grant 87/92–120 of the Interministrial Commission of Science Policy, Belgium.

Abbreviations: CK = creatine kinase; CK_m = mitochondrial creatine kinase; Cr/CrP = creatine/creatine-phosphate; μH^+ = electrochemical potential; IMM = inner mitochondrial membrane; OMM = outer mitochondrial membrane; ANT = adenine nucleotide translocase; EC = energy charge; ADM = adriamycin; DNP = dinitrophenol; NMG = non matricular granule; HK = hexokinase.

References

1. Adams, V., Bosch, W., Schlegel, J., Wallimann, T. & Brdiczka, D. (1989): Further characterization of contact sites from mitochondria of different tissues: topology of peripheral kinases. *Biochim. Biophys. Acta* **981**, 213–225.

2. Altschuld, R.A. (1980): Interaction between mitochondrial CK and phosphorylation. In *Heart creatine kinase: the integration of isozymes for energy distribution*, eds. W.E. Jacobus & J.S. Ingwall, pp. 127–132. Baltimore: Williams & Wilkins.

3. Ardail, D., Privat, J.P., Egret-Charlier, M., Levrat, C., Lerme, F. & Louisot, P. (1990): Mitochondrial contact sites. Lipid composition and dynamics. *J. Biol. Chem.* **265**, 18797–18802.

4. Atkinson, D.E. (1968): The energy charge of the adenylate pool as a regulatory parameter. Interaction with feedback modifiers. *Biochemistry* **7**, 4030–4034.

5. Baba, N., Kim, S. & Farrell, E.C. (1976): Histochemistry of creatine phosphokinase. *J. Mol. Cell. Cardiol.* **8**, 599–617.

6. Barnard, T. & Ruusa, J. (1979): Matrix granules in soft tissue. I. Elemental composition by X-ray microanalysis. *Exp. Cell Res.* **124**, 339–347.

7. Bessman, S.P. & Carpenter, C.L. (1985): The creatine–creatine phosphate energy shuttle. *Ann. Rev. Biochem.* **54**, 831–862.

8. Beyer, K. & Klingenberg, M. (1985): ADP/ATP carrier protein from beef heart mitochondria has high amounts of tightly bound cardiolipin, as revealed by ^{31}P nuclear magnetic resonance. *Biochemistry* **24**, 3821–3826.

9. Biermans, W. (1991): Mitochondrial contact sites: a dynamic compartment for creatine kinase activity. *PhD thesis*, University of Antwerp.

10. Biermans, W., Bakker, A. & Jacob, W. (1990): Contact sites between inner and outer mitochondrial membranes: a dynamic microcompartment for creatine kinase activity. *Biochim. Biophys. Acta* **1018**, 225–228.

11. Biermans, W., Bernaert, I., De Bie, M., Nys, B. & Jacob, W. (1989): Ultrastructural localisation of creatine kinase activity in the contact sites between inner and outer mitochondrial members of rat myocardium. *Biochim. Biophys. Acta* **974**, 74–81.

12. Brdiczka, D., Knoll, G., Riesinger, I., Weiler, U., Klug, G., Benz, R. & Krause, J. (1986): Microcompartmentation at the mitochondrial surface: its function in metabolic regulation. In *Myocardial and skeletal muscle bioenergetics*, ed. N. Brautbar, pp. 55-70. New York: Plenum Press.

13. Brooks, S.P.J. & Suelter, C.H. (1987a): Association of chicken mitochondrial creatine kinase with the inner mitochondrial membrane. *Arch. Biochem. Biophys.* **257**, 122–132.

14. Brooks, S.P.J. & Suelter, C.H. (1987b): Compartmented coupling of chicken heart mitochondrial creatine kinase to the nucleotide translocase requires the outer mitochondrial membrane. *Arch. Biochem. Biophys.* **257**, 144–153.

15. Cheneval, D. & Carafoli, E.(1988): Identification and primary structure of the cardiolipin-binding domain of mitochondrial creatine kinase. *Eur. J. Biochem.* **171**, 1–9.

16. Cullis, P.R., De Kruijff, B., Hope, M.J., Nayar, R. & Schmid, S.L. (1980): Phospholipids and membrane transport. *Can. J. Biochem.* **58**, 1091–1100.

17. Davis, E.J. & Davis-Van Thienen, W.I.A. (1978): Control of mitochondrial metabolism by the ATP/ADP ratio. *Biochem. Biophys. Res. Commun.* **83**, 1260–1266.

18. De Kruijff, B., Verkleij, A.J., Leunissen-Bijvelt, J., Van Echteld, C.J.A., Hille, J. & Rijnbout, H. (1982): Further aspects of the Ca^{2+}-dependent polymorphism of bovine heart cardiolipin. *Biochim. Biophys. Acta* **693**, 1–12.

19. Eppenberger, H.M., Perriard, J.C. & Wallimann, T. (1983): Analysis of creatine kinase isoenzymes during muscle differentiation. In *Isoenzymes: current topics in biological and medical research,* Vol. 7 eds. M. Rattazi, J.C. Scandalios & C.S. Whitt, pp. 19–38. New York: Alan R. Liss Inc.

20. Erickson-Viitanen, S., Viitanen, P., Geiger, P.J., Yang, W.C.T. & Bessman, S.P.(1982a): Compartmentation of mitochondrial creatine phosphokinase. I. Direct demonstration of compartmentation with the use of labeled precursors. *J. Biol. Chem.* **257**, 14395–14404.

21. Erickson-Viitanen, S., Geiger, P.J., Viitanen, P. & Bessman, S.P. (1982b): Compartmentation of mitochondrial creatine phosphokinase. II. The importance of the outer mitochondrial membrane for mitochondrial compartmentation. *J. Biol. Chem.* **257**, 14405–14411.

22. Hackenbrock, C.R. (1968): Chemical and physical fixation of isolated mitochondria in low-energy and high-energy states. *Proc. Natl. Acad. Sci. USA* **61**, 598–605.

23. Hertsens, R. (1986): Ultrastructureel onderzoek van het hartmitochondrion bij veranderende metabole toestanden. *PhD thesis,* University of Antwerp.

24. Hertsens, R. & Jacob, W. (1987): Freeze-fracture study of heart mitochondria in the condensed or orthodox state. *Biochim. Biophys. Acta* **894**, 507–514.

25. Jacob, W.A. & Hertsens, R. (1984): Intermembrane space in heart mitochondria correlated to the energy state and calcium loading. In *Electron Microscopy 1984 Proc. 8th European Congress in Electron Microscopy* Vol. 7 eds. A. Csanady, P. Rohlich, & D. Szabo, pp. 1873–1874. Kecskemit, Hungary: Petöfi Nyomda.

26. Jacob, W.A. & Hertsens, R. (1986): Interrelation between matrix granules and inner and outer membranes of myocardial mitochondria upon energisation. *4th EBEC Short Reports,* pp. 208. Cambridge: Cambridge University Press.

27. Jacob, W.A., Hertsens, R., Van Bogaert, A. & De Smet, M. (1990): Mitochondrial matix calcium ions regulate energy produced in myocardium: a cytochemical study. In *Proc. XIIth Int. Congress Electron Microsc.,* Vol. 2 eds. L.D. Peachy & D.B. Williams, pp. 166–167. San Francisco: San Francisco Press.

28. Jacobs, H., Heldt, H.W. & Klingenberg, M. (1964): High activity of creatine kinase in mitochondria from muscle and brain and evidence for a seperate mitochondria isoenzyme of creatine kinase. *Biochem. Biophys. Res. Commun.* **16**, 516–521.

29. Jacobus, W.E. & Lehninger, A.L. (1973): Creatine kinase of rat heart mitochondria. Coupling of cratine phosphorylation to electron transport. *J. Biol. Chem.* **248**, 4803–4810.

30. Jacobus, W.E. & Saks, V.A. (1973): Creatine kinase of heart mitochondria: changes in its kinetic properties induced by coupling to oxidative phosphorylation. *Arch Biochem. Biophys.* **219**, 167–178.

31. Kammermeier, H. (1987): Why do cells need phosphocreatine and a phosphocreatine shuttle.*J. Mol. Cell. Cardiol.* **9**, 115–118.

32. Klug, G., Krause, J., Östlund, A.K., Knoll, G. & Brdiczka, D. (1984): Alterations in liver mitochondrial function as a result of fasting and exhaustive exercise. *Biochim. Biophys. Acta* **764**, 272–282.

33. Knoll, G. & Brdiczka, D. (1983): Changes in freeze-fractured mitochondrial membranes correlated to their energetic state. Dynamic interactions of the boundary membranes. *Biochim. Biophys. Acta* **733**, 102–110.

34. Kottke, M., Adams, V., Riesinger, I., Bremm, G., Bosch, W., Brdiczka, D., Sandri, G. & Panfili, E. (1988): Mitochondrial boundary membrane contact sites in brain: points of hexokinase and creatine kinase location, and control Ca^{2+} transport. *Biochim. Biophys. Acta* **935**, 87–102.

35. Kottke, M., Adams, V., Walliman, T., Nolam, V.K. & Brdiczka, D. (1991): Location and regulation of octameric mitochondrial creatine kinase in the contact sites. *Biochim. Biophys. Acta* **1061**, 215–225.

36. Lindblom, G. & Rilfors, L.(1989): Cubic phases and istopic structures formed by membrane lipids – possible biological relevance. *Biochim. Biophys. Acta* **988**, 221–256.

37. Lipskaya, T.Y. & Rybina, I.V. (1987): [Properties of creatine kinase from skeletal muscle mitochondria] Svoistva kreatinkinazy iz mitokhondrii skeletnykh myshts. *Biokhimiya* **52**, 690–700.

38. Lipskaya, T.Yu., Trofimova, M.E. & Moiseeva, N.S. (1989): Kinetic properties of the octameric and dimeric forms of mitochondrial creatine kinase and physiological role of the enzyme. *Biochem. Int.* **19**, 609–613.

39. Lipskaya, T.Yu. & Trofimova, M.E. (1989): Study on heart mitochondrial creatine kinase using a cross-linking bifunctional reagent. II. The binding sites for the creatine kinase octamer on mitochondrial membranes have different properties. *Biochem. Int.* **18**, 1149–1159.

40. Marcillat, O., Goldschmidt, D., Eichenberger, D. & Vial, C.(1987): Only one of the two interconvertible forms of mitochondrial creatine kinase binds to heart mitoplasts. *Biochim. Biophys. Acta* **890**, 233–241.

41. Müller, M., Moser, R., Cheneval, D. & Carafoli, E. (1985): Cardiolipin is the membrane receptor for mitochondrial creatine phosphokinase. *J. Biol. Chem.* **260**, 3839–3843.

42. Nicolay, K., Timmers, R., Spoelstra, E., van der Neut, R., Fok, J.J., Huigen, Y.M., Verkleij, A.J. & De Kruijff, B. (1984): The interaction of adriamycin with cardiolipin in model rat liver mitochondrial membranes. *Biochim. Biophys. Acta* **778**, 359–371.

43. Nys, B., Hertsens, R., Van Espen, P. & Jacob, W. (1988): Electron spectroscopic imaging of calcium in matrix granules myocardial mitochondria. *Biol. Cell* **63**, 27a.

44. Nishiki, K., Erecinska, M., Wilson, D.F. & Cooper, S. (1978): Evaluation of oxidative phosphorylation in hearts from euthyroid, hypothyroid, and hypothyroid rats. *Am. J. Physiol.* **235**, C212–C219.

45. Quemeneur, E., Eichenberger, D., Goldschmidt, D., Vial, C., Beauregard, G. & Potier, M. (1988): The radiation inactivation method provides evidence that membrane-bound mitochondrial creatine kinase is an oligomer. *Biochem. Biophys. Res. Commun.* **153**, 1310–1314.

46. Quemeneur, E., Eichenberger, D. & Vial, C. (1990): Immunological determination of the oligomeric form of mitochondrial creatine kinase *in situ*. *FEBS Lett.* **262**, 275–278.

47. Riesinger, I., Knoll, G., Brdiczka, D., Denis-Pouxviel, C., Murat, J.C. & Probst, I. (1985): Defects of mitochondrial membrane structure in tumour cells. *Eur. J. Cell Biol.* **36**, 53.

48. Saks, V.A., Kuznetsov, A.N., Kupriyanov, V.V., Miceli, M.V. & Jacobus, W.E. (1985): Creatine kinase of rat heart mitochondria. The demonstration of functional coupling to oxidative phosphorylation in an inner membrane-matrix preparation. *J. Biol. Chem.* **260**, 7757–7764.

49. Saks, V.A., Rosenshtrauk, L.V., Smirnov, V.N. & Chazov, E.I. (1978): Role of creatine phosphokinase in cellular function and metabolism. *Can. J. Physiol. Pharmacol.* **56**, 691–706.

50. Sandermann, H. (1978): Regulation of membrane enzymes by lipids. *Biochim. Biophys. Acta* **515**, 209–237.

51. Schlame, M. & Augustin, W. (1985): Association of creatine kinase with rat heart mitochondria: high and low affinity binding sites and the involvement of phospholipids. *Biomed. Biochim. Acta* **44**, 1083–1088.

52. Schlegel, J., Wyss, M., Eppenberger, H.M. & Wallimann, T. (1990): Functional studies with the octameric and dimeric form of mitochondrial creatine kinase. Differential pH-dependent association of the two oligomeric forms with the inner mitochondrial membrane. *J. Biol. Chem.* **265**, 9221–9227.

53. Schlegel, J., Wyss, M., Schurch, U., Schnyder, T., Quest, A., Wegmann, G., Eppenberger, H.M. & Wallimann, T. (1988a): Mitochondrial creatine kinase from cardiac muscle and brain are two distinct isoenzymes but both form octameric molecules. *J. Biol. Chem.* **263**, 16963–16969.

54. Schlegel, J., Zurbriggen, B., Wegmann, G., Wyss, M., Eppenberger, H.M. & Wallimann, T. (1988b): Native mitochondrial creatine kinase forms octameric structures. I. Isolation of two interconvertible mitochondrial creatine kinase forms, dimeric and octameric mitochondrial creatine kinase: characterization, localization, and structure-function relationships. *J. Biol. Chem.* **263**, 16942–16953.

55. Schnyder, T., Engel, A., Lustig, A. & Wallimann, T. (1988): Native mitochondrial creatine kinase forms octameric structures. II. Characterization of dimers and octamers by ultracentrifugation, direct mass measurements by scanning transmission electron microscopy, and image analysis of single mitochondrial creatine kinase octamers. *J. Biol. Chem.* **263**, 16954–16962.

56. Van Venetië, R. & Verkleij, A.J. (1982): Possible role of non-bilayer lipids in the structure of mitochondria. A freeze-fracture electron microscopy study. *Biochim. Biophys. Acta* **692**, 397–405.

57. Verkleij, A.J. (1984): Lipidic intramembranous particles. *Biochim. Biophys. Acta* **779**, 43–63.

58. Wallimann, T., Schnijder, T., Schlegel, J., Wyss, M., Wegmann, G., Rossi, A., Hemmer, W., Eppenberger, H. & Quest, A. (1989): Subcellular compartimentation of creatine kinase isoenzymes, regulation of CK and octameric structures of mitochondrial CK: important aspects of the phosphoryl-creatine circuit, In *Progress in clinical and biological research, Vol. 315, Muscle energetics* eds. R.J. Paul, G. Eczinga & K. Yamada, pp. 159–176, New York: Alan R. Liss Inc.

59. Wynants, J. & Van Belle, H. (1985): Single-run high-performance liquid chromatography of nucleotides, nucleosides, and major purine bases and its application to different tissue extracts. *Anal. Biochem.* **144**, 258–266.

Guanidino Compounds in Biology and Medicine, eds. by P.P. De Deyn, B. Marescau, V. Stalon and
I.A. Qureshi. ©1992 John Libbey & Company Ltd., pp. 175–179.

Chapter 25

Alteration of muscle metabolism and contractile properties in response to creatine depletion

Y. OHIRA[1], T. WAKATSUKI[1], N. ENDO[1], K. NAKAMURA[2], T. ASAKURA[2],
K. IKEDA[3], T. TOMIYOSHI[4] and M. NAKAJOH[3]

[1]*Department of Physiology and Biomechanics, National Institute of Fitness and Sports, Kanoya City 891-23;*
[2]*Department of Neurosurgery,* [3]*Department of Radiology, Faculty of Medicine, and* [4]*University Hospital, University
of Kagoshima, Kagoshima City, 890, Japan*

Summary

Effects of creatine depletion and loading on metabolic and contractile properties of fast and slow skeletal muscles were studied in rats. Depletion of high-energy phosphates in muscles was induced by feeding the creatine analogue β-guanidinopropionic acid (β-GPA, 1 per cent) for approximately 9 weeks. Food for some rats was switched from either control or β-GPA to creatine-containing diet (1 per cent) for 1–2 weeks. Isometric contractile properties of soleus and extensor digitorum longus (EDL) were measured both *in situ* and *in vitro*. The concentration of high-energy phosphates was measured both biochemically in freeze-clamped muscle and by ^{31}P-nuclear magnetic resonance spectroscopy. Metabolite contents were also measured in the freeze-clamped samples. Furthermore, some enzyme activities and muscle fibre type were analysed biochemically and histochemically, respectively. Shift of speed-related contractile properties toward slow-type associated with an improved resistance to fatigue was induced by chronic creatine depletion. In response to creatine feeding, the contractile characteristics were reversed within 10 days. It was suggested that increased endurance capacity of muscles with depleted high-energy phosphates may be due to the elevated mitochondrial respiratory capacity and glycogen content. Shift of contractile properties to slow-twitch type was due to increased one-half relaxation time. These data suggest an important role of high-energy phosphates in the regulation of metabolic and contractile properties of muscles.

Introduction

Mitochondrial biogenesis is stimulated when the high-energy phosphates are depleted in animals by feeding the creatine analogue β-guanidinopropionic acid (β-GPA)[16,21] or by exposure to cold[17]. Mitochondrial enzyme activities in skeletal muscles sampled from these animals are greater than in control[21]. Following 22 days of 1 per cent β-GPA diet, increased cytochrome c mRNA is also found in rat muscles with inherently low cytochrome c mRNA[7]. It is also reported that the contractile characteristics of these muscles are shifted toward slow-twitch type[11,14]. However, the precise mechanism of such changes in metabolic and contractile properties of muscle is still unclear. The current study was performed to investigate the effects of creatine, which influences the levels of adenosine triphosphate (ATP) and phosphocreatine (PC)[1,2,10,11,13,14], on the contractile properties and metabolism of fast-twitch extensor digitorum longus (EDL) and slow-twitch soleus muscles in rats.

Materials and methods

Newly weaned male Wistar rats were randomly divided into two groups. Rats in the control group were fed powdered diet (CE-2, Nihon CLEA, Tokyo). The same diet, but containing β-GPA (1 per cent), was fed to the experimental group. They were pair-fed and the amount of food supplied, which was completely eaten within approximately 12 h, was gradually increased following the growth. From week four to the end of the experiment, each rat was fed 20 g daily. Water was supplied *ad libitum*. Temperature and humidity in the animal room with a 12:12 h light:dark cycle were maintained at approximately 23 °C and 55 per cent, respectively.

Effects of β-GPA feeding on creatine level in blood and enzymatic characteristics and fibre type in skeletal muscles were investigated. Rats were anaesthetized by i.p. injection of sodium pentobarbital (5 mg/100 g body weight). Blood was withdrawn from the jugular vein into a heparinized syringe. Blood was centrifuged and the supernatant was used for creatine determination[24]. Soleus, plantaris, and EDL muscles were sampled. The mid-portion of the muscle was mounted on a cork by using OCT compound and quickly frozen in isopentane cooled in liquid nitrogen. Serial cross-sections were cut at 10 μm thickness in a cryostat. These sections were stained qualitatively for myosin ATPase at an alkaline (pH, 8.75) and acid (pH, 4.35) pre-incubation according to the method of Nwoye *et al.*[15]. Tibialis anterior was used for determination of the activities of citrate synthase[23], succinate dehydrogenase[22], malate dehydrogenase[5], cytochrome oxidase[25,] creatine kinase[18], and lactate dehydrogenase[20].

Effects of creatine depletion and loading on high-energy phosphates and metabolites in plantaris muscle were measured at rest. The diet for five rats from each group was switched from either control or β-GPA diet to food containing 1 per cent creatine for 1–2 weeks. The concentration and distribution of phosphorus compounds in the dorsal thigh muscles of anaesthetized rat were evaluated by ^{31}P-nuclear magnetic resonance (NMR) spectroscopy (BEM 170/200, Otsuka Electronics, Tokyo). For biochemical analyses, plantaris in rat anesthetized with i.p. injection of sodium pentobarbital was carefully exposed keeping the nerve and blood supplies intact. The animal was allowed to recover for approximately 10 min on a heating pad covering the muscles with gauze moistened with Krebs–Ringer solution. Plantaris was freeze-clamped with a pair of aluminium tongs cooled in liquid nitrogen. The muscle samples were pulverized and homogenized as was reported in our previous study[17]. A portion of the whole homogenate was used for determination of glycogen[19]. Lactate[4], ATP[9], PC[8], and inorganic phosphate[3] were measured in the supernatant of the homogenate.

The contractile properties of soleus and EDL were measured both *in situ* and *in vitro*. Before the initiation of *in situ* experiment, right soleus and left EDL were dissected out and wet weight was measured. Since the muscle weight is increased due to elevated fluid content after electrical stimulation, these data taken from the resting contralateral side were used as the muscle weight in the experimental side. The left soleus and right EDL were exposed and stimulating apparatus composed of two stainless steel wires were placed around both right and left sciatic nerves at the gluteal region. The rat was placed in prone position on a table within a temperature-regulated chamber (37 °C). The left hindlimb was fixed to the table by inserting a needle through the knee. The distal tendon of soleus was connected to an isometric transducer. The apparatus placed around the left sciatic nerve was connected to an electrical stimulator. The muscles were stimulated electrically through the sciatic nerve. The optimum muscle length and stimulus intensity with 100 μs pulse duration which produce the maximum twitch tension were adjusted by checking the tension using an oscilloscope and paper recorder. The muscles were stimulated at 50 Hz (60 trains/min) and tension was recorded for 5 min. The anode and cathode in every pulse were alternated to avoid polarization at the electrodes. The fatigability of right EDL was then measured in the same way.

For *in vitro* experiment, soleus and EDL were dissected out and the distal tendon was fixed to a muscle holder and the proximal tendon was connected to a force transducer. Muscle was immersed in Krebs–Ringer solution (pH 7.2) gassed continuously with 95 per cent O_2 and 5 per cent CO_2. Muscle was stimulated through two stainless steel wires placed around the muscle. Isometric tension production in response to the electrical stimulation at 1, 10, 20, 50, 80, 100, 140, and 200 Hz with 3-min interval was recorded on both tape and paper recorders. The peak tension, time-to-peak tension (TPT), one-half relaxation time (1/2 RT), and the rate of tension production (dp/dt) in a twitch curve, summation pattern of tension, and the maximum tetanic tension were examined.

Results and discussion

The concentration of blood plasma creatine in β-GPA rats was approximately 52 per cent less than in controls ($P < 0.01$) but creatine in muscle was only 4 per cent of control ($P < 0.001$)). As was reported previously[1,2,10,11,13,14], high-energy phosphates in skeletal muscles were depleted. Evaluation of phosphorus compounds by ^{31}P-NMR spectroscopy also revealed this phenomenon. The level of PC in resting plantaris of β-GPA rats was only 8 per cent of that in control. The decrease in ATP was minor but still significant (–40 per cent, $P < 0.001$). The glycogen content, on the contrary, was increased by approximately 100 per cent in these muscles.

Distribution of slow-twitch oxidative fibres was increased following β-GPA feeding. The data obtained from the determination of contractile properties supported this histochemical transformation of fibre type. The β-GPA muscles tended to be slower than in control muscles as was reported by Mainwood *et al.*[11]. Such tendency was reversed within 10 days after creatine supplementation. In these muscles, the PC content was elevated to approximately 50 per cent of normal and the level of phosphorylated β-GPA was decreased. The TPT and 1/2 RT of soleus muscle tended to decrease even in control-diet group in response to creatine feeding. These data suggest that the characteristic of speed-related contractile properties is shifted toward slow-type when creatine content is lowered and toward fast-type in response to creatine loading, although the mechanism is still unclear.

Moerland *et al.*[14] found a change in the pattern of myosin isoform in hindlimb muscles of mice fed β-GPA. In soleus, the distribution of intermediate myosin was reduced and slow myosin was increased by approximately 50 per cent following β-GPA feeding. In EDL, β-GPA feeding resulted in a decreased fast isomyosin FM_1 (LC_3f homodimer) and an increased FM_3 (LC_1f homodimer). A shift of myosin isoform from fast V_1 to slower V_2 and V_3 in response to β-GPA feeding was also found in cardiac muscle[12]. In the current study, clear effects of creatine depletion and/or loading on the speed-related contractile properties were found in soleus 1/2 RT. Significant change in TPT, found by Mainwood *et al.*[11], was observed neither in EDL nor in soleus. These results may suggest that calcium uptake into sarcoplasmic reticulum may be influenced by creatine.

Endurance capacity of muscles, especially in EDL, was improved in rats fed β-GPA. Elevated mitochondrial enzyme activities found in tibialis anterior in our study and reported earlier by Shoubridge *et al.*[21], which increase mitochondrial respiratory capacity, may explain why the resistance to fatigue was improved in β-GPA muscles. Elongated mitochondria containing straightened layers of cristae were also found in rat muscles fed β-GPA[16]. Although the activities of cytoplasmic enzymes were inhibited in these muscles, greater glucose uptake[6] and hexokinase activity[21], reported elsewhere, may also help to increase the oxidative capacity of muscles. Another factor for the improved endurance capacity might be the elevated glycogen content in muscle.

The endurance capacity of muscles was decreased toward control level following 10 days of creatine supplementation. Although the resting high-energy phosphates were increased and mitochondrial enzyme levels were still elevated, glycogen content was normalized by creatine supply. Glucose uptake into muscles was also decreased to the control level in response to creatine loading[6]. These results may affect the endurance capacity of muscle. Further histochemical and biochemical examinations are currently performed to investigate the role of creatine in the contractile properties.

In conclusion, contractile properties and metabolic characteristics were studied in rat skeletal muscles with depleted and/or supplemented creatine. High-energy phosphates, especially PC, were depleted following 1 per cent β-GPA feeding for approximately 9 weeks. The characteristics of muscles were shifted toward slow-twitch type mainly due to the increased 1/2 RT in β-GPA rats. These muscles had greater resistance to fatigue than controls. Data suggest that improved endurance capacity may be caused by elevated mitochondrial respiratory capacity and glycogen content in muscles. Such alterations of contractile properties were reversed within 10 days of creatine feeding. Although the resting high-energy phosphates were increased toward control level and mitochondrial enzyme activities were still elevated, glycogen content was normalized by creatine supply. These data suggest an important role of high-energy phosphates in the regulation of metabolic and contractile properties of muscles.

Acknowledgements

This study was, in part, supported by the Inamori Foundation and the fund for basic experiments oriented to space station utilization, from the Institute of Space and Astronautical Science, Japan.

References

1. Fitch, C.D., Jellinek, M., Fitts, R.H., Baldwin, K.M. & Holloszy, J.O. (1975): Phosphorylated β-guanidino-propionate as a substrate for phosphocreatine in rat muscle. *Am. J. Physiol.* **228**, 1123–1125.

2. Fitch, C.D., Jellinek, M. & Mueller, E.J. (1974): Experimental depletion of creatine and phosphocreatine from skeletal muscle. *J. Biol. Chem.* **249**, 1060–1063.

3. Guynn, R.D., Veloso, D. & Veech, R.L. (1972): Enzymatic determination of inorganic phosphate in the presence of creatine phosphate. *Analyt. Biochem.* **45**, 277–285.

4. Hohorst, H.J. (1965): Determination of L-lactate with LDH and DPN. In *Methods of enzymatic analysis*, ed. H.U. Bergmeyer, pp. 266–270. New York: Academic Press.

5. Holloszy, J.O., Oscai, L.B., Don, I.J. & Molé, P.A. (1970): Mitochondrial citric acid cycle and related enzymes: adaptive response to exercise. *Biochem. Biophys. Res. Commun.* **40**, 1368–1373.

6. Ishine, S., Ohira, Y., Tabata, I., Kurata, H. & Takekura, H. (1988): Effects of creatine depletion on glucose uptake in mice. *Jpn. J. Phys. Fit. Sports Med.* **37**, 570.

7. Lai, M.M. & Booth, F.W. (1990): Cytochrome c mRNA and α-actin mRNA in muscles of rats fed β-GPA. *J. Appl. Physiol.* **69**, 843-848.

8. Lamprecht, W., Stein, P., Heinz, F. & Weisser, H. (1974): Creatine phosphate. In *Methods of enzymatic analysis*, 2nd edn, ed. H.U. Bergmeyer, pp. 1777–1781, New York: Academic Press.

9. Lamprecht, W. & Trautschold, I. (1974): ATP determination with hexokinase and glucose-6-phosphate dehydrogenase. In *Methods of enzymatic analysis*, 2nd edn, ed. H.U. Bergmeyer, pp. 2101–2110, New York: Academic Press.

10. Mainwood, G.W., Alward, M. & Eiselt, B. (1982): The effect of metabolic inhibitors on the contraction of creatine-depleted muscle. *Can. J. Physiol. Pharmacol.* **60**, 114–119.

11. Mainwood, G.W., Alward, M. & Eiselt, B. (1982): Contractile characteristics of creatine-depleted rat diaphragm. *Can. J. Physiol. Pharmacol.* **60**, 120–127.

12. Mekhfi, H., Hoerter, J., Lauer, C., Wisnewsky, C., Schwartz, K. & Ventura-Clapier, R. (1990): Myocardial adaptation to creatine deficiency in rats fed with β-guanidinopropionic acid, a creatine analogue. *Am. J. Physiol.* **258** (*Heart Circ. Physiol.* **27**), H1151–H1158.

13. Meyer, R.A., Brown, T.R., Krilowicz, B.L. & Kushmerick, M.J. (1986): Phosphagen and intracellular pH changes during contraction of creatine-depleted rat muscle. *Am. J. Physiol.* **250** (*Cell Physiol.* **19**), C264–C274.

14. Moerland, T.S., Wolf, N.G. & Kushmerick, M.J. (1989): Administration of a creatine analogue induces isomyosin transitions in muscle. *Am. J. Physiol.* **257** (*Cell Physiol.* **26**), C810–C816.

15. Nwoye, L., Mommaerts, W.F.H.M., Simpson, D.R., Seraydarian, K. & Marusich, M. (1982): Evidence for a direct action of thyroid hormone in specifying muscle properties. *Am. J. Physiol.* **242**, R401–R408.

16. Ohira, Y., Kanzaki, M. & Chen, C.-S. (1988): Intramitochondrial inclusions caused by depletion of creatine in rat skeletal muscles. *Jpn. J. Physiol.* **38**, 159–166.

17. Ohira, M. & Ohira, Y. (1988): Effects of exposure to cold on metabolic characteristics in gastrocnemius muscle of frog (*Rana pipiens*). *J. Physiol.* **395**, 589–595.

178

18. Oscai, L.B. & Holloszy, J.O. (1971): Biochemical adaptations in muscle. II. Response of mitochondrial adenosine triphosphatase, creatine phosphokinase, and adenylate kinase activities in skeletal muscle to exercise. *J. Biol. Chem.* **246**, 6968–6972.

19. Passonneau, J.V. & Landerolale, V.R. (1974): A comparison of three methods of glycogen measurement in tissues. *Analyt. Biochem.* **60**, 405–412.

20. Pesce, A., McKay, R.H., Stolzenbach, F., Cahn, R.D. & Kaplan, N.O. (1964): Comparative enzymology of LDH. *J. Biol. Chem.* **239**, 1753–1761.

21. Shoubridge, E.A., Challiss, R.A.J., Hayes, D.J. & Radda, G.K. (1985): Biochemical adaptation in the skeletal muscle of rats depleted of creatine with the substrate analogue β-guanidinopropionic acid. *Biochem. J.* **232**, 125–131.

22. Singer, T.P. (1974): Determination of the activity of succinate, NADH, choline, and α-glycerophosphate dehydrogenase. *Methods Biochem. Anal.* **22**, 123–175.

23. Srere, P.A. (1969): Citrate synthase. *Methods Enzymol.* **13**, 3–5.

24. Tanzer, M.L. & Gilvarg, C. (1959): Creatine and creatine kinase measurement. *J. Biol. Chem.* **234**, 3201–3204.

25. Wharton, D.C. & Tzagoloff, A. (1967): Cytochrome oxidase from beef heart mitochondria. *Methods Enzymol.* **10**, 245–250.

Guanidino Compounds in Biology and Medicine, eds. by P.P. De Deyn, B. Marescau, V. Stalon and I.A. Qureshi. ©1992 John Libbey & Company Ltd., pp. 181–186.

Chapter 26

Phosphocreatine, exercise, protein synthesis, and insulin

Samuel P. BESSMAN and Chandra MOHAN

Department of Pharmacology and Nutrition, University of Southern California, School of Medicine, Los Angeles, CA 90033, USA

Summary

The metabolic effects of insulin and of exercise are similar. Energy transfer in striated muscle is through the creatine phosphate shuttle. Isozymes of creatine phosphokinase (CPK) are attached near the myosin ATPase sites and the ATP translocase of the mitochondria. Myofibrillar ATPase converts ATP to ADP. ATP is generated by the attached CPK, converting creatine phosphate to creatine. Creatine diffuses to mitochondrial CPK where it is converted to phosphocreatine from nascent ATP, and the phosphocreatine returns to the myofibril. This process serves a twofold function. First, it supplies energy to the myofibril, and second, creatine stimulates the mitochondria to regenerate ATP (acceptor effect).

The acceptor role of creatine is similar to the mitochondrial effect of insulin. A parsimonious paradigm in which insulin attaches hexokinase to mitochondria, providing an acceptor stimulation, was proposed and much supporting data have been developed. Insulin stimulates mitochondrial Krebs cycle activity in isolated cells or tissues, reaching a maximum sustained rate within 45 to 60 s. This is accompanied by increased incorporation of ^{14}C substrate counts into protein, apparently due to the increased delivery of ATP from mitochondria to microsomes.

Creatine phosphate can stimulate microsomal protein synthesis more effectively than ATP. Creatine liberation, which occurs during exercise, causes increased influx of creatine phosphate to microsome. Exercise is therefore, equivalent to insulin stimulation of protein synthesis, both causing increased delivery of phosphate energy to the microsome.

Introduction

The growing conclusion among investigators that excess insulin is a major factor in the development of atherosclerosis makes it necessary to clarify the role of exercise in lowering the blood glucose. New understanding of the different mechanisms by which metabolic processes are affected by insulin and exercise now permits us to make some educated guesses about the preventive value of exercise in diabetes type I and type II. One fact is clear, although insulin affects almost all tissues, exercise acts only on the mechanically active organs – the heart and the skeletal muscle.

Mitochondrial action of insulin: the outboard motor concept

The intracellular action of insulin ('post receptor action') has been found to focus on the mitochondria, the primary energy centres of all cells[11]. About 95 per cent of the energy of the cell is generated by the mitochondria which have been characterized as small 'outboard motors' which are connected to various anabolic areas throughout all cells (Fig. 1). Any process which removes ATP as it is formed by the mitochondrion and converts it to ADP stimulates mitochondrial oxidation and concomitant generation of ATP. The enzymes which do this are called kinases. The two major enzymes which act

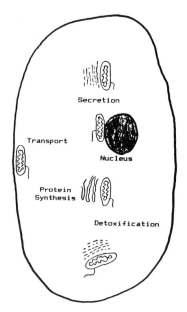

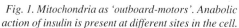

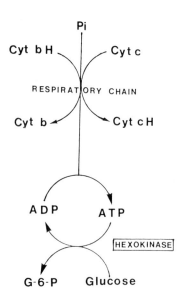

Fig. 1. *Mitochondria as 'outboard-motors'. Anabolic action of insulin is present at different sites in the cell.*

Fig. 2. *The acceptor role of hexokinase in the electron transport system of the mitochondrion. G-6-P, glucose-6-phosphate; Cyt, cytochrome.*

on mitochondria in this way are hexokinase and creatine kinase. Insulin stimulates these little motors, wherever they lie within the cell by attaching hexokinase to them. This results in a large acceleration of generation of ATP by the mitochondria (Fig. 2). This ATP is delivered most effectively to anabolic centres adjacent to the mitochondria. For example, those mitochondria adjacent to the microsomal network provide energy for protein synthesis, those adjacent to the Golgi apparatus deliver energy for secretion, and those adjacent to the cell membrane provide energy for transport of glucose and other substances into the cell. Consistent with this hypothesis is the well-known fact that insulin stimulates the uptake of glucose most effectively in fat cells in which mitochondria lie close to the plasma membrane, but in muscle and liver cells the effect of insulin on membrane transport would be less than in fat cells, because only a small proportion of their mitochondria are adjacent to the plasma membrane.

A unified hypothesis for the anabolic actions of insulin

In the past, it was very difficult to understand how insulin could work, because it has so many major effects, on glucose transport across the cell membrane, on glycogen synthesis, on protein synthesis, on CO_2 formation, on fat synthesis, on RNA synthesis, on oxidation, and on phosphorylation of most water-soluble vitamins. Bessman *et al.*[6] provided evidence that insulin increases turnover of ATP and almost all other phosphorylated intermediates in muscle tissue. It is not necessary to postulate many different pathways (second messengers) through which insulin might work. The dependence of insulin action on uninhibited mitochondrial function makes it clear that the manifold effects of insulin result from its stimulation of the *availability of the ATP which is necessary for every action of insulin on the cell. The physiological role of insulin is to accelerate the energy generation process in all susceptible cells in the same way.* The notion that the ATP supply from mitochondria is a controlling factor in endergonic processes has been proposed by others as a mechanism of changes in anabolic activity[1,2,9].

On the surface it would seem that when hexokinase attachment to mitochondria stimulates oxidative phosphorylation, all of the ATP generated might be consumed in the hexokinase reaction, leaving little or no ATP for other anabolic processes. Experiments of Bachur[3] showed that when respiration of liver mitochondria was enhanced by the addition of yeast hexokinase (with a very low K_m for ATP) over a wide range of concentrations, increased ATP was provided for a secondary synthesis which had a high K_m for ATP. On the basis of these experiments the hexokinase – mitochondrial binding theory was elaborated to include the 'cafeteria' principle[5] of preferred mitochondrial binding sites for many ATP-utilizing reactions, all of which profit from accelerated Krebs cycle turnover.

Exercise, acceptor hypothesis and the creatine phosphate shuttle

How does this relate to exercise? Based on the acceptor hypothesis, it was proposed that exercise produces effects on muscle metabolism resembling those produced by insulin[4,5]. From the acceptor hypothesis developed the concept of the creatine phosphate shuttle, which deals with energy control and distribution in muscle, heart and brain. Although the ultimate energy compound used for skeletal and cardiac muscle contraction is ATP, the primary transport medium of energy in striated muscle is creatine phosphate. This process is now called the *creatine phosphate shuttle*. In addition to the insulin – hexokinase process which is required at rest, the muscle mitochondrion has attached to it the enzyme creatine phosphokinase which catalyses reversible reactions to produce creatine phosphate and ADP at the mitochondrial surface, or creatine plus ATP at the myofibril. The energy for contraction is formed in the mitochondrion and immediately reacted with creatine to form creatine phosphate. It has been shown *in vitro* that free creatine added to mitochondria of muscle and heart accelerates oxidation and energy generation proportional to the free creatine added (Fig. 3). In striated muscle, not only the myofibrils are benefited by the creatine phosphate which comes from the mitochondria whenever the creatine resulting from contraction is brought into contact with them, but also the glucose uptake[13] and protein synthesis systems[8]. On the cell membrane there are bound creatine phosphokinase enzymes which use creatine phosphate for fueling glucose transport, and on the microsomes which synthesize muscle protein there is bound creatine phosphokinase which uses creatine phosphate to fuel protein synthesis. Thus when the muscle contracts, creatine phosphate is liberated, which goes to the mitochondria where creatine phosphate is generated. Some of this extra energy is used for glucose transport and protein synthesis. When the muscle is at rest there is no acceleration of creatine phosphate formation. The only provision of energy for transport of glucose is by the insulin hexokinase system. Carpenter *et al.*[8] provided evidence for the involvement of creatine phosphate shuttle in muscle protein synthesis. Using 2,4-dinitrofluorobenzene (DNFB), an inhibitor of creatine phosphokinase, they were able to show a significant inhibition (70 per cent) of protein synthesis in diaphragm muscle. This was not a direct effect of the inhibitor on any step of microsomal protein synthesis as shown by the fact that the same concentration of DNFB which caused 70 per cent inhibition of protein synthesis in diaphragm muscle had no inhibitory effect on protein synthesis in isolated hepatocytes, which

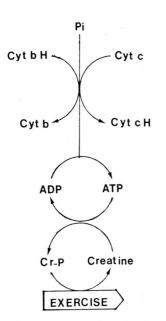

Fig. 3. The acceptor role of exercise. Cr-P, creatine phosphate.

possess no creatine phosphate shuttle. The involvement of creatine phosphate shuttle in muscle contraction is depicted in Fig. 4. The mitochondrial component of the shuttle lies within the intermembrane space and the myofibrillar component lies within the myofibril which includes ATPase of the light chains and the creatine phosphokinase associated with the A-band. The transport of energy between the site of its generation (mitochondrion) and site of its utilization (myofibril) takes place in the form of creatine–creatine phosphate shuttle. Energy (creatine phosphate) carried by this shuttle through the intervening space can be utilized by the microsomal system for protein synthesis (Fig. 5).

The mitochondria distributed throughout the muscle cells function as small outboard motors and are affected both by insulin and creatine generated by contraction. Exercising the myofibrils generates free creatine and causes an increase in glucose transport and an increase in protein synthesis. This is why exercise lowers the blood glucose, and at the same time, increases protein synthesis (growth) in the muscle. It is the result of more rapid production and delivery of energy to all anabolic sites in muscle. This process in the muscle of accelerating energy availability appears grossly to be similar

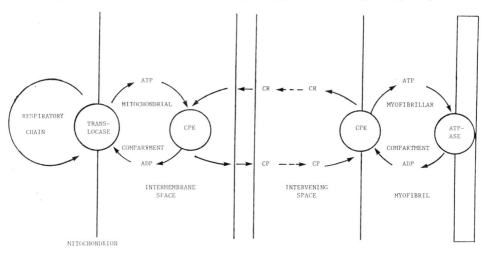

Fig. 4. Organization of the creatine phosphate shuttle within the inner membrane and the intermembrane space of the mitochondrion. CP, creatine phosphate; CPK, creatine phosphokinase; CR, creatine.

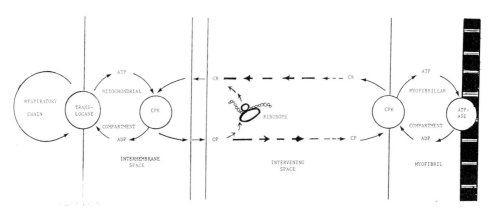

Fig. 5. Organization of the creatine phosphate shuttle and its relation to the protein synthesizing system in the muscle.

to the effect of insulin in all cells, which also results in more rapid production and distribution of energy. There is a major difference however. Insulin acts on all cells but the exercise effect occurs only in striated muscle. The diminution in blood glucose is produced by the movement of glucose into the muscle cells, and the protein synthesis effect is produced only in muscle cells. The effects of exercise are limited to changes in the metabolism only of the muscle. The only peripheral effect of exercise is the drop in blood glucose.

Diabetes, insulin and exercise

The first thing that one should remember is that insulin is necessary for normal growth and repair, two processes which are more or less abnormal in the insulin-deprived individual. The second is that exercise only acts like insulin in striated muscle. It does not release insulin or insulin-like substances into the circulation. The third, and perhaps most important finding, is that insulin itself, in excess, is the cause of most, if not all, of the complications of diabetics which are the result of vascular and neuronal degeneration.

In type I diabetes, there is an absolute deficiency of insulin. Since the only sign of normal 'control' of diabetes is the level of blood glucose, any activity which lowers the blood glucose diminishes the apparent 'need' for insulin. If a great deal of exercise maintains a normal glucose level there will be adequate and even increased growth of muscle but little insulin availability for all the other tissues. For type I diabetic patients who receive large doses of insulin exercise might be a boon, for it would diminish overzealous insulinization.

Type II diabetes, which represents more than 90 per cent of cases, is a different problem. These patients have much higher than normal levels of insulin in the blood, at least in the early phase of their disease. They secrete larger than normal amounts of insulin in response to a sugar stimulus. Since the high levels of insulin cause vascular complications it might be desirable to normalize the blood glucose by exercise. It should help to prevent, or at least delay, the onset of complications.

Physiological and pathological significance of the creatine phosphate shuttle

A progressive decline in mitochondrial creatine kinase activity has been shown in muscular dystrophy in chicken model[10]. The dystrophic chicks show greater resistance to loss of creatine kinase from mitochondria caused by phosphate. Brautbar *et al.*[7] have suggested that the heart and muscle disease resulting from phosphate depletion probably results from a defective creatine phosphate shuttle. They showed that rats maintained on a phosphate deficient diet become deficient in myocardial inorganic phosphate, creatine phosphate, and adenine nucleotides and exhibit significantly lower levels of total and mitochondrial creatine kinase. A significantly reduced activity of creatine kinase and creatine phosphate shuttle has been reported in diabetic rat hearts[12].

The delivery of energy (creatine phosphate) to protein synthesizing sites is increased by exercise. This would result in hypertrophy and increased activity of the creatine phosphate shuttle. Hypertensive cardiac hypertrophy could occur by a similar mechanism. The increased peripheral vascular resistance in hypertension would increase cardiac contraction resulting *pari passu* in increased protein synthesis and enlarged myocardium.

References

1. Aprille, J.R. & Asimakis, G.K. (1980): Postnatal development of rat liver mitochondria: State 3 respiration, adenine nucleotide translocase activity and the net accumulation of adenine nucleotides. *Arch. Biochem. Biophys.* **201**, 564–575.

2. Aprille, J.R., Yaswen, P. & Rulfs, J. (1981): Acute postnatal regulation of pyruvate carboxylase activity by compartmentation of mitochondrial adenine nucleotides. *Biochim. Biophys. Acta* **675**, 143–147.

3. Bachur, N. (1961): The effect of endogenous substances in the Krebs cycle. *PhD Thesis*, College Park, Maryland: University of Maryland.

4.   Bessman, S.P. (1960): Diabetes mellitus: observations, theoretical and practical. *J. Pediatrics* **56**, 191–203.

5.   Bessman, S.P. (1966): A molecular basis for the mechanism of insulin action. *Am. J. Med.* **40**, 740–749.

6.   Bessman, S.P., Borrebaek, B., Geiger, P.J. & Ben-Or, S. (1978): Mitochondrial creatine kinase and hexokinase – two examples of compartmentation predicted by hexokinase mitochondrial theory of insulin action. In *Microenvironments and cellular compartmentation*, eds. P. Srere & R.W. Estabrook, pp. 111–128. New York: Academic Press.

7.   Brautbar, N., Baczynski, R., Carpenter, C., Moster, S., Geiger, P., Finander, P. & Massry, S.G. (1982): Impaired energy metabolism in rat myocardium during phosphate depletion. *Am. J. Physiol.* **242**, F699–704.

8.   Carpenter, C.L., Mohan, C. & Bessman, S.P. (1983): Inhibition of protein and lipid synthesis in muscle by 2,4-dinitrofluorobenzene, an inhibitor of creatine phosphokinase. *Biochem. Biophys. Res. Comm.* **111**, 884–889.

9.   Lindberg, O. & Ernster, L. (1954): *Protoplasmatologia, Handbuch Der Protoplasmaforschung*, Band 3, pp. 133–136. Vienna, Austria: Verlag.

10.   Mahler, M. (1979): Progressive loss of mitochondrial creatine phosphokinase activity in muscular dystrophy. *Biochem. Biophys. Res. Comm.* **88**, 895–906.

11.   Mohan, C., Geiger, P.J. & Bessman, S.P. (1989): The intracellular site of action of insulin: the mitochondrial Krebs cycle. *Curr. Topics Cell. Regul.* **30**, 105–142.

12.   Savabi, F. (1988): Mitochondrial creatine phosphokinase deficiency in diabetic rat heart. *Biochem. Biophys. Res. Comm.* **154**, 469–475.

13.   Saks, V.A., Ventura-Clapier, R., Huchua, Z.A., Preobrazhensky, A.N. & Emelin, I.V. (1984): A further evidence for compartmentation of adenine nucleotides in cardiac myofibrillar and sarcolemmal coupled ATPase-creatine kinase systems. *Biochim. Biophys. Acta* **803**, 254–264.

Guanidino Compounds in Biology and Medicine, eds. by P.P. De Deyn, B. Marescau, V. Stalon and
I.A. Qureshi. ©1992 John Libbey & Company Ltd., pp. 187–194.

Chapter 27

Bioenergetic engineering with guanidino compounds. Loading tissues with extended-range synthetic thermodynamic buffers

James B. WALKER

Department of Biochemistry and Cell Biology, Rice University, PO Box 1892, Houston, Texas 77251, USA

Summary

Tissues of vertebrates that contain creatine kinase can be loaded *in vivo* or *in vitro* with massive amounts of a variety of synthetic analogues of creatine-P that differ widely in their kinetic and thermodynamic properties. Hearts and skeletal muscles of animal fed cyclocreatine were found to sustain ATP levels markedly longer than controls during subsequent ischaemic episodes; the delay in ATP depletion was manifested by a corresponding delay in onset of rigor in both tissues. This effect was specific for cyclocreatine. Its mechanism appears to be at least in part a consequence of the ability of cyclocreatine-P to continue to regenerate ATP from ADP and conserve the total adenylate pool during late stages of ischaemia, long after coexisting creatine-P reserves have been exhausted.

Introduction

Rapid muscular movements aid in the capture of prey and escape from predators. In both lower and higher vertebrates fast-twitch glycolytic muscles have the highest concentrations of creatine-P and creatine kinase, a phosphagen system that rapidly regenerates ATP and thermodynamically buffers (Stucki) transient changes in the ATP/free ADP ratio in cytosol, while furnishing inorganic phosphate (Pi) for glycolysis, oxidative phosphorylation, and various regulatory functions. Fast-twitch muscles also have high levels of pH buffers to counteract the acidity produced during anaerobic contractions. Comparisons between thermodynamic buffering and pH buffering are given in Table 1.

The creatine-P phosphagen system is thermodynamically poised to buffer the cytosolic ATP/free ADP ratio at a very high value, and therefore can support a series of extremely rapid contractions for a short period of time. However, in an evolutionary trade-off, this phosphagen system cannot continue its thermodynamic buffering function at the lower ATP/free ADP ratios that occur after a large number of rapid attack or escape manoeuvres, or during ischaemic episodes resulting from clamping of tissues by the jaws of a predator. During ischaemia the ATP levels decrease until rigor complexes eventually form between actin and myosin in both cardiac and skeletal muscles, Ca^{2+} ions accumulate, and loss of membrane integrity results in cell death.

The primary purpose of this report is to describe our laboratory's attempts to find a nontoxic synthetic phosphagen: (i) that can be accumulated in large amounts by tissues; and (ii) whose thermodynamic and kinetic properties are such that during ischaemia it can continue to regenerate ATP and buffer the cytosolic phosphorylation potential long after coexisting creatine-P reserves have been depleted. The model systems employed in this search have included: Ehrlich ascites tumour cells[3], mouse leg muscle[4], mouse brain[23,24], rat heart[2,12], chick heart[2,16], chick breast muscle[8,17], and chick embryo brain[23,25], heart[15], and skin[15].

Table 1. Thermodynamic buffering *vs* pH buffering in tissues

	pH buffering	Thermodynamic buffering
Component buffered	H⁺ ion concentration	ATP/free ADP ratio
Natural buffers	Pi, proteins, carnosine	Creatine-P, arginine-P
Buffer capacity	Buffer concentrations	Phosphagen concentration
Buffer kinetics	Instantaneous	Enzyme controlled; k_{cat}, K_m, phosphagen concentration
Buffering range	pK_a of each buffer	$\Delta G^{o\prime}$ of hydrolysis of each phosphagen in cell
Cell protection during ischaemia	Want low pK_a buffer to counteract acidity due to glycolysis	Want phosphagen with low $\Delta G^{o\prime}$ to continue to regenerate ATP late in ischaemia

Results and discussion

Table 2. Methods for altering thermodynamic buffering in tissues

1. Slow the kinetic responses of the phosphagen buffering system by introducing creatine-P analogues that have lower k_{cat}/K_m values with creatine kinase and that also displace endogenous creatine-P. This is equivalent to decreasing the effective concentration of creatine kinase in tissues, and thus is useful for evaluating the physiological importance of the 'creatine-P shuttle'. The nontoxic N-ethylguanidinoacetate is recommended for such an evaluation; its use is equivalent to 97 per cent inhibition of creatine kinase within cells, when it completely replaces endogenous creatine.

2. Increase the thermodynamic buffering capacity by loading tissues with additional creatine-P and/or synthetic phosphagens.

3. Extend the thermodynamic buffering range by loading tissues with creatine-P analogues capable of buffering the ATP/free ADP ratio at lower cytosolic phosphorylation potentials, where the creatine-P system can no longer function. Useful for delaying onset of rigor-contracture, and maintaining ion pumps and cell membrane integrity for longer periods of ischaemia. Several different synthetic phosphagens can be loaded into the same cells.

Table 2 lists some of the ways in which the thermodynamic buffering characteristics of tissues can be experimentally altered by loading *in vivo* or *in vitro* with appropriate synthetic analogues of creatine-P. Creatine analogues might also be expected to interact *in vivo* with all the presently established reactions of creatine depicted in Fig. 1, and our laboratory has reported a number of such studies[14,18,20,21]. Structures of those creatine analogues that have been fed to animals are given in Fig. 2. We have found that many of those creatine analogues were accumulated as their N-phosphorylated derivatives by a variety of tissues *in vivo* and *in vitro*. Coy D. Fitch and coworkers at St Louis University were the first to study the effects of dietary loading of animals with 3-guanidinopropionate-P, and George L. Kenyon and coworkers at the University of California were the first to synthesize cyclocreatine and find that it is an excellent substrate for creatine kinase *in vitro*.

Kinetic properties of synthetic phosphagens

Since the kinetic properties of these thermodynamic buffers are enzyme-controlled (Table 1), it was important to assay a variety of creatine analogues (Fig. 3) and creatine-P analogues (Fig. 4) for their

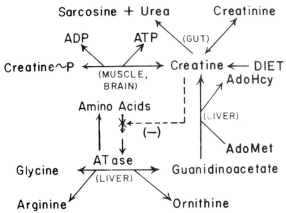

Fig. 1. Known reactions involving creatine (from J. Biol. Chem.)[14].

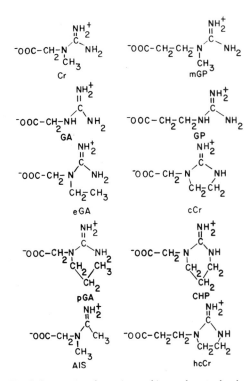

Fig. 2. Structures of creatine and its analogues that have been fed to animals. Abbreviations: Cr, creatine; mGP, N-methyl-3-guanidinopropionate; GA, guanidinoacetate; GP, 3-guanidinopropionate; eGA, N-ethylguanidinoacetate; cCr, cyclocreatine; pGA, N-propylguanidinoacetate; CHP, 1-carboxymethyl-2-iminohexahydropyrimidine; AIS, N-acetimidoylsarcosine; hcCr, homocyclocreatine (from J. Biol. Chem.)[14].

relative kinetic reactivities with rabbit muscle creatine kinase *in vitro* at pH 7.0[13,14]. We found that N-ethylguanidinoacetate-P is the most kinetically reactive synthetic phosphagen, and homocyclocreatine-P is the most kinetically unreactive synthetic phosphagen yet found to be accumulated by tissues (Fig. 4)[13,14].

Thermodynamic properties of creatine-P analogues

The mechanism by which thermodynamic buffering of the ATP/free ADP ratio by phosphagens occurs is shown in Table 3. The method we devised for calculating the midpoint thermodynamic buffering potential of synthetic phosphagens, relative to creatine-P, is illustrated in Table 4[1]. Our method was recently adopted by Ellington to calculate the midpoint buffering ranges of a series of invertebrate phosphagens[7]. Cyclocreatine-P[1,10] and homocyclocreatine-P[13] (cf. Fig. 5) are unique in having the lowest midpoint thermodynamic buffering potentials (analogous to having a low pK_a), with the natural phosphagen arginine-P having the next lowest buffering potential. Theoretically, in cytosol containing both cyclocreatine-P and creatine-P under steady-state near-equilibrium conditions, the cyclocreatine-P/cyclocreatine ratio should always be 30-fold greater than· the creatine-P/creatine ratio[1,10]. This relationship has been confirmed in Ehrlich ascites tumour cells[3], mouse brain[23], and chick breast muscle[8,13].

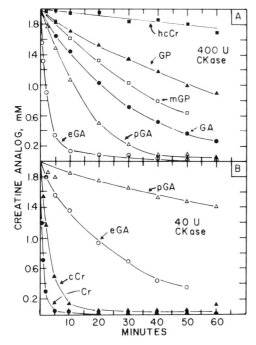

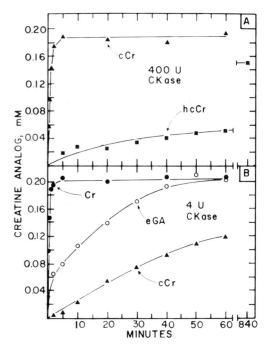

Fig. 3. Relative rates of reaction of creatine analogues as substrates for two different concentrations of creatine kinase (CKase) from rabbit muscle at pH 7.0. Abbreviations as in Fig. 2. (From Arch. Biochem. Biophys.)[13].

Fig. 4. Relative rates of reaction of creatine-P analogues as substrates for two different concentrations of creatine kinase from rabbit muscle at pH 7.0. Abbreviations as in Fig. 2. (From Arch. Biochem. Biophys.)[13].

Table 3. Thermodynamic buffering of the adenylate system

$$ATP^{4-} + H_2O \rightleftharpoons ADP^{3-} + Pi^{2-} + H^+ \tag{1}$$

$$X\text{-}P^{2-} + ADP^{3-} + H^+ \rightleftharpoons X + ATP^{4-} \tag{2}$$

$$\text{Net: } X\text{-}P^{2-} + H_2O \rightleftharpoons X + Pi^{2-} \tag{3}$$

In cells, reactions 1 and 3 are normally displaced 10^8 to 10^{10} from equilibrium; reaction 2 is in near-equilibrium in the steady-state and during ischaemia (usually). Reaction 3 is a formal, not actual, reaction. $\Delta G^{o\prime}$ in kcal/mole for reaction 3: creatine-P, -10.3; N-ethylguanidinoacetate-P, -10.3; cyclocreatine-P, -8.3; homocyclocreatine-P, -8.3; arginine-P, -9.0. Note that Pi does not arise directly from X-P but from ATP.

Table 4. Determination of $\Delta G^{o\prime}$ of hydrolysis of synthetic phosphagens

$$\text{Creatine-}P^{2-} + ADP^{3-} + H^+ \rightleftharpoons \text{creatine} + ATP^{4-} \tag{1}$$

$$X + ATP^{4-} \rightleftharpoons X\text{-}P^{2-} + ADP^{3-} + H^+ \tag{2}$$

$$\text{Net: Creatine-}P^{2-} + X \rightleftharpoons \text{creatine} + X\text{-}P^{2-} \tag{3}$$

Determine concentrations for reaction 3 at equilibrium: (i) *in vitro* with creatine kinase; (ii) *in vivo* after cell loading. Results obtained:

$$\frac{(\text{cyclocreatine-P})}{(\text{cyclocreatine})} = \frac{(30)(\text{creatine-P})}{(\text{creatine})} = \frac{(\text{homocyclocreatine-P})}{(\text{homocyclocreatine})} = \frac{(30)(\text{N-ethylguanidinoacetate-P})}{(\text{N-ethylguanidinoacetate})}$$

Calculate $\Delta\Delta G^{o\prime}$ for reaction 3; from this can calculate $\Delta G^{o\prime}$ of hydrolysis of X-P^{2-}, using $\Delta G^{o\prime}$ for creatine-P hydrolysis of -10.3 kcal/mole as reference standard[1,7,10].

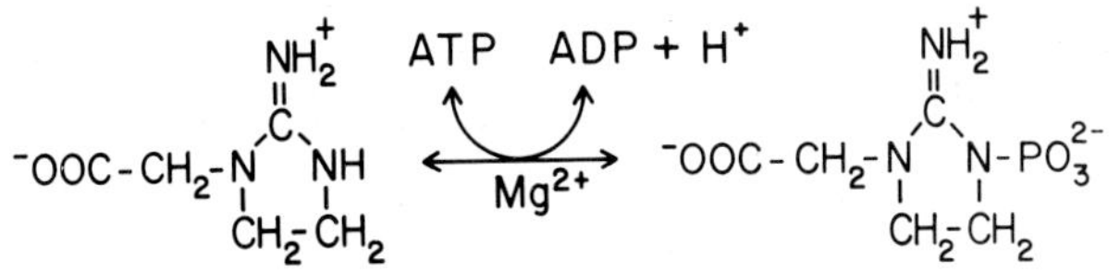

$$\text{}^-OOC-CH_2-\underset{\underset{CH_2-CH_2}{|}}{\overset{\overset{NH_2^+}{\|}}{\overset{C}{\underset{}{N}}}}\overset{}{NH} \quad \underset{Mg^{2+}}{\overset{ATP \quad ADP + H^+}{\rightleftharpoons}} \quad \text{}^-OOC-CH_2-\underset{\underset{CH_2-CH_2}{|}}{\overset{\overset{NH_2^+}{\|}}{\overset{C}{\underset{}{N}}}}\overset{}{N-PO_3^{2-}}$$

Fig. 5. Reversible formation of cyclocreatine-P by creatine kinase. Homocyclocreatine-P has one more -CH$_2$ - in its side chain[13].

Prior cyclocreatine feeding helps sustain ATP levels in ischaemic heart and skeletal muscles

Manual palpation[2], and later more quantitative methods[4,9,12], showed that prior cyclocreatine feeding delayed rigor onset almost twofold during subsequent ischaemia in chick heart[2], mouse leg muscle[4], and rat heart[12]. Hearts of rats made hypothyroid or hyperthyroid also showed rigor delay in response to cyclocreatine[11]. Cyclocreatine feeding resulted in a marked ATP-sustaining activity during subsequent ischaemia in rat heart[12], chick breast muscle (Fig. 6)[17], and in chick heart (Fig. 7)[16]. In all these tissues ATP-sparing action could be correlated with cyclocreatine-P accumulation and its utilization during ischaemia. Accumulation from the diet is slow because serum levels of the analogue seldom exceed 1 mM[8]; intravenous injection or infusion[6] can speed the process.

In vitro accumulation of synthetic phosphagens can be made quite rapid by employing high concentrations of the creatine analogues in the incubation medium[3,15,23]. We have found that intracellular Pi levels in such cells and tissues can be readily manipulated by varying Pi levels in the incubation medium during cyclocreatine-P accumulation.

Cyclocreatine feeding markedly increased the total high energy phosphates in mouse brain[23], resulting in interesting metabolic changes before and during ischaemia[24]. Feeding of N-methyl-3-guanidino-

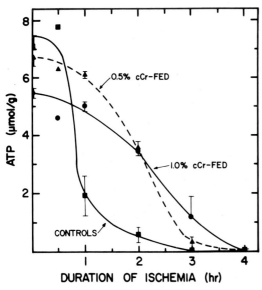

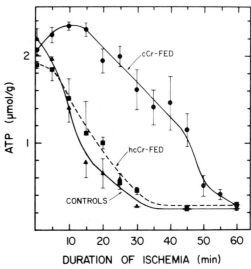

Fig. 6. Changes in ATP levels during total ischaemia in breast muscle of chicks previously fed a control diet, 0.5 per cent cyclocreatine, or 1.0 per cent cyclocreatine for 10–19 days. All chicks were injected with isoproterenol 2 h prior to killing to lower glycogen levels. (From J. Biol. Chem.)[17].

Fig. 7. Changes in ATP levels during total ischaemia in excised hearts of chicks previously fed a control diet, 1 per cent cyclocreatine (cCr) or 5 per cent homocyclocreatine (hcCr) for 10–14 days. (From Arch. Biochem. Biophys.)[16].

propionate likewise increased the total phosphagens in chick brain[14]. Recently transgenic mice engineered to express BB creatine kinase in their liver have been shown to accumulate cyclocreatine-P in liver following cyclocreatine feeding[5]. Assay problems involved in studying tissue extracts containing creatine-P plus synthetic phosphagens have now been overcome with the aid of creatine amidinohydrolase[14].

On mechanisms of the cyclocreatine effect

During ischaemia ATP can be regenerated from ADP by only two means: glycolysis and utilization of phosphagens, accompanied by formation of Pi. Our strategy for elucidating the mechanism of the ATP-sparing action of cyclocreatine feeding during subsequent ischaemia was to determine: (i) the specificity of the effect, i.e. whether any other creatine analogue exhibited ATP-sparing activity; (ii) if a relationship existed between the amount of cyclocreatine-P accumulated and ATP-sparing activity; (iii) what structural or functional properties of cyclocreatine or its phosphate were essential for the effect; and (iv) the importance of any change in glycogen metabolism and glycolysis resulting from cyclocreatine feeding.

Our findings were as follows: (i) among all the compounds whose structures are shown in Fig. 2, only feeding of cyclocreatine resulted in ATP-sustaining activity and/or delay in onset of rigor during ischemia; (ii) the extent of ATP-sustaining activity was linked to the concentration of cyclocreatine-P in tissues, both during feeding of cyclocreatine and after its removal from the diet, although during accumulation a lag period often could be detected which might signal an additional adaptive process; (iii) both cyclocreatine-P and homocyclocreatine-P have the same planar cycloguanidine ring and the same midpoint thermodynamic buffering potential, but accumulation of homocyclocreatine-P by tissues did not result in ATP-sparing activity, nor was homocyclocreatine-P utilized for regeneration of ATP during ischaemia[16] (cf. Fig. 4); (iv) glycogen levels were greatly increased in leg muscle by feeding 3-guanidinopropionate[4], in rat hearts by feeding cyclocreatine[12], and in chick hearts by feeding either cyclocreatine or homocyclocreatine[16], but only cyclocreatine feeding resulted in ATP-sustaining activity during ischaemia.

Suggested mechanisms of the cyclocreatine effect are considered in Table 5.

Table 5. Possible mechanisms of the cyclocreatine effect during ischaemia

Apparently ruled out: Effects due to increased total tissue phosphagens per se; increased glycogen reserves; hyper- and hypothyroidism; increased cellular dianions; pharmacological effects of phosphorylated cycloguanidine ring.

Not ruled out: Effects of changed Pi levels; adaptive effects on endocrine systems; effects on cAMP metabolism, Ca^{2+} levels or storage, free radicals, other cation complexes; effects of neuronal accumulation of cyclocreatine-P; pharmacological effects due to free cyclocreatine formation during ischaemia (cycloguanidine ring is common in drugs).

Application of Occam's razor:
Cyclocreatine-P is the only kinetically competent phosphagen that is thermodynamically poised to continue to regenerate ATP from ADP and conserve the total adenylate pool at the lower cytosolic phosphorylation potentials and more acid pH characteristic of prolonged ischaemic episodes, and thereby to delay onset of localized or general rigor-contracture and eventual cell death due to ATP depletion.

Possible evolutionary precursors of phosphagens in prokaryotes

In many microorganisms including streptomyces, polyphosphates serve as high-energy phosphate reserves. An interesting phosphotransferase system also occurs in streptomyces bacteria that synthesize antibiotics containing guanidinated inositol derivatives (Fig. 8)[19]. During streptomycin biosynthesis the phosphate adjacent to a guanidino group of streptidine-6-P (Fig. 8) is ultimately removed outside the plasma membrane by streptomycin-6-P phosphatase, an enzyme that can readily transfer this phosphate to hydroxyl groups adjacent to basic nitrogenous groups[22]. Streptomycin-6-P can thus theoretically substitute for ATP outside the cell, possibly phosphorylating cell wall components or inactivating aminoglycoside antibiotics produced by other species[22].

192

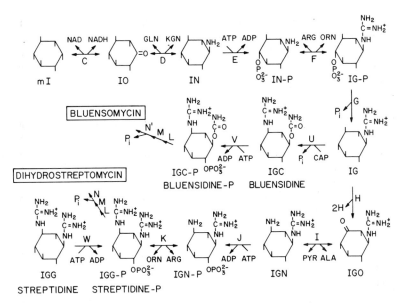

Fig. 8. Enzymatic reactions involved in biosynthesis of guanidinated inositol moieties of two antibiotics by streptomyces bacteria. Cofactors for reaction C are those for streptomyces that can utilize myo-inositol(mI) as the major carbon and energy source. (From J. Bacteriol)[19].

Acknowledgements

This research was supported by the Robert A. Welch Foundation. The experiments described were performed by the following students: Garth Griffiths, Tom Annesley, Dennis Woznicki, Jeff Roberts and David Turner. This paper is dedicated to the memory of Dennis T. Woznicki, Ph.D., M.D.

References

1. Annesley, T.M. & Walker, J.B. (1977): Cyclocreatine phosphate as a substitute for creatine phosphate in vertebrate tissues. Energetic considerations. *Biochem. Biophys. Res. Commun.* **74**, 185–190.

2. Annesley, T.M., Griffiths, G.R. & Walker, J.B. (1978): Anoxic hearts containing cyclocreatine-3-P, an analog of creatine-P, have delayed onset of rigor. *Federation Proc.* **37**, 1608.

3. Annesley, T.M. & Walker, J.B. (1978): Formation and utilization of novel high energy phosphate reservoirs in Ehrlich ascites tumor cells. *J. Biol. Chem.* **253**, 8120–8125.

4. Annesley, T.M. & Walker, J.B. (1980): Energy metabolism of skeletal muscle containing cyclocreatine phosphate. Delay in onset of rigor mortis and decreased glycogenolysis in response to ischemia or epinephrine. *J. Biol. Chem.* **255**, 3924–3930.

5. Brosnan, M.J., Chen, L., Van Dyke, T.A. & Koretsky, A.P. (1990): Free ADP levels in transgenic mouse liver expressing creatine kinase. Effects of enzyme activity, phosphagen type, and substrate concentration. *J. Biol. Chem.* **265**, 20849–20855.

6. Elgebaly, S.A., Allam, M.E., Rossomando, E.F., Cordis, G.A., Forouharg, F., Farghaly, A. & Kreutzer, D.L. (1990): Cyclocreatine inhibits the production of neutrophil chemotactic factors from isolated hearts. *Am. J. Pathol.* **137**, 1233–1241.

7. Ellington, W.R. (1989): Phosphocreatine represents a thermodynamic and functional improvement over the other muscle phosphagens. *J. Exp. Biol.* **143**, 177–194.

8. Griffiths, G.R. & Walker, J.B. (1976): Accumulation of analog of phosphocreatine in muscle of chicks fed 1-carboxymethyl-2- iminoimidazolidine (cyclocreatine). *J. Biol. Chem.* **251**, 2049–2054.

9. Jacobstein, M.D., Gerken, T.A., Bhat, A.M. & Carlier, P.G. (1989): Myocardial protection during ischemia by prior feeding with the creatine analog: cyclocreatine. *J. Am. Coll. Cardiol.* **14**, 246–251.

10. LoPresti, P. & Cohn, M. (1989): Direct determination of creatine kinase equilibrium constants with creatine or cyclocreatine as substrates. *Biochim. Biophys. Acta* **998**, 317–320.

11. Roberts, J.J. & Walker, J.B. (1981): Glycogen content and onset of rigor in hearts from cyclocreatine-fed and hyperthyroid rats. *Federation Proc.* **40**, 1626.

12. Roberts, J.J. & Walker, J.B. (1982): Feeding a creatine analogue delays ATP depletion and onset of rigor in ischemic heart. *Am. J. Physiol.* **243**, H911–H916.

13. Roberts, J.J. & Walker, J.B. (1983): Synthesis and accumulation of an extremely stable high-energy phosphate compound by muscle, heart, and brain of animals fed the creatine analog, 1-carboxyethyl-2-iminoimidazolidine (homocyclocreatine). *Arch. Biochem. Biophys.* **220**, 563–571.

14. Roberts, J.J. & Walker, J.B. (1985): Higher homolog and N-ethyl analog of creatine as synthetic phosphagen precursors in brain, heart, and muscle, repressors of liver amidinotransferase, and substrates for creatine catabolic enzymes. *J. Biol. Chem.* **260**, 13502–13508.

15. Turner, D. M. (1986): Metabolic studies of creatine analogs; ATP-sparing during ischemia in heart and skeletal muscle of chicks fed cyclocreatine. *PhD Thesis*, Rice Univ., Univ. Microfilms Int.

16. Turner, D.M. & Walker, J.B. (1985): Relative abilities of phosphagens with different thermodynamic or kinetic properties to help sustain ATP and total adenylate pools during ischemia. *Arch. Biochem. Biophys.* **238**, 642–651.

17. Turner, D.M. & Walker, J.B. (1987): Enhanced ability of skeletal muscle containing cyclocreatine phosphate to sustain ATP levels during ischemia following *beta*-adrenergic stimulation. *J. Biol. Chem.* **262**, 6605–6609.

18. Walker, J.B. (1979): Creatine: biosynthesis, regulation, and function. *Advances Enzymol.* **50**, 177–242.

19. Walker, J.B. (1990): Possible evolutionary relationships between streptomycin and bluensomycin biosynthetic pathways: Detection of novel inositol kinase and 0-carbamoyltransferase activities. *J. Bacteriol.* **172**, 5844–5851.

20. Walker, J.B., Griffiths, G.R. & Overton, P. (1976): Cyclocreatine, a synthetic energy messenger in heart, brain, and muscle. *Federation Proc.* **35**, 1745.

21. Walker, J.B. & Hannan, J.K. (1976): Creatine biosynthesis during embryonic development. False feedback suppression of liver amidinotransferase by N-acetimidoylsarcosine and 1-carboxymethyl-2- iminoimidazolidine (cyclocreatine). *Biochemistry* **15**, 2519–2522.

22. Walker, J.B. & Skorvaga, M. (1973): Streptomycin biosynthesis and metabolism. Phosphate transfer from dihydrostreptomycin 6-phosphate to inosamines, streptamine, and 2-deoxystreptamine. *J. Biol. Chem.* **248**, 2441–2446.

23. Woznicki, D.T. & Walker, J.B. (1979): Formation of a supplemental long time-constant reservoir of high energy phosphate by brain *in vivo* and *in vitro* and its reversible depletion by potassium depolarization. *J. Neurochem.* **33**, 75–80.

24. Woznicki, D.T. & Walker, J.B. (1980): Utilization of cyclocreatine phosphate, an analogue of creatine phosphate, by mouse brain during ischemia and its sparing action on brain energy reserves. *J. Neurochem.* **34**, 1247–1253.

25. Woznicki, D.T. & Walker, J.B. (1988): Utilization of the synthetic phosphagen cyclocreatine phosphate by a simple brain model during stimulation by neuroexcitatory amino acids. *J. Neurochem.* **50**, 1640–1647.

Guanidino Compounds in Biology and Medicine, eds. by P.P. De Deyn, B. Marescau, V. Stalon and I.A. Qureshi. ©1992 John Libbey & Company Ltd., pp. 195–204.

Chapter 28

Perinatal development of the creatine kinase system in mammalian heart

Renée VENTURA-CLAPIER[1], Jacqueline A. HOERTER[1], Andrei KUZNETSOV[2], Zaza KHUCHUA[2] and Joseph F. CLARK[1]

[1]*Laboratoire de Physiologie Cellulaire Cardiaque, INSERM U-241, Université Paris-Sud, Bât 443 91405 Orsay Cedex, France and* [2]*Laboratory of Bioenergetics, USSR Cardiology Research Center, 3 Cherepkovskaya Street 15A, Moscow 121552, USSR*

Summary

The creatine kinase (CK) pathway provides an efficient system of energy transfer between mitochondria and myofilaments. It involves specific localization of CK isoenzymes at sites of energy production and utilization. Perinatal development of functional activity of myofibrillar CK (MM-CK bound to myofibrils of Triton X-100 treated fibres) and mitochondrial CK (mt-CK bound to mitochondria of saponin-treated fibres) was followed in LV of foetal and adult guinea-pigs and in LV of rabbits ranging in age from 2 days before birth to adulthood. While MM-CK is expressed in rabbits before birth, its fixation and influence on contractile activity appeared progressively after birth; creatine-stimulated respiration was absent at birth and followed the same pattern of increase as myofibrillar-bound CK. Thus, although the main CK-isoforms are expressed at different stages, compartmentation of CK isoenzymes appears in parallel during postnatal development in rabbit. In guinea-pig both isoforms were present and functionally active in myofibrils and mitochondria before birth. This is in line with the known high degree of maturation at birth of the guinea-pig heart as compared to the rabbit heart. Compartmentation of creatine kinase is not linked to the increase in work load and pO_2 occurring at birth. It participates in the complex organization of the highly differentiated mammalian cardiac cell during cellular maturation.

Introduction

Foetal development is accompanied by a sharp increase in total CK activity[15]. The relative content of the different CK isoenzymes varies during perinatal development. The major isoform present during early foetal life is the BB-isoform. Foetal differentiation and maturation of muscle cell is characterized by a decrease in the BB isoform and an increase in the specific muscular form of CK, MM-CK, in skeletal muscle as well as in the cardiac cell. The replacement of B mRNA by M mRNA as the predominant species takes place at an earlier developmental stage in heart when compared to skeletal muscle [31]. The synthesis of MM-CK always precedes that of mt-CK. In heart, mt-CK isoform appears at the end of the foetal life in precocious animals or during postnatal development[15]. Thus, the two main isoforms that are specifically linked to the sites of production (mt-CK) and the sites of utilization (MM-CK) appear at different times during development. However, the presence of the specific isoforms of CK does not illustrate their localization and function in intracellular structures. Studies with subcellular fractionation or histochemical localization have revealed that CK isoenzymes are present in cytosol or bound to intracellular structures at the sites of energy production and energy utilization, such as plasma membranes, sarcoplasmic reticulum, nuclei, myofibrils and mitochondria[29]. It is known in skeletal muscle that the activity pattern determines the fibre type and the metabolic profile. Slow and sustained activity is associated with oxidative metabolism whereas rapid and short activity is associated with glycolytic activity. The organization of the CK system appears different in these two types of muscle: there is an abundance of cytosolic enzymes in glycolytic muscle and compartmentation of isoenzymes in oxidative muscle.

In the myofibrils, CK has been identified as the MM-type isoenzyme and was shown to be localized in the middle of the A-band, i.e. the M-line[37], as well as on the entire thick filament[23]. MM-CK is found near the location of ATPase activity[26] and has been shown to be biochemically and functionally coupled to the myosin ATPase[2,8,27,32,35,36,37]. In cardiac muscle, the MM-CK can rephosphorylate virtually all the intramyofibrillar ADP generated during maximal myosin ATPase activity[27,37] and contraction[36]. Furthermore, myosin ATPase reaction preferentially uses ATP supplied by the CK reaction rather than cytosolic ATP[2,27]. Cytosolic MM-CK is present during foetal life in rat heart, but maturation of the M-line and the presence of CK in myofibrils are detected only after birth[1,6].

In mitochondria, mt-CK is present at the outer surface of the inner membrane. The increased rate of PCr production from mitochondrial ATP has been explained by CK coupling to the ATP/ADP translocase which was suggested to supply ATP directly to CK. This process removes ADP, and thus drives the CK reaction in the reverse direction of PCr synthesis[2,16,17,28]. In cardiac cells, it has been proposed that in addition to the role of energy reservoir for contraction, the various locations of the CK isoenzymes permit microcompartmentation and functional coupling near the sites of energy production and consumption[2,16,28,37].

The role of energy transfer in adult mammalian cardiac cells implies CK isoenzymes compartmentation on their specific fixation sites and functional coupling with ATPases or translocase. Indeed, cellular maturation is accompanied by an increased structural complexity characterized by the development of intracellular organelles and compartments. We have thus followed the functional appearance of the compartmentation of the CK isozymes by using a physiological approach. An extensive study of the functional development of the creatine kinase system in perinatal rabbit heart has already been performed[11]. Here we compare the perinatal development of the CK system in two animals differing in the precocity of their maturational process: guinea-pig which is a precocious animal and rabbit having a postnatal maturation.

Materials and method

Animals

New Zealand rabbits obtained from the 'Institut National de la Recherche Agronomique' were used for this study. Gravid female guinea-pigs weighing 900–1100 g were used at 60–65 days of pregnancy and non-pregnant female guinea-pigs weighing 400–600 g were taken as control. All experimental animals were anaesthetized with intravenous or intraperitoneal pentobarbital sodium according to the recommendations of the Institutional Animal Care Committee (INSERM, Paris) and hearts were rapidly removed. Fibre bundles were dissected from left ventricular papillary muscles.

Mechanical experiments

Triton X-100 treatment induces complete disruption and vesicularization of all cellular membranes resulting in removal of the cytosolic and membrane associated fractions of proteins[36]. Fibres were incubated for 1 h in a relaxing solution (pCa 9, see solutions below) containing 1 per cent Triton X-100 to solubilize the membranes. Force was measured as described previously[20] using a transducer (model AE 801, Aker's Microelektroniks, Horten, Norway). All experiments were performed at 22 °C.

Relaxing (pCa 9, solution A) and activating (pCa 4.5, solution B) solutions were prepared as described previously[36]. Solutions were calculated to contain (mmol) EGTA 10, pCa 9, imidazole 30, pH 7.1, Na^+ 30.6, Mg^{2+} 3.16, and dithiothreitol 0.3; ionic strength was adjusted to 0.16 M with potassium acetate. Rigor solutions were obtained by mixing two solutions of pMgATP 6 and 2.5 in the absence of PCr and pMgATP 6 and 4 in the presence of PCr, at pCa 9.

Rigor tension relaxation experiments were carried out by bathing the fibres in low calcium solutions (pCa 9) containing decreasing amounts of MgATP in the presence or in the absence of PCr. Data were fitted using the Hill equation:

$$T = [MgATP]^{n_H} / (K + [MgATP]^{n_H})$$

where T is relative tension, n_H is the Hill coefficient and K is a dissociation constant. The pMgATP for half maximal relaxation, $pMgATP_{50} = (-\log_{10}K) / n_H$ was calculated for each experiment using linear regression analysis. CK efficacy was calculated as the difference between $pMgATP_{50}$ value in the absence of PCr minus $pMgATP_{50}$ value in the presence of PCr.

Mitochondrial respiration experiments

For determination of mitochondrial respiratory parameters in saponin-skinned fibres we used the method described previously[33], with minor changes. Conditions of the treatment and composition of the incubation medium used made it possible to obtain skinned fibres with functionally intact mitochondria[33]. Fibre bundles were incubated with intense shaking for 30 min in solution S (see below) containing 50 µg/ml of saponin. Bundles were then washed for 10 min in solution R (see below) without high energy phosphates. All the procedures were carried out at 4 °C. Solution S and R contained (mM): EGTA 10 (pCa 7), free Mg^{2+} 3, taurine 20, dithiothreitol 0.5, and imidazole 20 (pH 7.0). Ionic strength was adjusted to 0.16 M by addition of potassium 2-(N-morpholino) ethane sulphonate. Solution S also contained 5 mM MgATP and 15 mM PCr. In place of high energy phosphates, solution R contained 5 mM glutamate, 2 mM malate, 3 mM phosphate and 2 mg/ml fatty acid free bovine serum albumin.

The respiratory rates were determined using a Clark electrode (Yellow Spring Instruments Co., Yellow Springs, Oh., USA) in 3 ml solution R at 22 °C with continuous stirring by sequential substrates addition. The solubility of oxygen at 22 °C was taken to be 460 ng atoms/ml. Respiration rates were expressed as ng atoms of oxygen/min/mg dry wt.

Biochemical determinations

Total CK activity was measured in the left ventricle after Ultra-Turrax homogenization and CK extraction. All enzymatic determinations were performed at pH 7.1 and 22 °C in a reaction medium containing (mM): ADP 0.5, NADP 0.8, glucose 20, AMP 10 (to inhibit myokinase) and PCr 10 in the presence of 2 IU/ml hexokinase and glucose-6P-dehydrogenase at pCa 9.

Isoenzyme fractionation was determined using agarose electrophoresis (1 per cent) performed at 200 V and 4 °C for 1 h. Individual isoenzymes were observed by incubating the gel with a staining solution soaked paper for 30 min. Staining solution contained (mM): MES 22, Mg: acetate 50, glucose 70, AMP 15, NAC 120, ADP 9, NADP 9, PCr 120, 18 IU/ml hexokinase and 6 IU/ml G6PDH at a pH of 7.4. Individual isoenzyme bands can be visualized by observing the fluorescence of NADPH.

Statistical analysis

The results are presented as mean ± SEM. Significant changes were assessed by Student's *t*-test. A value of $P \leq 0.05$ was accepted as the level of significance.

Results

Creatine kinase isoenzymes

Figure 1 shows that the foetal guinea-pig heart at 60 days of gestation exhibits the four creatine kinase isoenzymes. The isoenzymic pattern is close to the adult one. By contrast, in rabbit heart, 1 day after

CK isoenzymes

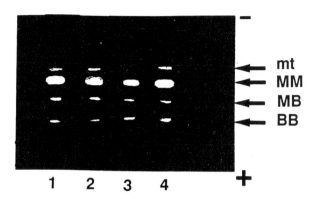

Fig. 1. CK isoenzymes pattern after electrophoresis on agarose gel of foetal and adult guinea-pig heart and newborn and adult rabbit heart. Buffer in mM: Tris/HCl 60, diethyl barbital 10, sodium barbital 50, EGTA 1, DTT 1, 0.1 per cent Triton X-100 (pH 9.0); 200 V, 1 h at 4 °C.
1, Foetal guinea-pig heart (60 days); 2, Adult guinea-pig heart; 3, Newborn rabbit heart (1 day); 4, Adult rabbit heart.

birth there is no mitochondrial CK present. In all animals MM-CK is present whether or not mt-CK is expressed.

Total CK activity was not different from adult, neither in 1-day-old rabbit: 355 ± 20 (n = 10) *vs* $< 372 \pm 14$ IU/g wwt in adult (n = 13), nor in guinea-pig heart 235 ± 23 IU/g wwt (n = 3) in foetus *vs* 356 ± 46 IU/g wwt (n = 5) in adult.

Functional properties of myofibrillar creatine kinase

Since MM-CK is expressed before birth in all mammalian species it is interesting to know when CK is associated with the contractile proteins and able to effectively rephosphorylate MgADP produced by the contractile activity. The next series of experiments was undertaken to study the influence of bound CK on the relaxation of rigor tension.

A stepwise decrease in MgATP in the absence of calcium and PCr, leads to the appearance of rigor tension. In the adult, the presence of PCr in the medium shifts the dependency of the rigor tension on MgATP concentration towards much lower MgATP concentrations[32,35]. This shift is due to local ADP rephosphorylation by bound CK, and can be taken as an index reflecting the functional state of bound CK. Figure 2 shows rigor tension as a function of MgATP concentration in the absence of PCr in the adult rabbit heart and in the presence of 12 mM PCr in 1-day-old and adult rabbit hearts (Fig. 2A) and in foetal guinea-pig heart in the presence and absence of PCr (Fig. 2B). The addition of PCr shifted the sensitivity to MgATP towards lower concentrations in the adult rabbit but not in the newborn. However, foetal guinea-pig heart already exhibited a large left shift of the pMgATP/tension relationship. It can be seen from Fig. 3 that myofibrillar CK efficacy was very low at birth in rabbits (0.26 ± 0.08, n = 4) and increased in the first 2 postnatal weeks to reach adult levels 17 days after birth (1.45 ± 0.06, n = 7). MM-CK is thus present in the cell before birth in the rabbit but not associated with the contractile proteins. In guinea-pig heart however, a few days before birth, MM-CK is present and functionally active: myofibrillar CK efficacy was 1.70 ± 0.03 (n = 3) in foetus compared to the value obtained previously in adult guinea-pig (1.79 ± 0.05, n = 5)[34].

Functional activity of mitochondrial creatine kinase

To study the appearance of the functional activity of CK in mitochondria during development, mitochondrial function was determined in saponin skinned fibres. Maximal respiration rate (V_{max}) is

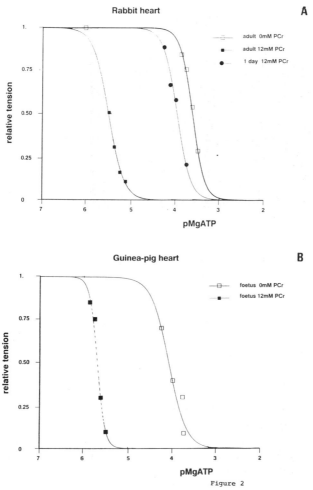

Fig. 2. Normalized pMgATP/tension relationships in adult and 1-day-old rabbits (A) and in foetal guinea-pig (B). Rigor tension was induced by decreasing MgATP concentration in the absence (open symbols) and in the presence (closed symbols) of 12 mM PCr at pCa 9. Data were fitted using the Hill equation.

obtained in the presence of both 1 mM ADP and 20 mM PCr. Respiration can be submaximally stimulated after addition of 0.1 mM ADP. Subsequent addition of 20 mM creatine, which activates mitochondrial CK and increases the ADP concentration in the vicinity of the inner mitochondrial membrane, enhanced respiration rate in the adult rabbit heart but not in the newborn (Fig. 4A). A significant stimulation of respiration is already observed in foetal guinea-pig (Fig. 4B). In guinea-pig hearts the percentage of increase of respiration following creatine addition amounted to 55.6 ± 7.1 per cent in foetal heart while it was 94.2 ± 8.2 per cent in adult guinea-pig. This enhancement indicates the functional efficacy of mt-CK. Creatine stimulated respiration was followed during development in the rabbit. It appeared immediately after birth and increased gradually to reach adult values 17 days after birth (Fig. 5).

Discussion

We have compared the appearance of CK function in myofibrils and mitochondria of guinea-pig and rabbit heart during foetal life and the early postnatal period. In myofibrils the ATP regenerating

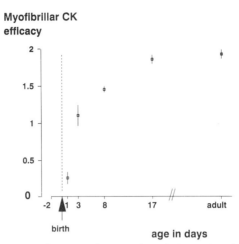

Fig. 3. Myofibrillar creatine kinase efficacy as a function of age during perinatal development in the rabbit. CK efficacy is calculated as the difference between the $pMgATP_{50}$ without PCr minus $pMgATP_{50}$ with PCr (see Fig. 3).

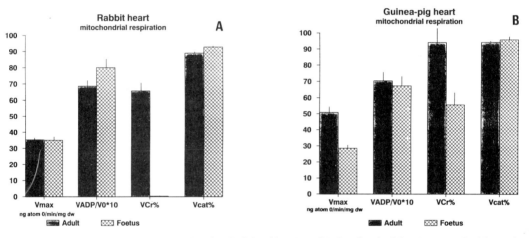

Fig. 4. Mitochondrial respiration rate in foetal and adult rabbit (A) and in foetal and adult guinea-pig (B). Maximal respiration rates (V_{max}) was measured in the presence of both 20 mM creatine and 1 mM ADP. V_{ADP}/V_0 (acceptor control ratio) is the ratio of V_{max} to basal respiration rate in the absence of nucleotide. V_{Cr} is the percentage of increase in respiration rate after addition of 20 mM creatine in the presence of 0.1 mM ADP. V_{cat}, an index of mitochondrial membrane integrity, is the percentage of inhibition of respiration by 35 μM carboxyatractyloside.

potential of bound CK was assessed by the ability of myofibrillar CK to relax rigor tension in Triton X-100 skinned fibres. In mitochondria, the activity of mitochondrial CK was assessed by the ability of creatine to stimulate mitochondrial oxygen consumption in saponin skinned fibres. The results show that the two isoforms associated with myofibrils and mitochondria are expressed at different stages during development in the rabbit. However, the functional activity and binding appear in parallel in the two organelles: before birth in guinea-pig and during the first 2 weeks after birth in rabbit.

Birth is associated with the combination of increased circulating pO_2 and changes in peripheral vascular resistance which increases the work load imposed on the left ventricle. Perinatal maturation of cardiac cells is characterized by a shift from glycolytic to oxidative metabolism and major changes in contractility. Such an adaptation involves both morphological and enzymatic complex modifica-

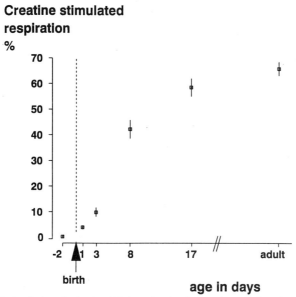

Fig. 5. Percentage of stimulation of mitochondrial respiration by creatine as a function of age in the rabbit left ventricle (see Fig. 4).

tions. Rabbits are born in an immature state, being naked, blind with sealed eyelids, deaf and inactive, while newborn guinea-pigs have hair, open eyelids and can hear and move actively. This contrast in development at birth corresponds to differences in the extent to which the cardiac muscle cells of the respective neonates are differentiated. During perinatal period, cell length and diameter increase as well as mitochondrial and myofibrillar volume; sarcoplasmic reticulum develops and during the late hypertrophic growth of the cell, T tubular system differentiates as an invagination of the sarcolemma. At the end of gestation, the guinea-pig cardiac cell already has a developed sarcoplasmic reticulum and T-tubules and corresponds to the second week after birth in rat and rabbit development [10,13,30].

Early foetal heart depends on high glycolytic and glycogenolytic activity for its ATP turnover[14,18]. At this stage of development, CK is also predominantly cytosolic since no binding to mitochondria or myofibrils was observed. Foetal cytosolic CK could play a role of temporal and spatial buffering of adenine nucleotides[21] as in white skeletal muscle. At birth, as metabolic requirements increase, oxidative metabolism becomes predominant in the cardiac cell. Although total CK activity is nearly constant during perinatal development, qualitative changes occur in the CK system: a mitochondrial isoform is expressed and binding of specific isoforms takes place at the sites of energy synthesis and energy utilization. Compartmentation of the CK system, a characteristic of aerobic muscles, allows rapid integration of metabolism[16] and quick signalling from myofibrils to mitochondria. It also permits rapid adaptation of energy production to energy utilization, a highly necessary process in cardiac tissue.

The most striking observation was that creatine kinase binding and functional coupling appeared in parallel in mitochondria and myofibrils of guinea-pig and rabbit heart, having a prenatal and postnatal maturation respectively. In both animals, the functional activity of mt-CK seems to parallel its expression in the cell. MM-CK binding to myofibrils is not concomitant to its expression in cardiac cells, and seems to parallel the expression of mt-CK.

Considerable species variation exists in mt-CK development. mt-CK appears before birth in precocious animals like the lamb and guinea-pig, and after birth in the mouse and rabbit[15]. In the rat heart, its specific activity still increases from weanling to adulthood[7,8]. Perry *et al.*[24] showed that in rabbit heart mt-CK appears after birth and that the enzymatic activity and CK reaction velocity measured by ^{31}P-magnetization transfer increased as mt-CK increased between 3 and 18 days of age. These results show that respiration was stimulated by creatine in the adult rabbit heart but not in the foetal rabbit heart. After birth, a marked increase in creatine-stimulated respiration occurs reaching the adult value by 17 days. However in guinea-pig heart, mt-CK is already present before birth and functionally coupled to the translocase. This strongly suggests that the increase in pO_2 *per se* is not a major determinant of mt-CK expression.

Myocardial contractility is well known to increase during perinatal development[22]. These quantitative changes are accompanied by a progressive organization of contractile proteins into characteristic cross-striated myofibrils. MM-CK, the major constituent of CK in adult heart, is present as a protein associated with the M-band in cardiac and skeletal muscle but may be also associated with the A-band[23] or with the I-band[37]. Despite the fact that MM-CK is present in high amounts in foetal life, its functional coupling to myofibrillar ATPase appears before birth in guinea-pig heart and after birth in rabbit. At birth in the rat heart, MM-CK is found diffuse in the cytosol and no evidence of M-band[1] nor of CK bound to the myofibrils is observed[1,6]; myofibrillar CK activity is found later during postnatal development[8]. We do not know which factors are involved in CK binding to intracellular structures. The changes in blood pressure and load at birth, do not appear to be the determinants of CK myofibrillar binding[11]. The appearance of myofibrillar CK could be associated with the maturation of myofibrils and appearance of the M-line as shown in the rat heart after birth[6]. At the present time there seems to be only one MM-CK gene. However, Boheler & Dillmann[4] have shown that in rat heart two isoforms of MM-CK were translated *in vitro* and that only one of these isoforms decreased with cardiac hypertrophy suggesting that the transcriptional or translational control of CK is much more complex than previously believed. It is tempting to speculate that a different MM-CK isoprotein differing in its binding properties to myofibrils would be expressed in parallel with the expression of mt-CK.

Thus CK isoenzymes binding and functional coupling to myofibrils and mitochondria appears at a time of intracellular transformation and therefore seems to be more closely related to the general process of cell maturation. At that time, the cell starts its hypertrophic growth and structures involved in excitation–contraction coupling increase their complexity: enhanced calcium release and calcium ATPase activity of the sarcoplasmic reticulum provide an efficient system of calcium handling[13,22]. Mitochondrial volume, surface density of cristae and cytochrome content increases[30,13]. Similarly, CK binding to intracellular structures increased the efficiency of systems involved in ATP turnover. Both changes in Ca handling and energetic turnover contribute to increase myocardial efficiency.

The differentiation and maturation of specialized cells leads to a hierarchy of spatial organization that is structured, integrated and compartmentalized; the appearance of integrated multienzyme systems constitutes a high degree of compartmentation[9]. This compartmentation of reactions allows an increased efficacy since the reactions are not limited by random diffusion of products and substrates and the intermediate reactants of a metabolic chain are directed toward the active sites. Association of CK with myosin ATPase or adenine nucleotide translocase represents such a two-step enzyme system. The importance of the CK-compartmentation in cardiac tissue compared with the buffering effect is illustrated in creatine phosphate depleted hearts, with the small amount of PCr present being able to maintain: high CK fluxes, rapid turnover of ATP and ADP[19], as well as sustained level of mechanical function[12]. Long-term creatine deficiency induces cardiac hypertrophy and isomyosin shifts towards more economical contractile activity; despite these adaptative changes the heart is not able to sustain normal contractile activity[20]. Thus CK-compartmentation is an important metabolic pathway in adult mammalian cardiac cell. This compartmentation occurs during the phase of cellular maturation and could be essential for a rapid perinatal adaptation of the myocardium to increased metabolic demand.

At the present time we do not know what signals are responsible for CK binding to intracellular structures. Events associated with birth *per se*, increased load or pO_2, can be excluded since compartmentation already exists in foetal guinea-pig. It is tempting to speculate that such binding can be hormonally controlled. CK activity has been shown to be hormonally regulated in many tissues. Testosterone regulates the plasticity of CK (mRNa) in levator ani muscle[5]. CK appears at an earlier age in the cerebellum of the rabbit which is a 'perinatal brain developer' compared to the rat, a 'postnatal brain developer'; the developmental differences in CK basal activity and responsiveness to vitamin D metabolites are correlated with differences in cellular growth rates[3]. Creatine kinase activity is stimulated by steroid hormones like oestrogens in rat uterus[25]. It is possible that a common signal associated with cardiac cell growth and maturation triggers the synthesis of mitochondrial and synthesis and/or binding of myofibrillar CK.

Acknowledgements

We thank P. Lechene for engineering assistance, I. Murat and V. Saks for helpful and constructive discussions and G. Vassort for constant support. Supported in part by grants from the AFM (Association Française contre les Myopathies). Z. Khuchua and A. Kuznetsov were supported by the French-Soviet exchange program 'Médecine et Techniques Médicales' between INSERM (France) and the Academy of Medical Sciences (USSR).

References

1. Anversa, P., Olivetti, G., Bracchi, P.G. & Loud, A.V. (1981): Postnatal development of the M-band in rat cardiac myofibrils. *Circ. Res.* **48**, 561–568.

2. Bessman, S.P. & Carpenter, C.L. (1985): The creatine–creatine phosphate energy shuttle. *Ann. Rev. Biochem.* **54**, 831–862.

3. Binderman, I., Harel, S., Earon, Y., Tomer, A., Weisman, Y., Kaye, A.M. & Somjen, D. (1988): Acute stimulation of creatine kinase activity by vitamin D metabolites in the developing cerebellum. *Biochim. Biophys. Acta* **972**, 9–16.

4. Boheler, K.R. & Dillmann, W.H. (1988): Cardiac response to pressure overload in the rat: the selective alteration of *in vitro* directed RNA translation products. *Circ. Res.* **63**, 448–456.

5. Boissonneault, G., Chapdelaine, P. & Tremblay, R.R. (1990): Actin and creatine kinase messenger RNAs in rat Levator Ani and Vastus muscles as a function of androgen status. *J. Appl. Physiol.* **68**, 1548–1554.

6. Carlsson, E., Kjorell, U. & Thornell, L.E. (1982): Differentiation of the myofibrils and the intermediate filament system during postnatal development of the rat heart. *Eur. J. Cell. Biol.* **27**, 62–73.

7. Dowell, R.T. (1986): Mitochondrial component of the phosphoryl-creatine shuttle is enhanced during rat heart perinatal development. *Biochem. Biophys. Res Commun.* **141**, 319–325.

8. Dowell, R.T. (1987): Phosphorylcreatine shuttle enzymes during perinatal heart development. *Biochem. Med. Metabol. Biol.* **37**, 374–384.

9. Hervagault, J.F. & Thomas, D. (1985): Theoretical and experimental studies on the behavior of immobilized multienzyme systems. In *Organized multienzyme systems: catalytic properties*, ed. G. Welch, pp. 381–418. London: Academic Press.

10. Hirakow, R. & Gotoh, T. (1975): *A quantitative ultrastructural study on the developing rat heart developmental and physiological correlates of cardiac muscle*, eds. M. Lieberman & T. Sano. New York: Raven Press.

11. Hoerter, J., Kuznetsov, A. & Ventura-Clapier, R. (1991): Functional development of the creatine kinase system in perinatal rabbit heart. *Circ. Res.* **69**, (in press).

12. Hoerter, J., Lauer, C., Vassort, G. & Gueron, M. (1988): Sustained function of normoxic hearts depleted in ATP and phosphocreatine: a ^{31}P-NMR study. *Am. J. Physiol.* **255**, C192–C201.

13. Hoerter, J., Mazet, F. & Vassort, G. (1981): Perinatal growth of the rabbit cardiac cell: possible implications for the mechanism of relaxation. *J. Mol. Cell. Cardiol.* **13**, 725–740.

14. Hoerter, J. & Opie, L. (1978): Perinatal changes in glycolytic function in reponse to hypoxia in the incubated or perfused rat heart. *Biol. Neonate* **33**, 144–161.

15. Ingwall, J.S., Kramer, M.F. & Friedman, W.F. (1980): Developmental changes in heart creatine kinase. In *Heart creatine kinase*, eds. W.E. Jacobus & J.S. Ingwall, pp. 9–17. Baltimore/London: Williams & Wilkins.

16. Jacobus, W.E. (1985): Respiratory control and the integration of heart high-energy phosphate metabolism by mitochondrial creatine kinase. *Ann. Rev. Physiol.* **47**, 707–725.

17. Jacobus, W.E. & Saks, V.A. (1982): Creatine kinase of heart mitochondria: changes in its kinetic properties induced by coupling to oxidative phosphorylation. *Arch. Biochem. Biophys.* **219**, 167–178.

18. Jarmakani, J.M., Nagatomo, T., Nakazawa, M. & Langer, G.A. (1978): Effect of hypoxia on myocardial high-energy phosphates in the neonatal mammalian heart. *Am. J. Physiol.* **235**, H475–H481.

19. Kupriyanov, V.V., Steinschneider, A.Y., Rhuge, A.Y., Kapelko, V.I., Lakomkin, V.L., Smirnov, V.N. & Saks, V.A. (1984): Regulation of energy flux through the creatine kinase reaction *in vitro* and in perfused rat heart. ^{31}P-NMR studies. *Biochem. Biophys. Acta* **805**, 319–331.

20. Mekhfi, H., Hoerter, J.A., Lauer, C., Wisnewsky, C., Schwartz, K. & Ventura-Clapier, R (1990): Myocardial adaptation to creatine deficiency in rats fed with β-guanidinopropionic acid, a creatine analogue. *Am. J. Physiol.* **258**, H1151–H1158.

21. Meyer, R.A., Sweeney, H.L. & Kushmerick, M.J. (1984): A simple analysis of the 'phosphocreatine shuttle'. *Am. J. Physiol.* **246**, C356–C377.

22. Nakanishi, T. & Jarmakani, J.M. (1984): Developmental changes in myocardial mechanical function and subcellular organelles. *Am. J. Physiol.* **246**, H615–H625.

23. Otsu, N., Hirata, M., Tuboi, S. & Miyazawa, K. (1989): Immunochemical localization of creatine kinase M in canine myocardial cells: most creatine kinase M is distributed in the A-band. *J. Histochem. Cytochem.* **37**, 1465–1470.

24. Perry, S.B., McAuliffe, J., Balschi, J.A., Hickey, P.R. Ingwall, J.S. (1988): Velocity of the creatine kinase reaction in the neonatal rabbit heart: role of mitochondrial creatine kinase. *Biochemistry* **27**, 2165–2172.

25. Reiss, N.A. & Kaye, A.M. (1981): Identification of the major component of the oestrogen-induced protein of rat uterus as the BB isozyme of creatine kinase. *J. Biol. Chem.* **256**, 5741–5749.

26. Saks, V.A., Chernousova, G.B., Vetter, R., Smirnov, V.N. & Chazov, E.I. (1976): Kinetic properties and the functional role of particulate MM-isoenzyme of creatine phosphokinase bound to heart muscle myofibrils. *FEBS Lett.* **62**, 293–296.

27. Saks, V.A., Khuchua, Z.A., Ventura-Clapier, R., Preobrazhensky, A.N. & Emelin, I.V. (1984): Creatine kinase in regulation of heart function and metabolism. 1. Further evidence for compartmentation of adenine nucleotides in cardiac myofibrillar and sarcolemmal coupled ATPase creatine kinase systems. *Biochim. Biophys. Acta* **803**, 254–264.

28. Saks, V.A., Rosenshtraukh, V.N., Smirnov, V.N. & Chazov, E.I. (1978): Role of creatine phosphokinase in cellular function and metabolism. *Can. J. Physiol. Pharmacol.* **56**, 691–706.

29. Sharov, V.G., Saks, V.A., Smirnov, V.N. & Chazov, E.I. (1977): An electron microscopic histochemical investigation of the localization of creatine phosphokinase in heart cells. *Biochim. Biophys. Acta* **468**, 495–501.

30. Smith, H.E. & Page, E. (1977): Ultrastructural changes in rabbit heart mitochondria during the perinatal period. Neonatal transition to aerobic metabolism. *Develop. Biol.* **57**, 109–117.

31. Trask, R.V. & Billadello, J.J. (1990): Tissue-specific distribution and developmental regulation of M-creatine and B-creatine kinase messenger RNAs. *Biochim. Biophys. Acta* **1049**, 182–188.

32. Veksler, V.I. & Kapelko, V.I. (1984): Creatine kinase in regulation of heart function and metabolism. II. The effects of phosphocreatine on the rigor tension of EGTA treated rat myocardial fibers. *Biochim. Biophys. Acta* **803**, 265–270.

33. Veksler, V.I., Kuznetsov, A.V., Sharov, V.G., Kapelko, V.I. & Saks, V.A. (1987): Mitochondrial respiratory parameters in cardiac tissues: a novel method of assessment by using saponin skinned fibers. *Biochim. Biophys. Acta* **892**, 191–196.

34. Ventura-Clapier, R., Mekhfi, H., Oliviero, P. & Swynghedauw, B. (1988): Pressure overload changes cardiac skinned fiber mechanics in rats, not in guinea pigs. *Am. J. Physiol.* **254**, H517–H524.

35. Ventura-Clapier, R., Mekhfi, H. & Vassort, G. (1987): Role of creatine-kinase in force development in chemically skinned rat cardiac muscle. *J. Gen. Physiol.* **89**, 815–837.

36. Ventura-Clapier, R., Saks, V.A., Vassort, G., Lauer, C. & Elizarova, G. (1987): Reversible MM-creatine kinase binding to cardiac myofibrils. *Am. J. Physiol.* **253**, C444–C455.

37. Wallimann, T. & Eppenberger, H.M. (1985): Localization and function of M-line-bound creatine kinase. M-band model and creatine phosphate shuttle. *Cell Muscle Mot.* **6**, 239–285.

Section IV

Clinical importance of creatine-creatinine,
phosphocreatine and creatine kinase

Guanidino Compounds in Biology and Medicine, eds. by P.P. De Deyn, B. Marescau, V. Stalon and
I.A. Qureshi. ©1992 John Libbey & Company Ltd., pp. 207–211.

Chapter 29

Creatine inhibits adenosine diphosphate- and collagen-induced thrombocyte aggregation

J. DELANGHE, M. De BUYZERE and G. BAELE

Central Laboratory, University Hospital Gent, De Pintelaan 185 (1B2), B-9000 Gent, Belgium

Summary

During the early phase of acute myocardial infarction (AMI), temporary increases of serum creatine values can be observed. The post-AMI rise in creatine levels (up to 70 mg/l) shows large interindividual variation and is independent of infarct size. Low creatine values during the early phase of AMI are associated with a higher cardiovascular risk. In view of the aspirin-like anti-inflammatory properties of creatine, the effect of creatinaemia on platelet aggregation was tested. Platelet-rich plasma from healthy volunteers (n = 11) was preincubated with creatine (final concentrations 100 and 200 mg/l) during 30 min at room temperature. Afterwards, ADP (4.27 mg/l) and collagen (1 mg/l) induced platelet aggregation was measured in the absence and the presence of creatine. At creatine concentrations of 100 mg/l, an inhibition of ADP-induced platelet aggregation was noted. However, large interindividual differences were observed. Changes in optical density after 5 min ($\Delta E5$) were 0.40 ± 0.33 (mean $\pm$ SD) *vs* 0.77 ± 0.16 for controls ($P < 0.05$). At a concentration of 200 mg/l, effects were comparable (change in optical density 0.39 ± 0.30). Similarly, collagen induced platelet aggregation was influenced by adding creatine: $\Delta E5$ 0.45 ± 0.35 (100 mg/l) and 0.36 ± 0.21 (200 mg/l) *vs* 0.74 $\pm$ 0.13 for controls ($P < 0.05$). These *in vitro* results suggest that the transient *in vivo* elevations of serum creatine concentrations observed during the early phase of AMI may as well inhibit intravasal thrombocyte aggregation.

Introduction

Creatine is a low-molecular mass compound (Mr 131), abundant in skeletal muscle and myocardial cells[18]. During the early phase of acute myocardial infarction (AMI), temporary increases of serum creatine values can be observed between the first and the seventh hour after infarction. The post-AMI rise in creatinaemia (up to 70 mg/l) shows large interindividual variation and is independent of infarct size. This finding has been used for early biochemical diagnosis of AMI[3-6]. Moreover, the released quantity of creatine largely exceeds the direct creatine losses from the infarcted myocardium, suggesting creatine loss from extra-cardiac tissues to contribute to the observed changes in creatine concentration[5].

Creatine possesses a number of pharmacological properties, such as anti-inflammatory effects, comparable to those of non-steroidal anti-inflammatory drugs[12]. Aspirin, a representative drug of the latter group, has been shown to be effective in the secondary prevention and the treatment of AMI[10]. In ischaemic conditions, creatine reduces intracellular breakdown of ATP[13] and positively influences muscle repair[15]. After AMI, temporary depression of platelet aggregation has been observed during the acute phase. This depression has been related with thrombotic coronary occlusion and the extent of coronary thrombus formation[11]. Spontaneous platelet aggregation *in vitro* was shown to be a useful marker for the prediction of coronary events and mortality in survivors of AMI[17].

Relationship between creatine release after AMI and the prevalence of further major cardiovascular complications and mortality is compared with the prognostic data of known risk factors such as infarct size and patients' age. In view of the aspirin-like anti-inflammatory properties of creatine, the effect of creatinaemia on platelet aggregation was tested.

Materials and methods

A group of 134 patients with retrosternal pain, suggestive of AMI were monitored. Among this group, 96 patients (76 males, age: 63 ± 11 y (mean $\pm$ SD) and 20 females, age: 74 ± 8 y) were finally diagnosed AMI. In 38 patients (24 males, age: 67 ± 8 y and 14 females, age: 76 ± 7 y), final diagnosis was unstable angina. Time interval between onset of symptoms (OOS) and hospital admission was 185 ± 170 min. Blood was sampled on admission, every 30 min during the first 4 h after admission, further on every hour during the next 4 h and finally 12, 16, 20 and 24 h after admission. Urine samples were taken on admission and were further obtained on spontaneous miction during the following 24 h. When insertion of a bladder catheter was necessary for medical reasons, urine collection was performed using the same sampling scheme as for serum samples. Serum and urine creatine were assayed using an enzymatic assay described earlier[4].

Blood was collected from the antecubital veins of healthy volunteers (n = 11; age: 30 ± 11 y) into 0.1 volume of 3.8 per cent of trisodium citrate and centrifuged at $250 \times g$ for 15 min to prepare platelet-rich plasma (PRP). PRP was diluted in 0.15 M sodium chloride until a platelet count of 300,000/mm^3 and preincubated with creatine (up to final concentrations 100 and 200 mg/l) during 30 min at room temperature. Afterwards, adenosine 5'-diphosphate (ADP; 4.27 mg/l) and collagen (1 mg/l) induced platelet aggregation tests were carried out in the absence or the presence of creatine in an Elvi 840 aggregometer (Elvi, Milan, Italy). Collagen reagent was purchased from Hormon Chemie (Munich, Germany), and adenosine 5'-diphosphate was purchased from Sigma Chemical Co (St Louis, MO, USA). Degree of inhibition of aggregation was determined by measurement of change of optical density 5 min after addition of the collagen or ADP (ΔE5).

Results

Post-AMI creatine release

Maximal post-AMI serum creatine concentrations were significantly different from reference range (6.1 ± 2.2 mg/l; $P < 0.005$)[3]. Table 1 summarizes creatine values in serum of AMI patients during the first 12 h after onset of symptoms. Two different patterns of serum creatine kinetics could be distinguished after AMI[5]. In 47 patients (49 per cent), single transient rises in creatinaemia of more than 1 mg/l were observed within the first 6 h (early peak) after onset of symptoms without further increases later on. In another 49 patients (51 per cent), an early peak was followed by a secondary serum creatine peak 20.9 ± 8.1 h after onset of symptoms.

Table 1. Creatine values in serum in 96 AMI patients during the first 12 h after onset of symptoms

Parameter	Value
Maximal serum creatine concentration (mg/l)	13.7 ± 13.3*
Time to peak for serum creatine[†] (min)	192 ± 122
Apparent half time of creatine (h)	5.9 ± 7.1
Time to peak for urine creatine[†] (min)	333 ± 180

Values are given as mean $\pm$ SD; *Significantly different from reference population ($P < 0.005$); [†]Time delay from onset of symptoms.

Creatine and infarct survival

A subgroup of 89 AMI-patients (70 males, 19 females, mean age 62 ± 11 y) who survived AMI for at least 2 months were followed further during 2 years. Three patients survived another AMI, 16 patients died from cardiovascular disease; 66 patients remained without AMI-recidive. Four patients died from non-cardiovascular causes and were excluded from the study. Infarction size was calculated using cumulative CK-MB release according to Roberts[14]. Data on infarction parameters are summarized in Table 2. The group without major cardiovascular events was about 7 years younger than the AMI survivor group with major events. In the latter group, infarction size was greater but showed a very large variation (range: 1 to 45 g CK-MB equivalent). Patients without cardiovascular events showed significantly higher creatinaemia and creatinuria during the early phase of the infarction, despite smaller infarction size. When post-AMI creatinaemia (0–12 h after OOS) exceeded 13 mg/l (twice the median of the reference range), incidence of death and major cardiovascular complications was only 3 out of 41 patients *vs* 16 out of 44 patients with low (< 13 mg/l) serum creatine values ($P < 0.005$, χ^2 analysis after Yates' correction).

Table 2. Infarction parameters, creatine, and outcome after acute myocardial infarction

	Patients without major cardiovascular complications	Patients with cardiovascular complications
Number	66	19
Patients' age (years)	61 ± 12	68 ± 9*
Maximal serum creatine (mg/l)	14.9 ± 7.3	9.7 ± 3.7*
Maximal urine creatine (mg/l)	45.7 ± 72.4	14.0 ± 9.7*
Infarction size (g CK-MB equivalent)	11.5 ± 13.1	17.3 ± 18.5*

Values are given as mean $\pm$ SD; *$P < 0.01$, Mann-Whitney U-test.

Creatine and platelet aggregation

In order to find out whether the protective effect of creatine is due to an inhibition of platelet aggregation, the effect of creatine addition to serum was tested *in vitro*. At final creatine concentrations of 100 mg/l, an inhibition of ADP-induced aggregation was noted. However, large interindividual $\Delta E5$ differences were observed: 0.40 ± 0.33 *vs* 0.77 ± 0.16 for controls ($P < 0.05$). At a concentration of 200 mg/l, effects were comparable: $\Delta E5$ 0.39 ± 0.30. Figure 1 shows the effect of

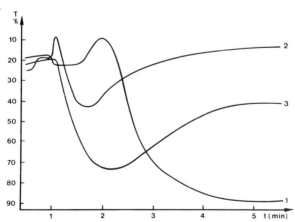

Fig. 1. Effect of creatine on ADP-induced platelet aggregation. Evolution of transmittance (T, per cent) as a function of time (t, min) after addition of ADP. In comparison with the control sample (1), addition of creatine at final concentrations of 200 mg/l (2) and 100 mg/l (3) results in an inhibition of platelet aggregation, followed by a disaggregation.

different creatine concentrations on ADP-induced platelet aggregation as a function of time. Similarly, collagen induced platelet aggregation was influenced by adding creatine: $\Delta E5$ 0.45 ± 0.35 (100 mg/l) and 0.36 ± 0.21 (200 mg/l) vs 0.74 ± 0.13 for controls ($P < 0.05$). Serial dilution of creatine in the range 10–200 mg/l showed that effects of creatine could already be observed at creatine concentrations of 20 mg/l.

Discussion

Acute myocardial infarction is accompanied by release of creatine in serum and urine. Although losses of creatine from the infarcted zone have been observed[5], rises in serum and urine creatine concentration after AMI were significantly higher than the expected changes in creatine concentrations. As expected, patients' age and infarction size are associated with higher incidence of cardiovascular events. However, also low creatine values during the early phase of AMI are associated with a higher cardiovascular risk. This phenomenon cannot be explained by the age difference between both groups as reference creatine values are age-independent[5]. Pharmacological effects of repetitive creatine release in patients showing important post-AMI creatinaemia may be partly responsible for the observed differences in incidence of cardiovascular events. Post-AMI creatine release may be considered as an additional parameter for predicting future major cardiovascular events after AMI. Both ADP- and collagen-induced platelet aggregation can be inhibited by creatine.

Like aspirin and thromboxane synthetase inhibitors[1], creatine is able to inhibit platelet aggregation. Platelet-inhibiting properties have as well been ascribed to structurally related compounds like imidazol and the poisonous guanidine derivatives arcaine (1,4-diguanidinobutane) and diguanidine diphenyl sulphone[2]. Effects may show large interindividual differences, as is the case for aspirin[16]. Effects of creatine on ADP-induced platelet aggregation can be demonstrated from concentrations of 20 mg/l, which are comparable to those observed after AMI. Effects of creatine on platelet count were negligible[8].

These in vitro results suggest that the transient in vivo elevations of serum creatine concentrations observed during the early phase of AMI may as well inhibit intravasal thrombocyte aggregation temporarily. Platelet aggregation plays an important role in the pathogenesis of AMI[9]. Useful effects of the administration of platelet aggregation inhibitors as an adjuvant to thrombolytic therapy have been demonstrated[10]. The association of high creatine release after myocardial infarction and long-term survival can be explained by the effects of creatine on platelet aggregation[7].

Acknowledgements

This work was supported by FGWO Grants N° 33002586 and 33008789

References

1. Bertele, V. & De Gaetano, G. (1982): Potentiation by dazoxiben, a thromboxane synthetase inhibitor, of platelet aggregation inhibitory activity of a thromboxane receptor antagonist and of prostacyclin. Eur. J. Pharmacol. **85**, 331–333.

2. De Gaetano, G., Vermylen, J. & Verstraete, M. (1970): Inhibition of platelet aggregation–its potential clinical usefulness: a review. In Hematological reviews, Vol. 2, ed. J. Anbrus, pp. 205. New York: Marcel Dekker.

3. Delanghe, J., De Buyzere, M., De Scheerder, I., Vogelaers, D., Van Den Abeele, A. & Wieme, R. (1986): Early diagnosis of acute myocardial infarction by enzymatic creatine determination. Clin. Chem. **32**, 1611.

4. Delanghe, J., Robbrecht, J., De Buyzere, M., De Scheerder, I., Vanhaute, O., Baert, M. & Thierens, H. (1988): Enzymatic creatine determinations as early marker for acute myocardial infarction diagnosis. Fresenius Z. Anal. Chem. **330**, 366–367.

5. Delanghe, J., De Buyzere, M., De Scheerder, I., Vogelaers, D., Vandenbogaerde, J., Van Den Abeele, A., Gheeraert, P. & Wieme, R. (1988): Creatine determinations as an early marker for acute myocardial infarction diagnosis. Ann. Clin. Biochem. **25**, 383–389.

6. Delanghe, J. & De Buyzere, M. (1991): Kreatin als frühzeitiger Marker für die Myokardinfarktdiagnostik. *Labor Medizin* **14**, 275–279.

7. Delanghe, J., De Buyzere, M., Leroux-Roels, G. & Clement, D. (1991): Can creatine predict further major cardiovascular complications after infarction? *Ann. Clin. Biochem.* **28**, 101–102.

8. Delanghe, J. (1989): Bijdrage tot de vroegtijdige biochemische diagnosestelling van het acuut myocardinfarct. *PhD Thesis*, University Gent.

9. De Wood, M., Spores, J., Notske, R. *et al.* (1980): Prevalence of total coronary occlusion during the early hours of transmural myocardial infarction. *N. Engl. J. Med.* **303**, 897–902.

10. ISIS-2 (Second International Study Of Infarct Survival) collaborative Group (1988): Randomised trial of intravenous streptokinase, oral aspirin, both, or neither among 17187 cases of suspected acute myocardial infarction; ISIS-2. *Lancet* **ii**, 349–360

11. Kawai, C. (1988): Acute myocardial infarction pathogenesis and implications for treatment: where do we go now? *Int. J. Cardiol.* **18**, 5–14.

12. Khanna, N. & Madan, B. (1978): Studies on the antiinflammatory activity of creatine. *Arch. Int. Pharmacodyn.* **231**, 340–350.

13. Millanvoye-Van Brussel, E., Freyss, M, Griffaton, G. & Lechat, P. (1984): Energy metabolism of cardiac cell cultures during oxygen deprivation: effects of creatine and arachidonic acid. *Biochem. Pharmacol.* **34**, 145–147.

14. Roberts, R., Henry, P. & Sobel. B. (1975): An improved basis for enzymatic estimation of infarct size. *Circulation* **52**, 743–754

15. Seraydarian, M., Artaza, L. & Abbott, B. (1974): Creatine and the control of energy metabolism in cardiac and skeletal muscle cells in culture. *J. Mol. Cell Cardiol.* **6**, 405–413.

16. Szczeklik, A., Gryglewski, R., Grodzinska, L., Musial, J., Serwonska, M. & Marcinkiwicz, E. (1979): Platelet aggregability, thromboxane A2 and malonaldehyde formation following administration of aspirin to man. *Thromb. Res.* **15**, 405–413.

17. Trip, M., Cats, V., van Capelle, F. & Vreeken, J. (1990): Platelet hyperreactivity and prognosis in survivors of myocardial infarction. *N. Engl. J. Med.* **322**, 1549–1554.

18. Walker, J. (1980): Creatine: biosynthesis, regulation, and function. *Adv. Enzymol.* **48**, 177–242.

Guanidino Compounds in Biology and Medicine, eds. by P.P. De Deyn, B. Marescau, V. Stalon and
I.A. Qureshi. ©1992 John Libbey & Company Ltd., pp. 213–222.

Chapter 30

Determination of creatine kinase isoforms in serum – clinical interest in patients with acute myocardial infarction

Jean-Paul CHAPELLE

*University of Liège, Department of Clinical Chemistry, University Hospital, B-35, Sart Tilman, B-4000 Liège,
Belgium*

Summary

Creatine kinase (CK) exists in human tissues as a dimer composed of two subunits, designated M and B. These subunits combine to produce three isoenzymes, termed MM, MB and BB. Moreover, multiple forms (isoforms) of CK-MM and CK-MB have been demonstrated; in particular, at least three MM isoforms (MM_1, MM_2, MM_3) and two MB isoforms (MB_1, MB_2) can be observed in blood as a result of a post-synthetic modification of the M subunits by serum carboxypeptidases. After the release of CK into the blood, the cellular or native isoforms MM_1 and MB_1 are rapidly turned into the modified or serum isoforms MM_2, MM_3 and MB_2. In this study, the MM and MB isoforms were determined at 1 h intervals during the early phase of myocardial infarction (MI) in 15 patients with very short hospitalization delays. MB_1, expressed as a percentage of the total CK-MB fraction, and the ratio of MB_1 to MB_2, reached their peak 1 h after admission; maximum values of MM_1 percentages and of the ratio of MM_1 to MM_3 were recorded after 6 h. Peak levels of the conventional markers, total CK and CK-MB, occurred after 12 h. The fast release of the native CK isoforms, especially MB_1, following myocardial damage was confirmed in 29 patients who underwent coronary bypass surgery, and in 15 patients who received intracardiac electric shocks for treatment of atrial fibrillation. We concluded from this study that the increase of MB_1 in serum is an early indication of myocardial necrosis and therefore, that its determination may be useful for the prompt diagnosis of MI. The other important clinical application of CK isoforms in MI patients is the assessment of successful myocardial reperfusion after thrombolytic therapy.

Introduction

Creatine kinase (CK, EC 2.7.3.2) exists in human tissues as a dimer composed of two subunits, designated M and B[8]. These subunits combine to produce three isoenzymes, termed MM, MB and BB. The various CK isoenzymes are expressed differentially in different tissues. In the human heart, 20–30 per cent of cytoplasmic CK is the MB isoenzyme, and the balance is the MM homodimer. The proportion of CK-MB in the heart exceeds that in any other tissue; in skeletal muscles, the MB isoenzyme represents only 1–4 per cent of the cytoplasmic CK. So the differential diagnosis between myocardial and muscular damage can be made on the basis of CK-MB, especially when expressed as a percentage of total CK[10].

Within the last few years, the presence of different forms of CK in the blood, other than the three classical isoenzymes, has been reported. Two of these atypical forms are macro-CK type 1 (CK-BB bound to immunoglobulin) and macro-CK type 2 (polymeric mitochondrial CK)[15]. Unlike atypical

CK forms, which only appear in serum on rare occasions, serum isoforms to CK-MM and CK-MB are part of the clearance process for CK and are present in all human sera; they should not be considered as variants.

In 1972, Smith[22] observed that the MM isoenzyme could be split into three distinct subbands using polyacrylamide gel electrophoresis; the presence of three MM isoforms in serum specimens was confirmed by Wevers *et al.*[23,24], by prolonging the usual time for standard electrophoresis on agarose gel. Using thin-layer isoelectric focusing (IEF), we determined the isoelectric points (pI) of these three forms as 6.88 (MM_1), 6.70 (MM_2) and 6.45 (MM_3). As to the MB isoenzyme, it appeared to consist of two isoforms, MB_1 and MB_2, with pI of 5.61 and 5.34, respectively[6].

The five major CK isoforms

Examination of human myocardial and skeletal muscles reveals the presence of the MM_1 and MB_1 isoforms[3,13]. These native or pure gene products are post-synthetically modified upon release into the circulation to produce two additional MM and one extra MB isoforms. Carboxypeptidase N is responsible for the removal of the C-terminal amino acid from MM_1 to produce modified MM_2 and MM_3 isoforms in a stepwise fashion[9,16]. Similarly, the cleavage of the C-terminal lysine of the single M subunit of MB_1 produces the modified MB_2 isoenzyme. However, a recent study using dogs showed that the B subunit undergoes C-terminal lysine cleavage by carboxypeptidase N, and that the second isoform of MB actually consists of a dimer in which both M and B subunits have been modified[1]. In this hypothesis, there would be an intermediate product between tissue and fully converted MB forms; but no such intermediate product has been reported till now, perhaps because of the insufficient resolution of the techniques used for the separation of the isoforms. Posttranslational modification of the B subunits should also give rise to at least 3 CK-BB, but no such heterogeneity has been reported in human serum.

The minor CK isoforms

Using high resolution techniques such as IEF[6], we have shown that serum contains at least two isoforms migrating cathodic to MM_1, with pI 7.29 and 7.10, and one isoform migrating anodic to MM_3, and designated as MM_4 (pI 6.25). Other reports suggested that the MM heterogeneity could be even more important[12,25]. Although these additional isoforms could be artefacts of IEF, that is due to denaturation of proteins or interaction of isoforms with ampholytes, some of them can be produced from the three major MM isoforms with oxidized glutathione in a reversible reaction[25]. Clearly, more work is needed to fully understand the clinical significance, if any, of these observations.

CK isoforms in human serum

In healthy individuals, the serum total CK (normal range of activity: 0–120 U/l) mainly consists of the MM isoenzyme. The three MM isoforms are present in normal serum, but MM_3, the final product, is found in proportions three times greater than the initial tissue form, MM_1. Using the most sensitive methods, small amounts of CK-MB can be detected in normal serum; the mean concentration is approximately 2 µg/l, with concentrations slightly higher in men than in women[5]. In normal individuals, an accurate determination of CK-MB isoforms is not easy, because the concentration of this fraction is low. Some authors have however reported that the MB_1/MB_2 ratio in normal serum is roughly equal to 1[17].

Purpose of the study

Important progress in the management of MI patients may result from a better knowledge of enzyme release from the jeopardized myocardium; CK isoforms, in particular, have two main potential clinical applications: early diagnosis of myocardial necrosis and non-invasive detection of coronary artery reperfusion. The main objective of the present study was to assess the clinical utility of CK isoforms as early markers of myocardial damage; in other words, to determine whether CK isoform analysis permits earlier detection of an MI than is possible using the conventional assay for total CK or total

CK-MB. For this purpose, we studied the evolution of the CK-MM and -MB isoforms during the early phase of MI by performing the assays at 1 h intervals in a group of patients with short hospitalization delays. We also investigated the kinetics of MM_1 and MB_1 appearance into the circulation following coronary bypass surgery (CBS) and after cardioversion.

Material and methods

Patient population

Our study population consisted of:

–15 MI patients (12 men and three women, mean age $\pm$ SD: 56 ± 8 years) who were admitted to the Coronary Care Unit (CCU), University Hospital, Liège, Belgium. Acute MI was diagnosed on the basis of a typical clinical history, electrocardiographic evidence, and a characteristic increase and decrease of total CK activity in serum. To be included in the study, patients had to reach the CCU within 4 h following the onset of chest pain and initial CK values had to be < 150 U/l. All patients received a fibrinolytic therapy (intravenous perfusion of 0.8–$1.0 \ 10^6$ units of streptokinase within 30 min) within 15 min following admission. Blood was sampled upon admission, at 1 h intervals during 6 h and after 8, 10, 12, 16, 20, 24, 28 and 32 h;

–30 male patients (mean age $\pm$ SD: 58 ± 9 years), who underwent elective coronary bypass grafting as an isolated surgical procedure at the University Hospital, Liège, Belgium. Saphenous vein or mammarial artery grafts were used in all patients; six patients received four grafts, 14 patients had three grafts, 10 patients two grafts, and one patient one graft giving an average number of grafts of 2.8 ± 0.8 (mean $\pm$ SD). There was no associated cardiac procedure. The operation was performed using cardiopulmonary bypass with a disposable membrane oxygenator primed with Haemacell®. In all cases, moderate hypothermia (28 °C) was used. All coronary anastomoses were completed during a single period of aortic cross-clamping, and myocardial protection was ensured by the infusion of cold potassium cardioplegic solution (4 °C) into the aortic root and by pericardial irrigation with iced saline. The duration of aortic cross-clamping averaged 37 ± 18 min (range 15–71 min); the mean duration of extracorporeal circulation (ECC) was 103 ± 48 min (range 41–246 min). Standard 12 lead electrocardiograms (ECG) were recorded before the operation, and repeated daily for 8 days after the operation. Twenty-nine patients followed a completely uneventful postoperative course; one case showed evidence of perioperative MI (new or deeper Q waves of at least 0.04 s duration)[2]. The 29 uncomplicated cases were allocated into two groups according to ECC duration. In the patients of group I (n = 15, ECC duration < 100 min), the mean duration was 68 ± 16 min. In the patients of group II (n = 14, ECC duration > 100 min), the mean duration was 142 ± 38 min. Peripheral blood specimens were taken preoperatively (A) and at 10 different times during the intervention: after anaesthesia (B); after sternotomy (C); after heparinization (D); 1 min (E) and 20 min (F) after the beginning of ECC; after rewarming of the patients to 30 °C (G); after pulmonary reperfusion (H); at the end of ECC (I); 15 min after the end of ECC (J); at the end of surgery, that is at patient's arrival in the intensive care unit. From this moment (time 0), blood was also drawn every 4th h during 36 h postoperatively and daily until day 5;

–15 patients (11 men and 4 women, mean age $\pm$ SD : 68 ± 15 years) who received intracardiac shocks for treatment of chronic atrial fibrillation (Department of Cardiology, Centre Hospitalier Hutois, Huy, Belgium). Intracardiac shocks were delivered from a multipolar USCI pacing catheter to the proximal ring electrode floating in the right atrium. The patients were subdivided into two groups according to the number of shocks delivered. The patients of group A (n = 9) received one shock per session and the energy delivered was < 400 J. The patients of group B (n = 6) received two or three shocks per session, for cumulated energies ranging from 440 to 1080 J. Venous blood samples were obtained 30 min before initiation of the procedure, and 1, 3, 6 and 24 h after.

For all patients, serum was separated promptly, kept at 4 °C and analysed within 24 h. Ethylene-diaminetetraacetate (EDTA) in a final concentration of 5 mmol/l was added to the samples devoted to the determination of MM and MB isoforms in order to avoid *in vitro* transformations[4].

Biochemical measurements

Total CK

We used an optimized spectrophotometric method[20] (Enzyline® CK NAC; BioMérieux, Lyon, France) to measure total CK activity (reference limits 0–120 U/l) with an automated analyser (ERIS 6170, Olympus-Eppendorf, Hamburg, Germany) at 37 °C. The activity of the total CK-MM isoenzyme was calculated by subtracting the activity of the CK-MB isoenzyme from that of total CK.

CK-MM isoforms

CK-MM isoforms were determined after separation by thin-layer isoelectric focusing on polyacrylamide gel, with specific CK staining, as previously described[6]. The proportions of the MM isoforms were determined using laser densitometry (Ultroscan XL, LKB, Bromma, Sweden).

Total CK-MB

Serum CK-MB (reference limits: 0–6 µg/l[5]) was measured in terms of mass by radial partition immunoassay using an automated analyser (Stratus, Baxter Dade, Miami, FL, USA). The conversion factor between mass (g/l) and activity (U/l) for CK-MB is approximately equal to 1[10].

CK-MB isoforms

CK-MB isoforms were determined after separation by discontinuous electrophoresis on polyacrylamide gel (PROTEAN TM II Slab Cell, BioRad Labs, Richmond, CA, USA). We applied 20 µl of serum and performed the electrophoresis at 4 °C for 5 h at 400 V. The CK subbands were made visible by specific CK staining. The proportions of the cathodal (MB1) and anodal (MB_2) isoforms were determined by laser densitometry (Ultroscan XL, LKB).

Results

Myocardial infarction

Table 1 shows the evolution of the proportions of the three major MM isoforms during the first 32 h following MI. In our series, the maximum percentage of MM_1 (52.4 per cent of the total CK-MM fraction) and the maximum value of the MM_1/MM_3 ratio (4.0) were obtained 6 h following admission. Maximum percentages of MM_2 were recorded after 24 h and MM_3 became the prominent isoform after 32 h. Total CK-MM reached its peaks after 12 h. At that time, MM_1 and the MM_1/MM_3 ratio had already decreased to 38.2 per cent and 1.9, respectively. The following hours were characterized by a rapid and regular decrease of the native isoform and of the MM_1/MM_3 ratio.

From 78.2 per cent of the total CK-MB fraction at the time of admission, MB_1 increased to a maximum value of 94.3 per cent after 1 h (Table 2). During the following hours, there was a gradual shift in which MB_1 decreased, while MB_2 progressively increased. MB_2, which represented only 5.7 per cent of the MB fraction after 1 h, became the predominant isoform after 16 h. The MB_1/MB_2 ratio (initial value: 3.6) reached its peak value (16.5) after 1 h; the ratio regularly decreased in the course of MI, a value close to 1 (equivalent proportions of MB_1 and MB_2) being recorded after 12 h. The ratio fell to 0.1 after 32 h.

Coronary bypass surgery

In our study population of patients who underwent CBS without complication, preoperative total CK activities were slightly above normal (171 ± 355 U/l, mean ± SD). A twofold increase in total CK activities was recorded during the intervention (Fig. 1). In the following hours, total CK regularly

Table 1. Evolution of the total CK-MM isoenzyme activity and of the three major MM isoforms in blood (mean values calculated in 15 patients with short admission delays). MM isoforms are expressed in percent of the total CK-MM fraction

	Time after hospital admission														
	0	1	2	3	4	5	6	8	10	12	16	20	24	28	32
Total CK-MM (U/l)	157	292	724	1166	1692	1818	1937	2239	2324	2593	1985	1750	1451	1157	1033
MM_1 (%)	36.7	38.9	44.1	45.9	47.2	49.5	52.4	45.8	41.5	38.2	32.9	26.0	21.3	16.8	14.1
MM_2 (%)	38.6	42.7	37.9	36.1	35.0	35.0	34.5	37.1	40.3	42.2	46.1	48.5	50.5	45.6	39.7
MM_3 (%)	24.7	18.4	18.0	18.0	17.8	15.5	13.1	17.1	18.2	19.6	21.0	25.5	28.2	37.6	46.2
MM_1/MM_3	1.5	2.1	2.4	2.5	2.7	3.2	4.0	2.7	2.3	1.9	1.6	1.0	0.8	0.4	0.3

Table 2. Evolution of the total CK-MB isoenzyme concentration and of the two MB isoforms in blood (mean values calculated in 15 patients with short admission delays). MB isoforms are expressed in percent of the total CK-MB fraction

	Time after hospital admission														
	0	1	2	3	4	5	6	8	10	12	16	20	24	28	32
Total CK-MB (µg/l)	3	18	51	115	143	166	180	191	208	225	173	143	112	73	45
MB_1 (%)	78.2	94.3	89.5	85.8	81.4	77.2	73.9	70.7	61.0	52.1	45.8	33.7	18.9	13.6	10.8
MB_2 (%)	21.8	5.7	10.5	14.2	18.6	22.8	26.1	29.3	39.0	47.9	54.2	66.3	81.1	86.4	89.2
MB_1/MB_3	3.6	16.5	8.5	6.0	4.4	3.4	2.8	2.4	1.6	1.1	0.8	0.5	0.2	0.2	0.1

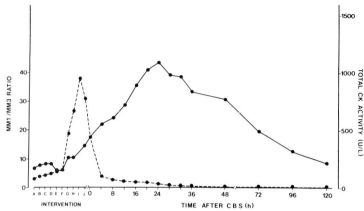

Fig. 1. Evolution of the mean serum total CK activity (—) and of the mean MM₁/MM₃ ratio (----) after CBS in the whole study population (n = 29). Measurement times during the intervention are coded as in Materials and Methods.

rose to reach its peak after 24 h (1084 ± 474 U/l); a slow decrease was observed afterwards, with a return to the initial values on day 5 (220 ± 312 U/l).

Preoperatively, MM_1 averaged 46.3 per cent of the MM fraction ; the mean MM_1/MM_3 ratio was 2.9. During the first steps of CBS, MM_3 increased moderately to a mean value of 52.9 per cent at the beginning of ECC. Similarly, the MM_1/MM_3 ratio increased to 6.1 (Fig. 1). During ECC, we observed a rapid rise in MM_1 (68.8 per cent after pulmonary reperfusion) resulting in a dramatic increase of the MM_1/MM_3 ratio to a value of 37.9 at the end of ECC. After a rapid fall to the initial values during the first post-operative hours, the MM_1/MM_3 ratio regularly decreased to the end of the investigation period. Mean values were 1.4, 1.0 and 0.5 after 24, 72 and 120 h, respectively. During the investigation

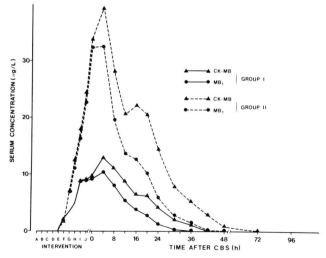

Fig. 2. Comparison of the evolution of the mean proportions of the serum MB₁ and MB₂ isoforms after CBS in the two groups of patients according to ECC duration. Measurement times during the intervention are coded as in Materials and Methods.

218

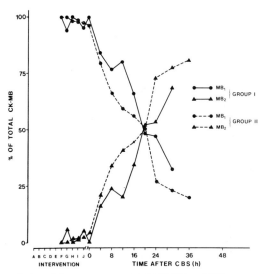

Fig. 3. Comparison of the evolution of the mean serum total CK-MB and MB₁ concentrations after CBS in the two groups of patients according to ECC duration. Measurement times during the intervention are coded as in Materials and Methods.

period, we did not find any significant differences in the proportions of the three MM isoforms and in the $MM_1/MM3$ ratio for the two groups constituted according to ECC duration.

In the majority of the cases, total CK-MB appeared either at the beginning of ECC (31.0 per cent of the patients) or at 20 min on bypass (44.8 per cent). In all patients, CK-MB was detected before the end of ECC. Peak levels occurred at 4 h postoperatively in the two groups (12.9 ± 6.9 µg/l in group I vs 39.5 ± 29.0 µg/l in group II, $P<0.05$). When CK-MB started to appear in the circulation, it entirely consisted of the MB_1 isoform. During the intervention, only traces of MB_2 were detected. In most cases, MB_2 appeared between the end of the intervention and the 8th postoperative hour. From this moment, there was a rapid shift in which MB_1 decreased whereas MB_2 increased comparatively. Equal proportions of MB_1 and MB_2 were recorded approximately 20 h after the intervention. In the following hours, MB_1 rapidly disappeared and represented only 25 per cent of the total MB fraction after 24 to 36 h. As shown in Fig. 2, there were no important differences between the proportions of the MB isoforms in the two groups constituted according to ECC duration.

In the two groups of patients, the rise in MB1, expressed in absolute concentration, was similar to that of total MB until the end of the intervention. In the first postoperative hours, the transformation of MB_1 into MB_2 led to a lower peak concentration for MB_1 comparatively to the total MB fraction (Fig. 3). MB_1 concentrations also returned earlier to normal than total MB levels. MB_1 concentrations were significantly higher in group II than in group I from the end of ECC to the 24th postoperative hour.

Cardioversion

In the overall population of patients who received intracardiac shocks, mean total CK levels increased from 36 ± 27 U/l 30 min before cardioversion to a maximum of 105 ± 68 U/l 6 h following the procedure ($P < 0.05$). In the nine patients of group A (one shock, energy < 400 J), MM_1 reached its peak (51.5 per cent of the total MM fraction) 3 h after cardioversion; in the six patients of group B (two to three shocks, cumulated energy 440–1080 J), MM_1 increased from 38.8 per cent 30 min before cardioversion to a maximum value of 62.5 per cent after 3 h. The maximum values of the MM_1/MM_3 ratio (4.5 in group A and 11.3 in group B) were also recorded after 3 h (Fig. 4).

Total CK-MB was not detected in the patients of group A during the entire investigation period. In

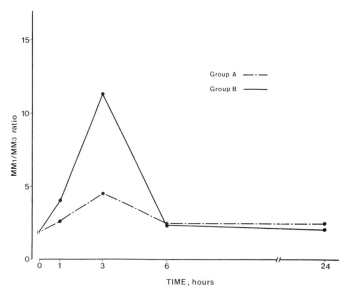

Fig. 4. Comparison of the evolution of the MM_1/MM_3 ratio after intracardiac electric shocks in the two groups of patients, according to the number of shocks delivered.

the patients of group B, CK-MB increased to a mean value of 12.5 ± 4.6 µg/l after 6 h. After 1 and 3 h, CK-MB entirely consists of the MB_1 isoform. After 6 and 24 h, the values of the MB_1/MB_2 ratio were 3.2 and 0.4, respectively.

Discussion

In this study, the changes in serum CK isoforms after MI were analysed by means of measurements performed at 1 h intervals during the acute phase of the disease. We showed that the tissue MM_1 isoform and the MM_1/MM_3 ratio rose above baseline in the first hours following MI and that both reached their peak 6 h after admission (on average, 8 h after onset of symptoms), that is 6 h earlier than total CK and CK-MB. After CBS, serum CK-MM mainly originates from muscles; anyhow, MM_1 increased in the circulation at the beginning of the intervention and the MM_1/MM_3 ratio reached its peak before the end of surgery. The rapidity of the rise of MM_1 after muscular trauma was confirmed by the study of the patients who received intracardiac shocks: the MM_1/MM_3 ratio reached its peak 3 h after the procedure was started.

CK-MB is absent – or present at very low levels – in normal individuals. In our series of MI patients, the initial total CK-MB concentrations were still normal or slightly increased. In contrast, MB_1 already represented 78 per cent of the total MB fraction at admission and increased to 94 per cent after 1 h. The MB_1/MB_2 ratio also peaked at that time. After CBS, MB_1 was also the first MB isoform to appear in the circulation. The highest total CK-MB peaks were recorded in the patients with the longest ECC duration, but ECC duration and the amounts of CK-MB released were without influence on the rate of conversion of MB_1 into MB_2. After cardioversion, CK-MB was only detected in the group of patients who received two or three electric shocks. During the first hours following the procedure, the total MB isoenzyme also entirely consisted of MB_1, and MB_2 only started to appear after 6 h. These investigations demonstrated that, after trauma to the heart, MB is released into the circulation under the MB_1 form.

Conclusions

CK isoforms are clinically useful in the management of an MI because they provide means for specifically monitoring the concentration of CK that has been recently released in the circulation. The native isoforms MM_1 and MB_1 have not yet undergone enzymatic cleavage, and therefore they represent a population of CK molecules that have been circulating for a relatively short time. Among patients with chest pain, this temporal specificity provides an enhanced basis for distinguishing the rapid CK release that occurs after an MI from baseline fluctuation in serum CK activity. Consequently, the first clinical application of CK isoforms is the early detection of myocardial necrosis. Although CK-MM represents the majority of total CK activity after MI, this isoenzyme is generally considered of secondary value for diagnostic purposes because of its lack of specificity. For this reason, and also because the MB_1/MB_2 ratio peaks earlier after MI than the MM_1/MM_3 ratio, MB_1 (or the MB_1/MB_2 ratio) is more promising as an early marker of MI than MM isoforms.

Using high-voltage electrophoresis to separate the isoforms[17], Puleo *et al.*[18] considered the test positive for an MI if the MB_1/MB_2 ratio exceeded 1.5; under these conditions, the false positive rate was in the range of 2–4 per cent. At 4 to 6 h after an MI, the sensitivity of this test for the diagnostic of an MI was 0.92. By comparison, the conventional assay for total CK-MB had a sensitivity in the same time period of only 0.5. By 6 to 8 hours post-MI the sensitivity of the isoform analysis was 1.0, compared with 0.71 for the conventional assay. The authors warned however that the sensitivity of the isoform assay appeared to decline 14–18 h after an MI, and that after this time the CK-MB assay was probably to be preferred.

Detection of myocardial reperfusion after thrombolytic therapy is the second important application of CK isoforms. Early intravenous perfusion with streptokinase is now widely used in AMI patients who reach the CCU within a short time after the attack, in order to recanalize the obstructed coronary artery and, consequently, to limit the infarcted area[21]. Successful fibrinolytic therapy is known to modify the kinetics of enzyme release, because the increased washing-out from the ischaemic area leads to an earlier appearance of the tissue markers in the patient's plasma[11,14]. Recent studies have shown that the analysis of CK isoforms is an earlier indicator and better discriminator of coronary reperfusion than either total CK or CK-MB. For example, the MM_1/MM_3 ratio displays earlier peaks in reperfused patients than in patients in whom thrombolytic therapy is either not undertaken or is unsuccessful. Successful reperfusion is characterized by peak ratios of the MM isoforms 4–6 h after acute onset; peak is delayed until about 6 h later in non-reperfused cases[26]. Differences in the rate of MM_1 decline can also be used to determine successful reperfusion[19]; a rate of decline that is, per hour, > 3.1 per cent is indicative of reperfusion whereas a rate of < 3.1 per cent is indicative of unsuccessful treatment. Recently, Christenson *et al.*[7] reported that the MB_1/MB_2 ratio allowed earlier detection of reperfusion than MB_1 and total CK-MB.

Today, isoform analysis by electrophoresis may allow reliable diagnosis of MI as early as 4 to 6 h following the cardiac event; they also have considerable potential for early assessment of coronary patency. However, because thrombolytic therapy needs to be initiated fewer than 3 h after an MI to be maximally effective, further efforts must be made to develop rapid, sensitive, and specific immunological methods for determining the native CK isoforms – particularly the cardiospecific MB_1 – and to report results on a stat basis.

References

1. Billadello, J.J., Fontanet, H.L., Strauss, A.W. & Abendschein, D.R. (1989): Characterization of MB creatine kinase isoform conversion *in vitro* and *in vivo* in dogs. *J. Clin. Invest.* **83**, 1637–1643.

2. Brewer, D.L., Bilbro, R.H. & Bartel, A.G. (1973): Myocardial infarction as a complication of coronary bypass surgery. *Circulation* **47**, 58–64.

3. Chapelle, J.P. (1984): Serum creatine kinase MM sub-band determination by isoelectric focusing. A potential method for the monitoring of myocardial infarction. *Clin. Chim. Acta* **137**, 273–281.

4. Chapelle, J.P., Bertrand, A. & Heusghem, C. (1981): The protection of creatine kinase MM sub-bands by EDTA during storage. *Clin. Chim. Acta* **115**, 255–262.

5. Chapelle, J.P. & El Allaf, M. (1990): Automated quantification of creatine kinase MB isoenzyme in serum by radial partition immunoassay, with use of the Stratus analyzer. *Clin. Chem.* **36**, 99–101.

6. Chapelle, J.P. & Heusghem, C. (1980): Further heterogeneity demonstrated for serum creatine kinase isoenzyme MM. *Clin. Chem.* **26**, 457–462.

7. Christenson, R.H., Ohman, E.M., Clemmensen, P., Grande, P., Toffaletti, J., Silverman, L.M., Vollmer, R.T. & Wagner, G.S. (1989): Characteristics of creatine kinase-MB and MB isoforms in serum after reperfusion in acute myocardial infarction. *Clin. Chem.* **35**, 2179–2185.

8. Dawson, D.M., Eppenberger, H.M. & Kaplan, N.O. (1975): Creatine kinase. Evidence for a dimeric structure. *Biochem. Biophys. Res. Commun.* **21**, 346–353.

9. Edwards, R.J. & Watts, D.C. (1984): Human 'creatine kinase conversion factor' identified as a carboxypeptidase. *Biochem. J.* **221**, 465–470.

10. El Allaf, M., Chapelle, J.P., El Allaf, D., Adam, A., Faymonville, M.E., Laurent, P. & Heusghem, C. (1986): Differentiating muscle damage from myocardial injury by means of the serum creatine kinase (CK) isoenzyme MB mass measurement/total CK activity ratio. *Clin. Chem.* **32**, 291–295.

11. Golf, S.W., Temme, H., Kempf, K.D., Bleyl, H., Brüstle, A., Bödeker, R. & Heinrich, D. (1984): Systemic short-term fibrinolysis with high dose streptokinase in acute myocardial infarction: time course of biochemical parameters. *J. Clin. Chem. Clin. Biochem.* **22**, 723–729.

12. Guslits, B.G. & Jacobs, H.K. (1983): Investigation of the heterogeneity of the MM isoenzyme of creatine kinase. *Clin. Chim. Acta* **130**, 55–69.

13. Heinbokel, N., Strivastava, L.M. & Goedde, H.W. (1982): Agarose gel isoelectric focusing of creatine kinase (EC 2.7.3.2) isoenzymes from different human tissue extracts. *Clin. Chim. Acta* **122**, 103–107.

14. Kwong, T.C., Fitzpatrick, P.G. & Rothbard, R.L. (1984): Activities of some enzymes in serum after therapy with intracoronary streptokinase in acute myocardial infarction. *Clin. Chem.* **30**, 731–734.

15. Lang, H. & Würzburg, U. (1982): Creatine kinase, an enzyme of many forms. *Clin. Chem.* **28**, 1439–1447.

16. Perryman, M.B., Knell, J.D. & Roberts, R. (1984): Carboxypeptidase-catalyzed hydrolysis of C-terminal lysine. Mechanism for *in vivo* production of multiple forms of creatine kinase in plasma. *Clin. Chem.* **30**, 662–664.

17. Puleo, P.R., Guadagno, P.A., Roberts, R. & Perryman, M.B. (1989): Sensitive, rapid assay of subforms of creatine kinase MB in plasma. *Clin. Chem.* **35**, 1452–1455.

18. Puleo, P.R., Guadagno, P.A., Roberts, R., Scheel, M.V., Marian, A.J., Churchill, D. & Perryman, M.B. (1990): Early diagnosis of acute myocardial infarction based on assay for subforms of creatine kinase-MB. *Circulation* **82**, 759–764.

19. Puleo, P.R., Perryman, M.B., Bresser, M.A., Rokey, R., Pratt, C.M. & Roberts, R. (1987): Creatine kinase isoform analysis in the detection and assessment of thrombolysis in man. *Circulation* **75**, 1162–1169.

20. Rosalki, S.B. (1967): An improved procedure for serum creatine phosphokinase determination. *J. Lab. Clin. Med.* **69**, 696–705.

21. Schwarz, F., Faure, A., Katus, H., Von Olshausen, K., Hofmann, M., Schuler, G., Manthey, J. & Kubler, W. (1983): Intracoronary thrombolysis in acute myocardial infarction: an attempt to quantitate its effect by comparison of enzymatic estimate of myocardial necrosis with left ventricular ejection fraction. *Am. J. Cardiol.* **51**, 1573–1578.

22. Smith, A.F. (1972): Separation of tissue and serum creatine kinase isoenzymes on polyacrylamide gel slabs. *Clin. Chim. Acta* **39**, 351–359.

23. Wevers, R.A., Delsing, M., Klein Gebbink, J.A.G. & Soons, J.B.J. (1978): Post-synthetic changes in creatine kinase isozymes (EC 2.7.3.2). *Clin. Chim. Acta* **86**, 323–327.

24. Wevers, R.A., Olthuis, H.P., Van Niel, J.C.C., Van Wilgenburg, M.G.M. & Soons, J.B.J. (1977): A study on the dimeric structure of creatine kinase (EC 2.7.3.2). *Clin. Chim. Acta* **75**, 377–385.

25. Williams, J., Williams, K.M. & Marshall, T. (1990): Heterogeneity of creatine kinase isoenzyme MM in serum in myocardial infarction: interconversion of the 'normal' and 'abnormal' sub-bands by glutathione. *Clin. Chem.* **36**, 775–777.

26. Wu, A.H.B. (1989): Creatine kinase isoforms in ischemic heart disease. *Clin. Chem.* **35**, 7–13.

Guanidino Compounds in Biology and Medicine, eds. by P.P. De Deyn, B. Marescau, V. Stalon and
I.A. Qureshi. ©1992 John Libbey & Company Ltd., pp. 223–230.

Chapter 31

Determination of creatine kinase activity and related problems in neuromuscular diseases

P. DIOSZEGHY

Department of Neurology and Psychiatry, Medical School, Debrecen, H-4012, Hungary

Summary

Serum creatine kinase (CK) activity has proved to be a useful diagnostic tool for the detection of muscle damage, as it is very sensitive and relatively specific to muscle. CK activity is extremely high in the sera of boys with Duchenne dystrophy, and a negative correlation was found between the age and the CK levels. The determination of CK activity is particularly important for genetic counselling in X-linked muscular dystrophy. Increased serum CK activity detects about 50–60 per cent of heterozygotes. However, carriers with familial predisposition to normal serum CK activity may exist.

The CK levels in the adult types of muscular dystrophy are only slightly increased or normal. The increase is more pronounced in the limb-girdle than in the facioscapulohumeral (FSH) type of muscular dystrophy. There are families with normal and others with elevated CK levels in FSH. In the presence of mononuclear infiltration in the muscle of patients with FSH, the serum CK activity is usually increased. This may cause some difficulties in regard to differentiating between FSH and polymyositis.

In myotonic diseases increased serum CK levels were found in cases of myotonia congenita with autosomal dominant inheritance.

In motorneuron diseases the serum CK concentration is moderately increased or normal. In spite of the rapid progression of muscle wasting in ALS the increase in serum CK activity is surprisingly small. The variable changes in serum CK level in different groups of patients with neuromuscular diseases may be the consequence of different types of muscle membrane damage.

Introduction

The clinical biochemistry of neuromuscular diseases concerns mainly the serum proteins originating from muscle. The main purpose is to find more specific and sensitive indicators of muscle damage[2]. Amongst the serum enzymes, creatine kinase (CK) has proved to be the most valuable and useful diagnostic tool for the detection of muscle damage, since it is generally considered highly sensitive and relatively specific to muscle[16]. It was first used in clinical practice by Ebashi *et al.*[3]. However, many factors have been reported to affect the serum CK levels, which may erode the specificity of serum CK measurements. Usually many factors influence its activity, like age[8,20,23], sex[14], hormonal status[1], puberty[14], race[9], muscle mass[15], occupational or exercise activity[13] and season[17].

CK may originate from other organs, not only from muscle, but the isoenzymes help to distinguish between the different sources. The human muscle contains CK-MM and less than 6 per cent CK-MB, while CK-BB is not detectable in muscle[18]. In a mature cross-striated muscle fibre the CK-MM is one of the protein components of the so-called M-line which is located at the centre of the sarcomer[24]. CK-MM is believed to form the primary m-bridges, which represent part if not all of the electrondense M-line structure, and which can be removed concomitantly with the removal of all measurable CK activity[25]. Recent data suggest a dual role for the M-line-bound CK-MM: attached to a strategically important site of the myofibril, CK-MM might function both as an enzyme involved in energy metabolism (ATP regeneration) and as a structural component in muscle types in which a high degree of filament-order maintenance is required during contraction[4]. Our intention in the study presented here was to survey the practical usefulness of serum CK determination in muscle clinic.

Patients and methods

A total of 169 patients with different neuromuscular diseases were studied (Table 1). In our facioscapulohumeral muscular dystrophy (FSH) cases 10 were familial and five sporadic. In addition, we have examined 96 female relatives of the patients with X-linked muscular dystrophies classifiable as definite (10) or possible (86) carriers according to the criteria outlined by Gardner-Medwin et al.[5]. The data of 46 (21 females and 25 males) healthy volunteers were used as controls. The mean + 2 SD value was taken as the upper normal limit. In carrier screening only the values of female subjects have been taken into consideration as controls.

Table 1. Diagnoses and the age of patients

Diagnosis	No. of patients	Age range (years)
Muscular dystrophies:		
Duchenne and Becker type	49	2–17
Facioscapulohumeral (FSH)	15	14–68
Limb girdle (LG)	10	12–59
Myotonic disorders:		
Myotonic dystrophy (MD)	6	18–53
Myotonia congenita (MC)	7	8–50
Myasthenia gravis (MG)	22	10–61
Neurogenic atrophies:		
Amyotrophic lateral sclerosis (ALS)	34	34–75
Spinal muscular atrophy (SMA)	15	17–74
Hereditary motor-sensory neuropathy (HMSN)	11	15–48

All patients were carefully examined clinically and electrophysiologically. Muscle biopsy was always done and fresh frozen cryostat sections were stained with a battery of histological and histochemical reactions. On frozen sections from the muscle of two patients with Becker (BMD) and six boys with Duchenne muscular dystrophy (DMD) dystrophin was detected immunohistochemically.

Venous blood was always drawn at 9 a.m. after 12 h of bed rest. Serum CK activity was measured by UV spectrophotometric method using monotest CK, NAC-activated (Boehringer).

The mean $\pm$ SD values were calculated. The results were statistically evaluated by Student's t-test and standard correlation; a probability level of $P < 0.05$ was considered statistically significant.

Results and discussion

Our knowledge about the regulation of serum enzyme activity is limited. The most important factors may be the following:

(1) Constant release of soluble proteins from muscle and other cells into the interstitial fluid, thence into lymph, then into plasma.

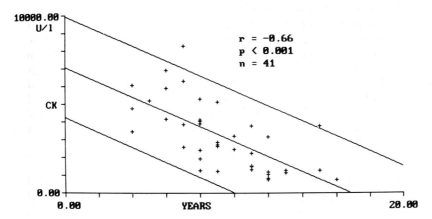

Fig. 1. Relationship between age vs serum CK activities in boys with Duchenne dystrophy.

(a) The release of enzymes from cells partly depends on cellular ATP, but other factors must also be involved.

(b) Binding of soluble enzymes to structural proteins of the muscle may modify release.

(2) Reuptake of proteins into the cells.

(3) Inactivation, degradation of enzymes in the plasma.

(4) Excretion.

The resulting serum enzyme activity is the net balance of all these processes.

It is well known that serum CK activity is extremely high in DMD and BMD type of muscular dystrophies. In our 49 patients with DMD and BMD the range of serum CK was found to be 730–19600 U/l (mean: 3542 ± 2384 U/l). A significant negative correlation was observed between the age of patients and the CK levels (Fig. 1). In pathological circumstances the most important factor of the increase in serum enzyme activity may be the damage of muscle membrane. Presumably two abnormalities may cause increased membrane permeability: (1) physical interruptions of membrane, or (2) functional abnormality of the mechanisms that normally regulate the outflow of soluble proteins from the sarcoplasm. In DMD the possible significance of morphological changes, the presence of

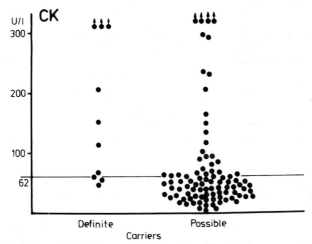

Fig. 2. Serum CK levels in carriers of X-linked muscular dystrophies. The thin line represents the upper normal limit.

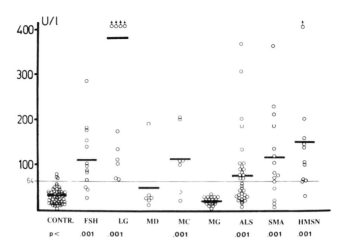

Fig. 3. CK activities in the sera of patients with different neuromuscular diseases. Abbreviations as in Table 1. The thick lines represent the mean values and the thin line shows the upper limit of normal range.

holes on the membrane surface were first described by Mokri and Engel[10]. Others demonstrated functional abnormality of the surface membrane on the basis of indirect evidence. In DMD serum enzyme activity declines after administration of thyroid hormones[19]. In hypothyroidism serum enzyme activity may be increased without any evidence of muscle necrosis. In patients with malignant hyperthermia CK activity may be increased between attacks without muscle weakness and histological abnormalities. These findings suggest a functional rather than morphological alteration. Another implication of importance of the functional disturbances is the existence of 'idiopathic CK-aemia': consistently elevated serum CK levels without family history of muscle diseases, malignant hyperthermia, the lack of muscle weakness and the absence of any electromyographical and histological abnormalities. The 'idiopathic CK-aemia' has an important practical implication: it bears upon the use of CK measurement in the screening of carriers of Duchenne and Becker type of muscular dystrophies and the identification of individuals at risk for malignant hyperthermia.

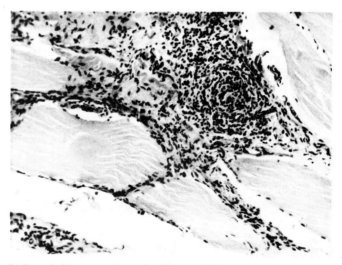

Fig. 4. Anterior tibial muscle biopsy specimen of an 18-year-old boy with FSH. A nodular dense infiltration is present (H & E staining).

The identification of the protein product of Duchenne gene, the dystrophin[6,7], that is a cytoskeletal protein attached to the muscle membrane, gave a new prospect to the study of the role of muscle membrane in the pathomechanism of muscular dystrophies. However, there is no direct connection between membrane permeability and the lack of this protein. In a patient with mild symptoms and signs of Becker type muscular dystrophy, who had nearly normal dystrophin reaction in the muscle membrane, we found extremely high serum CK activity. On the other hand, in patients with Duchenne type muscular dystrophy the dystrophin was missing from the membrane and their serum CK activities were the same or less than in the serum of the previously mentioned patient with Becker dystrophy.

Carrier detection of X-linked muscular dystrophies is particularly important for genetic counselling. Although the new results of molecular genetics made carrier determination more accurate, serum CK measurements have their proper importance even nowadays in carrier screening. The traditional screening method has been the assay of serum CK, which detects only about 52 per cent[12] of heterozygotes, and false-negative results have been obtained in 30 per cent of cases[22]. Three of our 10 definite carriers had normal CK, whereas the others showed elevated values (Fig. 2). There are different explanations for false-negative results. According to Sibert *et al.*[22] definite carriers with familial predisposition to normal serum CK activity do exist. Another reason is that CK activity in carriers decreases with age[21]. Consequently, the possibility of detecting carriers by CK measurement is the greater the younger the presumptive carrier. A total of 86 possible carriers were also studied. The serum CK levels were increased in 27 of them (Fig. 2). This group displayed distinctly high mean CK activity (229 ± 303 U/l, range: 63–1485 U/l) and might be designated as 'probable carriers'.

In adult types of muscular dystrophies the mean CK activities were significantly elevated (Fig. 3). The increase was more pronounced in limb girdle dystrophy (LG) than in FSH. The values were increased in all patients with LG but nearly 50 per cent of patients with FSH had values in the normal range. Consequently, in adults the highly increased serum CK values render the LG dystrophy probable.

In certain families with FSH the CK values were increased without exception, whereas in others they were normal in all affected members. In FSH we have found a special relationship between the serum CK activity and the histological findings. It was previously reported that in muscle specimens of a few patients with clinically typical FSH inflammatory cell infiltrates were present[11]. It is of interest that the histological changes in the affected members of a given family are identical in respect to the presence of mononuclear infiltrates (Fig. 4). This suggests, that the inflammatory infiltrations in FSH muscle may be genetically determined. Parallel with the findings of infiltrations we measured a slight but consequent increase in serum CK levels. This could be the result of a more prominent fibre necrosis

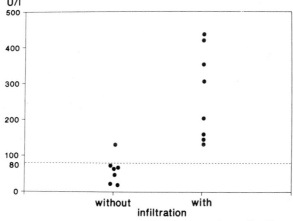

. Serum CK activities of FSH patients with and without mononuclear cell infiltrations in their muscles. The dotted line represents the upper normal limit.

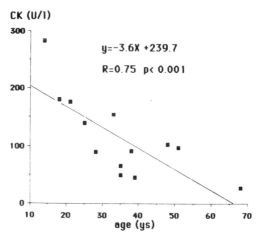

Fig. 6. Relationship between age vs serum CK activities in patients with FSH dystrophy.

in these patients compared to those without infiltrations, where the CK concentrations were in the normal range (Fig. 5). As it was predicted, progression was accelerated and the course of the disease was worse in patients with infiltrations and elevated CK levels. In the clinical practice difficulties may arise with regard to differentiation between sporadic cases of FSH with mononuclear cell infiltrates and polymyositis. When inflammatory cells are present, histopathology cannot promote the establishment of an exact diagnosis. The infiltrations are indistinguishable from the ones found in polymyositis and can be the source of diagnostic difficulties. Hereditary transmission, the characteristic clinical symptoms and signs (early and predominant facial muscle involvement, muscle weakness less marked than expected from the amount of atrophy), the lack of serological abnormalities but moderately elevated CK level may help to exclude polymyositis. Similarly as in X-linked dystrophies in the adult types of muscular dystrophies, the serum CK concentrations significantly decrease with age (Fig. 6). This tendency may be the consequence of waning muscle mass and/or decreasing muscle work during progression[23].

It was surprising to find that increased CK values were found in myotonia congenita: patients with high concentrations belonged to the variety with autosomal dominant inheritance, while the others with normal values proved to be autosomal recessive cases. On the contrary, in all but one of our patients with myotonic dystrophy the CK values were in the normal range (Fig. 3).

In neurogenic muscle atrophies the CK levels were only slightly increased or normal (Fig. 3). The most surprising is that the change in serum CK activity was the most moderate in the most severe and progressive disease, in amyotrophic lateral sclerosis (ALS). Although the mean value increased significantly in ALS, 56 per cent of the patients had normal CK values. In spinal muscular atrophy (SMA) and hereditary motor-sensory neuropathy (HMSN) the increased values were more frequent and the increase in mean values was also higher compared to ALS. In SMA, 2/3 of the patients had higher than normal CK concentrations, while in HMSN the normal values were exceptional. According to the severity of muscle damage and the rate of progression, a striking increase in serum CK activity could be predicted in ALS and SMA. However, compared to muscle dystrophies, it falls behind the expectation. In muscle dystrophies – first of all in the X-linked dystrophies – the abnormalities of muscle membrane are verified, they play an important role in the escape of enzymes. The characteristics of the muscle membrane will change after denervation, too, but in a different way than in muscle dystrophies. In denervated muscle membrane permeability concerning proteins remains unchanged or is only slightly altered. This may explain the mild increase in serum CK in ALS. It is conceivable, that the enzymes are inactivated, catabolized even inside the muscle, consequently they cannot get into the blood.

In conclusion, serum CK estimation is a useful method for the detection of muscle damage in primary and secondary skeletal muscle disorders. It is an important tool for carrier screening in X-linked muscular dystrophies in spite of the new results of molecular genetics.

References

1. Bundey, S., Edwards, J.H. & Insley, J. (1979): Carrier detection in Duchenne muscular dystrophy. *Lancet* **i**, 881.

2. Dioszeghy, P. & Mechler, F. (1989): The significance of simultaneous estimation of serum creatine kinase and myoglobin in neuromuscular diseases. *J. Neurol.* **235**, 174–176.

3. Ebashi, S., Toyokura, Y., Momoi, H. & Sugita, H. (1959): High creatine phosphokinase activity of sera of progressive muscular dystrophy. *J. Biochem.* **46**, 103–104.

4. Eppenberger, H.M. (1984): Expression of muscle-specific isoprotein form of creatine kinase. In *Neuromuscular diseases*, ed. G. Serratrice, pp. 5–8. New York: Raven Press.

5. Gardner-Medwin, D., Pennington, R.J. & Walton, J.N. (1971): The detection of carriers of X-linked muscular dystrophy gene. A review of some methods studied in Newcastle upon Tyne. *J. Neurol. Sci.* **13**, 459–474.

6. Hoffman, E.P., Brown, R.H. & Kunkel, L.M. (1987): Dystrophin: the protein product of the Duchenne muscular dystrophy locus. *Cell*, **51**, 919–928.

7. Hoffman, E.P., Fishbeck, K.H., Brown, R.H., Johnson, M., Medori, R., Loike, J.D., Harris, J.B., Waterson, R., Brooke, M., Sprecht, L., Kupsky, W., Chamberlain, J., Caskey, C.T., Shapiro, F. & Kunkel, L.M. (1988): Characterization of dystrophin in muscle-biopsy specimens from patients with Duchenne's or Becker's muscular dystrophy. *N. Engl. J. Med.* **318**, 1363–1368.

8. Lane, R.J.M. & Roses, A.D. (1981): Variation of serum creatine kinase levels with age in normal females – implications for genetic counselling in Duchenne muscular dystrophy. *Clin. Chim. Acta* **113**, 75–86.

9. Meltzer, H.Y., Donus, E., Greenhans, L., Davis, J.M. & Belmaker, R. (1978): Genetic control of human plasma phosphokinase activity. *Clin. Genet.* **13**, 321–326.

10. Mokri, B. & Engel, A.G. (1975): Duchenne dystrophy: Electron microscopic findings pointing to a basic or early abnormality in the plasma membrane of the muscle fiber. *Neurol.* **25**, 1111–1120.

11. Munsat, T.L., Piper, D., Cancilla, P. & Mednick, J. (1972): Inflammatory myopathy with facioscapulohumeral distribution. *Neurol.* **22**, 335–347.

12. Nicholson, L.V.B. (1981): Serum myoglobin in muscular dystrophy and carrier detection. *J. Neurol. Sci.* **51**, 411–426.

13. Nicholson, G.A., McLeod, J.G., Morgan, G., Meerkin, M., Cowan, J., Bretag, A., Graham, D., Hill, G., Robertson, E. & Sheffield, L. (1985): Variable distributions of serum creatine kinase reference values – relationship to exercise activity. *J. Neurol. Sci.* **71**, 233- 245.

14. Nicholson, L.V.B. & Walls, T.J. (1983): Variation of serum myoglobin levels in normal individuals. *J. Neurol. Sci.* **62**, 41–58.

15. Novak, L.P. & Tillery, G.W. (1977): Relationship of serum creatine phosphokinase to body composition. *Human Biology*, **49**, 375–380.

16. Pennington, R.J.T. (1977): Serum enzymes. In *Pathogenesis of human muscular dystrophies*. ed. L.P. Rowland, pp. 341–349. Amsterdam: Excerpta Medica.

17. Percy, M.E., Andrews, D.F. & Thompson, M.W. (1982): Serum creatine kinase in the detection of Duchenne muscular dystrophy carriers – effects of season and multiple testing. *Muscle Nerve* **5**, 58–64.

18. Prellwitz, W., Kapp, S., Neumeier, D., Knedel, M., Lang, H. & Heuwinkel, D. (1978): Isoenzyme der Kreatine-Kinase: Verteilungsmuster in der Skeletmuskulatur und im Serum bei Erkrankungen sowie Schädigungen der Muskulatur. *Klin. Wschr.* **56**, 559–565.

19. Rowland, L.P. (1980): Biochemistry of muscle membranes in Duchenne muscular dystrophy. *Muscle Nerve* **3**, 3–20.

20. Satapathy, R.K. & Skinner, R. (1978): Serum creatine kinase levels in normal females. *J. Med. Genet.* **16**, 49–51.

21. Scheuerbrandt, G. (1980): Screening for the early detection of Duchenne muscular dystrophy. In *Muscular dystrophy research*, eds. C. Angelini, G.A. Danieli & D. Fontanari, pp. 157–166. Amsterdam: Excerpta Medica.

22.   Sibert, J.R., Harper, P.S., Thompson, R.j. & Newcombe, R.G. (1979): Carrier detection in Duchenne muscular dystrophy. Evidence from a study of obligatory carriers and mothers of isolated cases. *Arch. Dis. Child.* **54**, 534–537.

23.   Smith, I., Elton, R.A. & Thomson, W.H.S. (1979): Carrier detection in X-linked recessive (Duchenne) muscular dystrophy – serum creatine phosphokinase values in premenarchal, menstruating, postmenopausal and pregnant normal women. *Clin. Chim. Acta,* **98**, 207–216.

24.   Wallimann, T., Pelloni, G., Turner, D.C. & Eppenberger, H.M. (1978): Monovalent antibodies against MM-creatine kinase remove the M-line from myofibrils. *Proc. Natl. Acad. Sci.* **75**, 4296–4300.

25.   Wallimann, T., Turner, D.C. & Eppenberger, H.M. (1977): Localization of creatine kinase isoenzymes in myofibrils I. Chicken skeletal muscle. *J. Cell. Biol.* **75**, 297–317.

Guanidino Compounds in Biology and Medicine, eds. by P.P. De Deyn, B. Marescau, V. Stalon and I.A. Qureshi. ©1992 John Libbey & Company Ltd., pp. 231–238.

Chapter 32

Creatine kinase and its isoenzymes as indices for muscle damage

P.R. BÄR and G.J. AMELINK

Department of Neurology and Janus Jongbloed Research Centrum, University of Utrecht, The Netherlands

Summary

Creatine kinase is an enzyme present in many tissues, from which it may leak into the bloodstream. When tissue is damaged, for example the heart during an acute myocardial infarct, or skeletal muscle after strenuous exercise or as a consequence of a muscle disease, a higher than normal activity of creatine kinase can be observed in the circulation. This phenomenon is often used to detect tissue damage at an early stage to start treatment as soon as possible (heart) or to diagnose conditions which lead to muscle degeneration (excessive exercise, muscular dystrophies). Analysis of the isoenzymes of creatine kinase may yield useful extra information as to the source of the creatine kinase activity in the circulation. In this review we will discuss several ways in which provoked muscle damage caused by a standardized amount of exercise can be used to diagnose or recognize several disorders, and to assess the resistance of muscle against mechanical and metabolic stress. First, the general principle and the validity of this approach will be demonstrated in an animal exercise model in which exercise conditions and drugs can be tested for their influence on exercise-induced muscle damage. Second, an exercise test for humans will be discussed that has been used to assess damage due to different causes: during disease, after training or as a side effect of certain drugs. In the conclusion we will reflect on the use of these tests and the combination with modern, non-invasive methods.

Introduction

Creatine kinase (CK) is an ubiquitous enzyme that catalyses the transfer of a phosphate group between the two energy-rich substances phosphocreatine (PCr) and ATP. PCr is the energy store that is used first during sudden, intense exercise. ATP is formed from PCr and ADP during the first few seconds, while the enzymatic machinery of the cell gets ready to produce more ATP by increasing the flux through the glycolysis and the Krebs cycle. Muscle especially, on which this type of sudden demand is more or less normal, contains very high levels of CK in its cytoplasm, very near to where the action is, i.e the myofibrils[30].

CK is a dimeric molecule with a molecular weight of 82.000 and consists of two subunits (each MW 41.000): CK-M, the muscle type, and CK-B, the brain type. Thus, three different combinations exist, resulting in three recognizable isoenzymes – CK-MM, CK-MB and CK-BB. Many tissues show a more or less characteristic isoenzyme pattern: skeletal muscle contains more than 98 per cent CK-MM, although under certain conditions (e.g. after endurance training or in neuromuscular disease) CK-MB and CK-BB may be present[7,28,29,33]. Heart muscle contains about 20 per cent CK-MB, and is thus the most important source of CK-MB in the circulation. The early diagnosis of an acute myocardial infarct (AMI) is often confirmed by measuring CK-MB in serum. Brain contains almost exclusively CK-BB. CK-BB from brain does not easily enter the circulation because of the blood–brain barrier. Only under certain conditions (disease, strenuous exercise), can CK-BB be found in both muscle and the circulation[7]. The isoenzyme profile in human serum reflects this tissue distribution of CK isoenzymes: in serum more than 97 per cent is CK-MM (from skeletal muscle) and only traces of CK-MB and CK-BB are found. This profile is different in the rat as rat blood platelets contain CK-BB (see below).

When CK enters the circulation, it will be attacked by a circulating enzyme, carboxypeptidase N[38]. This enzyme removes a lysine molecule from the C-terminus of the M-subunit (producing M⁻). In this way, a set of isoforms can be formed: starting from CK-MM (also called MM$_1$) CK-MM (MM$_2$) and CK-M⁻M⁻ (MM$_3$) can be formed, and starting from CK-MB (MB$_1$), which has only one M-unit to modify, only one isoform is known, CK-M⁻B (MB$_2$). As the removal of lysine results in a changed behaviour in an electric field, these isoforms can all be separated. The fact that isoforms are formed only after entry of MM$_1$ or MB$_1$ into the bloodstream has been used to estimate the time of an AMI by measuring the ratio MM$_3$/MM$_1$. This aspect will be discussed in more detail elsewhere (Chapter 30). For more details on the clinical biochemistry of CK we refer the reader to a recent review[19].

Animal studies

Exercise, especially strenous and unaccustomed exercise, may lead to an increased activity of CK and other enzymes in serum[13,25]. Since the first report on CK activity in 1965 by Vejjaajiva and Teasdale[37] a large number of studies have been published on this subject[32]. Generally, CK activity peaks 24–48 h after prolonged exercise such as long distance running (e.g. marathon), and can reach values up to thousands of U/l. After eccentric exercise even of relative short duration much larger elevations have been described[31]. Interestingly, after eccentric exercise the CK activity peaks even later, usually not before 96 h after exercise. It is predominantly CK-MM that is responsible for this post-exercise increase in CK activity[33]. However, after long distance running CK-MB and CK-BB may also contribute to the increase in CK activity. It is generally accepted that in the absence of AMI all three isoenzymes are derived from skeletal muscle sources, because an increased ratio of CK-MB and CK-BB (which is virtually absent in normal skeletal muscle) is found in skeletal muscle of trained long-distance runners[7]. This is thought to reflect continuous degeneration and regeneration of muscle in these persons.

In animals CK elevations after exercise have also been well documented especially in rats, after eccentric exercise[8], after concentric exercise[2,3,6] and in an *in vitro* model[4]. Interestingly, in rats the CK-isoenzyme pattern in serum or plasma is quite different from the pattern observed in humans: 85–90 per cent of the CK activity in sedentary rats is CK-BB, probably derived from platelets, as rat platelets contain considerable amounts of CK-BB[3]. The remaining 10–15 per cent consists almost entirely of CK-MM. Although the isoenzyme patterns at rest do not differ between the sexes, after exercise CK-MM increases almost sevenfold in males, whereas in females only a twofold increase is seen (Fig. 1). In both sexes only a 40 per cent increase in CK-BB is seen[3]. This is in keeping with the sex-linked difference in CK(-MM) release after exercise in humans, which has been shown to be larger in males in several studies[15,33,35]. To test the hypothesis that oestrogens are involved in this sex-linked difference several studies were undertaken in an *in vivo* model of a running rat and an *in vitro* model of isolated rat soleus muscles. It was shown that ovariectomy in female rats enhanced the exercise-induced increase in CK[2]. Oestradiol treatment of males attenuated the exercise-induced CK response[12]. These observations were confirmed in an *in vitro* model with isolated soleus muscles of male and female rats: pretreatment of male rats with oestradiol attenuated the CK efflux *in vitro*, whereas soleus muscles of ovariectomized female rats showed an augmented CK release that could be curtailed by oestradiol supplemention[4]. Recently it has been shown that pretreatment with tamoxifen, a (partial) oestrogen antagonist, also diminishes the release of CK from rat soleus muscle *in vitro*[20,21].

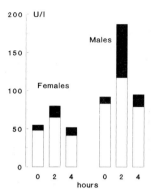

Fig. 1. CK activity (units/litre) in serum after 2 h running exercise of rats: the total height of the bars represents the total CK-activity, the hatched part represents CK-BB activity and the black part CK-MM activity. Measurements were done before (0), immediately after (2) and 2 h after exercise (4).

In the exercising rat paradigm we used muscle specific CK-MM release after exercise to study the possible beneficial effect of dantrolene sodium on exercise-induced muscle damage. Dantrolene sodium is a muscle relaxant that affects the calcium flux over the membrane of the sarcoplasmic reticulum (SR). We found that treatment with dantrolene sodium reduced the CK(-MM) increase after exercise. The treated animals also showed significantly less histological evidence of muscle damage. From this study it was concluded that calcium release from the SR is important in the pathogenesis of exercise-induced muscle damage[5]. The effect of vitamin E deficiency on exercise-induced muscle damage was also studied. A vitamin E deficient diet was found to result in an augmented CK(-MM) release after exercise especially in male rats. In male rats the increase in post-exercise plasma CK activity coincided with a marked increase in histological signs of muscle damage[6].

Clinical studies

The use of CK as a marker of muscle damage has a long history and it is an important part of the diagnostic procedure for patients with muscle weakness. A (very) high CK activity is found in serum, in myopathic disorders such as myositis and muscular dystrophies. In neurogenic disorders no or small to moderate increases in CK are usually found. We will not dwell long on this aspect, as it is treated in more detail elsewhere in this volume (Chapter 31). Rather we will give examples of the use of a diagnostic exercise test that was originally developed to detect Duchenne (DMD) carriers. The identification of the DMD-gene product, dystrophin[24], has led to new ways to detect patients and carriers. However, carrier detection was and is still helped by the fact that carriers display the same defect as DMD boys in some of their muscle fibres and, thus, show a higher than normal CK level at rest. Unfortunately, the sensitivity of this method is 50 to 70 per cent at best. We therefore attempted to develop a test based on the assumption that the existing but small difference in muscle membrane strength between non-carriers and carriers may be enlarged by exercise. This was shown to be the case for CK leakage after exercise, but the level of exercise necessary to obtain the desired separation between the two populations was extreme, and therefore unpractical as a standard test[23]. We had reasons to believe that myoglobin would be a better, i.e. more sensitive marker of muscle membrane permeability[9,11] and tested this hypothesis using a milder exercise regimen. The results are described below.

Myoglobin is the oxygen carrier in skeletal and heart muscle, to which it gives their red colour, and is not found outside these tissues. It is a small molecule (MW 17.500) and located mainly in the I-band region (as O_2-shuttle) and also near the Z-lines and near the outer membrane of the mitochondria[27]. Its release into the circulation seems to precede that of CK, which may be caused by its smaller size. When released into the circulation it is rapidly cleared by the kidneys and excreted via the urine. This rapid appearance and disappearance seems to make myoglobin a good marker for acute muscle damage, rather than for chronic states of muscle damage, such as in certain neurological diseases[1]. Kagen[26] was the first to see the use of this early release in recognising AMI, and subsequently it has been used as a more general marker for muscle damage. Borleffs *et al.*[14] tested whether myoglobin could be used to follow the efficacy of treatment of patients with polymyositis: they measured both CK and myoglobin during treatment with corticosteroids and found that CK activity in serum was the better index for this purpose. With its relatively long half-life in serum, CK does not disappear as quickly as myoglobin. The authors concluded that myoglobin is the better index for acute muscle damage: it appears within 30 min after exercise[10], but disappears rapidly. CK, in contrast, appears later after exercise[9], but remains in the circulation longer. Thus, CK is a good indicator for chronic states of muscle damage, and as such better suited to follow up, e.g. the effect of drug treatment or training on muscle damage.

Driessen *et al.*[17] measured both CK and myoglobin in an exercise test, originally set up for DMD carrier detection. From Fig. 2 it can be seen that with this relatively mild exercise protocol, the CK leakage of DMD carriers is only slightly larger than that of non-carriers. The spread in the results would hamper correct detection of carriers, or in a broader sense, it would hamper the discrimination between the population with normal, physiological CK leakage and that of people with an increased permeability of the muscle membrane. Figure 2 shows that for myoglobin this test confirms the earlier statement: it is a better marker for acute muscle damage and gives a far better separation of the two populations than CK[17]. For DMD carrier detection, with the sophisticated methods that are now available, this test may still be used as a prescreening. More importantly, it may very well be used in more general instances where a question arises about increased permeability or enhanced suscepti- bility of the muscle membrane to damage.

Another example of the use of leakage of CK and myoglobin from muscle was found in a series of patients with chronic progressive external ophthalmoplegia (CPEO), a mitochondrial disease. CPEO patients often experience exercise intolerance. They are diagnosed based on their history and a combination of laboratory findings: an electromyograph with signs of myopathy, ragged red fibres in a muscle biopsy and (under EM) abnormalities in the mitochondria, and sometimes a higher than normal lactate level in serum. Interestingly, CK activity at rest is normal or only mildly elevated. Seven CPEO patients (with normal CK in rest) were asked to perform a long-term exercise test[15],

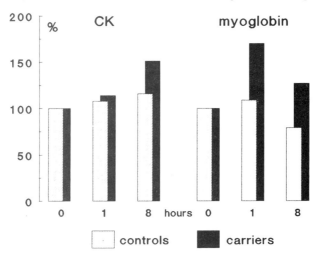

Fig. 2. The activity of CK (left part) and the concentration of myoglobin (right part) in serum of 54 control women (hatched bars) and 23 carriers of Duchenne's muscular dystrophy (black bars), measured before (0), and 1 and 8 h after a 45-min bicycle exercise. The changes in CK and myoglobin are given as per cent change with respect to the values immediately before exercise.

consisting of a 2-h cycle ride on an ergometer at 50 per cent of each individual's maximal capacity. Before, during and after this exercise test blood samples were taken and analysed for a series of metabolites from the energy metabolism (glucose, FFAs, ketone bodies, lactate, pyruvate), and for markers of muscle damage (CK, myoglobin and aldolase). Moreover, muscle biopsies were taken to analyse the carnitine level, and the functional, phosphorylative capacity of mitochondria. Patients and controls showed almost identical values for most indices studied. The only parameters that differed were the lactate production during exercise and the leakage of muscle proteins. Lactate production was higher than in controls, indicating that oxidative phosphorylation could not cope with the demands made on the system during exercise. There was no correlation between the degree of mitochondrial dysfunction and lactate production, which may point to adaptation which varies between individuals, dependent on training, life style etc. However, the leakage of CK and myoglobin was significantly correlated to the mitochondrial dysfunction ($P < 0.001$ for both CK and myoglobin). Figure 3 shows the correlation between the remaining capacity to oxidize pyruvate (x-axis) of the seven patients, and the CK release 24 h after exercise (two patients had a similar rest activity of 98 per cent and are represented by one triangle in Fig. 3). The authors concluded that, in view of this correlation, the mitochondrial defect is probably the cause of the post-exercise muscle damage[16].

Several drugs may cause a rise in serum CK activity[28]. A class of drugs that has received a lot of attention in the last few years are the so called statins – drugs that inhibit the activity of a key enzyme in the biosynthesis of cholesterol, hydroxymethyl glutaryl CoA-reductase[22]. Examples of these drugs are simvastatin, lovastatin and pravastatin. The last drug differs from the other two in chemical structure: it contains an open lactone ring rather than a closed ring structure and is therefore hydrophilic instead of lipophilic. Pravastatin is a more selective drug that is reported to have fewer side effects when compared to lova- and simvastatin. In patients receiving lovastatin or simvastatin, mild rises in CK activity in serum have been reported. Two cases of myositis causing discontinuation of treatment have been reported for patients using simvastatin[18]. Until now, no reports on adverse effects of pravastatin have appeared. Lovastatin, in combination with other drugs (gemfibrozil, niacin

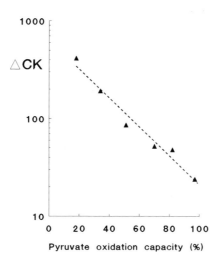

Fig. 3. The release of CK from skeletal muscle of seven patients with chronic progressive external ophthalmoplegia is given as the difference between the activity 24 h after exercise minus the activity immediately before exercise (CK, y-axis), and represented as function of the capacity of each individual's muscle mitochondria to oxidize pyruvate (x-axis). This capacity was measured in a mitochondrial fraction prepared from a muscle biopsy and is given as percentage of the capacity measured in control mitochondria (= 100 per cent). NB: the rightmost triangle represents two patients with identical values.

or cyclosporin), has been reported by several groups to cause severe rhabdomyolysis[39]. The reason why these drugs affect the muscle membrane is not known, but deserves attention.

The effects on the muscle are apparently often subclinical, comparable to the situation in the DMD carrier group. To determine a possible weakness of the muscle membrane, Smit *et al.*[36] tested 12 male patients that had been treated for hypercholesterolaemia for at least 1 year in an adapted version of the DMD carrier exercise test. The patients' age was 44 ± 10 years, their serum cholesterol level 7.1 ± 1.1 mmol/l. As controls, 17 normolipaemic controls were tested. They all exercised for 45 min on an ergometer at 2 watt per kg body weight, and had been asked not to perform exercise 48 h before the test. Blood samples were taken before the test, and 1, 2, 4 and 8 h after exercise. The samples were assayed for CK and myoglobin. Two things can be concluded from Table 1. First, both myoglobin and CK values were slightly higher in the patients, even at rest. Second, both CK and myoglobin rose with time after the test, but with a different temporal profile: myoglobin peaked already 1 h after the test, whereas CK rose slowly in time[36]. The pattern is, thus, very similar to that obtained with the DMD-carriers: a slow rise in CK which is only significant at one time point, and a fast rise in myoglobin. Myoglobin remained slightly elevated, at least for 8 h after the test. No observations were done at 24 h after exercise: most likely CK would still have been higher than control values[15], but myoglobin would probably have been cleared from the circulation[10]. These findings are consistent with an increased muscle permeability during simvastatin treatment. So, by using a provocation test, a subclinical (and maybe innocent) disorder, caused in this case by simvastatin, can be uncovered at an early stage.

Table 1. The values of creatine kinase (units per litre) and myoglobin (ng per ml) in the serum of 12 male hypercholesterolaemic patients, treated with simvastatin, and of 17 normolipaemic controls before and after an exercise test.

Time	Creatine kinase, U/l		Myoglobin, ng/ml	
	Patients	Controls	Patients	Controls
Before test	87 ± 6	66 ± 11	46 ± 3	36 ± 2
1 h after test	92 ± 4	71 ± 10	66 ± 7	38 ± 4
2 h after test	101 ± 6	72 ± 10	62 ± 7	36 ± 4
4 h after test	108 ± 7	78 ± 11	62 ± 8	36 ± 4
8 h after test	121 ± 11	74 ± 9	59 ± 10	31 ± 3

Conclusions

The release of CK from muscle during and after exercise is a physiological phenomenon indicating transient loss of integrity of the muscle membrane. Under extreme conditions and in susceptible subjects massive leakage of muscle enzymes and myoglobin may occur, accompanied by muscle degeneration, sometimes leading to rhabdomyolysis. The latter condition can be life-threatening and is therefore a medical emergency. Despite numerous studies on the release of CK after exercise little is known about the exact mechanism underlying this release. However, several factors are known to influence the process. Metabolic factors, such as a shortage of ATP, seem to be important as shown in the study of Driessen *et al.*[17] and as reflected by the increased susceptibility to rhabdomyolysis of patients with certain metabolic myopathies. Mechanical factors, directly affecting the membrane, also seem of importance as especially eccentric exercise causes large (delayed) CK elevations. Eccentric exercise is known to generate high forces at relatively low metabolic cost. As we have shown damage is also under hormonal control[2,12]. The effect of vitamin E deficiency points to a possible role of free radicals and lipid peroxidation, since free radical production and lipid peroxidation increase during exercise[6]. Calcium, either from the extracellular space (entering after mechanical tearing or via stretch activated cation channels), or from the sarcoplasmic reticulum, probably plays a key role as mediator of membrane-damaging processes. Calcium can activate proteases which may attack the Z-lines or the cytoskeleton, and activate phospholipase A2 (PLA2) which affects the membrane in several ways.

It liberates FFAs from membrane phopholipids, leaving lysophospholipids behind, which weaken the membrane and can act as detergent. Membrane lipids and unsaturated FFAs can undergo lipid peroxidation leading to a destructive chain reaction. Arachidonic acid, a precursor of prostaglandin and leukotriene synthesis, is released in relatively high amounts by PLA2. Since so many different factors influence the degree of muscle damage, it seems likely that different mechanisms are involved, as evidenced by the fact that the time course of enzyme release after exercise depends very much on the type of exercise (eccentric *vs* concentric) involved[13].

A relatively new technique that may help shed light on the time course of events in muscle after exercise is magnetic resonance (MR): MR imaging can be used to visualize, for instance, oedema, and MR spectroscopy (MRS) can be used to follow the energy metabolism – phosphor-MRS yields information on ATP, PCr and intracellular pH, and proton-MRS provides information on lactate production in living tissue[34]. MRI and MRS are non-invasive and allow one to 'look inside the muscle' during exercise. Whatever the exact mechanism underlying muscle damage is, the evaluation of CK and its isoenzymes in serum can be very useful in monitoring muscle (membrane) damage in neuromuscular or heart disease, after exercise or training, or that due to toxic effects (e.g. drugs). Additionally, exercise as a way to provoke CK release can be used to detect subclinical damage. Measuring CK isoenzymes and myoglobin as extra markers may raise the sensitivity of such tests.

Acknowledgements: During the past few years a great many people have been working on parts of the work presented here. Naming them means undoubtedly forgetting one or two but we thank Drs. Kamp, Winckers, the students E.J. Heyting, R. Adang, R. Kolsters, B. Rikken, G. van Steyn, H. Mossel, M. Harfterkamp, C. van der Kallen, O. van Dobbenburg, R. Koot, B. van Berckel, W. Balemans, P. Bosman; the anonymous 'controls' and the mothers of Duchenne patients who performed most of the real 'work'; and the excellent biotechnicians G.J. Vosmeer and C. Brand.

References

1. Adornato, B.T., Kagen, L.J. & King Engel, W.K. (1978): Myoglobinaemia in Duchenne muscular dystrophy patients and carriers: a new adjunct to carrier detection. *Lancet* **ii**, 499–501.

2. Amelink, G.H. & Bär, P.R. (1986): Exercise-induced muscle damage in the rat: effects hormonal manipulation. *J. Neurol. Sci.* **76**, 61–68.

3. Amelink, G.J., Kamp, H.H. & Bär, P.R. (1988): Creatine kinase isoenzyme profiles after exercise in the rat: sex-linked differences in leakage of CK-MM. *Pfl. Arch. Eur. J. Physiol.* **412**, 417–421.

4. Amelink, G.J., Koot, R., Erich, W.B.M., Van Gijn, J. & Bär, P.R. (1990): Sex-linked variation in creatine kinase release, and its dependence on oestradiol, can be demonstrated in an *in vitro* rat muscle preparation. *Acta Physiol. Scand.* **138**, 115–124.

5. Amelink, G.J., Van der Kallen, C.J.H., Wokke, J.H.J. & Bär, P.R. (1990): Dantrolene sodium diminishes exercise-induced muscle damage in the rat. *Eur. J. Pharmacol.* **179**, 187–192.

6. Amelink, G.J., Van der Wal, W.A.A., Van Asbeck, B.S. & Bär, P.R. (1991): Sex-linked differences and exercise-induced muscle damage in the rat: the effect of vitamin E deficiency. *Pfl. Arch. Eur. J. Physiol.* **419**, 304–309.

7. Apple, F.S., Rogers, M.A., Casal, D.C., Sherman, W.M. & Ivy, J.L. (1985): Creatine kinase-MB isoenzyme adaptations in stressed human skeletal muscle of marathon runners. *J. Appl. Physiol.* **59**, 149–153.

8. Armstrong, R.B., Ogilvie, R.W. & Schwane, J.A. (1983): Eccentric exercise-induced injury to rat skeletal muscle. *J. Appl. Physiol.* **54**, 80.

9. Bär, P.R., Amelink, G.J. & Jennekens, F.G.I. (1985): Muscle protein leakage in healthy females and Duchenne carriers: myoglobin and aldolase preceed creatine kinase. *J. Neurol.* **S232**, 172.

10. Bär, P.R., Driessen, M.F., Rikken, B. & Jennekens, F.G.I. (1985): Muscle protein leakage after strenuous exercise: sex and chronological differences. *Neurosci. Lett.* **S22**, 288.

11. Bär, P.R. & Amelink, G.J. (1986): Serum myoglobin after exercise: a useful parameter for the detection of Duchenne carriers. *Muscle Nerve* **9**(5S), 200.

12. Bär, P.R., Amelink, G.J. & Blankenstein, M.A. (1988): Prevention of exercise-induced muscle membrane damage by oestrogen. *Life Sci.* **42**, 2677–2681.

13. Bär, P.R., Amelink, G.J., Jackson, M.J., Jones, D.A. & Bast, A. (1990): Aspects of exercise-induced muscle damage. In *Sports, medicine and health*, ed. G.P.H. Hermans, pp. 1143–1148. Amsterdam: Elsevier Science.

14. Borleffs, J.C.C., Derksen, R.H.W.M. & Bär, P.R. (1987): Serum myoglobin and creatine kinase concentrations in patients with polymyositis or dermatomyositis. *Ann. Rheum. Dis.* **46**, 173–175.

15. Brooke, M.H., Carroll, J.E., Davis, J.E. & Hagberg, J.M. (1979): The prolonged exercise test. *Neurology* **29**, 636–643.

16. Driessen, M.F., Bär, P.R., Scholte, H.R. & Hoogenraad, T.U. (1987): A striking correlation between muscle damage after exercise and mitochondrial dysfunction in patients with chronic progressive external ophthalmoplegia. *J. Inherit. Metabol. Dis.* **10**, 252–255.

17. Driessen-Kletter, M.F., Amelink, G.J., Bär, P.R. & Van Gijn, J. (1990): Myoglobin is a sensitive marker of increased muscle membrane vulnerability. *J. Neurol.* **237**, 234–238.

18. Emmerich, J. & Aubert, I. *et al.* (1990): Efficacy and safety of simvastatin. A one-year study in 66 patients with type II hyperlipoproteinaemia. *Eur. Heart J.* **11**, 149–155.

19. Jones, M.G. & Swaminathan, R. (1990): The clinical biochemistry of creatine kinase. *J. Int. Fed. Clin. Chem.* **2**, 108–113.

20. Koot, R., Amelink, G.J., Blankenstein, M.A. & Bär, P.R. (1990): Tamoxifen and oestrogen both protect the rat muscle membrane against physiological damage. *J. Steroid Biochem.* **36**, 18S.

21. Koot, R.W., Amelink, G.J., Blankenstein, M.A. & Bär, P.R. (1991): Tamoxifen and oestrogen both protect the rat muscle membrane against physiological damage. *J. Steroid Biochem.* **40**, 689–695.

22. Grundy, S.M. (1988): HMG-CoA reductase inhibitors for treatment of hypercholesterolemia. *N. Engl. J. Med.* **319**, 24–33.

23. Herrmann, F.H., Spiegler, A.W.J. & Wiedemann, G. (1982): Muscle provocation test. A sensitive method of discrimination between carriers and noncarriers of Duchenne muscular dystrophy. *Hum. Genet.* **61**, 102–104.

24. Hoffman, E.P., Brown, B.H. & Kunkel, L.M. (1987): Dystrophin:the protein product of the Duchenne muscular dystrophy gene. *Cell* **51**, 919–928.

25. Hortobagyi, T. & Denahan, T. (1989): Variability in creatine kinase: methodological, exercise and clinically related factors. *Int. J. Sports Med.* **10**, 69–80.

26. Kagen, L.J., Scheidt, S. & Butt, A. (1975): Myoglobinaemia following acute myocardial infarction. *Am. J. Med.* **58**, 177–182.

27. Kawai, H., Nishino, H. & Saito, S. (1987): Localisation of myoglobin in human muscle cells by immunoelectron microscopy. *Muscle Nerve* **10**, 144–149.

28. Lott, J.A. & Abbott, L.B. (1986): CK isoenzymes. *Clin. Lab. Med.* **6**, 547–576.

29. Nanji, A.A. (1983): Serum creatine kinase isoenzymes: a review. *Muscle Nerve* **6**, 83–90.

30. Neumeier, D., (1981): Subcellular distribution of creatine kinase isoenzymes. In *Creatine kinase isoenzymes*, ed. H. Lang, pp. 110–115. Berlin, Heidelberg: Springer-Verlag.

31. Newham, D.J., Jones, D.A. & Edwards R.H.T. (1986): Plasma creatine kinase changes after eccentric and concentric contractions. *Muscle Nerve* **9**, 59–63.

32. Noakes T.D. (1987): Effects of exercise on serum enzyme activities in humans. *Sports Med.* **4**, 245–267.

33. Rogers, M.A., Stull, M.A. & Apple, F.S. (1985): Creatine kinase isoenzyme activities in men and women following a marathon race. *Med. Sci. Sports Exerc.* **17**, 679–682.

34. Rodenburg, J.B., De Boer, R.W. & Bär, P.R. (1991): Muscle damage: a combined MRI and MRS study. *Basic and Applied Myology*, Perspectives for the 90's, eds. U. Carraro & S. Salmons, pp. 155–162. Padova: Uni Press.

35. Shumate, J.B., Brooke, M.H., Carroll, J.E. & Davis, J.E. (1979): Increased serum creatine kinase after exercise: a sex linked phenomenon. *Neurology* **29**, 902–904.

36. Smit, J.W.A., Amelink, G.J., Bär, P.R., De Bruin, T.W.A. & Erkelens, D.W. (1990): Increased human myoglobin creatine kinase serum levels after standardized exercise during simvastatin treatment. *Eur. J. Clin. Invest.* **20**, A31.

37. Vejjaajiva A. & Teasdale G.M. (1965): Serum creatine kinase and physical exercise. *B. M. J.* **1**, 1653–1654.

38. Wu A.H.B. (1989): Creatine kinase isoforms in ischaemic heart disease. *Clin. Chem.* **35**, 7–12.

39. Three reports in *N. Engl. J. Med.* (1988): 318, 46–48; two reports in *Lancet* (1989): **ii**, 1097–1098

Guanidino Compounds in Biology and Medicine, eds. by P.P. De Deyn, B. Marescau, V. Stalon and
I.A. Qureshi. ©1992 John Libbey & Company Ltd., pp. 239–248.

Chapter 33

Phosphocreatine as effective drug in clinical cardiology

V.A. SAKS, V.I. KAPELKO, M.Y. RUDA, M.L. SEMENOVSKY[1] and E. STRUMIA[2]

*Cardiology Research Centre, 3rd Cherepkovskaya, 15A, Moscow 121552, USSR; [1]Institute of Transplantology and
Artificial Organs, Moscow, USSR; [2]Schiapparelli-Searle Medical Research Department, Torino, Italy*

Summary

A review is given of the pharmacological effects of exogenous phosphocreatine in experimental conditions and the
efficacy of phosphocreatine in clinical cardiology is highlighted. Data are discussed that seem to indicate that exogenous
phosphocreatine contributes to the protection of cellular membranes from irreversible ischaemic damage and prevents
calcium overload of cells or at least slows down these processes helping the cells to overcome the ischaemic period and
to recover when the blood flow is restored. Exogenous phosphocreatine has been shown to be efficacious in some
multinational clinical trials during cardiac surgery and after myocardial infarction.

Introduction

Physiology function of energy carrier is fulfilled by the phosphocreatine (PCr) and creatine kinase system in heart, skeletal muscle, brain, retina and spermatozoa[7,14,21,23]. However, the last decade has produced new information showing that phosphocreatine possesses significant pharmacological activity[4,20]. It is already used in several countries for treatment of acute ischaemic diseases, mostly myocardial infarction and intraoperational ischaemia.

The cellular damage occurring during myocardial ischaemia is extensively studied and well known in general[4,6,10]. If reperfusion is started after a short period of ischaemia all changes observed may be completely reversed. However, if the ischaemic period is too long, the reperfusion does not help to restore the heart function but causes further damage (reperfusion injury). The mechanisms of irreversible injury are as following: (1) due to activation of phospholipases the membrane phospholipids are being degraded, this damaging the integrity of cellular membranes and producing harmful intermediates – lysophosphoglycerides, which are extremely arrhythmogenic; (2) the metabolic products accumulate in the cells – phosphate, lactate, protons etc. disturbing osmolarity and functioning of intracellular structures and systems; (3) due to complete disappearance of PCr adenine nucleotides start to degrade up to adenosine and inosine, these leaving the cells easily and degrading further into xantine and hypoxantine, the central reaction being the sarcolemmal 5′-nucleotidase reaction: AMP → Pi + adenosine; (4) as a result of destruction of cellular membrane and also due to inhibition of Na, K-ATPase and accumulation of sodium, massive overloading of cells by calcium is observed, this resulting in mitochondrial damage and hypercontraction of myofibrils; (5) due to changes in membrane structure and accumulation of reducing equivalents, at reperfusion lipid peroxidation occurs that further damages cellular membranes enhancing calcium overload and death of cells.

Thus, it is most important to protect the cellular membranes from irreversible ischaemic damage and prevent calcium overload of cells, or at least to slow down the rate of these processes to help the cells overcome the ischaemic period and recover normally when the blood flow is restored. It seems that exogenous phosphocreatine serves well for these purposes. This paper gives a general review of experimental and clinical data on this topic.

Pharmacological effects of exogenous phosphocreatine in experiments

The main effect of phosphocreatine added extracellularly is significant protection of heart function as estimated both by much better recovery of developed tension and by quicker decrease of end diastolic pressure.

Figure 1 shows a good example of such a protection: in the control only 10 per cent of contractile force is restored after 60 min of hypoxia of rat heart papillary muscle but in the presence of 10 mM of PCr the functional recovery is more than 60 per cent, and thus PCr gives a sixfold protective effect[16]. What is very important is the complete relaxation (decrease of resting tension to pre-ischaemic value) and also the absence of reperfusion contracture. This effect is one of the most important since the relaxation phase determines heart filling and in this way cardiac work.

It is especially important that the protective effect of PCr is related to the peculiarities of its intact molecular structure. Indeed, creatine and inorganic phosphate, its components, do not afford any protection[9,11,12]. Protection is not afforded either by structural analogues of PCr, such as phospho-arginine, and is not related to Ca^{2+}-chelation[9,13]. Ronca et al.[22,24], and also Conorev et al.[3] have shown that functional protection is afforded by PCr also against another type of injury – oxidative damage caused by perfusion of isolated heart by H_2O_2, 90 µM. This damage simulates that caused by lipid peroxidation due to formation of active oxygen radicals during post-ischaemic reperfusion.

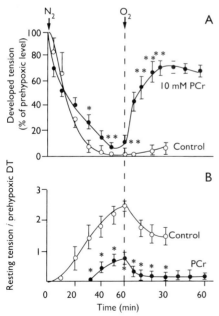

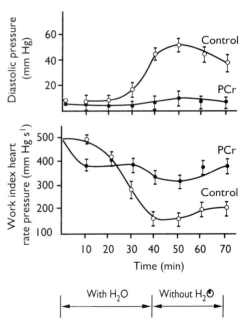

Fig. 1. Recovery of developed tension (A) and resting tension (B) of rat papillary muscle after 60 min of hypoxia in the presence of PCr. From Seppet et al.[16] with permission.

Fig. 2. Effect of PCr on contractility of isolated rat heart subjected to H_2O_2-induced oxidative stress. A – changes in diastolic pressur; B – changes in cardiac work index. From Conorev et al.[3] with permission.

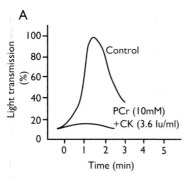
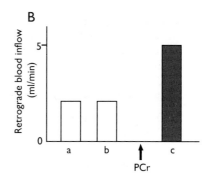

Fig. 3. (A) Inhibition of ADP-induced aggregation of platelets by the PCr-creatine kinase system. (B) The effect of PCr on retrograde inflow of blood into ischaemic zone. (a)(b) – 5 and 40 min after ligation of coronary artery; (c) after 15 min of PCr infusion.

The oxidative stress first induces elevation of diastolic pressure due to lipid peroxidation by H_2O_2-generated radicals (Fig. 2). This damages sarcolemma and results in calcium overload and finally in contracture. In the presence of PCr the development of contracture is completely prevented (Fig. 2A), and also the decline of work index in the presence of PCr is much less (Fig. 2B). This kind of effect shows that exogenous PCr may effect the cellular surface membrane – sarcolemma, which is the target of H_2O_2 attack. However, when used in combination with an antioxidant – tocopherol-phosphate (TPP) – the effects are additive.

Functional protection given by PCr is concomitant with significantly improved metabolic recovery of the heart at post-ischaemic reperfusion. This was clearly demonstrated first in experiments with isolated perfused rat hearts by using ^{31}P-NMR spectroscopy for detection of intracellular metabolites. A very remarkable decrease of creatine kinase leakage from the hearts with PCr as compared to controls was also observed[13,18]. The latter effect is again directly pointing to the protection of cardiac sarcolemma against ischaemic damage.

The ability of PCr to decrease the size of the ischaemic zone has been reported and this is most probably due to an increase in collateral blood flow and improvement in microcirculation[1,17]. It is not excluded that PCr directly influences endothelial cells in blood vessels. However, improvement in microcirculation is mostly the result of the direct effect of PCr on platelet aggregation (Fig. 3A) on a simple principle of ADP removal in the presence of high activity of creatine kinase. Besides, Piacenza, Strumia et al.[2,14] have shown that by improving the energy status of erythrocytes, PCr elevates their plasticity and haemorrheological properties, enhancing their osmotic resistance. These membrane effects of PCr on erythrocytes may increase the rate of their movement in capillaries. In physiological experiments all these effects manifest as an increase in retrograde blood flow into the zone of regional ischaemia (Fig. 3B)[14].

Besides disturbances of contractile function, very often consequences of both myocardial ischaemic (infarction) and post-ischaemic reperfusion are cardiac rhythm disorders – arrhythmias, including life-threatening ventricular fibrillations. The latter are among the main causes of sudden death, an increasing factor of mortality from cardiovascular diseases. In many experimental studies exogenous PCr was found to eliminate ventricular arrhythmias related to ischaemia or anoxia[5,8,19] that may be directly related to its effect on the membranes and lipid metabolism.

Direct determinations of penetration of PCr from extracellular space into cardiac cells showed that this process is very slow[11]. It is not easy to assume that the metabolic effects of PCr can be explained by its direct influence on the energy metabolism of the cell. This may be one of the minor effects of exogenous PCr. The major effects of exogenous PCr are explained by the phenomenon of stabilization of the sarcolemma by PCr in ischaemic heart cells. To some extent preservation of ATP can also be explained by inhibition of 5′-nucleotidase of cardiac sarcolemma[11].

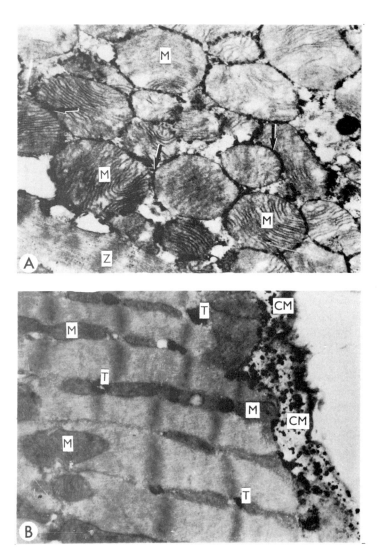

Fig. 4. (A) Heart myocardium after 35 min of ischaemia and 20 min of reperfusion with addition of lantha-num of which particles can be detected on mitochondrial membrane (arrow). (B) The same as A, but 10 mM PCr was added into solution before and during ischaemia. Lanthanum stays outside the cells. From Sharov et al.[18] with permission.

A clear direct demonstration of the stabilization of ischaemic sarcolemma by PCr was first obtained by Sharov *et al.*[17,19] in experiments with colloidal lanthanum: the particles of lanthanum penetrated easily the sarcolemma of control cardiomyocytes at reperfusion after 30 min of ischaemia but stayed in the extracellular space if experiments were carried out in the presence of PCr (Fig. 4).

Further important demonstration of the effect of exogenous PCr on the surface membrane was found in studies on phospholipid metabolism in ischaemic heart.

In 1978 Corr *et al.*[4] described an elevation of the products of the phospholipid degradation, lysophosphoglycerides (LPG), mostly of lysophosphatidylcholine (LPC) and lysophosphatidyletha-nolamine (LPE), in the ischaemic zone of the rabbit heart. Accumulation of LPG in the ischaemic zone may be considered to be one of the major factors of the electrical instability of the ischaemic myocardium and also related to the disintegration of the sarcolemmal structure. Besides the loss of natural components of the membrane phospholipid molecules, and because of the detergent activities of LPC its accumulation destroys membrane further as, is demonstrated in the upper part of Fig. 5. Agents which are able to depress the lysophosphoglyceride accumulation in ischaemic myocardium should also protect myocardium from electrophysiological and structural damage. Figure 5 demonstrates this activity of PCr: it shows the elevation in the lysophosphatidylcholine (LPC) after 8 min of total ischaemia.

However, this elevation of LPC was completely suppressed by PCr. This is a very significant observation since two important conclusions can be made: (1) PCr directly effects the metabolism of phospholipids in membrane; (2) this results in elimination of accumulation of highly arrhythmogenic intermediates of phospholipid metabolism, that may be a basis of antiarrhythmic effects of PCr[14].

All these experimental data very logically led to the experiments with isolated sarcolemmal vesicles. Into these vesicles spin-labelled fatty acid derivates were inserted to measure the mobility of phospholipid molecules within the membrane. The mobility was determined from the width of the spectral line in EPR spectrum (Fig. 6). These measurements showed that in the presence of PCr the mobility of phospholipid molecules is lower (structural parameter S is higher) at any temperature. The direct interaction of PCr with sarcolemmal phospholipids may be explained by a hypothesis of zwitterionic interactions (Fig. 7). All data on the mechanism of the protective effect of PCr on ischaemic tissue described above can be summarized by the scheme shown in Fig. 8.

Biochemical evidence gives reasons to believe that the mechanism of this protective action is complex and includes at least four components (Fig. 8): (1) inhibition of accumulation of lysophosphoglycerides in ischaemic myocardium and preservation of the structure of cardiac cells' sarcolemma; (2)

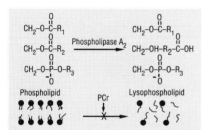

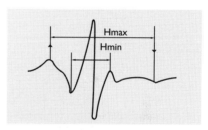

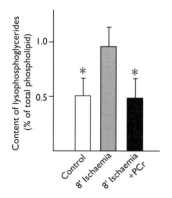

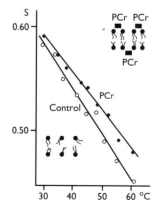

Fig. 5. Inhibition of accumulation of lysophosphoglycerides in ischaemic rat myocardium by PCr. LPC-lysophosphatidylcholine.

Fig. 6. EPR spectrum of spin-labelled 5-oxylstearate in the cardiac sarcolemmal preparations. From the spectrum a structural order parameter S was calculated. The values of S as a function of temperature are shown for control and in the presence of PCr, 10 mM.

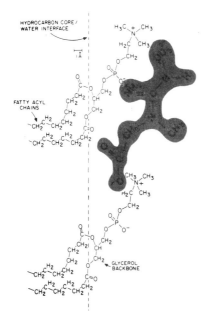

Fig. 7. Hypothetical zwitterionic interaction of PCr with polar heads of phospholipid molecules in the interphases of sarcolemma.

extracellular inhibition of platelet aggregation via removal of ADP in the extracellular creatine kinase reaction and increasing plasticity of red blood cells; (3) some extent of PCr penetration into the cells and participation in the PCr energy transport system by maintaining high local ATP levels is possible; (4) inhibition of adenine nucleotide degradation at the step of 5′ nucleotidase reaction at cardiac cells' sarcolemma. The total result of the action of these four components is a delay (sometimes very significant) of irreversible sarcolemmal damage, preservation of adenine nucleotide pool and better functional recovery, strong antiarrhythmic action of PCr. Also some indications for limitation of infarct size exist[14].

It is difficult not to agree with Dr. Ronca that PCr, which is present inside of normal cardiomyocytes in high concentration should be considered as a neutral factor stabilizing cellular membrane[22]. Its disappearance in ischaemia is one of the reasons of membrane destabilization and damage. Addition of PCr extracellulary may restore this stabilizing factor.

By protecting membranes from degradation, PCr

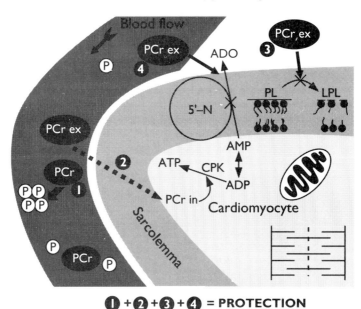

❶ + ❷ + ❸ + ❹ = PROTECTION

Fig. 8. Mechanism of the protective action of PCr on ischaemic myocardium. (1) Inhibition of platelet aggregation, extracellular mechanism; (2) Possible penetration into the cells and participation in PCr shuttle minor component; (3) Inhibition of accumulation of lysophosphoglycerides major component; (4) Inhibition of 5′-nucleotidase of sarcolemma.

may make phospholipid less accessible for active oxygen species and in this way exert an antioxidative effect. On the other hand, antioxidants reduce the content of these species. That explains why PCr and antioxidants (TPP) potentiate the effect of each other and afford additive protection.

The results of clinical trial of PCr

PCr is produced in the form of pharmacological preparation Neoton by Schiapparelli in Italy. Positive results from pharmacological studies formed a basis for a wide clinical trial of Neoton in Italy, USSR and several other countries. Given below are the results of clinical use of Neoton in: (1) cardiac surgery; (2) treatment of acute myocardial infarction.

Cardiac surgery

The first physician to use Neoton for intraoperational protection of heart was Professor M. Semenovsky, Moscow, who published his work in 1987[15]. The protocol of Neoton use was to add it to cardioplegic solution in a concentration of 10 mM and to start its administration before operation. By now, more than 500 patients have been investigated according to this protocol[2]. Clinical analyses showed that the use of Neoton both in crystalloid and blood cardioplegic solutions at a concentration of 10 mM resulted in much better restoration of sinus rhythm and decreased frequency of fibrillations, and much less inotropic support was necessary in the postoperative period[15]. In controls, the use of cardioplegia saved significant amounts of ATP and PCr which content decreased by 25 per cent by the end of the operation. In the group with Neoton the high energy compounds stayed, however, at the preoperational level[15]. In accordance with the results of experimental research, ultrastructural analysis showed good preservation of sarcolemmal structure in the Neoton group[15]. This was clearly seen with the use of the lanthanum method, even despite a high degree of heterogeneity of the human heart material studied. Very similar results were obtained by an Italian group of cardiac surgeons directed by Professor Luigi D'Alessandro that operated upon several hundreds of patients with Neoton cardioplegia with improved postoperative functional recovery[20].

Professor Semenovsky has studied the effect of Neoton in cardioplegic solution in intraoperational protection of heart in experiments on dogs in the control group and in animals which were made hypertrophic. In both cases the use of Neoton gave positive results. This is important since it shows that the protection is given in the case of hypertrophic heart.

Myocardial infarction

Among the reasons for mortality from acute myocardial infarction (AMI) are arrhythmias turning into fibrillation and heart rupture. Studies of the development of AMI showed that after occlusion of coronary artery the ischaemic zone is converted into a necrotic one within 20–30 h, if the reperfusion is not initiated, for example by using thrombolytic means[6,10]. Development of AMI may be influenced by increasing collateral blood flow into the ischaemic zone or by improving microcirculation that helps the drug to get into the infarcted area and to afford cellular protection for ischaemic cardiomyocytes, or to cardiomyocytes in non-infarcted areas with increased energy demand[6]. The results of experimental research *in vivo* described in the previous part gave us hope that Neoton may be useful in these early hours of myocardial infarction. The trial protocol was worked out and used by Professor M. Ruda *et al.*[13]. Patients with a first myocardial infarction (6 h after the attack and without congestive heart failure) were included in a randomized placebo-controlled study.

Phosphocreatine was infused in the following way:

1st day – (immediately after control examination), 2 g bolus intravenous injection, followed by intravenous infusion with a rate of 4 g/h during 2 h; 2nd–6th day – 2 g intravenous injection twice a day (placebo – 0.9 per cent NaCl water solution)[10].

At present more than 1000 patients have been treated according to this protocol, mostly in Moscow and St Petersburg Cardiology Centres and in Italy. In all studies both control and treated groups were carefully selected to be similar in basic parameters such as age, character of diseases and treatment. In rather good accordance with the experimental data, the clinical trial showed that PCr can be used as an effective drug for treatment of acute myocardial infarction to eliminate life-threatening arrhythmias, to improve relaxation of the heart in diastole (decrease of ischaemic contracture), decrease the frequency of development of heart failure and cardiac rupture, with a possible decrease of mortality rate.

In the series of investigations performed by Professor Ruda *et al.* electrocardiographic data demonstrated that Neoton infused according to the protocol described above resulted in more rapid positive dynamics of precordial cartograms as compared to controls. At the second hour of observation the tendency for lowering of the total elevation of ST segment was already revealed in the PCr group (see Fig. 9A).

The effect of PCr was seen very clearly when its antiarrhythmic activity was studied by Holter monitoring. 24 h monitoring showed an average number of extra ventricular beats of around 700 in the PCr group and around 2500 in the control group (Fig. 9B).

The effect of PCr is mostly due to a decrease in the frequency of complex rhythm disturbances. Ventricular fibrillation was not seen in PCr group. As has already been observed by Fagbemi *et al.*[5] the antiarrhythmic effect was very often seen both during its intravenous infusion and long after cessation of its administration, when according to pharmacokinetic data its concentration in blood becomes zero. We may explain that by transient accumulation of PCr in the heart and also by preservation of the cardiomyocytes' membrane structure and electrophysiological characteristics due to stabilization of phospholipid membrane in acute phase of AMI and in particular due to suppression of accumulation of lysophosphatidylcholine.

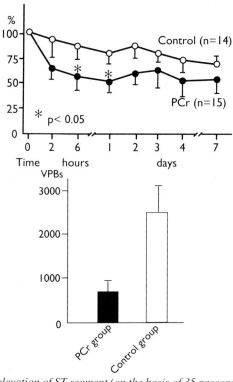

Fig. 9. (A) Dynamics of total elevation of ST segment (on the basis of 35 precordial leads). Effects of Neoton infusion: statistically significant decrease of ST segment elevation is seen; (B) Holter 24 h monitoring data showing reduction of ventricular premature beats by PCr. From Ruda et al.[13] with permission.

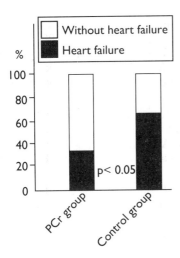

Fig. 10. The frequency of development of heart failure (Killip II–III class) during hospitalization in control and Neoton groups. From Ruda et al.[13] with permission.

The most important observation in this trial, besides antiarrhythmic action, was improvement in postinfarctional clinical course as seen in diminished frequency of development of heart failure (Fig. 10)[13].

In the PCr group one observed an appearance of left ventricular failure necessitating administration of glycosides and diuretics (class II and III according to Killip) in 33 per cent of the cases, and in 57 per cent in the control group during 1 month of observation ($P < 0.05$). The mechanism of this phenomenon may include both better preservation of severely ischaemic myocardium and also improvement of the metabolic and functional state of nonischaemic heart muscle.

In summary, more than 5 years of clinical trial of Neoton in cardiac surgery and for treatment of acute myocardial infarction have collected sufficient amount of data to make a firm conclusion.

Exogenous phosphocreatine is a cardioactive drug which is effective in protection of myocardium of patients in early hours of infarction and from intraoperational ischaemic damage. Its application in doses of several g (3 g/l cardioplegic solution and about 10 g per day intravenously) ensures significantly better functional recovery in postinfarction period or at reperfusion.

References

1. Afonskaya, N.I., Samarenko, M.B. & Ruda, M. Ya. (1989): PCr in the treatment of patients with acute myocardial infarction. *Sov. Med. Rev. A. Cardiol.* **2**, 169–202.

2. Borgoglio, R., Piacenza, G. & Osella, R. (1982): Effects of exogenous PCr on the metabolic and contractile state of the hypoxia-perfused rabbit heart. In *Advances in studies on heart metabolism*, eds. C.H. Caldarera & P. Harris, pp. 287–292. Bologna: CLUEB.

3. Conorev, E.A., Sharov, V.G. & Saks, V.A. (1991): Improvement in contractile recovery of isolated rat heart after cardioplegic ischemic arrest with endogenous phosphocreatine: involvement of antiperoxidative effect? *Cardiovasc. Res.* **25**, 164–171.

4. Corr, P.B., Gross, R.W. & Sobel B.E. (1984): Amphipathic metabolites and membrane dysfunction in ischemic myocardium. *Circ. Res.* **55**, 135–154.

5. Fagbemi, O., Kane, K.A. & Parratt, J.R. (1982): PCr suppresses ventricular arrhythmias resulting from coronary artery ligation. *J. Cardiovasc. Pharmacol.* **4**, 53–58.

6. Hearse, D.J. (1990): Ischemia, reperfusion and the determinants of tissue injury. *Cardiovasc. Drugs Therapy* **4**, 777–776.

7. Jacobus, W.E. (1985): Respiratory control and the integration of heart high-energy phosphate metabolism by mitochondrial creatine kinase. *Ann. Rev. Physiol.* **47**, 707–725.

8. Marshall, R.J. & Parratt, J.R. (1974): Reduction of ventricular arrhythmias follwoing acute coronary artery ligation on the dog after administration of PCr. *Naunyn Schmiedebergs Arch. Pharmacol.* **281**, 437–441.

9. Parratt, J.R. & Marshall, R.J. (1990): The response of isolated cardiac muscle to acute anoxia: protective effect of adenosine triphosphate and PCr. *J. Pharm. Pharmacol.* **26**, 427–433.

10. Piper, H.M. (1990): *Pathophysiology of severe ischemic myocardial injury.* Dordrecht, Netherlands: Kluwer Academic Publishers.

11. Preobrazhensky, A.N., Javadov, S.A. & Saks, V.A. (1986): Study of the hypothetical mechanism of protective effect of phosphocreatine in ischemic mycocardium. *Biochimia* **51**, 675–683.

12. Robinson, L.A., Braimbridge, M.V. & Hearse, D.J. (1984): PCr: an additive mycocardial protective agent in cardioplegia. *J. Thorac. Cardiovasc. Surg.* **87**, 190–200.

13. Ruda, Ya. M., Samarenko, M.B., Afonskaya, N.I. & Saks, V.A. (1988): Reduction of ventricular arrhythmias by PCr (Neoton) in patients with acute myocardial infarction. *Am. Heart J.* **116**, 393–397.

14. Sakis, V.A., Bobkov, Y.G. & Strumia, E. (1987): *Creatine phosphate: biochemistry, pharmacology and clinical efficiency.* Torino: Edizioni Minerva Medica.

15. Saks, V.A., Sharov, V.G., Kupryanov, V.V., Kryzhanovky, S.A., Semenovksy, M.L., Mogilevsky, G.M., Lakomkin, V.L., Shteinshneider, A.Ya., Preobrazhensky, A.N., Javadov, S.A., Anukhovsky, E.P., Beskrovnova, N.N., Rosenshtraukh, L.V. & Kaverina, N V (1987): PCr and related compounds: protective action on the ischemic mycocardium. In *Myocardial metabolism*, eds. A.M. Katz & V.N. Smirnov, pp. 446–480. Switzerland: Harwood Academic Publishers GmbH.

16. Saks, V.A., Rosenschtraukh, L.V., Smirnov, V.N. & Chazov, E.I. (1978): Role of creatine phosphokinase in cellular function and metabolism. *Can. J. Physiol. Pharmacol.* **56**, 691–706.

17. Semenovsky, M.L., Shumakov, V.I., Sharov, V.G., Mogilevsky, G.N., Asamolovsky, A.V., Machotina, L.A. & Saks, V.A. (1987): Protection of ischemic myocardium by exogenous PCr. II. Clinical, ultrastructural, and biochemical evaluations. *J. Thorac. Cardiovasc. Surg.* **94**, 762–769.

18. Seppet, E.K., Eimree, M.A., Kallikorm, A.P. & Saks, V.A. (1988): Effect of exogenous PCr on heart muscle contractility modulated by hyperthyroidism and extracellular calcium concentration. *J. Appl. Cardiol.* **3**, 369–380.

19. Sharov, V.G., Afonskaya, N.I., Ruda, M.Y., Cherpachenko, N.M., Posin, E.Ya., Markosian, R.A., Chepeleva, I.I., Samarenko, M.B. & Saks, V.A. (1986): Protection of ischemic myocardium by exogenous PCr (Neoton): pharmacokinetics of PCr reduction of infarct size, stabilization of sarcolemma of ischemic cardiomyocytes and antithrombotic action. *Biochem. Med.* **35**, 101–114.

20. Sharov, V.G., Saks, V.A., Kupriyanov, V.V., Lakomkin, V.L., Kapelko, V.I., Steinshneider, A.Ya. & Javadov, S.A. (1987): Protection of ischemic myocardium by exgenous PCr: I. Morphologic and phosphorus 31-nuclear magnetic resonance studies. *J. Thorac. Cardiovasc. Surg.* **94**, 749–761.

21. Sharov, V.G., Bescrovnova, N.N., Kryzhanovsky, S.A., Bobkov, I.G., Saks, V.A. & Kaverina, N.G. (1989): Ultrastructure of purkinje cells in the subendocardium and false tendons in early experimental myocardial infarcation complicated by fibrillation in the dog. *Virchows Arch. B. Cell. Pathol.* **57**, 131–139.

22. Tronconi, L. & Saks, V.A. (1990): *Proceedings of International Meeting on PCr in cardiology and cardiac surgery.* Pavia: Universita di Pavia.

23. Walliman, T., Schnyder, T., Schlegel, J., Wyss, M., Wegmann, G., Rossi, A-M., Hemmer, W., Eppenberger, H.M. & Quest, A.F.G. (1989): Subcellular compartmentation of creatine kinase isoenzymes, regulation of CK and octametric structure of mitochondrial CK: important aspects of the phosphoryl-creatine circuit. In *Progress in clinical and biological research, 'Muscle energetics'*, 315, eds. R.J. Paul, G. Elzinga & K. Yamada, pp. 159–176. New York: Alan R. Liss.

24. Zucchi, R., Poddighe, R., Limbruno, U., Mariani, M., Ronca- Testoni, S. & Ronca, G. (1989): Protection of isolated rat heart from oxidative stress by exogenous PCr. *J. Mol. Cell. Cardiol.* **21**, 67–73.

Guanidino Compounds in Biology and Medicine, eds. by P.P. De Deyn, B. Marescau, V. Stalon and
I.A. Qureshi. ©1992 John Libbey & Company Ltd., pp. 249–251.

Chapter 34

The crucial role of creatine kinase
for cardiac pump function

V.I. KAPELKO, V.A. SAKS, E.K. RUUGE, V.V. KUPRIYANOV, N.A. NOVIKOVA,
V.L. LAKOMKIN, A.Y. STEINSCHNEIDER and V.I. VEKSLER

*Institute of Experimental Cardiology, Russia Cardiology Research Centre, 3rd Cherepkovskaya St. 15A, 121552
Moscow, Russia*

Summary

The pump function as well as isovolumic pressure development of the isolated rat heart after acute or chronic inhibition
of the flux through creatine kinase (CK) were studied. The acute CK inhibition by iodoacetamide (0.5 mM) was followed
by complete disappearance of the aortic output in working mode while left ventricular (LV) isovolumic pressure
decreased by only 16%. The difference is related to the fact that pump function is highly dependent on LV filling which
has dropped due to elevated LV diastolic pressure and stiffness. Similar but less prominent changes have been observed
at chronic CK inhibition after 6-wk feeding with guanidinopropionate, an inhibitor of creatine influx. The predominant
deterioration of LV distensibility both at acute and chronic inhibition of the flux through CK may be explained by the
ability of myofibrillar CK to maintain high rate of ADP rephosphorylation and detachment of actomyosine bonds.

Introduction

The energy transport in myocardial cells from mitochondria to sites of its utilization is now
believed to be accomplished both by creatine kinase (CK)-dependent transformation of ATP
into phosphocreatine (PCr) and back, the PCr pathway[1], and by direct diffusion of ATP through
the myoplasm, the adenylate pathway[2]. For evaluation of the relative importance of both pathways
of energy supply we studied changes in the cardiac pump function under conditions both of acute CK
inhibition by iodoacetamide (IAA) and of chronic depletion of the flux through CK[3].

Materials and methods

Isolated rat hearts were perfused either in isovolumic or working modes at 37 °C with standard
Krebs–Henseleit solution[3] to which pyruvate (5 mM) was added. The cardiac output and coronary
flow were measured and left ventricular (LV) pressure was monitored on a Gould 2600S recorder
through a needle inserted into the LV cavity through its wall. Cardiac volume × pressure work was
calculated as the product of cardiac output and mean pressure in aortic chamber. The LV diastolic
stiffness index was calculated by dividing the difference between the end and minimal diastolic
pressures by LV filling volume during diastole. At steady state, the latter value was equal to the stroke
volume. LV filling for a given diastole could be monitored by the time course of LV diastolic pressure.
The time difference between stable filling pressure and pressure area was the product of the difference
and the duration of LV filling. The content of phosphate compounds participating in energetic

metabolism as well as the flux through CK were estimated by ^{31}P-NMR technique. More detailed information has been given elsewhere[3].

Acute inhibition of creatine kinase (CK) activity was caused by adding iodoacetamide (IAA, 0.5 mM) to perfusate for 15 min retrograde perfusion followed by the same period for its washout[2]. A sustained reduction of the flux through CK was produced by feeding rats a creatine-free diet to which guanidino-propionate (GP), an inhibitor of creatine entry into cardiomyocytes, was added[3].

Results and discussion

The IAA-treatment resulted in almost complete CK inhibition, its activity dropping to less than 1 per cent. Mild depletion of PCr content, from 39.0 ± 1.8 to 21.8 ± 1.6 mCM/g dry weight, was due to residual CK activity. Similar depletion of ATP content was also observed, from 18.2 ± 1.1 to 12.6 ± 0.9 μM/g, while Pi content doubled from 7.5 ± 1.2 to 15.0 ± 2.0 μM/g.

The cardiac pump function was greatly depressed. Most of these hearts could not even maintain a low level of resistance pressure (80 cm water) due to decreased LV systolic pressure (66 ± 8 mmHg) as compared to control value (105 ± 6 mmHg). This was combined with a rise in both LV minimal and end-diastolic pressures by 7 mmHg, a deep fall in LV filling pressure area and an elevation of LV diastolic stiffness from 20 ± 4 to 106 ± 15 mmHg/ml.

In isovolumic mode, a similar rise in LV end-diastolic pressure, from 15 ± 1 to 32 ± 2 mmHg, was observed. However, LV systolic pressure after IAA treatment (125 ± 5 mmHg) was only 16 per cent less ($P < 0.05$) than control value (149 ± 8 mmHg). Thus the ability to develop pressure was decreased after IAA treatment distinctly less than cardiac pump function. Obviously, this difference is due to deteriorated distensibility of LV myocardial fibres in the latter mode.

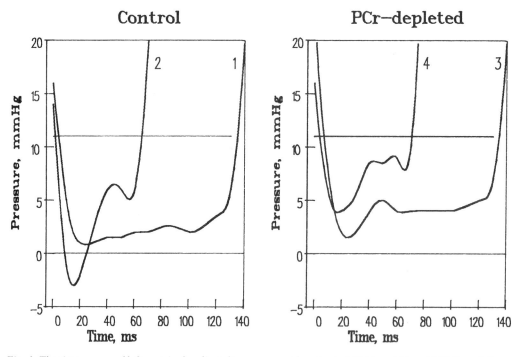

Fig. 1. *The time course of left ventricular diastolic pressure at heart rate 240 (1 and 3) and 360 (2 and 4) per min. The upper horizontal lines show the stable level of the left atrial pressure.*

Prolonged GP treatment caused distinct myocardial depletion of both creatine and PCr content. The latter fell to low levels in various groups ranging 2.7–7.9 µM/g from the normal value 29.7 ± 2.5 µM/g. In contrast, myocardial ATP content as well as CK activity were similar in both groups. Calculated ATP-PCr flux and ATP turnover rates, 400 ± 77 and 198 ± 30 µM/min/g, were lower after GP-treatment than the corresponding control values, 602 ± 65 and 330 ± 22 µM/min/g ($P < 0.01$).

The cardiac pump function at low level of resistance pressure (80 cm water) was similar in both groups in spite of slightly decreased LV systolic pressure, 100 ± 4 *vs* 116 ± 3 mmHg ($P < 0.05$). Diastolic indices were changed more apparently after GP treatment, diastolic stiffness increased from 46 ± 4 to 65 ± 7 mmHg/ml ($P < 0.05$) and LV filling pressure time area decreased from 0.95 ± 0.05 to 0.59 ± 0.08 mmHg*s ($P < 0.01$).

A gradual rise in aortic resistance up to complete disappearance of aortic outflow was followed by a more steep elevation of LV diastolic stiffness in PCr-deficient hearts up to 130 ± 17 mmHg/ml, while in control hearts this value rose only to 78 ± 9 mmHg/ml. Cardiac work increased in both groups, however, its maximal value in the GP-treated group (2312 ± 348 mmHgml/min) was 40 per cent less than in the control group (3881 ± 330, $P < 0.01$).

Another functional load, namely an increase in heart rate induced by electrostimulation of right atrium, was followed by a rise in both LV minimal and maximal diastolic pressures in PCr-deficient heart (Fig. 1B). This was combined with a diminution of LV filling pressure area and a half of these hearts could not maintain pump function at rates higher than 390 per min. In contrast, LV minimal diastolic pressure dropped in control hearts (Fig. 1A) and all hearts could pump at a rate of 420/min and more.

The predominant deterioration of LV distensibility both at acute and chronic inhibition of the flux through CK may be explained by the ability of CK bound to myofibrils to maintain functional compartmentation of adenine nucleotides, normal Ca^{2+} sensitivity of myofibrils and low myofibrillar stiffness[4]. This provides adequate supply of ATP for myosin ATPase to elicit dissociation of actomyosin cross bridges. Thus, in addition to recently postulated involvement of PCr pathway to energy supply for maximal cardiac work[3], this system seems to play the crucial role in maintenance of normal myofibrillar relaxation and distensibility. This renders PCr absolutely necessary for normal cardiac pump function.

Reference

1. Bessman, S.P. & Geiger, P.J. (1981): Transport of energy in muscle: the phosphorylcreatine shuttle. *Science* **211**, 448–452.

2. Fossel, E.T. & Hoefeler, H. (1987): Complete inhibition of creatine kinase in isolated perfused rat hearts. *Am. J. Physiol.* **252**, (*Endocrinol. Metab.* **15**), E124–E130.

3. Kapelko, V.I., Kupriyanov, V.V., Novikova, N.A., Lakomkin, V.L., Steinschneider, A.Y., Severina, M.Y., Veksler, V.I. & Saks, V.A. (1988): The cardiac contractile failure induced by chronic creatine and phosphocreatine deficiency. *J. Mol. Cell. Cardiol.* **20**, 465–479.

4. Ventura-Clapier, R., Saks, V.A., Vassort, G., Lauer, C. & Elizarova, G.V. (1987): Reversible MM-creatine kinase binding to cardiac myofibrils. *Am. J. Physiol.* **253** (*Cell. Physiol.* **22**), C444–C455.

Section V
Guanidino Compounds in Renal Insufficiency

Guanidino Compounds in Biology and Medicine, eds. by P.P. De Deyn, B. Marescau, V. Stalon and I.A. Qureshi. ©1992 John Libbey & Company Ltd., pp. 255–260.

Chapter 35

Urinary guanidinoacetic acid: don't shoot the messenger

Burton D. COHEN and Harini PATEL

Bronx-Lebanon Hospital, 1276 Fulton Avenue, Bronx, NY 10456, USA

Summary

Urinary guanidinoacetic acid (UGAA), most of which originates in renal tubular epithelium, reflects sharply and precisely the renal mass. It is, therefore, proposed as a messenger of actual or impending tubular necrosis. Unfortunately, it also reflects the state of protein malnutrition and since conditions leading to ATN are often associated with nutritional deficiency, its message is garbled. Simple oral protein supplementation with gelatine, however, can overcome the nutritional element and restore UGAA as a messenger with a warning. Neither glomerular filtration nor tubular reabsorption appear to be affected by the amino acid supplement and quantitative urine output in spot samples is most helpful when reported in relation to urinary creatinine.

Introduction

In terrestrial animals the kidney serves, like the canary in a coal mine, as a signal of hypoxia. The combination of plasma skimming coupled with the countercurrent configuration of the vasa recti produce in the medulla an 80 per cent reduction in oxygen tension in the face of which the cells of the thick ascending loop of Henle actively transport massive quantities of sodium against a considerable diffusion gradient.

Aggravating this already highly vulnerable situation are the demands of tubulo-glomerular balance. To prevent salt loss, the chemoreceptors of the macula densa monitor sodium delivery out of the proximal nephron and, when the limit for distal nephron conservation is sensed, they feed angiotensin into the afferent arterioles to reduce filtration and distal salt delivery.

The problem arises when cytotoxins, concentrated in the nadir of the loop, disrupt the sodium pumps causing excessive salt delivery, feedback vasoconstriction and further hypoxia leading to tubular necrosis. Obviously, the most useful way to minimize morbidity is to remove the cytotoxins through early recognition of ischaemia or renal angina.

Currently, the search for a messenger includes monitoring blood levels of cytotoxins or watching for the indices of uraemia. Both are late warnings, however, and correlate poorly with toxicity. Of interest are tubular specific enzymes and proteins in urine which might serve a function comparable to the role of myocardial enzymes in coronary disease.

Guanidinoacetic acid (GAA) is manufactured by tubular epithelium as shown in Fig. 1. Synthesis involves the transamidination of glycine via arginine and the principal, though not the exclusive, site of this reaction is the renal tubular epithelium. Production of GAA, therefore, reflects the integrity of the renal tubules and competes for substrate with urea synthesis and proteogenesis both of which also utilize arginine and glycine. Recently, a series of reports have appeared using the decline or

METABOLIC PATHWAY OF GUANIDINIUM COMPOUNDS

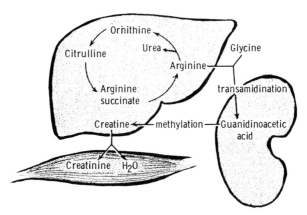

Fig. 1. Guanidinoacetic acid formation in the kidney.

disppearance of urinary guanidinoacetic acid (UGAA) as an early marker of both acute and chronic renal dysfunction[5,6,7,9,10,11]. The following study is a further exploration of UGAA as a messenger with an early warning.

Materials and methods

UGAA was measured by high performance liquid chromatography using an alkali-ninhydrin reagent for the post-column derivatization of guanidines[3,4]. A small volume of urine was diluted 1:10 with buffer at pH 2.2 and filtered through a 0.2 µm membrane filter. The diluted filtrate was then injected onto the column. Separation of the guanidino compounds was accomplished via a strong cation-exchange column using appropriate sodium citrate buffers. Detection was achieved by fluorometry following post-column reaction at 50 °C with two reagents: 0.75 N sodium hydroxide and 0.6 per cent ninhydrin. Urinary creatinine was determined by a standard kinetic method employing a Beckman autoanalyser.

Results and discussion

Figure 2 shows a plot of UGAA versus serum urea nitrogen (SUN) in a patient with pneumonia followed for 6 days on 80 mg gentamycin intravenously every 6 h. The patient developed an episode of acute tubular necrosis (ATN) from which he recovered after a month's follow-up. The rise in SUN, followed daily, was apparent 3 days after the level of UGAA fell to zero, a sentinel signal.

Thirty-seven patients receiving aminoglycosides were followed for 6–65 days. Ten developed ATN from which all recovered. UGAA fell to sentinel levels at an average of 7 days before the serum urea or creatinine rose significantly.

There are, however, problems as illustrated in Table 1. Here we compare spot urines collected from hospitalized subjects who are divided into two groups: sick patients with pneumonia *vs* stable individuals admitted for elective procedures on the orthopaedic, gynaecological and ophthalmological services. UGAA is significantly reduced in the seriously ill and the reason becomes apparent when we examine the results shown in Table 2. Here are three small groups, two of whom are normal healthy individuals and the third, severe uraemics. Obviously, the loss of kidney tissue to less than 10 per cent, as reflected in the glomerular filtration rate (GFR), leads to a loss of production of GAA and a fall in UGAA comparable to that seen in Table 1. However, a reduction in dietary substrate, presumably glycine and arginine, as achieved by committed vegetarians also results in a fall in UGAA. Interestingly, the fractional excretion of GAA, which normally reflects no tubular absorption, appears to parallel the GFR, e.g. the less delivered the greater the per cent reabsorbed.

Table 1. The role of nutritional status in urinary guanidinoacetic acid concentration;
UGAA in sickness and health hospitalized subjects

	Controls	Pneumonia
N	31	26
UGAA	53.4 ± 33.6*	4.3 ± 3.4*

*mg/g CR ± 1 SD.

Table 2. The effect of nutritional status compared to kidney disease on urinary guanidinoacetic acid output

	SGAA mg/dl	UGAA mg/d	FEGAA %	GFR ml/min
Controls (6)	0.027 ± 0.008	40 ± 3	99 ± 25	125 ± 49
Uraemics (5)	0.028 ± 0.016	3 ± 3	67 ± 21	11 ± 4
Vegetarians (6)	0.035 ± 0.005	18 ± 6	57 ± 36	76 ± 19

FE, Fractional excretion; GFR, glomerular filtration.

It is, of course, these very sick, malnourished patients with pneumonia who are exposed to cytotoxins and at the greatest risk to develop ATN. If UGAA is at the sentinel level at the outset, its usefulness as a messenger is compromised.

To overcome this problem, we tried providing substrate. We compared, in 19 severely ill patients with pneumonia, the effect on UGAA of substrate loading with either arginine HCl intravenously

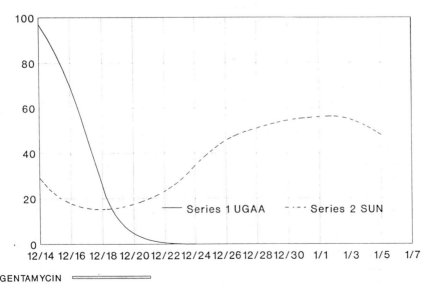

Fig. 2. Daily spot determinations of serum urea and urine guanidinoacetic acid (GAA) in a patient on anti-biotics for pneumonia. The Y-axis plots urea in mg/dl and GAA in mg/10g creatinine.

ARG∗ and GLY∗∗ SUPPLEMENTATION
SPOT URINE INCREMENTS IN mgm GAA/gm CR

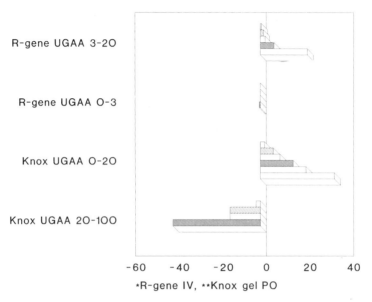

∗R-gene IV, ∗∗Knox gel PO

Fig. 3. Plot of delta UGAA following amino acid loading comparing patients with low initial values (< 20 mg/g CR) to those with higher presenting concentrations.

UGAA CONCENTRATION
MGMS PER GM CREATININE

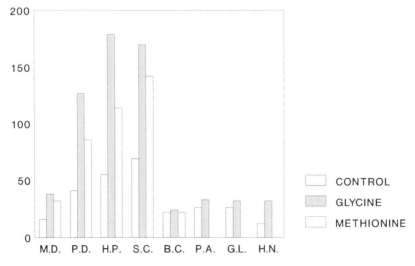

Fig. 4. Mean daily spot urine concentrations of GAA expressed in relation to creatinine after amino acid feeding.

(R-gene) at 50 g/day for one dose or glycine by mouth (Knox gelatine) at 6 g/day for 3–6 days. The increment in UGAA over the period of study is plotted in Fig. 3. Where UGAA was initially between 3 and 20 mg per g creatinine (CR), the output increased with either agent. Where malnutrition was so bad that UGAA was less than 3 mg/g CR, arginine had no effect. Presumably, the bulk of that given goes to the production of urea, the demand for which is 20-fold greater than that for either creatine synthesis or proteogenesis. Where no evidence of malnutrition was evident (UGAA was initially over 20 mg/g CR), UGAA was essentially unaffected by substrate loading.

Table 3 shows the effect of similar glycine feeding in controls compared to patients with chronic renal failure (CRF) and end-stage renal disease (ESRD) presented in terms of 24-h GAA output. The presence of renal disease inhibits totally the response to added substrate indicating that UGAA will still deliver the message warning of renal failure even in the face of artificial substrate supplementa-tion. Table 4 compares the 24-h GAA output in eight normal subjects on a vegetarian diet following 1 control week, 1 week on glycine 2 g TID and 3 days on methionine 0.5 g TID. There is a marked increase in daily GAA output (and presumably production) following glycine which is not reflected in concentration. Serum levels and fractional excretion are also unaffected by glycine while meth-ionine causes an absolute decrease which would be anticipated as a result of increased methylation[1,2].

Table 3. The effect of glycine feeding on UGAA output in the presence of moderate to severe renal disease

			Glycine and GAA (glycine* 2 g TID)	
		(GFR)	PreRx	Glycine
Urine GAA, mg/d	Controls	(92)	27.5	56.2
	CRF	(35)	10.1	11.5
	ESRD	(9)	2.3	1.8
Serum GAA, mg/dl			0.032	0.033
FE GAA, %			74	83

*Knox gelatine.

Table 4. A comparison of UGAA 24-h output *vs* spot concentrations following amino acid supplements

	UGAA output *vs* concentration			
	UGAA mg/d	UGAA mg/dl	SGAA mg/dl	FEGAA %
Control	21 ± 4	3.1 ± 1.8	0.028 ± 0.008	88 ± 38
Glycine*	60 ± 30	3.5 ± 2.1	0.030 ± 0.009	89 ± 38
Methionine	36 ± 11	4.8 ± 1.7	0.022 ± 0.002	–

*Knox gelatine.

The mean change in UGAA in spot samples presented as mg/g CR is shown in Fig. 4 which reflects the same change after glycine and methionine as demonstrated with the 24-h collection suggesting that spot samples are best analysed in relation to creatinine rather than volume. This is not surprising since urine creatinine is more constant in relation to time than is urine volume.

Figure 5 diagrams the relationships demonstrated. Glycine appears to be effective in restoring UGAA in malnourished individuals. Despite evidence that GAA shares membrane transport with glycine[12] and creatine[8], fractional excretion was unaffected by dietary supplements suggesting that the quan-tities given were totally metabolized.

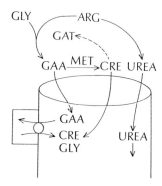

Fig. 5. Renal tubular handling of the products and substrates of creatine synthesis. GLY = glycine, ARG = arginine, GAT = glycine arginine transamidinase, GAA = guanidinoacetic acid, MET = methionine, CRE = creatine.

References

1. Cohen, B.D. (1972): Aberrations of the urea cycle in uremia. In *Uremia: pathogenesis, diagnosis and therapy*, eds. R. Kluthe, G. Berylne & B. Burton, pp. 1. Stuttgart: Georg Thieme Verlag.

2. Cohen, B.D. & Patel, H. (1987): Methionine and the control of creatine synthesis. *Ann. N.Y. Acad. Sci.* **494**, 329–331.

3. Hiraga, Y. & Kinoshita, T. (1981): Post-column derivatization of guanidino compounds in high-performance liquid chromatography using ninhydrin. *J. Chromatography* **226**, 43–51.

4. Hiraga, Y. & Kinoshita, T. (1985): High-performance liquid chromatographic analysis of guanidino compounds using ninhyrin reagent. II. Guanidino compounds in blood of patients on hemodialysis therapy. *J. Chromatography* **342**, 269–275.

5. Koller, A., Comess, T.D. & Natelson, S. (1975): Evidence supporting a proposed mechanism explaining the inverse relationship between guanidinoacetate and guanidino-succinate in human urine. *Clin. Chem.* **21**, 235–242.

6. Nakayama, S., Kiyateake, I., Shirokane, Y. & Koide, H. (1989): Effect of antibiotic administration on urinary guanidinoacetic acid excretion in renal disease. In *Guanidines 2*, eds. A. Mori, B. Cohen & H. Koide, pp. 313. New York: Plenum Press.

7. Sasaki, M., Takahara, K. & Natelson, S. (1973): Urinary guanidinoacetate/guanidinosuccinate ratio: an indicator of kidney dysfunction. *Clin. Chem.* **19**, 315–321.

8. Sims, E.A.H. & Seldin, D.W. (1949): Reabsorption of creatine and guanidinoacetic acid by the renal tubules. *Am. J. Physiol.* **157**, 14–20.

9. Tofuku, Y., Muramoto, H., Kuroda, M. & Takeda, R. (1985): Impaired metabolism of guanidinoacetic acid in uremia. *Nephron* **41**, 174–178.

10. Takano, Y., Gejyo, F., Shirokane, Y., Nakajima, N. & Arakawa, M. (1988): Urinary excretion rate of guanidinoacetic acid in essential hypertension. *Nephron* **48**, 167–168.

11. Takano, Y., Aoike, I., Gejyo, F. & Arakawa, M. (1989): Urinary excretion rate of guanidinoacetic acid as a new marker in hypertensive renal damage. *Nephron* **52**, 273–277.

12. Van Pilsum, J.F. & Canfield, T.M. (1962): Transamidinase activities *in vitro* of kidneys from rats fed diets supplemented with nitrogen-containing compounds. *J. Biol. Chem.* **237**, 2574–2577.

Guanidino Compounds in Biology and Medicine, eds. by P.P. De Deyn, B. Marescau, V. Stalon and
I.A. Qureshi. ©1992 John Libbey & Company Ltd., pp. 261–268.

Chapter 36

Factors affecting urinary excretion of creatinine

M. LAVILLE, N. POZET, D. FOUQUE, A. HADJ-AïSSA, J.P. FAUVEL and P. ZECH

Département de Néphrologie, Université Claude-Bernard Lyon I, Hôpital Edouard-Herriot, 69437 Lyon Cedex 03, France

Summary

While generally viewed as a sensitive and reliable marker of changes in renal function, plasma creatinine levels are strongly influenced by changes in either creatinine generation or excretion which are sometimes independent of renal function. The authors review the confounding factors in the assessment of creatinine excretion: inadequate urine collection and storage; changes in endogenous generation of creatinine (physical exercise, increased muscle breakdown, age, body lean mass); dietary load of creatinine; changes in extrarenal routes of excretion and changes in tubular secretion of creatinine due to urine flow rate, renal insufficiency, and drug effects. As a consequence, plasma creatinine levels should be disregarded for accurate follow-up of patients with renal failure, especially when a therapeutic intervention is being evaluated. Moreover, changes in plasma levels should not lead to invasive diagnostic procedures until extrarenal factors modifying creatinine metabolism, especially drug-related, have been eliminated.

Introduction

Serial measurements of glomerular filtration rate (GFR) are an important part of the follow-up of patients with renal diseases. Unfortunately, the cost and the time necessary for reference methods such as inuline or isotope clearances are incompatible with routine clinical use. Since creatinine is produced at a constant rate and cleared almost exclusively by glomerular filtration, it has been considered for a long time as a suitable marker of GFR for day-to-day assessment of renal function. Even if this is true in a majority of patients with stable clinical status, a large discrepancy between creatinine clearance and the true value of GFR could be observed in several circumstances, such as rapid changes in endogenous creatinine generation or dietary intake. On the other hand, in patients with renal insufficiency, tubular capacity of creatinine secretion is unequivocally preserved, depending on the extent of tubular cell damage. Both circumstances could result in an overestimation of GFR when measured by creatinine clearance only, hence increasing the risk of unadapted clinical decisions[21]. Conversely, creatinine secretion could be blocked by drugs interfering with tubular transport mechanisms. Whatever the direction, changes in plasma creatinine levels and creatinine clearance dissociated from changes in GFR impair the assessment of the rate of progression of renal insufficiency and the evaluation of therapeutic interventions. In addition, creatinine excretion is often used as a standardization parameter for serial or between-patient comparisons of urinary protein and calcium, which will be impaired by acute changes in creatinine excretion.

Metabolism of creatinine

The predominant site of creatine biosynthesis is liver, where creatine is formed from guanidinoacetate through an irreversible, non-rate-limiting, reaction. The limiting step is the synthesis of guanidino-acetate from glycine and arginine by the transamidinase, which takes place mainly in the kidney. Creatine ingested from meat represses transamidinase, which is conversely stimulated by testosterone. Creatine released into the circulation is actively concentrated by muscle for 90–98 per cent, and other tissues. Muscle creatine and phosphocreatine dehydrate to creatinine at a rate of 1.1 per cent and 2.64 per cent per day, respectively. Creatinine is then excreted in urine by glomerular filtration and proximal tubular secretion, through both organic cation and anion transport systems. Because the creatine pool is large, small changes in turnover rates result in large differences in urinary creatinine excretion.

Factors affecting urinary excretion of creatinine

Even in healthy subjects, 24 h-creatinine excretion is far from constant, since its coefficient of variation reaches 15 per cent on daily measurements for 1 week. Since urinary excretion of creatinine can be modified by either changes in creatinine generation or changes in renal handling, it could be expected that the relationships between creatinine clearance and GFR will be less disturbed by the former than by the latter. However, creatinine secretion can increase to adapt to a creatinine load whatever the source. There are no clinically relevant differences between increased endogenous production and exogenous modifications of creatinine metabolism, and they will be described successively. We will not review possible interference between either endogenous compounds or drugs with methods used for creatinine measurement.

Inadequate urine collection and storage

Inaccurate 24 h-urine collection is obviously the main confounding factor in assessment of creatinine excretion. In circumstances where creatine production is increased, a large increase in urinary creatinine could suggest the occurrence of spontaneous interconversion during urine storage. It was found that conversion rates of up to 25 per cent per day are obtained in urine maintained at pH 4 and at temperatures above 30 °C[14]. The conversion is negligible at 4 °C or after freezing, and at pH between 6 and 10, interconversion reaches an equilibrium. Errors due to this phenomenon are overall of little clinical importance, since conversion rates reach significant levels in relatively unusual storage conditions. These findings emphasize the need for accurate urine collection and storage during studies on creatinine metabolism.

Changes in endogenous generation of creatinine

Physical exercise

Extremely strenuous exercise, and to a smaller extent emotional stress, can increase urinary output of creatinine by 5–10 per cent[16]. Following a 56 km foot race, a sustained 48 h increase in creatinine clearance was described in association with an increase in plasma levels of creatinine related to increased creatinine production[18]. However it is not clear whether the increase in creatinine clearance due to increased tubular secretion, was only related to the endogenous load of creatinine or also to the release of amino-acids from damaged muscle cells.

Increased muscle breakdown

Both fever and severe trauma are accompanied by an increase in creatinine excretion up to 100 per cent in the acute phase[16]. This could be due to increased muscle breakdown from stimulated interleukin-1, tumour necrosis factor and other acute phase protein secretion[1,12].

Age

It is well known that creatinine clearance decreases with age, on the average from 10 ml/min every 10 years after the fourth decade. However, at similarly decreased levels of GFR, the elderly have

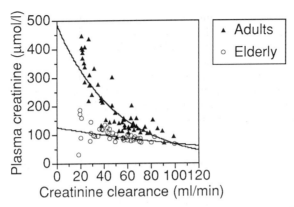

Fig. 1. Relationship between plasma creatinine level and creatinine clearance, in 58 adults (aged 19 to 65, mean 50.0 ± 10.4, ▲) and 42 elderly (aged 66.1 to 94.8, mean 78.9 ± 6.9, ○) with normal to severely impaired renal function.

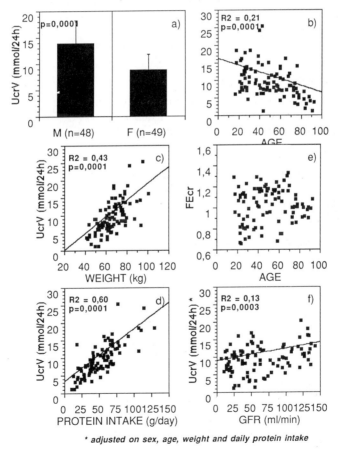

* adjusted on sex, age, weight and daily protein intake

Fig. 2. Relationship between daily excretion of creatinine and: (a) sex; (b) age; (c) weight; (d) protein intake; (e) relationship between age and tubular secretion of creatinine expressed as fractional excretion (Ccr/Cin) and adjusted on both GFR and daily protein intake; (f) relationship between GFR and daily excretion of creatinine adjusted on sex, age, weight and protein intake.

plasma levels of creatinine much lower than adult patients (Fig. 1)[15]. This is due to decreased creatinine production from decreased body lean mass, as shown by a 50 per cent lower urinary creatinine excretion (Fig. 2b). Creatinine production decreases from on average 20 mg/kg/day before 30 years to 12 mg/kg/day after 80 years. The decrease in meat intake with age could also play a role, but there is no decrease in tubular secretion of creatinine compared to adults with similar GFR values (Fig. 2e). Accordingly, GFR in elderly can be predicted from plasma creatinine only through the use of formulas including age, weight and sex[9,19].

Body lean mass

In healthy subjects, creatinine excretion depends on sex, body size and diet protein and creatinine content (Fig. 2a, 2c, 2d)[5,13,16,17]. In immobilized acutely ill patients or in patients with neurological diseases, body lean mass could be less than expected according to age. In such patients, Cockroft's formula could overestimate creatinine clearance by 46 per cent[10], but it is not known whether this overestimation also applies to GFR.

Dietary load of creatinine

The contribution of exogenous creatinine intake should not be underestimated, since creatinine excretion was shown to increase up to 50 per cent when subjects fed with low-creatinine formulas for 11 weeks were returned to *ad libitum* diet[5,17]. The creatine and creatinine content of meat is on the average 350 mg and 19 mg per 100 g, the latter being increased to 80 mg by cooking[24]. The renewal of interest in the renal handling of an exogenous load of creatinine came from the description of renal functional reserves elicited by acute protein loads[6]. Whereas initial studies using creatinine clearance claimed increases in GFR as large as 60 per cent following a meat meal, it was found to be less than 30 per cent using inulin clearance[22]. These results highlight the fact that creatinine secretion could adapt in a few minutes to an acute load. Even in subjects fed with a creatine-free diet, the intake of the amino acid precursors arginine and glycine, is followed by an increase in urinary creatinine excretion[16].

Changes in extrarenal routes of excretion: endogenous degradation

Creatinine excretion was found to be diminished in patients with chronic renal failure despite stable plasma levels and normal body lean mass and diet creatinine content (Fig. 2f). Since the work of Jones & Burnett[20] using labelled creatinine, it is generally admitted that 16 to 66 per cent of creatinine formed could be metabolized or excreted via extrarenal routes. This was further studied by Mitch and co-workers[25,26] who measured a constant extrarenal metabolism of, on the average, 26 per cent of daily creatinine production, and showed that creatinine accumulation into body fluids was not involved in decreased creatinine excretion. Rather, they suggested that besides creatinine degradation by the gut flora, direct conversion of creatinine into creatine could account for a significant part of extrarenal creatinine metabolism.

Changes in tubular secretion of creatinine

Urine flow rate

In 25 healthy kidney donors studied by our group, creatinine clearance rose from 129 ± 17 to 137 ± 18 ($P < 0.01$) when urine flow rate increased from 0.54 ± 0.14 to 1.56 ± 1.32 ml/min following an oral 200 ml water load[28]. Accordingly, creatinine clearance measured during the period of high urine flow rate was significantly higher than simultaneously measured inulin clearance. We also observed an acute increase in creatinine clearance following iv administration of a loop diuretic, ethacrynic acid (Fig. 3). In healthy volunteers, fractional creatinine clearance (Ccr/Cin) rose from 1.05 in the dehydrated state induced by furosemide infusion (urine flow rate: 8.4 ml/min), to 1.45 after both oral and iv rehydration (urine flow rate: 23.2 ml/min). Accordingly, tubular secretion of creatinine was

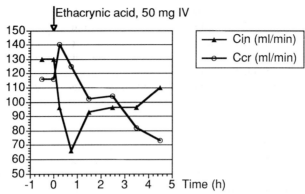

Fig. 3. Concomitant changes in creatinine (○) and inulin (▲) clearances following acute diuresis induced by ethacrynic acid, 50 mg iv, in healthy volunteers.

estimated at 47 per cent of filtered creatinine in a hydrated state, whereas in a dehydrated state there was more likely to be a simultaneous reabsorption[31].

Renal insufficiency

An increase in the tubular component of creatinine excretion is well characterized in chronic renal insufficiency. However it is not known whether this is due only to an increased secretion induced by increased plasma levels[23], or also to a decreased reabsorption related to tubule alterations and elevated urine flow rate[31]. The extent of the dissociation between creatinine clearance and GFR depends on both the degree of renal insufficiency and the primary disease. There is indeed an inverse correlation between fractional tubular secretion of creatinine and GFR, with a significant increase in fractional clearance from 1.19 at GFR > 80 ml/min to 1.92 at GFR < 40 ml/min[30], or from 0.97 at GFR > 110 ml/min to 1.48 at GFR < 30 ml/min (unpublished result).

For instance, tubular dysfunction is prominent in lupus nephritis, where fractional clearances of creatinine up to 2.21 have been observed[30]. In two patients with advanced renal insufficiency, the amount of secreted creatinine accounted for up to 70 per cent of the total urinary creatinine[2]. In a similar way, increased fractional excretion of β2 microglobulin and DMSA was found during exacerbation of proliferative lupus nephritis[32]. Interestingly, cimetidine could be able to reduce creatinine secretion by injured tubular cells, and improve the accuracy of creatinine clearance as a marker of GFR in lupus patients[29] as well as in other glomerulopathic patients[30].

In patients with diabetic nephropathy, conversely, there is a dissociation between the tubular handling of creatinine and β2 microglobulin, since creatinine clearance appears to be a better predictor of GFR than β2 microglobulin clearance[11].

Drugs modifying tubular handling of creatinine

Since creatinine is an amphoteric compound which undergoes active tubular secretion by both organic cation and anion transport systems[4], urinary excretion can be modified by drugs interfering with either cation transport system (i.e. cimetidine) or anion transport system (i.e. probenecid), or both.

Trimethoprim. Administration of trimethoprim 80 to 160 mg twice daily, alone or in combination with sulphamethoxazole, was found to result in a short-term and fully reversible 20 per cent increase in plasma creatinine levels in 21 patients and two healthy volunteers. There were no concomitant changes in iothalamate clearance, indicating an effect of trimethoprime on tubular secretion of creatinine[4]. This effect was further enhanced by oral creatinine loading, and attributed to a competition between creatinine and trimethoprim at the level of the base transport system. This effect could be masked in chronic renal failure by reduction of tubular secretory capacity.

Probenecid. During a systematic comparison of several markers of GFR in healthy volunteers, it was found that probenecid 1 g intravenously reduced by 22 per cent the renal clearance of inulin, but significantly reduced by 34 per cent and 63 per cent the fractional clearances of iothalamate and creatinine, respectively[27]. However the creatinine clearance which was 40 per cent higher than inulin clearance at the basal state, remained significantly different even after probenecid administration. In rats, probenecid did not alter the renal extraction of ^{51}Cr-EDTA, but reduced significantly by 80 per cent the extraction of PAH and by 40 per cent that of ^{125}I iothalamate indicating competition mainly at the level of organic anion transport system.

Cimetidine. Administration of cimetidine 0.8 to 2.0 g/day was found to result in a sustained increase of plasma creatinine levels in 10 per cent of patients, until the drug was discontinued. In healthy volunteers, cimetidine 800 mg orally did not change plasma creatinine concentration, but significantly decreased creatinine clearance from 117 to 90 ml/min, even in maximal diuresis conditions[8]. In healthy volunteers, cimetidine 5 mg/kg iv induced a significant decrease in both endogenous and exogenous creatinine clearances without concomitant changes in inulin clearance[7]. This was attributed to a 50 per cent decrease in tubular secretion of creatinine by competition at the level of the organic bases transport system. In glomerulopathic patients with a mean GFR of 40 ml/min, Shemesh[30] found that a 300 mg cimetidine infusion decreased fractional clearance of creatinine from 1.7 to 1.2. This interaction was further used in patients with lupus nephritis, in order to improve the accuracy of GFR estimation by creatinine clearance[29].

Non steroidal anti-inflammatory drugs. A decrease in creatinine clearance has been reported following administration of a number of NSAID[3]. This change is assumed to represent a decrease in GFR secondary to increased renal vascular resistance predominantly at the preglomerular level. However, there is a possibility for this decrease to result in part from decreased tubular secretion of creatinine, since salicylate could compete for the organic cation transport system.

Table 1. Relative magnitude of extrarenal factors affecting urinary creatinine excretion

Factor	Magnitude
Creatine to creatinine interconversion during storage	+25%
Normal daily variation	± 4 to 8%
Very strenuous exercise	+ 5 to 10%
Switching from meat diet to creatinine free diet	−10 to 30%
Acute meat load	+ 30%
Menstrual cycle	+ 10 to 15%
Severe infection or trauma	+ 20 to 100%
Age over 65 years	− 50%
Extrarenal metabolism	−25%
Acute increase in urine flow rate	+40%
Tubular secretion/drug interferences	± 30%

Conclusion

While generally viewed as a reliable marker of even slight changes in renal function, plasma creatinine level is strongly influenced by changes in either creatinine generation or excretion, which are sometimes independent of renal function (see Table 1). As a consequence, first, plasma creatinine levels should be disregarded for accurate follow-up of patients with renal failure especially when a therapeutic intervention is being evaluated; and second, acute changes in plasma levels should not lead to aggressive diagnostic procedures until extrarenal factors modifying creatinine metabolism, especially drug-related, have been eliminated.

References

1. Baracos, V., Rodemann, P., Dinarello, C.A. & Goldberg, A.L. (1983): Stimulation of muscle protein degradation and prostaglandin release by leukocytic pyrogen (Interleukin-1). *N. Engl. J. Med.* **308**, 553–558.

2. Bastl, C., Katz, M.A. & Shear, L. (1977): Uremia with low serum creatinine – an entity produced by marked creatinine secretion. *Am. J. Med. Sci.* **273**, 289–292.

3. Bergamo, R., Cominelli, F., Kopple, J.D. & Zipser, R.D. (1989): Comparative acute effects of aspirin, diflunisal, ibuprofen and indomethacin on renal function in healthy man. *Am. J. Nephrol.* **9**, 460–463.

4. Berglund, F., Killander, J. & Pompeius, R. (1975): Effect of trimethoprim-sulfamethoxazole on the renal excretion of creatinine in man. *J. Urol.* **114**, 802–808.

5. Bleiler, R.E. & Schedl, H.P. (1962): Creatinine excretion: variability and relationships to diet and body size. *J. Lab. Clin. Med.* **59**, 945–955.

6. Bosch, J.P., Saccagi, A. Lauer, A. *et al.* (1983): Renal functional reserve in humans. Effect of protein intake on glomerular filtration rate. *Am. J. Med.* **75**, 943–950.

7. Burgess, E., Blair, A., Krichman, K. & Cutler, R.E. (1982): Inhibition of renal creatinine secretion by cimetidine in humans. *Renal Physiol.* **5**, 27–30.

8. Burland, W.L., Gleadle, R.I., Mills, J.G., Sharpe, P.C. & Wells, A.L. (1977): The effect of cimetidine on renal function. In *Proceedings of the Second International Symposium on Histamine H2-Receptors Antagonists*, eds. W.L. Burland & M. Alison Simkins, pp. 67–73. Amsterdam. Oxford, Excerpta Medica.

9. Cockroft, D.W. & Gault, M.H. (1976): Prediction of creatinine clearance from serum creatinine. *Nephron* **16**, 31–41.

10. Drinka, P.J. (1987): Estimating creatinine clearance from serum creatinine in chronically immobilized nursing home residents. *Nephron* **47**, 310–311.

11. Feehally, J., Taverner, D., Burden, A.C. & Walls, J. (1983): Predictors of renal function in diabetic and non-diabetic renal disease. *Clin. Chim. Acta* **133**, 169–175.

12. Flores, E.A., Bistrian, B.R., Pomposelli, J.J., Dinarello, C.A., Blackburn, G.L. & Istfan, N.W. (1989): Infusion of tumor necrosis factor/cachectin promotes muscle catabolism in the rat. *J. Clin. Invest.* **83**, 1614–1622.

13. Forbes, G.B. & Bruining, G.J. (1976): Urinary creatinine excretion and lean body mass. *Am. J. Clin. Nutr.* **29**, 1359–1366.

14. Fuller, N. & Elia, M. (1988): Factors influencing the production of creatinine: implications for the determination and interpretation of urinary creatinine and creatine in man. *Clin. Chim. Acta* **175**, 199–210.

15. Hadj-Aïssa, A., Dumarest, C., Maire, P. & Pozet, N. (1990): Renal function in the elderly. *Nephron* **54**, 364–365.

16. Heymsfield, S.B., Arteaga, C., McManus, C., Smith, J. & Moffitt, S. (1983): Measurement of muscle mass in humans: validity of the 24-hour creatinine method. *Am. J. Clin. Nutr.* **37**, 478–494.

17. Hoogwerf, B.J., Laine, D.C. & Greene, E. (1986): Urine C-peptide and creatinine (Jaffé method) excretion in healthy young adults on varied diets: sustained effect of varied carbohydrate, protein, and meat content. *Am. J. Clin. Nutr.* **43**, 350–360.

18. Irving, R.P., Noakes, T.D., Burger, S.C., Myburgh, K.H., Querido, D. & Van Zyl Smit, R. (1990): Plasma volume and renal function during and after ultramarathon running. *Med. Sci. Sports Exerc.* **22**, 581–587.

19. Jelliffe, R.W. (1973): Creatinine clearance: a bedside estimate. *Ann. Intern. Med.* **79**, 604–605.

20. Jones, J.D. & Burnett, P.C. (1974): Creatinine metabolism in humans with decreased renal function: creatinine deficit. *Clin. Chem.* **20**, 1204–1212.

21. Labeeuw, M., Fouque, D., Hadj-Aïssa, A., Laville, M., Zech, P. & Pozet, N. (1990): Est-il encore possible d'utiliser la créatinine pour suivre l'évolution de l'insuffisance rénale. *Néphrologie* **11**, 231–235.

22. Laville, M., Hadj-Aïssa, A., Pozet, N., Le Bras, J.H., Labeeuw, M. & Zech, P. (1989): Restrictions on use of creatinine clearance for measurement of renal functional reserve. *Nephron* **51**, 233–236.

23. Levey, A.S., Berg, R.L., Gassman, J.J., Hall, P.M., Walker, W.G. & the MDRD Study Group (1989): Creatinine filtration, secretion and excretion during progressive renal disease. *Kidney Int.* **36**, S73–S80.

24. Mayersohn, M., Conrad, K.A. & Achari, R. (1983): The influence of a cooked meat meal on creatinine plasma concentration and creatinine clearance. *Br. J. Pharmacol.* **15**, 227–230.

25. Mitch, W.E., Collier, V.U. & Walser, M. (1980): Creatinine metabolism in chronic renal failure. *Clin. Sci.* **58**, 327–335.

26. Mitch, W.E. & Walser, M. (1978): A proposed mechanism for reduced creatinine excretion in severe chronic renal failure. *Nephron* **21**, 248–254.

27. Odlind, B., Hällgren, R., Sohtell, M. & Lindström, B (1985): Is ^{125}I iothalamate an ideal marker for glomerular filtration? *Kidney Int.* **27**, 9–16.

28. Pozet, N., Labeeuw, M., Kaffa, I., Hadj-Aïssa, A., Cochat, P., Zech, P. & Traeger, J. (1985): Clairance de la créatinine à débit urinaire faible. *Néphrologie* **6**, 78.

29. Roubenoff, R., Drew, H., Moyer, M., Petri, M., Whiting-O'Keefe, Q. & Hellmann, D.B. (1990): Oral cimetidine improves the accuracy and precision of creatinine clearance in lupus nephritis. *Ann. Intern. Med.* **113**, 501–506.

30. Shemesh, O., Golbetz, H., Kriss, J.P. & Myers, B.D. (1985): Limitations of creatinine as a filtration marker in glomerulopathic patients. *Kidney Int.* **28**, 830–838.

31. Sjöstrom, P.A., Odlind, B.G. & Wolgast, M. (1987): Extensive tubular secretion and reabsorption of creatinine in humans. *Scand. J. Urol. Nephrol.* **22**, 129–131.

32. Ter Borg, E.J., de Jong, P.E., Meijer, S.S. & Kallenberg, C.G.M. (1991): Tubular dysfunction in proliferative lupus nephritis. *Am. J. Nephrol.* **11**, 16–22.

Guanidino Compounds in Biology and Medicine, eds. by P.P. De Deyn, B. Marescau, V. Stalon and
I.A. Qureshi. ©1992 John Libbey & Company Ltd., pp. 269–274.

Chapter 37

Serum creatinine and creatinine clearance in preterm neonates

N. GORDJANI[1], H. PETRI[2], R. BURGHARD[3], R. PALLACKS[2], J. LEITITIS[4] and
M. BRANDIS[4]

[1]*Klinikum der Albert-Ludwigs-Universität, Universitäts-Kinderklinik, Mathildenstrasse 1, D–7800 Freiburg,
Germany;* [2]*University of Marburg, Department of Paediatrics, Deutschhausstr. 12, 3550 Marbug, Germany;*
[3]*Children's Hospital Memmingen, Germany;* [4]*University of Freiburg, Department of Paediatrics, Germany*

Summary

In order to assess renal function in the early period of adaptation to extrauterine life 61 otherwise healthy, appropriate
for gestational age, preterm newborns were studied on postnatal days 1, 2, 4–5, 8–10 and 28–32. Patients were divided
into two groups of 34 very premature neonates (gestational age: 27–32 weeks) and 27 moderately premature neonates
(gestational age: 33–37 weeks). Serum creatinine and creatinine in timed urine samples was measured according to a
kinetic approach to the method of Jaffe. Creatinine clearance was calculated according to the formula: urine creatinine
×urine output/serum creatinine [ml/min], and referred to body surface area. Estimated creatinine clearance was assessed
by the formula: $l \times k$/serum creatinine, where l is body length in cm and k is 29 representing a factor depending on
muscle mass, and compared to the calculated creatinine clearance. In addition creatinine concentration was measured
in 171 specimens of amniotic fluid from healthy women with normal pregnancies.

Serum creatinine was higher, and both calculated and estimated creatinine clearance lower in very preterm infants.
Creatinine excretion was not different between the two groups. Calculated creatinine clearance correlated significantly
with the estimated creatinine clearance from the second postnatal day onwards. Gestational age correlated with estimated
and calculated creatinine clearance. The correlation with estimated creatinine clearance however was stronger than with
calculated creatinine clearance.

Creatinine concentration in amniotic fluid increased significantly from the 16th to the 38th week of gestation. Towards
the end of gestation markedly higher values of creatinine concentrations in amniotic fluid than in corresponding maternal
serum were found.

It is concluded that estimation of creatinine clearance by using body length and serum creatinine provides a useful
method of assessing neonatal glomerular function in preterm newborns even in the first week of life. It correlates more
strongly with gestational age and therefore seems to be more accurate than the calculated creatinine clearance. On the
other hand calculated creatinine clearance is more difficult to obtain because timed urine has to be collected. As a marker
of foetal glomerular function the concentration of creatinine in amniotic fluid increased significantly during gestation
to almost twice the maternal serum creatinine levels.

Introduction

With advancing medical progress more immature infants can be treated and survive in
paediatric intensive care units. Rapid postnatal changes in fluid and mineral homeostasis
and severe haemodynamic changes in the presence of immature kidney function predispose
these neonates to develop fatal imbalances. Morphological maturation of the kidneys, i.e. induction
of new nephrons, is not completed before the 32nd week of gestation. Therefore very preterm neonates

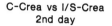

C-Crea vs I/S-Crea
2nd day

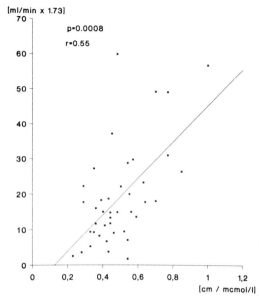

Table 2. Correlation between gestational age and estimated and calculated creatinine clearance in the first 10 days of life

	P		r	
	Calc-ulated	Esti-mated	Calc-ulated	Estimated
1	0.004	0.0182	0.53	0.42
2	0.023	0.0009	0.23	0.53
4–5	0.021	0.0001	0.35	0.78
8–10	0.016	0.0004	0.51	0.73

Lower P values and higher correlation coefficients (r) demonstrate stronger correlation between gestational age and estimated rather than calculated creatinine clearance.

Fig. 3. Significant correlation between the ratio: body length/serum creatinine (l/SCcr) and calculated creatinine clearance (Ccr) on the second day of life is indicated. Creatinine clearance is represented on the y-axis and l/Scr is shown on the x-axis. The correlation coefficient r is 0.55.

Creatinine concentrations in amniotic fluid are shown in Fig. 4. Values increased significantly more than twofold from 6.9 mg/l in the 16–17th week of gestation to 15.9 mg/l at term in the 37th–38th week.

Creatinine concentrations in amniotic fluid and corresponding maternal serum demonstrated an increasing difference during gestation with significantly lower maternal serum levels.

Discussion

Serum creatinine on the first day of life reflects maternal levels and thus is similar in moderately and very preterm infants. In order to prevent patent ductus arteriosus botalli and severe respiratory problems fluid intake is restricted to 60–80 ml/kg body weight in preterm newborns. Thus the increase of Scr in both groups on the second day was due to haemoconcentration following an intended weight loss. Accordingly estimated, in contrast to calculated, Ccr decreased slightly during this period. The following increase reflected rapid glomerular maturation in both groups. As expected moderately preterms displayed a markedly higher Ccr. Estimated and calculated Ccr in very immature neonates after 4 weeks equalled those of moderately immature neonates at 10 days of life although conceptional age (i.e. gestational age plus postnatal age) was slightly lower in very preterm newborns in this period. This may be an expression of postpartal acceleration of glomerular function as an adaptive reaction to preterm birth. Adaptive acceleration of renal tubular rather than glomerular function has been reported by others[1,2]. Already from the second day on, calculated Ccr correlated significantly with estimated Ccr showing the adequate validity of this parameter for the assessment of renal function even in the early period of postnatal adaptation to extrauterine life. A positive correlation of gestational age and calculated Ccr existed, but was even stronger with estimated Ccr. This may be due to inaccuracies in collection of urine samples which is difficult especially in preterm neonates weighing less than 1000 g. Urinary concentrations of creatinine were highly variable ranging from 250 to 14000 μmol/l and so was urinary excretion of creatinine even if correlated to body surface area. Additional factors which are more constant in human serum than in urine, such as pH and unknown chromogen

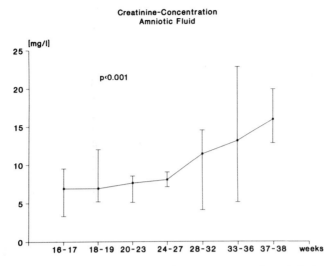

Fig. 4. Creatinine concentration in mg/l in amniotic fluid increases significantly during gestation from the 16th to 38th week. Values are given as median and range.

substances, may disturb measurement of creatinine in urine, especially in routine analysis, leading to misjudgement of renal function.

Within each group the value of k was not the same for all patients. It may even not remain the same for a given patient during a major change in body habitus. In this study it was not possible to carry out many serial measurements in the same patient. Thus the individual values for k could not be determined. Schwartz and associates[11] examined a large group of premature infants for a sufficient long period of 15 months and were able to obtain serial values of the same patient. They could show that 77 per cent of k values were between 17.6 and 44. In the presence of severe malnutrition, e.g. in anorectic adolescents, the reliability of the estimation of Ccr from Scr and body length is limited, and more anthropometric measurements such as arm muscle volume may be necessary[4]. Intrauterine maturation of renal glomerular function could be demonstrated by a steady and significant increase of creatinine concentration in amniotic fluid. Corresponding values in maternal serum were markedly lower from the 32nd week of gestation showing active foetal secretion of creatinine into amniotic fluid. However the variability was considerable and no reliable estimation of foetal glomerular function could be derived. Tubular rather than glomerular parameters have been advocated for the assessment of foetal renal function[5,6].

These data demonstrate that the estimation of Ccr using body length and Scr is a reliable method of assessing glomerular function in appropriate for gestational age very premature newborns in the first week of life. Estimation of Ccr correlated more strongly with gestational age than calculation of Ccr and may even be more accurate in clinical practice. In addition it is easier to obtain because no timed urine has to be collected. Premature neonates with a gestational age of less than 33 weeks have reduced glomerular function as compared to preterm newborns with a gestational age of more than 32 weeks. Creatinine concentration in amniotic fluid increases in the course of gestation indicating maturation of foetal glomerular function. However variability is considerable and thus no reliable estimation of foetal GFR is possible by the sole determination of this parameter.

Acknowledgements

The authors thank Mrs. C. Vestweber for valuable technical assistance in the preparation of the manuscript.

References

1. Al-Dahhan, J., Haycock, G.B., Chantler, C. & Stimmler, L. (1983): Sodium homeostasis in term and preterm neonates. *Arch. Dis. Child.* **58**, 335–342

2. Arant, B.S. (1978): Developmental patterns of renal functional maturation compared in the human neonate. *J. Pediatr.* **92**, 705–712.

3. Brion, L.P., Fleischmann, A.R., McCarton, C. *et al.* (1986): A simple estimate of glomerular filtration rate in low birth weight infants during the first year of life: noninvasive assessment of body composition and growth. *J. Pediatr.* **109**, 698–707.

4. Brion, L.P., Boeck, M.A., Gauthier, B. *et al.* (1989): Estimation of glomerular filtration in anorectic adolescents. *Pediatr. Nephrol.* **3**, 16–21.

5. Burghard, R., Pallacks, R., Gordjani, N. *et al.* (1987): Microproteins in amniotic fluid as an index of changes in fetal renal function during development. *Pediatr. Nephrol.* **1**, 574–580.

6. Glick, P.L., Harrison, M.R., Golbus, M.S. *et al.* (1985): Management of the foetus with congenital hydronephrosis II. Prognostic criteria and selection for treatment. *J. Pedratr. Surg.* **20**, 376–387.

7. Gordjani, N., Burghard, R., Leititis, J.U. & Brandis, M. (1988): Serum creatinine and creatinine clearance in healthy neonates and prematures during the first 10 days of life. *Eur. J. Pediatr.* **148**, 143–145.

8. Ross, B., Cowett, R.M. & Oh, W. (1977): Renal function of low birth weight infants during the first two months of life. *Pediatr. Res.* **11**, 1162–1164.

9. Schwartz, G.J., Haycock, G.B., Edelmann, C.M. Jr. *et al.* (1976): A simple estimate of glomerular filtration rate in children derived from body length and plasma creatinine. *Pediatrics* **58**, 259–263.

10. Schwartz, G.J., Feld, L.G. & Langford, D.J. (1984): A simple estimate of glomerular filtration rate in full-term infants during the first year of life. *J. Pediatr.* **104**, 849–854.

11. Schwartz, G.J., Brion, L.P. & Spitzer, A. (1987): The use of plasma creatinine concentration for estimating glomerular filtration rate in infants, children and adolescents. *Pediatr. Clin. North. Am.* **34**, 571–590.

Guanidino Compounds in Biology and Medicine, eds. by P.P. De Deyn, B. Marescau, V. Stalon and
I.A. Qureshi. ©1992 John Libbey & Company Ltd., pp. 275–280.

Chapter 38

Urinary excretion rate of guanidinoacetic acid as a sensitive indicator of early-stage nephropathy

Makoto ISHIZAKI, Hiroshi KITAMURA[1], Yoshio TAGUMA[1], Kazumasa AOYAGI[2]
and Mitsuharu NARITA[2]

*Kidney Centre, Eijinkai Nagano Hospital, 4–29 Higashi-machi, Furukawa 989–61, Miyagi, Japan; [1]Kidney Centre,
Sendai Shakaihoken Hospital, 3–16–1 Tsutsumimachi, Sendai 981, Japan; [2]Department of Internal Medicine,
University of Tsukuba, 1–1–1 Tennohdai, Tsukuba 305, Japan*

Summary

The serum creatine (CRT) concentration controls the renal production of guanidinoacetic acid (GAA) as well as its excretion into the urine. We measured blood and urine GAA and CRT in patients with early-stage nephropathy and investigated whether the urinary excretion rate of GAA, with plasma CRT levels considered, might be an early indicator of renal dysfunction. The subjects consisted of 30 patients with various types of nephropathy but Ccr of ≥ 70 ml/min and 14 healthy controls. Taking into account plasma CRT which was correlated with urinary GAA and CRT, we investigated GAA urinary excretion dynamics in the control and nephropathy groups from urinary GAA/urinary CRT. This value was 9.7 ± 8.1 in the former and significantly lower (1.0 ± 0.8) in the latter ($P < 0.001$). Conclusively, urinary GAA/urinary CRT is extremely useful as a noninvasive indicator of early-stage nephropathy.

Introduction

Guanidinoacetic acid (GAA) is mainly synthesized in the kidney[15], and serves as a precursor of creatine which is indispensable to the energy metabolism of the muscle. In 1963, Bonas *et al.*[1] first reported that the urinary excretion of GAA was reduced in patients with renal failure as compared with controls. This was followed by a similar report by Sasaki *et al.*[18]. Kadono *et al.*[12] stated that the normal value for urinary GAA excretion was difficult to determine because of great individual and physiological variations in normal humans. They also demonstrated by creatine (CRT) loading experiments that the serum concentration of CRT controlled the urinary excretion of its precursor GAA. We therefore determined the plasma and urinary concentrations of GAA and CRT in patients with various early-stage nephropathies, and examined whether the urinary excretion rate of GAA could be used as an indicator of early-stage nephropathy.

Subjects and methods

Thirty patients (16 men, 14 women) aged 15–66 years (mean: 35.6 years), who were admitted to Sendai Shakai-Hoken Hospital for renal disorders, participated in the study. Creatinine clearance (Ccr) in these patients was over 70 ml/min. Biopsy specimens were obtained from all patients to confirm histopathologically the presence or absence of nephropathy. Of the patients with nephropathy, nine (30 per cent) were receiving 20–30 mg/day oral prednisolone at the time of blood and urine sampling. The normal control group comprised 14 healthy adults (13 men, 1 woman) aged 25–48 years (mean: 34.4 years). Every urine specimen was a part of a 24-h urine collection. Blood was collected in conjunction with the measurement of 24-endogenous creatinine clearance (Ccr).

The concentration of GAA in urine was determined using an enzymatic method[19,20] with a slight modification. The concentrations of GAA and CRT in plasma were determined by a high performance liquid chromatography (HPLC) apparatus assembled according to Hung et al.[11].

Urinary CRT (u-CRT) was measured by an enzymatic method (Toyobo, Japan).

Results

Histopathological diagnosis in 30 cases of nephropathy (Table 1)

Table 1. Histopathological diagnosis in the studied patients with nephropathy. Renal biopsy specimens were obtained from all patients as soon as possible after admission

Type	No. of cases
IgA nephropathy	16
Benign nephrosclerosis	3
Minimal change nephrotic syndrome	3
Membranous nephropathy	3
Purpura nephritis	1
Acute glomerulonephritis	1
Interstitial nephritis	1
Membranoproliferative glomerulonephritis	1
Hepatic glomerulosclerosis	1
Total	30

IgA nephropathy was diagnosed in 16 cases, benign nephrosclerosis, minimal change nephrotic syndrome and membranous nephropathy in three cases each, and purpura nephritis, acute glomerulonephritis, interstitial nephritis, membranoproliferative glomerulonephritis and hepatic glomerulosclerosis in one case each.

Comparison of laboratory findings between the normal and nephropathy groups (Table 2)

Table 2. Laboratory findings in this study (mean ± SD)

		Controls (n = 14)	Patients (n = 30)
Ccr	(ml/min)	97.7 ± 14.7	89.3 ± 21.6
Urinary excretion of GAA	(mg/day)	88.8 ± 41.9	59.6 ± 44.7*
Plasma GAA	(μg/dl)	41.0 ± 9.8	30.4 ± 9.5**
CGAA/Ccr		1.31 ± 0.52	1.44 ± 1.07
Urinary excretion of CRT	(mg/day)	19.8 ± 22.2	142 ± 190***
Plasma CRT	(μg/dl)	451 ± 187	675 ± 331*
CCRT/Ccr		0.02 ± 0.02	0.12 ± 0.13***

$*P < 0.05$, $**P < 0.01$, $***P < 0.001$ vs controls;
Ccr = creatinine clearance; GAA = guanidinoacetic acid; CGAA = GAA clearance; CRT = creatine; CCRT = CRT clearance.

There were no significant differences in Ccr or the ratio of GAA clearance (CGAA) to Ccr between the normal and nephropathy groups. Urinary GAA (u-GAA) was 59.6 ± 44.7 mg/day for the nephropathy group and 88.8 ± 41.9 mg/day for the normal group, with the former being significantly

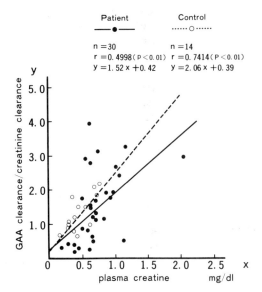

Fig. 1. Correlation between plasma creatine and GAA clearance/creatinine clearance.

lower ($P < 0.05$). Plasma GAA (p-GAA) was 30.4 ± 9.5 μg/dl for the nephropathy group and 41.0 ± 9.8 μg/dl for the normal group, with the former being significantly lower ($P < 0.01$).

The excretion of CRT in 24-h urine (u-CRT) was 142 ± 190 mg/day for the nephropathy group and 19.8 ± 22.2 mg/day for the normal group, with the value for the nephropathy group significantly higher at $P < 0.001$. Plasma CRT (p-CRT) was 675 ± 331 μg/dl for the nephropathy group and 451 ± 187 μg/dl for the normal group, with a significantly higher ($P < 0.05$) value for the nephropathy group.

The ratio of CRT clearance (CCRT) to Ccr was significantly higher for the nephropathy group at $P < 0.001$.

Correlation between p-CRT and CGAA/Ccr

p-CRT was positively correlated with CGAA/Ccr in both the normal and nephropathy groups, although the correlation was stronger in the former. When the p-CRT value exceeded 0.30 mg/dl in the normal group and 0.38 mg/dl in the nephropathy group, the CGAA/Ccr value exceeded 1, and GAA was secreted from the renal tubular cells. When the p-CRT value was low, reabsorption through the renal tubules occurred (CGAA/Ccr < 1) (Fig. 1).

Interrelationships among p-GAA, p-CRT, u-GAA, u-CRT and u-GAA/u-CRT

In both the normal and nephropathy groups, significant positive correlations were detected between p-CRT and u-GAA (Fig. 2), p-CRT and u-CRT as well as between u-GAA and u-CRT. A significant negative correlation was found between u-CRT and u-GAA/u-CRT. A significant negative correlation

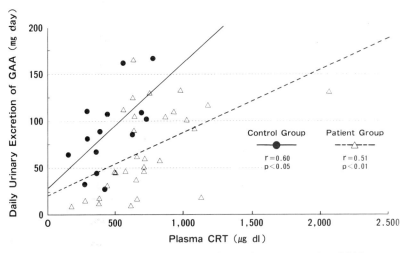

Fig. 2. Correlation between plasma CRT and urinary excretion of GAA.

was present between p-CRT and u-GAA/u-CRT in the normal group, but not in the nephropathy group (Table 3).

Table 3. Coefficients of correlation in the patients with nephropathy (normal controls)

	u-GAA	u-CRT	u-GAA/u-CRT
p-CRT	0.51**	0.63***	−0.27
	(0.60)*	(0.88)***	(−0.75)***

*$P < 0.05$; **$P < 0.01$; ***$P < 0.001$;
p-CRT was significantly correlated with u-GAA, u-CRT in both groups, and u-GAA/u-CRT in the normal group. However, it was not correlated with u-GAA/u-CRT in the nephropathy group.

Comparison of u-GAA/u-CRT in the nephropathy and normal groups (Fig. 3)

The u-GAA/u-CRT value was 1.0 ± 0.8 for the nephropathy group (n = 30) and 9.7 ± 8.1 for the normal group (n = 14). It was significantly lower for the nephropathy group at $P < 0.01$. With 1.0 and less considered as abnormal u-GAA/u-CRT values, 18 cases (60 per cent) of the nephropathy group showed abnormal values.

Discussion

GAA is synthesized mainly in the kidney by glycine amidinotransferase (GAT) with glycine and arginine as substrates[9,15]. The GAA synthesized in the kidney is converted to CRT in the liver through the action of GAA methyltransferase[3], and incorporated into the muscle. CRT is indispensable to the energy metabolism of the muscle, but there is a saturation limit to its uptake by muscle[5]. Moreover, the plasma CRT level is subject to physiological variation due to diet[13,23], hormone levels[22], drugs[2,10,17] and CRT antagonists[6,7,8,16,21].

When plasma CRT is elevated, the reabsorption of CRT through the renal tubules is suppressed and the excess CRT is excreted through the kidney[12]. Meanwhile, the plasma CRT concentration sends negative feedback signals to the GAA-producing enzyme GAT to suppress the production of GAA in the renal parenchymal cells, by which mechanism the plasma CRT concentration is controlled[4,14,21]. Thus, we examined whether or not it is possible to detect early kidney parenchymal injury from changes in the urinary excretion of GAA corrected for the plasma CRT concentration.

We selected as subjects only patients without subjective symptoms but who showed haematuria and/or proteinuria at periodical medical examination, and were admitted for closer examination. The Ccr level was above 70 ml/min in all patients. The types of nephropathy were determined histopathologically.

While the u-GAA level depends on p-CRT, p-CRT and p-GAA are present in the blood in extremely low concentrations that are below the detection limit of the enzyme method. Consequently, HPLC is necessary for the determination of p-CRT and p-GAA concentrations, which is not practical as a routine laboratory procedure.

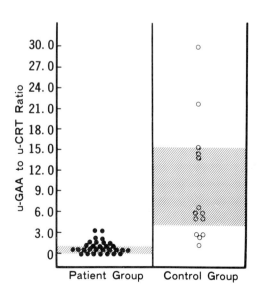

Fig. 3. Comparison of u-GAA/u-CRT in the nephropathy and normal groups.

On the other hand, the urinary levels of CRT and GAA are within the range of sensitivity of the enzyme method, and can be measured in routine laboratories.

Our patients were all inpatients. The p-CRT value for this nephropathy group was significantly higher than that for the normal group ($P < 0.05$). This may be because the consumption of CRT decreased as a result of reduced physical activity during hospitalization, so that the muscle was already saturated with CRT and the excess CRT was released into the blood resulting in raised p-CRT levels. This elevation of p-CRT is speculated to have stimulated the negative feedback control of renal GAT to suppress GAA production in the kidney, with both p-GAA and u-GAA reduced in the nephropathy group. Furthermore, the elevation of p-CRT suppressed the reabsorption of CRT in the renal tubules, thereby accelerating the urinary excretion of CRT. Although p-CRT was positively correlated with u-GAA and u-CRT in both the normal and nephropathy groups, the correlation coefficient was higher for the normal group. Moreover, p-CRT was negatively correlated with u-GAA/u-CRT in the normal group, but not correlated in the nephropathy group. These findings indicate that the negative feedback mechanism from CRT to renal GAT is not insignificantly affected by the renal cell injury.

On the other hand, u-GAA is reduced in various nephropathies. Since it is subject to negative feedback control by p-CRT, p-CRT must be taken into consideration in studies of variation in the urinary excretion of GAA. The enzyme method is inadequate for the measurement of p-CRT, the level of which is extremely low. We therefore examined variations in the urinary excretion of GAA corrected for p-CRT using u-GAA/u-CRT which is determined enzymatically.

In the 30 patients with mild nephropathy in the present study, u-GAA/u-CRT was 1.0 ± 0.8, *vs* 9.7 ± 8.1 in the normal group. The value was significantly lower for the nephropathy group than that for the normal group at $P < 0.001$. If u-GAA/u-CRT below 1.0 is regarded as abnormal, 60 per cent of the patients had some abnormality. Unlike renal biopsy, u-GAA/u-CRT can be evaluated repeatedly without any invasiveness.

Consequently, u-GAA/u-CRT is thought to be a sensitive and useful indicator of early stage nephropathy.

Acknowledgements

The authors wish to thank Dr. Shirokane for skilful technical assistance and Ms. E. Sakai for excellent secretarial help.

References

1. Bonas, J.E., Cohen, B.D. & Natelson, S. (1963): Separation and estimation of certain guanidino compounds: application to human urine. *Microchem. J.* **7**, 63–77.

2. Bourdakos, N. & Wolf, S. (1962): Creatine and muscular dystrophy. Relationship of creatine–creatinine metabolism to diet and drugs. *Arch. Neurol.* **6**, 439–450.

3. Cantoni, G.L. & Vignos, P.J. Jr. (1954): Enzymatic mechanism of creatine synthesis. *J. Biol. Chem.* **209**, 647–659.

4. Fitch, C.D., Hsu, C. & Dinning, J.S. (1960): Some factors affecting kidney transamidinase activity in rats. *J. Biol. Chem.* **235**, 2362–2364.

5. Fitch, C.D. & Shields, R.P. (1966): Creatine metabolism in skeletal muscle: I. Creatine movement across muscle membranes. *J. Biol. Chem.* **241**, 3611–3614.

6. Fitch, C.D., Shields, R.P., Payne, W.F. & Dacus, J.M. (1968): Creatinine metabolism in skeletal muscle: III. Specificity of the creatine entry process. *J. Biol. Chem.* **243**, 2024–2027.

7. Fitch, C.D., Jellinek, M. & Mueller, E.J. (1974): Experimental depletion of creatine and phosphocreatine from skeletal muscle. *J. Biol. Chem.* **249**, 1060–1063.

8. Fitch, C.D. & Chevli, R. (1980): Inhibition of creatine and phosphocreatine accumulation in skeletal muscle and heart. *Metabolism* **29**, 686–690.

9. Funahashi, M., Kato, H., Shiosaka, S. & Nakagawa, H. (1981): Formation of arginine and guanidinoacetic acid in the kidney *in vivo*. Their relations with the liver and their regulation. *J. Biochem.* **89**, 1347–1356.

10. Hoberman, H.D., Sims, E.A.H. & Engstrom, W.W. (1948): The effect of methyltestosterone on the rate of synthesis of creatine. *J. Biol. Chem.* **173**, 111–116.

11. Hung, Y., Kai, M., Nohta, H. & Ohkura, Y. (1984): High performance liquid chromatographic analysis for guanidino compounds using benzoin as a fluorogenic reagent. *J. Chromatogr.* **305**, 281–294.

12. Kadono, K., Irie, A., Kushiro, H., Kodama, J., Satani, M., Hayashi, C. & Miyai, K. (1982): Study of guanidinoacetic acid metabolism as a measure for evaluation of renal function: metabolism of interference with creatine. *Igaku No Ayumi* **121**, 419–421.

13. Kim, G.S., Chevli, K.D. & Fitch, C.D. (1983): Fasting modulates creatine entry into skeletal muscle in the mouse. *Experientia* **39**, 1360–1362.

14. McGuire, D.M., Gross, M.D., Van Pilsum, J.F. & Towle, H.C. (1984): Repression of rat kidney L-arginine: glycine amidinotransferase synthesis by creatine at a pretranslational level. *J. Biol. Chem.* **259**, 12034–12038.

15. McGuire, D.N., Gross, M.D., Elde, R.P. & Van Pilsum, J.F. (1986): Localization of L-arginine–glycine amidinotransferase protein in rat tissues by immunofluorescence microscopy. *J. Histochem. Cytochem.* **34**, 429–435.

16. Makanna, D.A., Fitch, C.D. & Fischer, V.W. (1980): Effects of β-guanidinopropionic acid on murine skeletal muscle. *Exp. Neurol.* **68**, 114–121.

17. Perkoff, G.T., Silber, R., Tyler, F.H., Cartwright, G.E. & Wintrobe, M.M. (1959): Studies in disorders of muscle: XII. Myopathy due to the administration of therapeutic amounts of 17-hydroxy-corticosteroids. *Am. J. Med.* **26**, 891–898.

18. Sasaki, M., Takahara, K. & Natelson, S. (1973): Urinary guanidinoacetate/guanidinosuccinate ratio: an indicator of kidney dysfunction. *Clin. Chem.* **19**, 315–321.

19. Shirokane, Y., Utsushikawa, M. & Nakajima, M. (1987): A new enzymic determination of guanidinoacetic acid in urine. *Clin. Chem.* **33**, 394–397.

20. Shirokane, Y., Nakajima, M. & Mizusawa, K. (1991): Easier enzymic determination of guanidinoacetic acid in urine. *Clin. Chem.* **37**, 478.

21. Van Pilsum, J.F. & Canfield, T.M. (1962): Transamidinase activities, *in vivo*, of kidneys from rats fed diets supplemented with nitrogen-containing compounds. *J. Biol. Chem.* **237**, 2574–2577.

22. Van Pilsum, J.F., Carlson, M., Boen, J.R., Taylor, D. & Zakis, B. (1970): A bioassay for thyroxine based on rat kidney transamidinase activities. *Endocrinology* **87**, 1237–1244.

23. Walker, J.B. (1960): Metabolic control of creatine biosynthesis: I. Effect of dietary creatine. *J. Biol. Chem.* **235**, 2357–2361.

Guanidino Compounds in Biology and Medicine, eds. by P.P. De Deyn, B. Marescau, V. Stalon and
I.A. Qureshi. ©1992 John Libbey & Company Ltd., pp. 281–285.

Chapter 39

^{31}P-NMR study for the energy metabolism of skeletal muscle in chronic haemodialysis patients

G. OGIMOTO, S. OZAWA, T. MAEBA, S. OWADA and M. ISHIDA

*The First Department of Internal Medicine, St. Marianna University School of Medicine, 2-16-1, Sugao,
Miyamae-ku, Kawasaki 216, Japan*

Summary

The muscle energy metabolism in chronic uraemic patients was evaluated by ^{31}P nuclear magnetic resonance (^{31}P-NMR) spectroscopic determination. Before and after haemodialysis, the relative concentrations of phosphocreatine (PCr), ATP and inorganic phosphate (Pi) in the bicipital femoral muscle at rest were computed from each peak area of the NMR spectrum. The intracellular pH was also calculated from the chemical shift of Pi from PCr in the spectrum. The mean PCr and ATP concentrations in uraemic patients (PCr; before 2114 ± 512.5, after 2195 ± 473.5, ATP; before 571 ± 123.3, after 625 ± 85.3) were significantly lower than those in control subjects (PCr; 2983 ± 437.6, ATP; 914 ± 166.8). The mean ratio of PCr/ATP in uraemic patients at the time before haemodialysis (3.72 ± 0.560) was significantly higher than in controls (3.28 ± 0.154). However, there were no significant changes of NMR parameters in response to haemodialysis. These findings suggest that haemodialysis does not directly affect the cellular bioenergetics, though the disturbance of energy production is present in the skeletal muscle of uraemic patients.

Introduction

Chronic uraemic patients undergoing maintenance haemodialysis often complain of muscle weakness and easy fatiguability. Though the causes of these symptoms have not been ascertained sufficiently, Isaacs[5] has demonstrated the existence of a primary myopathy in addition to the neurogenic myopathy. During the last 10 years, several factors have been incriminated in the myopathy of uraemia[1]. In this study, focusing on the disturbance of muscle energy metabolism, we measured high energy phosphate compounds as well as intracellular pH in the femoral muscle of uraemic patients before and after haemodialysis using ^{31}P nuclear magnetic resonance (^{31}P -NMR).

Patients and methods

Ten patients, six males and four females, undergoing maintenance haemodialysis were studied. All patients had received regular haemodialysis with bicarbonate dialysate three times a week, for a period of 4 h each time. The mean age was 53.3 years and the mean haemodialysis duration was 20.4 months. The causes of renal failure at the end stage were chronic glomerulonephritis in five patients, diabetic nephropathy in four patients, and unknown in one patient. Control subjects were five non-uraemic healthy male volunteers, mean aged 47.2 years.

For ^{31}P-NMR determination, a Philips Gyroscan S15 NMR apparatus was applied with a supercon-ducting magnet at a field of 1.5 Tesla giving an operating frequency of 25.89 MHz. Each patient was placed on the NMR apparatus under a fully relaxed condition, and examined before and after haemodialysis. A two-turn surface coil in a diameter of 14 cm was mounted over the left bicipital femoral muscle and a volume of $40 \times 100 \times 40$ mm was set up by the method of image selected *in vivo* spectroscopy (ISIS) as shown in Fig. 1. A pulse of 0.2 ms with 4 s of repetition time was given for the spectroscopy. The spectra were obtained from an average of 128 or 256 accumulations by a Fourier transform spectrometer in the function of spectroscopic mode.

The spectrum consisted of five major peaks as shown in Fig. 2. From the left, inorganic phosphate (Pi), phosphocreatine (PCr) and three peaks of ATP, γ-, α- and β-ATP were identified. Each area of the peaks was computed from integration of signals and expressed as arbitrary units. Three ATP peak

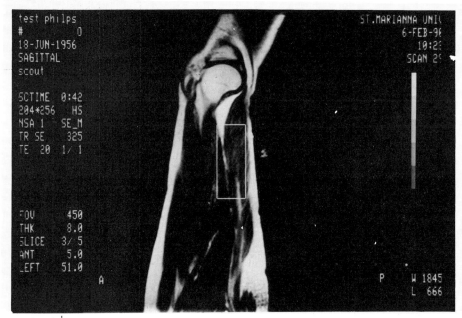

Fig. 1. ^{1}H-NMR image of the left bicipital femoral muscle for volume selection (ISIS method).

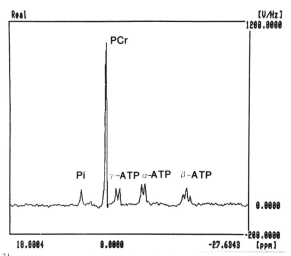

Fig. 2. ^{31}P-NMR spectrum obtained from the femoral muscle of a uraemic patient.

areas were averaged as the ATP area. The ratios of PCr/ATP, PCr/Pi and ATP/Pi were computed from each peak area. The intracellular pH was also calculated from the chemical shift difference (δ) between the PCr and Pi signals according to the equation[8]:

$$pH = 6.75 + \log [(\delta - 3.27) / (5.69 - \delta)].$$

In addition to the analyses of the spectra, arterial blood pH, as well as serum creatinine, urea nitrogen, Pi and potassium were measured before and after haemodialysis as biochemical parameters by conventional laboratory methods.

Statistical significance of differences of the data were evaluated by the paired and non-paired two-tailed *t*-test, and *P* values of less than 0.05 were regarded as significant.

Results

The data of biochemical parameters are shown in Table 1. All parameters, serum urea nitrogen, creatinine, Pi and potassium were significantly decreased after haemodialysis.

Table 1. Changes of serum parameters of uraemic patients in response to haemodialysis

	n	Urea-N (mg/dl)	Creatinine (mg/dl)	Pi (mg/dl)	Potassium (mEq/l)
Before HD	5	74 ± 19.2	9.1 ± 1.98	3.6 ± 0.45	4.1 ± 0.60
After HD	5	28 ± 5.40*	4.1 ± 0.63*	2.1 ± 0.50*	3.1 ± 0.23*

Mean ± SD; *$P < 0.01$.

Arterial blood pH and intracellular pH are shown in Table 2. After haemodialysis, arterial blood pH was distinctly increased from 7.344 to 7.436. However, there was no significant change in intracellular pH in response to haemodialysis.

Table 2. Arterial blood pH and intracellular pH

	n	Arterial blood pH	Intracellular pH
Uraemic patients	10		
Before HD		7.344 ± 0.048	6.977 ± 0.043
After HD		7.436 ± 0.032*	6.968 ± 0.038
Control subjects	5		7.020 ± 0.027

Mean ± SD; * $P < 0.01$.

The concentrations of phosphate compounds in the muscle are shown in Table 3. The mean areas of PCr and ATP in uraemic patients, both before and after haemodialysis, were significantly lower than those in control subjects. After haemodialysis, the mean areas of PCr and ATP were slightly increased and Pi was decreased compared to those before haemodialysis. These differences were not statistically significant.

Table 3. Concentrations of phosphate compounds in muscle of uraemic patients and control subjects

	n	PCr	ATP	Pi
Uraemic patients	10			
Before HD		2114 ± 512.5*	571 ± 123.3*	345 ± 139.0
After HD		2195 ± 473.5*	625 ± 85.3*	315 ± 85.3
Control subjects	5	2983 ± 437.6	914 ± 166.8	337 ± 109.3

Mean ± SD; *$P < 0.01$; *vs* control.

Table 4 shows the ratios of phosphate compounds in the muscle of uraemic patients and control subjects. The mean ratios of PCr/ATP (before haemodialysis) were significantly higher and ATP/Pi (both before and after haemodialysis) was significantly lower in uraemic patients than in control subjects. However, no significant changes were observed in these ratios in response to haemodialysis.

Table 4. Ratios of phosphate compounds in muscle of uraemic patients and control subjects

	n	PCr/ATP	PCr/Pi	ATP/Pi
Uraemic patients	10			
Before HD		3.72 ± 0.560*	6.74 ± 2.395	1.82 ± 0.560*
After HD		3.50 ± 0.449	7.12 ± 1.303	2.08 ± 0.550*
Control subjects	5	3.28 ± 0.154	9.55 ± 3.012	2.93 ± 0.985

Mean ± SD; *$P < 0.05$; *vs* control.

No correlation was found between the biochemical parameters and the NMR parameters in uraemic patients.

Discussion

^{31}P-NMR spectroscopy is a safe and non-invasive method useful for the study of energy metabolism in intact human muscle. In this study, we measured relative muscle concentrations of phosphocreatine (PCr), ATP and Pi as well as intracellular pH in the femoral muscle of chronic uraemic patients, and evaluated the changes of these values in response to haemodialysis. Since the myopathy in uraemia is prominent in the lower extremities, particularly at the proximal muscles[1], we selected the bicipital femoral muscle for the measurement of NMR spectroscopy.

It is commonly accepted that there are three steps in the bioenergetics of skeletal muscle: energy production, transport and utilization. In uraemia, abnormalities in all of these steps have been recognized by many investigations. In the present study, though the NMR spectra were varied both in uraemic patients and in control subjects, the relative concentrations of PCr and ATP were significantly lower in uraemic patients than in control subjects. Moreover, the ratio of PCr/ATP was significantly higher in uraemic patients. These results suggest that the energy production in the muscle is strongly impaired in uraemic patients. Histochemical studies have demonstrated a type II fibre atrophy in uraemic patients. The type II fibres contain high PCr and ATP concentrations[6]. These facts may support the findings presented in this study.

In contrast to our results, Garber[4] examined PCr and ATP levels in the skeletal muscle of uraemic rats enzymatically, and found no reduction in the total concentration of these high energy phosphates. His observation may indicate that mitochondrial energy production and oxidative phosphorylation are intact in uraemic rats. However, our current study of HPLC determination for PCr and ATP levels in the femoral muscle of uremic rats revealed extremely low PCr and high ATP concentrations (data not shown), which suggests the impaired energy transport from the form of ATP to the form of PCr.

^{31}P-NMR research for the effect of haemodialysis on the energy status of skeletal muscle was previously performed by Cardoso *et al.*[3]. They demonstrated that the muscle concentrations of both PCr and ATP were not significantly changed by haemodialysis. In agreement with their report, we found that the relative concentrations and the ratios of phosphate compounds were not changed in response to haemodialysis. Moreover, no significant change was observed in the intracellular pH, although the arterial blood pH was significantly increased after haemodialysis. Therefore, we may conclude that haemodialysis does not directly affect the bioenergetics of the skeletal muscle in chronic haemodialysis patients at rest.

There are some problems for the evaluation of the muscle bioenergetics in chronic haemodialysis patients. It is well known that Pi concentrations both in the extra and intracellular fluids are high in those patients. Since Pi is a permeable substance, it is easily dialysed into the dialysate. As the result, the ratios of PCr/Pi and ATP/Pi vary with haemodialysis. Therefore, it is not appropriate to evaluate the muscle bioenergetics using these ratios in chronic haemodialysis patients, nevertheless the ratio of PCr/Pi reflects the phosphorylation potential[9]. Previous contributions[2,7] have revealed the increased extracellular water quantity in chronic uraemia. Oh *et al.* indicated that among the reduced extracellular water component by haemodialysis, about 70 per cent was removed into dialysate and the remaining 30 per cent was translocated into the intracellular fluid. This increased water component in the muscle, less than 5 per cent of the body weight in our patients, may affect the density of muscle fibre and the PCr and ATP concentrations. When we notice these factors that affect the NMR parameters in the volume of interest, ^{31}P-NMR is useful in the study for the muscle bioenergetics in chronic haemodialysis patients.

References

1. Brautbar, N. (1983): Skeletal myopathy in uremia: abnormal energy metabolism. *Kidney Int.* **16**, S81–S86.

2. Broyer, M., Delaporte, C. & Maziere, B. (1974): Water, electrolytes and protein content of muscle obtained by needle biopsy in uremic children. *Biomedicine* **21**, 278–285.

3. Cardoso, M., Shoubridge, E., Arnold, D., Léveillé, M., Prud'homme, M., St-Louis, G. & Vinay, P. (1988): NMR monitoring of the energy status of skeletal muscle during hemodialysis using acetate. *Clin. Invest. Med.* **11**, 292–296.

4. Garber, A.J. (1978): Skeletal muscle protein and amino acid metabolism in experimental chronic uremia in the rat. *J. Clin. Invest.* **32**, 623–632.

5. Isaacs, H. (1969): Electromyographic study of muscular weakness in chronic renal failure. *S. Afr. Med. J.* 683–688.

6. Meyer, R.A., Brown, T.R. & Kushmerick, M.J. (1985): Phosphorous nuclear magnetic resonance of fast- and slow-twitch muscle. *Am. J. Physiol.* **248**, C279–C287.

7. Oh, M.S., Levison, S.P. & Canoll, H.J. (1975): Content and distribution of water and electrolytes in maintenance hemodialysis. *Nephron* **14**, 421–423.

8. Taylor, D.J., Bore, P.J., Styles, P., Gadian, D.G. & Radda, G.K. (1983): Bioenergetics of intact human muscle: a ^{31}P nuclear magnetic resonance study. *Mol. Biol. Med.* **1**, 77–94.

9. Zochonde, D.W., Thompson, R.T., Driedger, A.A., Strong, M.J., Gravelle, D. & Bolton, C.F. (1988): Metabolic changes in human muscle denervation: topical ^{31}P NMR spectroscopy studies. *Magn. Reson. Med.* **7**, 373–383.

Guanidino Compounds in Biology and Medicine, eds. by P.P. De Deyn, B. Marescau, V. Stalon and
I.A. Qureshi. ©1992 John Libbey & Company Ltd., pp. 287–292.

Chapter 40

Erythrocyte membrane fluidity in uraemic patients and the effect of guanidino compounds on erythrocytes

Manabu FUKASAWA, Takashi YASUDA, Chikako SHIBA, Tomoya FUJINO,
Makoto SUGIYAMA, Shigeru OWADA and Masashi ISHIDA

*The First Department of Internal Medicine, St Marianna University School of Medicine, 2-16-1, Sugao,
Miyamae-ku, Kawasaki 216, Japan*

Summary

To elucidate the changes in physical properties of erythrocyte membranes in uraemia and the effect of guanidino compounds on them, membrane fluidity of erythrocytes in uraemic patients and changes in membrane fluidity of intact erythrocytes in the presence of methylguanidine, guanidinosuccinic acid or guanidinopropionic acid were examined using spin labelled stearic acids.

The erythrocyte membrane fluidity of patients undergoing haemodialysis was significantly decreased at the hydrocarbon region immediately beneath the phospholipid head group and increased at the interior region of the membrane bilayer compared with that of controls. In patients undergoing continuous ambulatory peritoneal dialysis, the fluidity was significantly increased at the interior region of the membrane bilayer compared with controls.

There were no apparent effects of guanidino compounds on the membrane fluidity of erythrocytes, except for 10 mM of GPA which increased the fluidity at the interior region of membrane bilayer.

The pattern of change observed in uraemic patients was different from that in the presence of GPA, therefore it is conceivable that guanidino compounds are not responsible for the change of erythrocyte membrane fluidity observed in uraemic patients.

Introduction

Decrease of red cell life span and increase of haemolysis are characteristic features in patients with renal failure. Disturbance of the physical properties of erythrocyte membrane such as increased mechanical fragility and decreased deformability in uraemia are considered to be the causes of haemolysis[3,12].

In the uraemic state, accumulation of guanidino compounds, especially methylguanidine, guanidino-succinic acid and guanidinopropionic acid and their potentially toxic effects are thought to be the causes of the pathogenesis of the uraemic syndromes[1,6,17]. Several workers demonstrated that guanidino compounds were responsible for the abnormalities of erythrocyte function, including decreased deformability and depressed Na^+,K^+-ATPase activity of erythrocyte membranes[3,5,10].

The cell membrane has a crucial role in the maintenance of cellular function and membrane fluidity is one of its important physical properties[4]. Perturbation of membrane fluidity is known to elicit changes in membrane function, such as permeability and enzyme reaction[8,11,13]. Therefore, we measured the erythrocyte membrane fluidity of uraemic patients by electron spin resonance using a series of doxyl stearate spin probes and compared results with controls to elucidate the change of physical properties of the erythrocyte membrane in patients with uraemia. We also examined the effect of guanidino compounds on erythrocyte membrane fluidity.

Materials and methods

Patients

Patients consisted of 12 uraemic patients undergoing regular haemodialysis (HD) and eight patients undergoing continuous ambulatory peritoneal dialysis (CAPD). The mean durations of dialysis were 58.5 ± 55.4 months in HD patients and 35.8 ± 25.2 months in CAPD patients. The cause of uraemia was chronic glomerulonephritis in all patients and they had no history of diabetes mellitus. Controls consisted of 10 healthy volunteers.

Measurement of erythrocyte membrane fluidity with spin probe

Three kinds of spin-labelled stearic acid (doxyl stearic acid; DSA, Aldrich Chemical Co., Milwaukee, WI) were used for the measurement of membrane fluidity (Fig. 1). Each of them has a nitroxide radical ring at the 5th, 12th and 16th carbon position from the carboxyl group of the acyl-chain. These spin

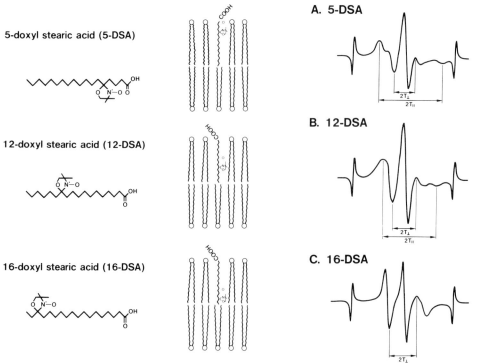

5-doxyl stearic acid (5-DSA)

12-doxyl stearic acid (12-DSA)

16-doxyl stearic acid (16-DSA)

A. 5-DSA

B. 12-DSA

C. 16-DSA

Fig. 1. Spin-labelled stearic acids and the location of nitroxide radical in the membrane bilayer.

Fig. 2. ESR spectra of nitroxide-labelled stearic acid in erythrocyte membrane. A: 5-doxyl stearic acid; B: 12-doxyl stearic acid; C: 16-doxyl stearic acid. 2T‖: outer hyperfine splitting, 2T⊥: inner hyperfine splitting.

288

probes were incorporated into the lipid bilayer of the membrane, and the amplitude of motion of these labels is sensitive to the flexibility of the membrane. As indicated in Fig. 1, 5-DSA reflects the hydrocarbon region immediately beneath the phospholipid head groups, and 16-DSA reflects the interior region of the membrane bilayer. Heparinized blood taken from controls and uraemic patients was centrifuged, and plasma and buffy coat were discarded. The erythrocytes were washed twice with Dulbecco's phosphate buffered saline without Ca and Mg (PBS). Five hundred μl of washed erythrocytes solution in which the haematocrit had been adjusted to 50 per cent with PBS was placed into a test tube, to which 10 μg of DSA in ethanol had been already added and evaporated to dryness. They were incubated for 15 min at 37 °C with gentle shaking to incorporate the spin labels into the cell membrane, then washed twice with PBS to remove the free spin label. Labelled erythrocytes were centrifuged at $1800 \times g$ for 5 min and were packed into a glass capillary tube. ESR spectra of the spin labelled membrane were recorded with a JEOL JES-RE1X ESR spectrometer (JEOL Ltd., Tokyo, Japan). Recorded temperature was controlled at 37 °C by a variable temperature controller (JEOL JES-DVT2). The instrumental parameters were as follows: scan range, 10 mT; centre field, 327 mT; time constant, 0.1 s; sweep time, 2 min; field modulation width, 0.4 mT; modulation frequency, 100 kHz; receiver gain 1×100; microwave power, 4 mW; microwave frequency, 9.2 GHz.

Figure 2 indicates the ESR spectra of each probe incorporated into the erythrocyte membranes, which show the anisotonic movements of spin-label. Order parameter (S), that is the empirical and quantitative estimation of membrane fluidity, was calculated from the inner and outer hyperfine splitting of the spectrum by the following equation as described by Gaffney[4].

Order parameter (S)

$$S = \frac{T\| - T\perp - C}{T\| + 2T\perp + 2C} \times 1.723$$

$$C = 1.4G - 0.053 (T\| - T\perp)$$

In the spectrum for 16-DSA, only the inner hyperfine splitting was well resolved, so the outer hyperfine splitting was taken as (44.5 − inner hyperfine splitting), as described by Gaffney[4]. The S value shifts between 1.0 to 0.0, and high values represent a relatively solid membrane, while low values represent high mobility.

The effects of guanidino compounds on erythrocyte membrane fluidity

To examine the effects of guanidino compounds, washed erythrocytes from healthy controls were incubated with 0.1 mM, 1 mM or 10 mM methylguanidine (MG), guanidinosuccinic acid (GSA) or guanidinopropionic acid (GPA) (Sigma Chemical Co., St Louis, MO) for 1 h at 37 °C before labelling. After incubation, they were labelled with spin probe, ESR spectra were recorded, and the order parameters were calculated as described above. Statistical significance was determined by Student's *t*-test and *P* values of less than 0.05 were regarded as significant.

Results

Erythrocyte membrane fluidity in uraemic patients

Order parameters of controls, HD patients and CAPD patients are shown in Fig. 3. The mean order parameter for 5-DSA was significantly higher in HD patients compared with controls (0.671 ± 0.004 and 0.676 ± 0.006, respectively). In CAPD patients, the mean order parameter for 5-DSA was not significantly changed compared to controls or HD patients (0.672 ± 0.006). There were no significant changes in mean order parameters for 12-DSA among controls, HD and CAPD patients (0.488 ± 0.008, 0.494 ± 0.009 and 0.494 ± 0.013, respectively). For 16-DSA, the mean order parameter of HD (0.320 ± 0.012) and CAPD patients (0.320 ± 0.020) were significantly lower than that of controls (0.337 ± 0.011).

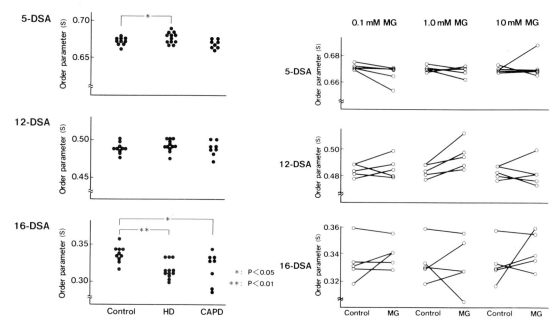

Fig. 3. Erythrocyte membrane fluidities in controls, patients undergoing haemodialysis (HD) and patients undergoing continuous ambulatory peritoneal dialysis (CAPD). *P < 0.05; **P < 0.01.

Fig. 4. Effect of methylguanidine (MG) on the order parameters for each spin probe in erythrocyte membrane.

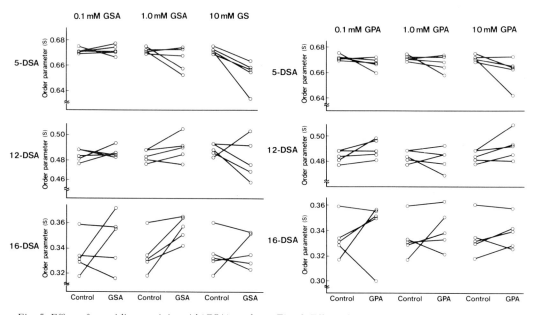

Fig. 5. Effect of guanidinosuccinic acid (GSA) on the order parameters for each spin probe in erythrocyte membrane.

Fig. 6. Effect of guanidinopropionic acid (GPA) on the order parameters for each spin probe in erythrocyte membrane.

Effect of MG, GSA or GPA on erythrocyte membrane fluidity

In the presence of MG, there was no apparent tendency towards change of membrane fluidity in all spin probes and at each concentration (Fig. 4). The order parameter for 5-DSA was significantly decreased by the incubation with 10 mM of GSA (Fig. 5). But no other apparent tendency to change was observed in the presence of GSA. The order parameter for 5-DSA was significantly decreased in the presence of 10 mM GPA (Fig. 6). There were no significant changes in order parameters for other probes incubated with GPA.

Discussion

The present study showed that there was no apparent effect of guanidino compounds on the membrane fluidity of intact erythrocyte, except for 10 mM GSA or 10 mM GPA, which increased the membrane fluidity of the hydrocarbon region immediately beneath the phospholipid head group. Since GSA is a strong acid, the concentration of 10 mM may decrease the pH of the sample solution, and change in pH is known to be one of the factors that affect fluidity of the erythrocyte membrane[15]. Therefore, the effect of GSA may be masked by the change in pH of the sample solution.

The membrane fluidity in the patients undergoing haemodialysis was significantly decreased at the hydrocarbon region immediately beneath the phospholipid head group, observed using 5-DSA, and increased at the interior region, observed using 16-DSA, compared with that of controls. In the patients undergoing CAPD, the fluidity was significantly increased at the interior region of the membrane bilayer compared with controls. These results are different from those of Komidori[9] who reported that only the motion parameter for 16-DSA increased compared with controls, and that there was no significant change in the mean order parameter for 5- and 12-DSA. It is difficult to say why the results were so different, but we need to measure the membrane fluidity of erythrocytes in a larger number of uraemic patients to clarify this difference. There are several factors affecting membrane fluidity of erythrocytes, that is the composition of phospholipids, cholesterol to phospholipid ratio, lipid to cholesterol ratio, degree of unsaturation length of phospholipid acyl chains, interaction between phospholipid and protein and so on[13,14,16]. Komidori[9] examined membrane lipids and found that there were significant decreases of the polyunsaturated to saturated acyl chain ratio and the phosphatidylcholine content in the erythrocyte membrane of the uraemic patients compared with that of controls, but no other significant changes were observed in other phospholipid classes or free cholesterol to phospholipid molar ratio. He also mentioned that there were no significant correlations between the change in erythrocyte membrane fluidity and the change in the contents of phosphatidylcholine or polyunsaturated to saturated acyl chain ratio.

Active oxygen is one of the important factors that contribute to the pathophysiological condition of uraemia. Grzelinska et al.[7] demonstrated that γ-irradiation of bovine erythrocyte membrane resulted in a decrease of membrane fluidity at the depth of C_{12} of membrane lipids but not at the depth of C_5. Bruch et al.[2] also reported that the effect of lipid peroxidation on fluidity was maximal in the membrane microenvironment sampled by 12-DSA, whereas other regions of the bilayer were less affected using sonicated soybean phospholipid vesicles. Therefore, the pattern of change in erythrocyte membrane observed in uraemic patients was different from that of lipid peroxidation. Guanidino compounds such as MG, GSA and GPA are accumulated in uraemic patients and their toxicity was thought to be responsible for some pathological conditions observed in uraemia[1]. One of them, MG is known to affect the biochemical functions of erythrocyte membrane. Erythrocyte deformability and Na^+,K^+-ATPase activity of erythrocyte membrane were reduced by exposure of erythrocytes to MG[10]. Kimelberg[8] reported that Na^+,K^+-ATPase activity was apparently influenced by the change in membrane fluidity. However, the present study did not demonstrate the effect of MG on erythrocyte membrane fluidity. Among the three tested guanidino compounds, GPA at 10 mM changed the membrane fluidity. But the concentration of 10 mM GPA is far greater than that of GPA in serum and erythrocyte in uraemic patients. The pattern of change in membrane fluidity in uraemic patients was different from that induced by GPA. Therefore, at present it is difficult to explain the changes in erythrocyte membrane fluidity observed in uraemic patients by the accumulation of the tested guanidino compounds. However, it is necessary to examine the effect of lower concentrations or more prolonged periods of incubation to clarify the effect of the guanidino compounds on the membrane fluidity of the intact erythrocyte.

References

1. Barsotti, G., Bevilacqua, G., Morelli, E., Cappelli, P., Balestri, P.L. & Giovannetti, S. (1975): Toxicity arising from guanidine compounds: role of methylguanidine as a uremic toxin. *Kidney Int.* **7**, S299–S301.

2. Bruch, R.C. & Thayer, W.S. (1983): Differential effect of lipid peroxidation on membrane fluidity as determined by electron spin resonance probes. *Biochem. Biophys. Acta* **733**, 216–222.

3. Forman, S., Bischel, M. & Hochstein, P. (1973): Erythrocyte deformability in uremic hemodialyzed patients. *Ann. Intern. Med.* **79**, 841–843.

4. Gaffney, B.J. (1975): Fatty acid chain flexibility in the membranes of normal and transformed fibroblasts. *Proc. Natl. Acad. Sci. USA* **72**, 664–668.

5. Giovannetti, S., Cioni, L., Balestri, P.L. & Biagini, M. (1968): Evidence that guanidine and some related compounds cause hemolysis in chronic uremia. *Clin. Sci.* **34**, 141–148.

6. Giovannetti, S., Biagini, M., Balestri, P.L., Navelesi, R., Giagnoni, P., de Mattetis, A., Ferro-Milone, P. & Perfetti, C. (1969): Uremia like syndrome in dogs chronically intoxicated with methylguanidine and creatinine. *Clin. Sci.* **36**, 445–452.

7. Grzelinska, E., Bartosz, G., Gwozdzinski, K. & Leyko, W. (1979): A spin-label study of the effect of gamma radiation on erythrocyte membrane. Influence of lipid peroxidation on membrane structure. *Int. J. Radiat. Biol.* **36**, 325–334.

8. Kimelberg, H.K. (1975): Alteration in phospholipid-dependent (Na$^+$-K$^+$)-ATPase activity due to lipid fluidity. *Biochem. Biophys. Acta* **413**, 143–156.

9. Komidori, K. (1988): Lower levels of erythrocyte membrane fluidity and changes in membrane lipids in uremia patients. *Fukuoka Univ. Med. J.* **79**, 314–325.

10. Mikami, H., Ando, A., Fujii, M., Okada, A., Imai, E., Kokuba, Y., Orita, Y. & Abe, H. (1985): Effect of methyl-guanidine on erythrocyte membranes. In *Guanidines*, eds. A. Mori, B.D. Cohen & A. Lowenthal, pp. 205–212. New York: Plenum Publishing Co.

11. Moriyama, T., Nakahama, H., Fukuhara, Y., Horio, M., Yanase, M., Orita, Y., Kamada,T., Kanashoro, M. & Miyake, Y. (1989): Decrease in the fluidity of brush-border membrane vesicles induced by gentamicin. A spin-labeling study. *Biochem. Pharmacol.* **38**, 1169–1174.

12. Rosenmund, A., Binswanger, U. & Straub, P.W. (1975): Oxidative injury to erythrocytes, cell rigidity, and splenic hemolysis in hemodialyzed uremic patients. *Ann. Intern. Med.* **82**, 460–465.

13. Schreier, S., Polnazek, C.F. & Smith, I.C.P. (1978): Spin labels in membranes: problems in practice. *Biochem. Biophys. Acta* **515**, 395–436.

14. Simon, I., Burns, C.P. & Spector, A.A. (1982): Electron spin resonance studies on intact cells and isolated lipid droplets from fatty acid-modified L 1210 murine leukemia. *Cancer Res.* **42**, 2715–2721.

15. Yamaguchi, T., Koga, M., Fujita, Y. & Kimoto, E. (1982): Effect of pH on membrane fluidity of human erythrocytes. *J. Biochem.* **91**, 1299–1304.

16. Yamaguchi, T., Kuroki, S., Tanaka, M. & Kimoto, E. (1982): Effects of temperature and cholesterol on human erythrocyte membranes. *J. Biochem.* **92**, 673–678.

17. Yatzidis, H., Oreopoulus, D., Tsaparas, N., Voudiceari, S., Stavroulaki, A. & Zestanakis, S. (1966): Colorimetric determination of guanidines in blood. *Nature* **212**, 1498–1499.

Guanidino Compounds in Biology and Medicine, eds. by P.P. De Deyn, B. Marescau, V. Stalon and I.A. Qureshi. ©1992 John Libbey & Company Ltd., pp. 293–299.

Chapter 41

The effect of guanidino compounds on the membrane fluidity of the cultured rat mesangial cell

Takashi YASUDA, Manabu FUKASAWA, Masayuki OHMINATO, Teruhiko MAEBA, Sadanobu OZAWA, Shigeru OWADA and Masashi ISHIDA

The First Department of Internal Medicine, St Marianna University, School of Medicine, 2-16-1, Sugao, Miyamae-ku, Kawasaki 216, Japan

Summary

The effects of the guanidino compounds methylguanidine, guanidinosuccinic acid and guanidinopropionic acid, at 1.0 and 10 mM, on the membrane fluidity of mesangial cells were examined by electron spin resonance using 5-, 12- or 16-doxyl stearic acid. The membrane fluidity of mesangial cells was reduced at the hydrocarbon region of the lipid bilayer immediately beneath the phospholipid head group and increased at the interior region of the membrane bilayer in the presence of methylguanidine or guanidinopropionic acid (1.0 mM and 10 mM). The above findings indicate that guanidino compounds affect the membrane fluidity of mesangial cells.

Introduction

Mesangial cells are one component of the glomerulus. The mesangium consists of mesangial cells and mesangial matrices, and sclerotic change of the mesangium is often observed in many glomerular diseases. According to the recent development of cell culturing techniques, there is increasing knowledge about the role of mesangial cell in the maintenance of glomerular function and in the progression of glomerular diseases[10].

The accumulation of guanidino compounds such as methylguanidine (MG), guanidinosuccinic acid (GSA) and guanidinopropionic acid (GPA) is observed in the uraemic state, and they are thought to be uraemic toxins[5]. There have been many studies on the toxic effects of guanidino compounds on the function of cells including erythrocytes, lymphocytes and platelets[1,4]. Few studies have reported on these toxic effects on somatic cells while some more reports have demonstrated effects on the nervous system[6].

Membrane fluidity is one of the important physical properties of cell membrane, and is related to cellular functions such as permeability, hormone receptor and enzyme activities in membrane, and physiological importance of membrane fluidity in regulatory intracellular functions has also been recognized[7].

Therefore, we examined the effect of MG, GSA and GPA on the membrane fluidity of mesangial cells using spin labelled stearic acids.

Materials and methods

Preparation of rat cultured mesangial cells

Cultured mesangial cells were prepared from isolated glomeruli from 5-week-old Wistar rats, and cultured with RPMI 1640 medium supplemented with 15 per cent foetal bovine serum, 300 µg/ml glutamine, 0.66 u/ml insulin, 100 u/ml penicillin and 100 µg/ml streptomycin, and incubated in a humidified incubator with 5 per cent CO_2 in air at 37 °C. The cultured cells were characterized as mesangial cells by their morphological and immunohistochemical properties [2]. All experiments were performed at confluency between five and 10 passages.

Spin label of mesangial cells

Three kinds of spin-labelled stearic acids; 5-, 12- and 16-doxyl stearic acid (DSA, Aldrich Chemical Co., Milwaukee, WI) were used for the labelling of mesangial cell membrane.

Mesangial cells were detached from the cultured dish using 0.05 per cent trypsin/0.02 per cent EDTA and washed once with the culture medium to stop the action of trypsin/EDTA and twice with Hank's balanced salt solution (HBSS). Ten micrograms of DSA in ethanol were added to a test tube and evaporated to dryness. Two ml of cell suspension (2×10^6 cells) in HBSS were put into the test tube, and incubated for 10 min at 37 °C to incorporate the spin-labelled stearic acid into the membrane bilayer. The labelled cells were washed once with HBSS.

Electron spin resonance spectrometry

The labelled cells were centrifuged at $600 \times g$ for 5 min, and the pellet was packed into a glass capillary tube. ESR spectra of the spin labelled membranes were recorded with a JEOL JES-RE1X ESR spectrometer (JEOL Ltd., Tokyo, Japan) at 37 °C controlled by a variable temperature controller (JES-DVT2, JEOL). The instrumental parameters were as follows: scan range, 10 mT; centre field, 327 mT; time constant, 0.1 s; sweep time, 2 min; field modulation width, 0.4 mT; modulation frequency, 100 kHz; receiver gain, 1×100; microwave power, 4 mW; microwave frequency, 9.2 GHz. Fig. 1 indicates the typical ESR spectra of each probe incorporated into mesangial cell membranes. In the case of 5-DSA, the order parameters were calculated from the inner and outer hyperfine splitting of the spectrum by the equation (1) and (2). In the case of 12- and 16-DSA, since the outerfine splitting was not resolved, the order parameters were calculated by equations (1), (2) and (3)[3].

Order parameter (S)

$$S = \frac{T\| - T\perp - C}{T\| + 2T\perp + 2C} \times 1.723 \tag{1}$$

$$C = 1.4G - 0.053\,(T\| - T\perp) \tag{2}$$

$$T\| = 44.5G - 2T\perp \tag{3}$$

In addition, the motion parameter was calculated from the spectra derived from 16-DSA according to the equation (4)[3].

Motion parameter(τ_0)

$$\tau_0 = 6.5 \cdot 10^{-10} \cdot W_0[(h_0/h_{-1})^{1/2} - 1] \tag{4}$$

Effect of DSA concentration

Two ml of cell suspension (2×10^6 cells) in HBSS were labelled with 5, 10, 20 or 40 µg of DSA by incubation for 10 min at 37 °C and the spectrum of the probe incorporated into the membrane was recorded. After spectrum measurement, cell viability was determined with trypan blue exclusion.

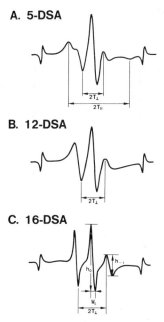

A. 5-DSA

B. 12-DSA

C. 16-DSA

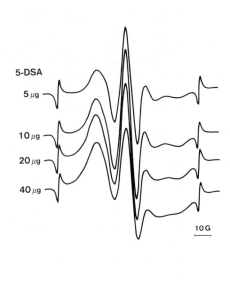

5-DSA

5 μg

10 μg

20 μg

40 μg

10 G

Fig. 1. ESR spectra of nitroxide-labelled stearic acid in mesangial cell membrane. A: 5-doxyl stearic acid; B: 12-doxyl stearic acid; C: 16-doxyl stearic acid. $2T\|$: outer hyperfine splitting; $2T\perp$: inner hyperfine splitting; h_o: central-field peak height; h_{-1}: high-field peak height; W_o: line width of the central peak.

Fig. 2. Effects of the concentration of 5-doxyl stearic acid on the ESR spectra. Mesangial cells were adjusted $2 \times 10^6/2$ ml and labelled with 5, 10, 20 or 40 μg of 5-DSA.

Effect of incubation after labelling on the spectrum

After the labelling of mesangial cell membrane with spin probes, ESR spectra were recorded at 15 min, 30 min, 60 min, 120 min and 180 min, and the order parameters were calculated from the spectra.

Effect of guanidino compounds

The labelled mesangial cells with 10 μg of each DSA were incubated with 1.0 or 10 mM MG, GSA or GPA for 15 min at 37 °C. After incubation with these guanidino compounds, ESR spectra of probes were recorded and cell viabilities were determined.

Effect of pH of the sample solution

The pH of the sample solution was adjusted to 6.5 or 5.0 with addition of hydrochloride, then the labelled mesangial cells were incubated in this solution for 15 min at 37 °C. After incubation, ESR spectra of probes were recorded and cell viabilities were determined.

Results

ESR spectrum of each probe incorporated into mesangial cell membrane and the effect of DSA concentration on the spectrum

Typical ESR spectra obtained with 5-, 12- and 16-DSA showed anisotonic movements of spin-label (Fig. 1). The ESR spectra labelled with various concentrations of 5-DSA are shown in Fig. 2. The signal amplitude increased as the concentration of DSA increased, but at 40 μg peak height became smaller than at 20 μg and the outerfine splittings were not resolved. The values of order parameters

295

were virtually identical for every concentration in each spin label in which a spectrum could be resolved (Table 1). Viabilities of the cells were decreased at 20 and 40 µg. Therefore, we used 10 µg of DSA for labelling.

Table 1. The order parameters (S) for each probe and viability of mesangial cells with different amounts of DSA

Dose (µg)	5-DSA		12-DSA		16-DSA	
	S	Viability, %	S	Viability, %	S	Viability, %
5	0.558	82.0	ND	79.6	0.360	84.5
10	0.560	83.9	0.488	80.3	0.354	82.8
20	0.559	78.3	0.490	80.6	0.358	80.5
40	ND	68.1	0.491	50.7	0.354	64.6

ND: not detected.

Effect of incubation time after labelling on ESR spectra

The time course of the spectra after labelling with 5-DSA is shown in Fig. 3. At 2 h after labelling, there was considerable reduction in amplitude of signal. Therefore, the ESR spectra were recorded within 1 h after introduction of the spin label. No change in the order parameter was observed over 1 h in each probe.

Effect of guanidino compounds on the mesangial cell membrane fluidity

In the presence of 1 mM MG or GPA, the order parameters for 5-DSA were increased, and the order parameters and motion parameters for 16-DSA were decreased (Fig. 4). Incubation with GPA decreased the order parameter for 12-DSA. Incubation with GSA decreased the order parameters for 5- and 16-DSA, but in one case, the motion parameter for 16-DSA increased.

Incubation with 10 mM MG or GPA increased the order parameters for 5-DSA, and decreased the motion parameters for 16-DSA. This effect was the same as that produced by 1 mM of MG or GPA (Fig. 5). In the presence of 10 mM GSA, the order parameters for 5-DSA were markedly decreased in two cases.

Cell viabilities did not differ in any experiment, except for cases incubated with GSA (Table 2). GSA is a strong acid, so the pH changed from 7.3 to 6.5 in 1 mM sample solution, and pH was 5.0 in the 10 mM sample solution.

Table 2. Effect of MG, GSA and GPA on the order parameters (S) for 5-, 12- and 16-DSA and cell viabilities

	Control	1 mM MG	1 mM GSA	1mM GPA	Control	10 mM MG	10 mM GSA	10 mM GPA
5-DSA								
S	0.593±0.001	0.600±0.001	0.585±0.001	0.605±0.005	0.603±0.004	0.614±0.007	0.547±0.038	0.621±0.012
Viability	89.0±1.72	79.4±5.23	74.1±3.14	76.4±3.86	79.6±3.95	76.9±4.61	44.9±14.8	78.1±5.20
12-DSA								
S	0.452±0.006	0.451±0.018	0.454±0.006	0.444±0.005	0.474±0.010	0.475±0.010	0.467±0.010	0.478±0.010
Viability	80.3±6.57	76.6±4.49	77.8±4.32	75.8±4.45	78.0±1.56	74.6±1.30	51.9±10.5	77.9±2.00
16-DSA								
S	0.306±0.006	0.294±0.005	0.300±0.008	0.294±0.005	0.321±0.004	0.310±0.010	0.335±0.008	0.316±0.002
Viability	77.4±3.90	80.0±4.24	75.9±1.88	78.4±0.84	77.7±1.33	79.2±3.39	44.8±8.70	76.7±1.60

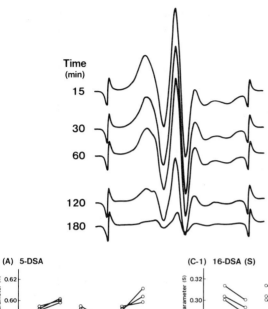

Time
(min)

15

30

60

120

180

Fig. 3. Time course of the ESR spectra. ESR spectra were recorded at 15, 30, 60, 120 and 180 min after labelling with 5-DSA.

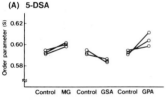

(A) 5-DSA

(B) 12-DSA

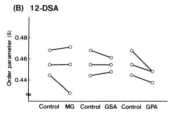

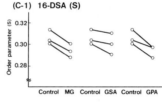

(C-1) 16-DSA (S)

(C-2) 16-DSA (τ_0)

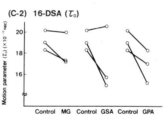

Fig. 4. Effect of 1 mM methyl-guanidine, guanidinosuccinic acid and guanidinopropionic acid on the order parameter for each spin probe and motion parameters for 16-DSA in mesangial cell membranes.

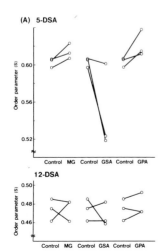

(A) 5-DSA

12-DSA

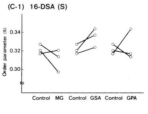

(C-1) 16-DSA (S)

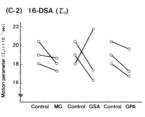

(C-2) 16-DSA (τ_0)

Fig. 5. Effect of 10 mM methylguanidine, guanidinosuccinic acid and guanidinopropionic acid on the order parameter for each spin probe and motion parameters for 16-DSA in mesangial cell membranes.

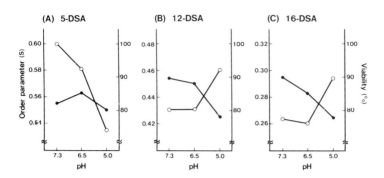

Fig. 6. Effect of pH on the order parameters and viabilities of the mesangial cells.

Effect of pH of the sample solution on the membrane fluidity and cell viability

With decrease in the pH of the sample solution, the order parameter for 5-DSA decreased and that for 12- and 16-DSA increased (Fig. 6). Cell viabilities decreased at pH 5.0.

Discussion

The mesangial cell membrane was more flexible than that of erythrocytes in all of the regions probed when compared with the results reported by Fukasawa *et al.* (*see* Chapter 40), and the order parameters of each probe were similar to that of fibroblast[3]. No magnetic interaction between closely adjacent nitroxide molecules of DSA was observed at the concentration used in this experiment, because the order parameters were virtually identical between 10 µg and 20 µg of each probe for 2×10^6 cells. There was a considerable reduction in amplitude of the signal at 2 h after labelling, presumably due to chemical reduction of the nitroxide group by cell components[9]. Therefore, the spectra were recorded within 1 h after introduction of the spin label in all experiments.

The change in pH of the sample solution induced a marked effect on the order parameters for each probe and cell viability[11]. Since there were changes in pH in the sample solutions of GSA, the effects of GSA on membrane fluidity were probably masked by the effect of the change of pH and cell viability. The use of another buffered system which does not affect the cell is necessary to clarify the effects of GSA. In the presence of MG and GPA (1 mM, 10 mM), the membrane fluidity of mesangial cells decreased at the hydrocarbon region of the membrane bilayer immediately beneath the phospholipid head group, and increased the fluidity of the interior region of the membrane bilayer. This pattern of change was similar to that observed in erythrocytes obtained from the patients undergoing haemodialysis (*see* Chapter 40). At present, it is difficult to make a detailed interpretation based on these data about how MG and GPA act on the membrane and change the membrane fluidity, and how cellular function is affected by the change in membrane fluidity. It is also difficult to interpret the discrepancy between the effect of guanidino compounds on erythrocytes and mesangial cells. However, one reason for this may be the individual heterogeneity of erythrocyte membrane components such as fatty acid composition and the ratio of cholesterol to phospholipid in individual persons[8].

Cultured cells, on the other hand, were quite homogeneous, so that the effects of guanidino compounds probably become more apparent than in erythrocytes. The present results indicate that membrane fluidity is affected by the presence of MG or GPA. These results suggest the possibility that the accumulation of guanidino compounds in uraemia changes the somatic cellular function through modification of cell membrane fluidity. To clarify the effect of guanidino compounds on the membrane fluidity, it is necessary to examine the effect at lower concentration of guanidino compounds and to study other cell lines.

References

1. Barsotti, G., Bevilacqua, G., Morelli, E., Cappelli, P., Balestri, P.L. & Giovannetti, S. (1975): Toxicity arising from guanidine compounds: role of methylguanidine as a uremic toxin. *Kidney Int.* **7**, S299–S301.

2. Foidart, J.B., Dechenne, C., Dubois, C., Deheneffe, J. & Mahieu, P. (1981): Tissue culture of isolated renal glomeruli: present and future. *Adv. Nephrol.* **10**, 267–292.

3. Gaffney, B.J. (1975): Fatty acid chain flexibility in the membranes of normal and transformed fibroblasts. *Proc. Natl. Acad. Sci. USA* **72**, 664–668.

4. Giovannetti, S., Cioni, P., Balestri, P.L. & Biagini, M. (1968): Evidence that guanidine and some related compounds cause hemolysis in chronic uremia. *Clin. Sci.* **34**, 141–148.

5. Giovannetti, S., Biagini, M., Balestri, P.L., Navelesi, R., Giagnoni, P., de Mattetis, A., Ferro-Milone, P. & Perfetti, C. (1969): Uremia like syndrome in dogs chronically intoxicated with methylguanidine and creatinine. *Clin. Sci.* **36**, 445–452.

6. Matsumoto, M. & Mori, A. (1976): Effects of guanidino compounds on rabbit brain microsomal Na^+-K^+ ATPase activity. *J. Neurochem.* **27**, 635–636.

7. Schreier, S., Polnazek, C.F. & Smith, I.C.P. (1978): Spin labels in membranes: problems in practice. *Biochem. Biophys. Acta* **515**, 395–436.

8. Shiga, T., Maeda, N., Suda, T., Kon, K. & Sekiya, M. (1979): The decreased membrane fluidity of *in vivo* aged, human erythrocytes. A spin label study. *Biochem. Biophys. Acta* **553**, 84–95.

9. Simon, I., Burns, C.P. & Spector, A.A. (1982): Electron spin resonance studies on intact cells and isolated lipid droplets from fatty acid-modified L 1210 murine leukemia. *Cancer Res.* **42**, 2715–2721.

10. Striker, G.E., Lange, M.A., MacKay, K., Bernstein, K. & Striker, L.J. (1987): Glomerular cells *in vitro*. *Adv. Nephrol.* **16**, 169–186.

11. Yamaguchi, T., Koga, M., Fujita, Y. & Kimoto, E. (1982): Effect of pH on membrane fluidity of human erythrocytes. *J. Biochem.* **91**, 1299–1304.

Guanidino Compounds in Biology and Medicine, eds. by P.P. De Deyn, B. Marescau, V. Stalon and
I.A. Qureshi. ©1992 John Libbey & Company Ltd., pp. 301–307.

Chapter 42

Effect of hyperbaric oxygen therapy on urinary methylguanidine excretion in patients with or without renal failure

Katsumi TAKEMURA, Kazumasa AOYAGI[1], Sohji NAGASE[1], Masako SAKAMOTO[1],
Toshiko ISHIKAWA[1] and Mitsuharu NARITA[1]

*Kamitsuga General Hospital, Grand Haitsu 805, 1-853 Shimoda-cho, Kanuma, Japan; [1]Department of Internal
Medicine, Tsukuba University, Tsukuba, Japan*

Summary

We have reported that methylguanidine (MG), implicated as a potent uraemic toxin, was a peroxidative product of creatinine (Cr) and also that active oxygen species play an important role in MG synthesis *in vitro*. On the other hand, there are many reports demonstrating that active oxygen species are biologically produced in hyperoxia. In this study, we investigate the effect of hyperoxia on MG synthesis in patients who were treated by hyperbaric oxygen therapy (HBO). Eighteen patients with cerebral infarction, ileus, etc. were selected for this study who had indications for HBO. HBO conditions were 100 per cent oxygen, 2 atmospheres absolute (ATA) or 3 ATA, 1 h. Patients who received HBO under 2 ATA were divided into three groups (group I: creatinine clearance (Ccr) < 10 ml/min, group II: $10 \leq$ Ccr < 50, group III: Ccr $\geq$ 50 ml/min). Case No. 18 received HBO under 3 ATA twice and was classified as group IV (Ccr $\geq$ 50 ml/min). Before and after HBO, urine and serum were collected and MG and Cr were determined. Urinary MG excretion increased significantly after HBO therapy in every group. Among the three groups (groups I–II), the increase of MG excretion after HBO rose dramatically in group I, but the ratio of urinary MG/serum Cr which was calculated as an indicator of the MG production rate from Cr was almost identical. The ratio of urinary MG/serum Cr increased markedly with exposure to oxygen of 3 ATA compared to 2ATA. These results suggest that the increase of MG excretion under HBO therapy reflects the peroxidative state in this condition and that the ratio of urinary MG/serum Cr is a useful indicator of the peroxidative state *in vivo*.

Introduction

A peroxidative state has been reported as a possibly important pathogenic factor in uraemia[3,6]. We have reported that methylguanidine (MG), implicated as a potent uraemic toxin, was a peroxidative product of creatinine (Cr) *in vitro*[8], and that active oxygen species play an important role in MG synthesis in isolated rat hepatocytes[1] and activated human neutrophils[10]. However, it remains unclear whether the peroxidative state affects MG synthesis *in vivo*. On the other hand, hyperoxia is very injurious to various tissues and is considered to be an extremely peroxidative state. Some investigators reported that active oxygen species are biologically produced in hyperoxia[2,4].

Hyperbaric oxygen (HBO) therapy is remarkably effective against hypoxic disorders such as carbon monoxide poisoning, ileus, gangrene, and cerebrovascular disorder. However, excessive hyperbaric oxygen pressure causes headache and/or convulsions. There have been some reports concerning increased production of free radicals under the hyperoxia caused by HBO therapy[7].

In this study, we investigated the effect of hyperoxia on MG synthesis in patients with or without renal failure under HBO therapy.

Materials and methods

Eighteen patients (seven males, eleven females), aged 25–82 years, treated by HBO therapy were studied. Their primary diseases were cerebral infarction, ileus, and dermal gangrene. Some of them had complications of diabetes mellitus or hypertension (Table 1).

Table 1. Subjects and clinical diagnosis

Case no.	Age	Sex	Clinical diagnosis	Serum creatinine (mg/dl)
1	74	F	Parkinson syndrome	0.5
2	82	F	Ileus	0.5
3	78	F	Cerebral infarction	0.3
4	65	M	Cerebral infarction	0.8
5	48	F	Diabetic gangrene	0.6
6	50	F	Soft tissue necrosis after operation	0.5
7	45	M	Cerebral infarction, DM	0.6
8	56	F	Oedema of right upper limb after mastectomy	0.5
9	72	M	Phlebothrombosis of left upper limb	0.8
10	71	F	Parkinson syndrome, cerebral infarction	0.5
11	77	M	Ileus, renal insufficiency	0.7
12	71	M	Cerebral infarction, HT, renal insufficiency	1.0
13	65	M	Cerebral infarction, HT, renal insufficiency	1.3
14	51	M	Cerebral infarction, HT, renal insufficiency	1.2
15	54	F	Cerebral infarction, HT, renal insufficiency	1.5
16	72	M	Cerebral infarction, DM, CRF	6.1
17	63	F	Cerebral infarction, DM, CRF	6.4
18	25	F	Pneumatosis intestinalis	0.7

Abbreviations: DM: diabetes mellitus. HT: hypertension. CRF: chronic renal failure (creatinine clearance < 10 ml/min); *renal insufficiency: 10 ml/min ≤ creatinine clearance < 50 ml/min.
Group I: cases No. 1–10; Group II: cases No. 11–15; Group III: cases No. 16–17; Group IV: case No. 18.

The conditions of HBO therapy were as follows: the hyperbaric chamber was flushed with 100 per cent oxygen and pressurized to 2 atmospheres absolute (ATA) for 10 min, and the oxygen pressure was maintained at 2 ATA for the following 60 min. Then the chamber was decompressed for 20 min. In case No. 18, hyperbaric oxygen pressure was 3 ATA.

The applied HBO system was the Model KHO–201 for one person (Kawasaki Engineering Company, Japan).

Patients were fasted overnight or had a low (Cr) and low MG meal (Cr: 7.2 mg, MG: 90 nmol/total meal) at 7 a.m. They received HBO therapy at 10 a.m.

Urine samples were collected for 2 h just before HBO therapy (8–10 a.m.) and during HBO therapy (10–11.30 a.m.). Blood samples were obtained from the patients just before and immediately after HBO therapy. In three cases who had a low Cr and low MG diet without HBO therapy, urine and blood samples were collected in the same manner as above to investigate the effect of a low Cr and low MG diet alone on urinary and serum concentration MG and Cr.

Urinary and serum Cr was determined by Cr analyser (IL-919, USA), MG was determined by high-performance liquid chromatography using 9,10-phenanthrenequinone for the post-labelling method as described previously[11]. Statistical analysis was performed with the paired Wilcoxon test.

Results

Patients who received HBO therapy under 2 ATA were divided into three groups according to their endogenous creatinine clearance (Ccr).

Group I consisted of two patients whose Ccr was below 10 ml/min.

Group II consisted of five patients whose Ccr was between 10 and 50 ml/min.

Group III consisted of 10 patients whose Ccr was more than 50 ml/min.

Case No. 18 received HBO under 3 ATA and formed group IV (Ccr ≥ 50 ml/min).

No significant difference in serum MG and Cr levels before and after HBO therapy was observed. Particularly, the serum concentrations of MG in patients without renal failure remained below the detection limit of the applied method (Table 2).

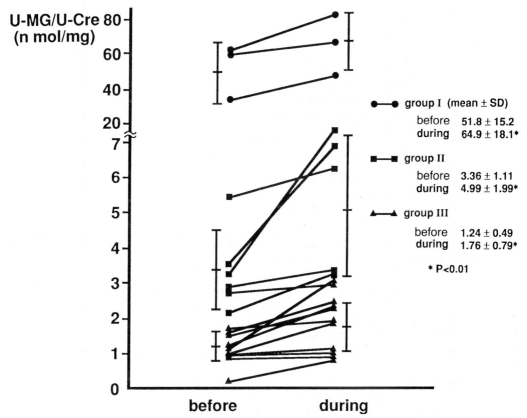

Fig. 1. Changes in the urinary MG excretion under hyperbaric oxygen therapy (HBO).
Conditions of HBO: 100 per cent O_2, 2 atmospheres absolute, 1 h.
Group I: creatinine clearance (Ccr) < 10 (ml/min);
Group II: 10 ≤ Ccr < 50 (ml/min);
Group III: Ccr ≥ 50 (ml/min).

Table 2. Levels of methylguanidine and creatinine in serum and urine of patients before and during hyperbaric oxygen therapy

Case no.	Before HBO				During HBO		At the end of HBO		
	U-MG nmol/ml	U-Cr mg/dl	S-MG nmol/ml	S-Cr mg/dl	U-MG nmol/ml	U-Cr mg/dl	S-MG nmol/ml	S-Cr mg/dl	Diet*
1	1.31	111	ND*	0.5	0.43	23	ND	0.5	+
2	0.17	18	ND	0.5	0.26	24	ND	0.5	–
3	0.03	14	ND	0.3	0.20	26	ND	0.4	+
4	0.39	32	ND	0.8	0.59	25	ND	0.8	–
5	0.20	18	ND	0.6	0.98	32	ND	0.6	+
6	0.50	32	ND	0.5	0.57	25	ND	0.5	+
7	1.00	60	ND	0.6	1.10	46	ND	0.7	+
8	0.22	23	ND	0.5	0.30	31	ND	0.5	+
9	0.55	33	ND	0.8	0.72	38	ND	0.8	+
10	0.09	10	ND	0.5	0.08	9	ND	0.5	+
11	0.46	21	ND	0.7	1.59	48	ND	0.8	–
12	1.02	35	ND	1.0	1.25	38	ND	1.1	+
13	0.39	14	ND	1.3	1.58	51	ND	1.3	+
14	2.48	70	ND	1.2	3.84	56	ND	1.1	–
15	0.79	24	ND	1.5	0.58	8	ND	1.5	+
	1.68	31	ND	1.7	1.00	16	ND	1.6	+
16	21.95	64	1.9	6.1	17.26	37	1.9	6.1	+
	16.75	28	2.2	6.4	16.25	25	2.2	6.4	+
17	11.66	19	1.2	6.4	12.44	15	1.1	6.4	+
18	0.24	32	ND	0.7	0.59	12	ND	0.7	–
	0.11	5	ND	0.7	0.76	16	ND	0.7	–

*Low creatinine and MG diet (Cr: 7.2 mg, MG: 90 nmol); ND = not detectable.

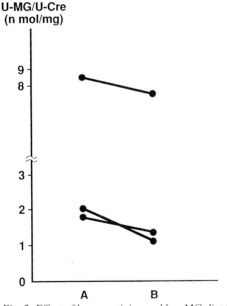

U-MG/U-Cre (n mol/mg)

Fig. 2. Effect of low creatinine and low MG diet (Cr: 7.2 mg, MG: 90 nmol) on urinary MG excretion without HBO therapy.
A: urine collected during 1–3 h after the diet; B: urine collected during 3–4.5 h after the diet.

The urinary excretion rate of MG is expressed by the value of urinary MG per urinary Cr. These values increase significantly during HBO therapy. In the group of patients with renal failure, the increase of MG excretion rate during HBO therapy was much more pronounced than that in the group without renal failure. In addition, the MG excretion rate before HBO therapy in the renal failure group was much higher than in the group without renal failure (Fig. 1).

Figure 2 indicates the change of urinary excretion rate of MG in the three cases who had a low Cr and MG diet without HBO therapy. As shown in this figure, the urinary excretion rate of MG in this group decreased as time went on. This result suggests that this diet had no effect on the increase of MG excretion during HBO therapy.

The urinary excretion rate of MG in case No. 18, who received HBO therapy twice under the condition of 3 ATA with 100 per cent O_2, rose markedly (from 1.5 mmol/mg to 4.8 mmol/mg) during HBO therapy in spite of a Ccr that was more than 50 ml/min.

Discussion

Under the circumstances of high oxygen pressure as HBO condition, the arterial PO_2 increases about 14- to 21-fold.

In this study, it was clear that the urinary excretion rate of MG elevated during HBO therapy and that a low Cr and MG diet had no effect on the increase of MG excretion.

In addition, we demonstrated that this increase was not caused by chemical response in urine under hyperbaric oxygen exposure. Although the results are not shown here, the MG and Cr levels of seven urine specimens did not change at all when placed in a HBO chamber under 2 ATA with 100 per cent O_2 for 2 h.

It has been reported that glomerular filtration rate (GFR) did not change, but renal plasma flow decreased and subsequently the filtration fraction rose under hyperbaric oxygen[9]. In this study, we did not measure GFR, but the serum Cr level did not change under the HBO therapy. Also, there is no report supporting the increase of GFR under hyperbaric oxygen. Therefore, we speculate that the increase of MG excretion was not caused by an increase of GFR but by stimulated synthesis of MG from Cr *in vivo* under HBO therapy.

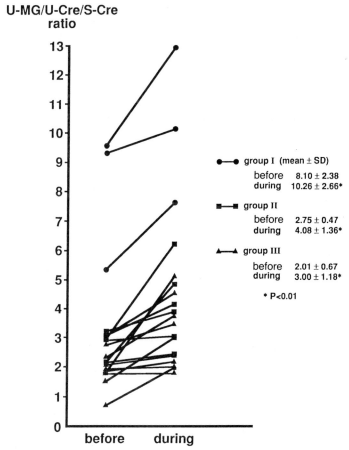

Fig. 3. Change in U-MG/U-Cr/S-Cr ratio under HBO therapy.
Conditions of HBO: 100 per cent O_2, 2 atmospheres absolute, 1 h;
Group I: Ccr < 10 ml/min; group II: 10 ≤ Ccr < 50 ml/min; group III: Ccr ≥ 50 ml/min.

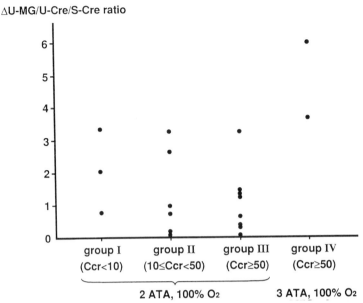

Fig. 4. Increase of U-MG/U-Cr/S-Cr ratio after HBO therapy in four groups.

The ratio of urinary MG excretion to serum Cr concentration (U-MG/U-Cr/S-Cr ratio) was calculated as an indicator of MG production rate from Cr. The U-MG/U-Cr/S-Cr ratio in the three groups (Group I–III) significantly increased during HBO therapy (Fig. 3).

Although the increase of MG excretion during HBO therapy rose dramatically in the group with renal failure, the increase of U-MG/U-Cr/S-Cr ratio was almost identical in these three groups. Also, the U-MG/U-Cr/S-Cr ratio increased markedly by exposure to oxygen pressure of 3 ATA as compared to that of 2 ATA (Fig. 4).

These results suggest that MG production rates correlate better with exposure to hyperbaric oxygen than with renal function.

There were a few deviant cases in each group showing large increases of the U-MG/U-Cr/S-Cr ratio during HBO therapy. The U-MG/U-Cr/S-Cr ratio increased markedly in one case of diabetic nephropathy in group I, two cases of nephrosclerosis in group II and one case of diabetic gangrene in group III. These results suggest that antioxidant activity may be low in patients with diabetes mellitus or nephrosclerosis, though further study is required to clarify this issue.

There are many reports indicating that the production of active oxygen rises under hyperbaric oxygen exposure[2,4] and that active oxygen is one of the most important factors of oxygen toxicity[5]. Based on these facts, while we have already reported that MG was synthesized by active oxygen species in vitro[8], we conducted this investigation in humans. As a result, it was clear that the increase of urinary MG excretion in the patients with or without renal failure under the HBO therapy reflects the peroxidative state under this condition. It is suggested that the U-MG/U-Cr/S-Cr ratio is a useful marker reflecting the production of active oxygen species in vivo.

Acknowledgements

These research findings were presented, in part, at the 32th Annual Meeting of the Japanese Congress of Nephrology in November 1989, in Hamamatsu and at the 12th Annual Meeting of the Japanese Guanidino Compounds Research Association in October 1989, in Sendai. We thank Ms. Junko Tanaka and Yukari Simozawa for their skilled technical assistance.

References

1. Aoyagi, K., Nagase, S., Narita, M. & Tojo, S. (1987): Role of active oxygen on methylguanidine synthesis in isolated rat hepatocytes. *Kidney Int.* **32**, S229–S233.

2. Chance, B., Sies, H. & Boveris, A. (1979): Hydroperoxide metabolism in mammalian organs. *Physiol. Rev.* **59**, 527–605.

3. Fillit, H., Elion, E., Sullivan, R., Sherman, R. & Zabriskie, J.B. (1981): Thiobarbituric acid reactive material in uremic blood. *Nephron* **29**, 40–43.

4. Freeman, B.A. & Crapo, J.D. (1981): Hyperoxia increases oxygen radical production in rat lungs and lung mitochondria. *J. Biol. Chem.* **256**, 10986–10992.

5. Fridovich, I. (1978): The biology of oxygen radicals. *Science* **201**, 875–880.

6. Kuroda, M., Asaka, S., Tofuku, Y. & Takeda, R. (1985): Serum antioxidant activity in uremic patients. *Nephron* **41**, 293–298.

7. Mano, Y., Akiba, C., Takano, N., Doi, N. & Shibayama, M. (1987): Research on hydroxyl radicals in plasma under hyperbaric oxygen exposure. *Jpn. J. Hyg.* **42**, 570–577.

8. Nagase, S., Aoyagi, K., Narita, M. & Tojo. S. (1986): Active oxygen in methylguanidine synthesis. *Nephron* **44**, 299–303.

9. Norman, J.G., Shearer, J.R., Napper, A.J., Robertson, I.M. & Smith, G. (1974): Action of oxygen on the renal circulation. *Am. J. Physiol.* **227** (**3**), 740–744.

10. Sakamoto, M., Aoyagi, K., Nagase, S., Ishikawa, T., Takemura, K. & Narita, M. (1989): Methylguanidine synthesis by active oxygen species derived from activated human neutrophils. *Jpn. J. Neprol.* **31**, 55–62.

11. Yamamoto, Y., Manji, T., Saito, A., Maeda, K. & Ohta, K. (1979): Ion exchange chromatography separation and fluorometric detection of guanidino compounds in physiologic fluids. *J. Chromat.* **162**, 327–340.

Guanidino Compounds in Biology and Medicine, eds. by P.P. De Deyn, B. Marescau, V. Stalon and
I.A. Qureshi. ©1992 John Libbey & Company Ltd., pp. 309–313.

Chapter 43

Dipyridamole decreased urinary excretion of methylguanidine increased by puromycin aminonucleoside *in vivo*

Kazumasa AOYAGI, Sohji NAGASE, Katsumi TAKEMURA, Shoji OHBA, and
Mitsuharu Narita

*Department of Internal Medicine, Institute of Clinical Medicine, University of Tsukuba, Tsukuba-city, Ibaraki, 305,
Japan*

Summary

We have reported that methylguanidine is formed from creatinine by active oxygen and that puromycin aminonucleoside
(PA) increased MG synthesis in isolated rat hepatocytes. We suggested that PA-induced proteinuria may be caused by
active oxygen. In the same isolated hepatocytes, we also reported that adenosine, its agonists and their potentiators
inhibited MG synthesis stimulated by PA. Moreover, it has been reported that adenosine potentiators ameliorate
PA-induced proteinura, experimental glomerulonephritis due to immune complex and human nephritis. In this paper,
we investigated the effect of PA and dipyridamole, an adenosine potentiator *in vivo*.

Twenty-eight rats were divided into four groups and kept in metabolic cages. PA (80 mg/kg) was administered to each
rat subcutaneously at the beginning of the experiment except in the control group. Dipyridamole was administered 50
mg/kg body weight or 200 mg/kg body weight per os after PA injection. Urinary excretion was determined by HPLC.
The amount of daily MG excretion in the urine of the control group, the PA group, the PA + 100 mg dipyridamole group
and PA + 400 mg dipyridamole group were 374 ± 159, 364 ± 113, 366 ± 67 and 325 ± 190 (mean ± SD) nmole/day at
the 1st day after PA injection; 521 ± 84, 593 ± 122, 389 ± 84 and 329 ± 128 nmoles/day at the 2nd day after PA injection;
680 ± 223, 995 ± 377, 560 ± 207 and 560 ± 67 nmoles/day at the 8th day after PA injection, respectively. In this
experiment, on the 2nd day after PA injection, a significant increase of proteinuria (from 11.1 ± 3.6 to 18.6 ± 5.4 mg/day,
mean ± SD, n = 7) was observed and this increase of protein excretion was inhibited by dipyridamole.

We have proposed that the MG/creatinine ratio may be a good marker of active oxygen generation *in vivo*. These results
suggested that PA increased active oxygen *in vivo* and dipyridamole decreased active oxygen generation *in vivo*. This
phenomenon may account for their action in kidney disease.

Introduction

We reported that methylguanidine (MG) is formed from creatinine by active oxygen[4,14] and also that puromycin aminonucleoside (PA) increased MG synthesis in isolated rat hepatocytes[2,3]. Therefore it was suggested that abnormal active oxygen generation may be a cause of proteinuria induced by PA. Again, in isolated rat hepatocytes, we demonstrated that adenosine and its agonists such as 2-chloroadenosine and [6]N-monomethyladenosine and their potentiators dipyridamole and zilazep hydrochloride inhibited MG synthesis[5,6,7]. Moreover, it has been reported that active oxygen scavengers decreased the proteinuria induced by PA in rats[9].

On the other hand, it has been reported that adenosine potentiators alleviate experimental glomeru-lonephritis[12], PA-induced proteinuria[11,13] and human glomerulonephritis[10,15].

In this paper, we investigated the effect of PA and also the effect of co-application of an adenosine potentiator, dipyridamole on urinary excretion of methylguanidine of rats.

Materials and methods

Four groups of male Sprague Dawley rats were considered. Each group consisted of seven rats. They were kept in metabolic cages for urine collection. PA (80 mg/kg) alone was injected subcutaneously in one group (PA group).

In two other groups, dipyridamole was administered soon after PA injection at a dosage of 50 mg/kg bid (PA + dipyridamole 100) or 200 mg bid (PA + dipyridamole 400) and continued until the end of the experiment as shown in Fig. 1. Methylcellulose without dipyridamole was administered per os to the control rats. The urine was collected in the flask from 0 to 24 h, from 24 to 48 h and from the 7th day to the 8th day after PA injection. To avoid *in vitro* conversion of creatinine to MG and bacterial contamination, 0.5 ml of a mixed solution containing 40 mg/ml glutathione, 4 mg/ml penicillin G and 4 mg/ml streptomycin was added to each flask and the flasks were chilled during urine collection. Dipyridamole was dissolved in 1 percent methylcellulose at the concentration of 20 mg/ ml. PA was dissolved in 0.9 per cent saline at the concentration of 22.8 mg/ml. Urinary protein precipitated by trichloroacetic acid was measured by Lowry method to avoid the interference of glutathione. Trichloroacetic acid (final concentration, 10 percent w/v) was added to the urine. MG in the supernatant was measured fluorometrically using HPLC and 9,10-phenanthrenequinone for post-labelling.

Results

Effect of dipyridamole on urinary protein excretion induced by puromycin aminonucleoside

Urinary protein excretion from 24 to 48 h after PA injections increased slightly in a statistically significant manner ($P < 0.01$) as shown in Table 1. Both 100 mg/day and 400 mg/day of dipyridamole inhibited the early phase increment of proteinuria by PA. Urinary protein excretion after 7 days of PA injection was slightly although not significantly decreased from 527 mg/day to 445 mg/day by 100 mg/kg/day of dipyridamole.

Table 1. Effect of puromycin and dipyridamole on urinary protein excretion

Condition	Urinary protein (mg/day, mean ± SD)		
	0–24 h	24–48 h	7–8th day
PA alone	11.1 ± 3.6^a	18.6 ± 5.4^d	527 ± 227^g
PA + dipyridamole (100)	10.6 ± 3.5^b	12.7 ± 3.6^e	445 ± 197^h
PA + dipyridamole (400)	11.2 ± 3.6^c	13.0 ± 5.5^f	505 ± 164^i

Values are expressed as mean ± SD of 7 rats. a *vs* b: NS, a *vs* d, g: P < 0.01, d *vs* e: P < 0.05, d *vs* f: NS, b *vs* e: NS, c *vs* f: e *vs* f: NS.

Effect of PA and dipyridamole on MG excretion in urine

MG excretion in urine collection from 24 h to 48 h increased significantly ($P < 0.05$) after PA injection as shown in Table 2. This increase was significantly prevented by both doses of dipyridamole. After 7 days of PA injection, MG excretion in urine markedly increased ($P < 0.01$) and this increase was inhibited by 100 mg/kg/day of dipyridamole ($P < 0.10$) and 400 mg/kg/day of dipyridamole ($P < 0.05$). In normal rats, urinary MG excretion increased significantly after being kept for 7 days.

Table 2. Effect of puromycin and dipyridamole on urinary MG excretion

Condition	Urinary MG excretion (nmol/day)		
	0–24 h	24–48 h	7–8th day
PA alone	364 ± 113^a	593 ± 122^e	995 ± 377^i
PA + dipyridamole (100)	366 ± 67^b	389 ± 84^f	560 ± 207^j
PA + dipyridamole (400)	325 ± 190^c	329 ± 128^g	560 ± 67^k
Control	374 ± 159^d	521 ± 84^h	680 ± 223^l

Values are expressed as mean ± SD of 7 rats. a *vs* b, c, d: NS, a *vs* e: $P < 0.05$, a *vs* i: $P < 0.01$, e *vs* f, g: $P < 0.05$, i *vs* j: $P < 0.10$, i *vs* k, l: $P < 0.05$, d *vs* l: $P < 0.05$, e *vs* h: NS, e *vs* f, g: $P < 0.05$.

Effect of PA and dipyridamole on other parameters

Body weight of the control group increased from 212.5 ± 2.8 g to 249.8 ± 5.2 g (mean ± SD) over 1 week in this experiment. Body weight increased from 200 ± 5 g to 227 ± 12.6 g in the PA-injected group, from 197 ± 4.8 g to 219 ± 12.2 g in the PA + 100 mg/kg/day dipyridamole group and from 201 ± 12.4 g to 230 ± 15.3 g in the PA + 400 mg/kg/day dipyridamole group.

Daily food intake by a rat in the control group, PA group, PA + dipyridamole (100) group, and PA + dipyridamole (400) group were 20.5 ± 2.5 g, 10.5 ± 4.6 g, 11.1 ± 4.6 g and 12.1 ± 2.5 g (mean ± SD) per day, respectively. Food intake by rats in the control group was significantly greater than that of PA-injected rats.

Discussion

Urinary protein excretion increased slightly on the second day after PA injection. Recently, this early phase increase of proteinuria by PA injection was also pointed out by other investigators. This increase of proteinuria was inhibited by dipyridamole administration. In our experiment, massive proteinuria on the 7th day after PA injection was not significantly inhibited by dipyridamole. This result is different from earlier findings[11,13]. We injected 80 mg/kg of PA subcutaneously. This dose of PA might have been too high to be inhibited by dipyridamole since decrease of MG excretion on 7th day by dipyridamole was not so strong ($P < 0.10$).

Urinary MG excretion was significantly increased 24 h after PA injection ($P < 0.05$) and 7 days after PA injection ($P < 0.01$). MG is easily excreted by the kidney, and is difficult to detect in the serum of normal subjects. So urinary excretion of MG reflects MG formation in the various organs. In our experiment, MG was not detected in the serum of any rat. We have reported that in various organs MG is formed from creatinine. Furthermore, we have reported that MG is formed by reactive oxygen in isolated rat hepatocytes as well as in activated human leukocytes[14]. We also demonstrated that under hyperbaric oxygen conditions which increase active oxygen generation *in vivo*, urinary excretion of MG/creatinine ratio increased (*see* Chapter 42). So the increase of urinary MG excretion

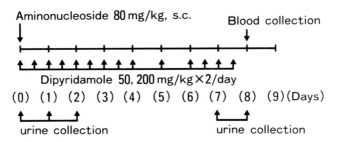

Fig. 1. Schedule of experiment.

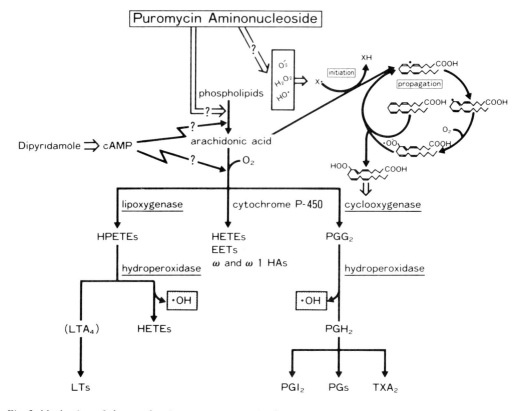

Fig. 2. Mechanism of abnormal active oxygen generation by puromycin aminonucleoside and its suppression by dipyridamole.

suggested the increase of active oxygen generation by puromycin aminonucleoside in the rats *in vivo*. Decrease of the urinary excretion of MG by the administration of dipyridamole to the PA-injected rats also suggested the decrease of active oxygen generation *in vivo* as well as in isolated rat hepatocytes *in vitro*. MG is formed in various organs, so dipyridamole might have inhibited active oxygen generation in the whole body. In the normal control rats, urinary MG excretion also increased during this experiment. The reason for this increase was not clear. The stress of being kept in the metabolic cages or growth of the rats may have been related to it. The food containing small amounts of MG could have been of influence as well. Food intake by normal rats was more than that of the PA-injected rats, and there is no difference in the food intake among the PA-injected rats with or without dipyridamole. Urinary MG excretion of PA-injected rats was significantly higher than that of normal rats.

Concerning the relation between adenosine and active oxygen generation, it has been reported that adenosine is a physiological modulator of superoxide anion generation in leukocytes[1]. In isolated rat hepatocytes, we have demonstrated that cyclic AMP may be an important intracellular transducer or suppressor of active oxygen generation. Endo *et al.* also reported that adenosine A_2 receptor which increases cyclic AMP was related to the inhibition of active oxygen generation in isolated rat glomeruli[16].

Concerning the mechanism of abnormal active oxygen generation in isolated rat hepatocytes, we reported that the arachidonic cascade might be an important source of hydroxyl radicals[8]. So we propose that adenosine potentiators, such as dipyridamole and zilazep inhibit abnormal active oxygen generation as shown in Fig. 2.

References

1. Aoyagi, K., Ohba, S., Narita, M. & Tojo, S. (1983): Regulation of biosynthesis of guanidinosuccinic acid in isolated rat hepatocytes and *in vivo. Kidney Int.* **24**, S224–S228.

2. Aoyagi, K., Nagase, S., Narita, M. & Tojo, S. (1987): Role of active oxygen on methylguanidine synthesis in isolated rat hepatocytes. *Kidney Int.* **32**, S229–S233.

3. Aoyagi, K., Nagase, S., Sakamoto, M., Narita, M. & Tojo, S. (1989): Puromycin aminonucleoside stimulates the synthesis of methylguanidine: a possible marker of active oxygen in isolated rat hepatocytes. In *Guanidines 2*, eds. A. Mori, B.D. Cohen & H. Koide, pp. 71–77. New York: Plenum Press.

4. Aoyagi, K., Nagase, S., Sakamoto, M., Narita, M. & Tojo, S. (1989): Active oxygen in methylguanidine synthesis by isolated rat hepatocytes. In *Guanidines 2*, eds. A. Mori, B.D. Cohen & H. Koide, pp. 79–85. New York: Plenum Press.

5. Aoyagi, K., Nagase, S., Sakamoto, M., Narita, M. & Tojo, S. (1989): Adenosine, adenosine analogues and their potentiators inhibit methylguanidine synthesis, a possible marker of active oxygen in isolated rat hepatocytes. In *Guanidines 2*, eds. A. Mori, B.D.Cohen & H. Koide, pp. 123–128. New York: Plenum Press.

6. Aoyagi, K., Nagase, S., Sakamoto, M., Takemura, K., Ishikawa, T. & Narita, M. (1989): Regulation of biosynthesis of methylguanidine, a marker of active oxygen generation in isolated rat hepatocytes. In *Medical, biochemical and chemical aspects of free radicals*, eds. O. Hayaishi, E. Niki, M. Kondo & T. Yoshikawa, pp. 137–140. Amsterdam: Elsevier.

7. Aoyagi, K., Nagase, S., Sakamoto, M., Takemura, K., Ishikawa, T. & Narita, M. (1990): Regulation of active oxygen generation in isolated rat hepatocytes by adenosine. *Heart and Vessels*, Suppl. **4**, 33.

8. Aoyagi, K. & Narita, M. (1991): Mechanism of abnormal active oxygen generation in tissue cells by puromycin aminonucleoside. *Acta Medica Biologica* **39**, 53–62.

9. Bruce, N., Cronstein, B.N., Karmer, S.B., Weissmann, G.D. & Hirschhorn, R. (1983): Adenosine: a physiological modulator of superoxide anion generation by human neutrophils. *J. Exp. Med.* **158**, 1160–1177.

10. Diamond, J.R., Bonventrre, J.V. & Karnovsky, S.B. (1986): A role for oxygen free radicals in aminonucleoside nephrosis. *Kidney Int.* **29**, 478–483.

11. Kan, K., Wada, T., Kitamoto, K., Konishi, K., Ozawa, Y., Kato, E. & Matsuki, S. (1974): Dipyridamole for proteinuria suppression: use in a patient with proliferative glomerulonephritis. *JAMA* **229**, 557–558.

12. Kimura, K., Endo, H. & Sakai, F. (1985): Suppressive effect of dipyridamole on the proteinuria of aminonucleoside nephrosis in rat. *J. Toxicol. Sci.* **4**, 1–10.

13. Koyama, A., Inage, H., Sano, M., Tojo, S., Neild, G.H. & Cameron, J.S. (1985): Platelet involvement in the nephritis of acute serum sickness in rabbits: protection by dipyridamole and FUT-175. *Clin. Exp. Immunol.* **61**, 388–396.

14. Nagase, S., Aoyagi, K., Narita, M. & Tojo, S. (1986): Active oxygen in methylguanidine synthesis. *Nephron*, **44**, 299–303.

15. Sakamoto, M., Aoyagi, K., Nagase, S., Ishikawa, T., Takemura, K. & Narita, M. (1989): Methylguanidine synthesis by active oxygen generated by stimulated human neutrophils. *Jap. J. Nephrol.* **31**, 851–858.

16. Tojo, S., Narita, M., Koyama, A., Sano, S., Suzuki, H., Tsuchiya, T., Tsuchida, H., Yamamoto, S. & Shishido, H. (1978): Dipyridamole therapy in the nephrotic syndrome. *Contrib. Nephrol.* **9**, 111–127.

Guanidino Compounds in Biology and Medicine, eds. by P.P. De Deyn, B. Marescau, V. Stalon and
I.A. Qureshi. ©1992 John Libbey & Company Ltd., pp. 315–316.

Chapter 44

Guanidino compounds in cerebrospinal fluids of patients with cerebrovascular disease and chronic renal failure treated by haemodialysis

K. ISODA, M. KINOSHITA, S. HIROSE, H. TOKUSHIMA, N. KASAHARA, S. KIM, G.
MATSUMOTO, T. SAGARA, Y. ITAKURA, H. TAMURA, R. NAGASAWA and T. MITARAI

4th Department of Internal Medicine, Saitama Medical Center, Saitama Medical School, Saitama 350, Japan

Introduction

Efforts have been made to detect causative substances related to the development of central and peripheral neuropathy in uraemic patients. Both small and middle molecular substances which are easily removed from the blood stream by dialysis have been thought to be candidate neurotoxins. However, distinct relationships between these substances and neurological symptoms have not been clarified in patients on long-term haemodialysis treatment. It is known that central and peripheral neuropathy such as involuntary movements, myoclonus, epileptic attacks and severe burning feet sensations are found among patients on 'so-called' appropriate dialytic therapy. Only a few studies have reported on the association between neuropathy and guanidino compound alterations in cerebrospinal fluid (CSF) in nondialysed patients with renal insufficiency[1,2,3].

In this study, we measured seven guanidino compounds in serum and CSF obtained from patients in renal failure and on long-term haemodialysis and from patients with cerebrovascular disease (CVD), with or without renal impairment, to elucidate the close relationship between neurological symptoms and concentrations of guanidino compounds in CSF.

Materials and methods

Patients

Twenty patients were selected in this study including 10 patients with CVD (average age 60.8 ± 11.6 years, eight male, two female), seven patients with chronic renal failure on a dialysis programme (average age 64.5 ± 4.6 years, two male, five female) and three subjects with CVD (average age 73.3 ± 4.7 years, one male, two female) treated by haemodialysis because of complication of renal failure.

Eleven of 13 patients with CVD were found to suffer from infarction by physical examination and computed tomography of the brain (brain CT). Two of 13 patients with CVD were diagnosed to suffer from hypertensive encephalopathy. The group of patients treated by haemodialysis was composed of six patients with diabetic neuropathy, two patients with chronic glomerulonephritis as original disease and two patients with cerebral infarction with renal failure.

Methods

Serum and CSF were collected from 20 patients and stored at –40 °C until analysis. Guanidinosuccinic acid (GSA), guanidinoacetic acid (GAA), guanidinobutyric acid (GBA), guanidinopropionic acid (GPA), guanidine (G), methylguanidine (MG) and creatinine (Cr) were estimated using high performance liquid chromatography with fluorescence detection. A stepwise pH gradient from 2 to 9 was used.

Results and discussion

Mean concentrations of GSA, GAA, G, MG and Cr in serum and CSF of the three groups are shown in Table 1. Concentrations of GSA in serum and CSF showed the highest levels in patients who suffered from CVD during a dialysis programme. GAA and MG were also detected in serum and CSF of patients with chronic renal failure on dialysis treatment, however those substances were not detectable in serum and CSF obtained from patients with CVD accompanied with renal failure. The reason of those results remained uncertain in this study. GPA and GBA were not detected in serum and CSF in all patients examined.

Table 1. Concentration of various guanidino compounds in serum and CSF from patients with cerebrovascular disease and these treated by haemodialysis and chronic renal failure on haemodialysis treatment

	CVD	CVD + HD	HD
sGSA	1.90 ± 0.33	8.00 ± 5.07	4.57 ± 2.57
cGSA	0.25 ± 0.23	2.43 ± 1.66	1.61 ± 1.03
sGAA	1.87 ± 0.35	ND	1.43 ± 0.54
cGAA	ND	ND	0.21 ± 0.19
sG	ND	0.60 ± 0.84	ND
cG	ND	0.32 ± 0.23	0.54 ± 0.45
sMG	ND	1.60 ± 1.13	2.80 ± 1.18
cMG	ND	0.30 ± 0.22	1.09 ± 0.02
sCTN	62.3 ± 11.0	512.6 ± 298.9	618.2 ± 149.7
cCTN	144.0 ± 135.0	180.2 ± 29.5	415.8 ± 273.0

s: Serum, c: CSF; unit: μmol/l; Values are mean ± SD; ND: not detected.

Episodes of myoclonus were seen in five of 20 patients, epileptic attacks in three of 20 and severe peripheral neuropathy in two of 20 patients. Concentrations of GSA in CSF were ranged from 16 to 98.9 per cent of the concentration in serum in patients who showed myoclonus, and from 23.4 to 63.3 per cent in patients with severe peripheral neuropathy. De Deyn et al.[1,2] reported in vivo and in vitro data on GSA. GSA was detectable in CSF in nondialysed patients with chronic renal insufficiency, and adding GSA at clinically significant concentrations into the medium of mouse spinal cord neurons in primary dissociated cell culture inhibited both GABA and glycine responses in a dose-dependent manner. These data suggested that guanidino compounds inhibit the inhibitory neurotransmitters GABA and glycine by blocking the chloride channel. The presence of guanidino compounds in CSF might indicate that those substances cross the blood–brain barrier in chronic renal failure and cerebrovascular disease. Further studies of the neurotransmitter systems of patients with chronic renal failure on long-term haemodialysis may lead to more understanding of the pathogenesis of uraemic central and peripheral neuropathy.

References

1. De Deyn, P.P., Marescau, B., Cuyckens, J.J., Van Gorp, L., Lowenthal, A. & De Potter, W.P. (1987): Guanidino compounds in serum and cerebrospinal fluid of non-dialyzed patients with renal insufficiency. Clin. Chim. Acta 167, 81–88.

2. De Deyn, P.P. & Macdonald, R.L. (1990): Guanidino compounds that are increased in cerebrospinal fluid and brain of uremic patients inhibit GABA and glycine responses on mouse neurons in cell culture. Ann. J. Neurol. 28, 627–633.

3. Marescau, B., De Deyn, P.P., Wiechert, P., Hiramatsu, M., Van Gorp, L., De Potter, W.P. & Lowenthal, A. (1989): Guanidino compounds in serum and cerebrospinal fluid of epileptic and some other neurological patients. In Guanidines 2, eds. A. Mori, B.D. Cohen & H. Koide, pp. 203–212. New York/London: Plenum Press.

Guanidino Compounds in Biology and Medicine, eds. by P.P. De Deyn, B. Marescau, V. Stalon and I.A. Qureshi. ©1992 John Libbey & Company Ltd., pp. 317–320.

Chapter 45

Influence of 'oral 1,25(OH)$_2$D$_3$ pulse therapy' on methylguanidine in haemodialysis patients with severe secondary hyperparathyroidism

S. TAKAHASHI, Y. KINOSHITA, M. YANAI, K. OKADA, T. KUNO, M. MAEJIMA, Y. NAGURA and M. HATANO

2nd Department of Internal Medicine, Nihon University School of Medicine, 30–1 Oyaguchi-kami, Itabashi-ku, Tokyo 173, Japan

Summary

Methyl guanidine (MG) is one of the guanidino compounds (GC) and investigations on it as a uraemic toxin have long been made. The correlation among MG, parathyroid hormone (PTH) which is one of the same uraemic toxins and vitamin D have not been clarified. We gave oral 1,25(OH)$_2$D$_3$ pulse therapy to patients with chronic renal failure for the treatment of secondary hyperparathyroidism (2° HPT) in an attempt to assess the effectiveness of this therapeutic regimen and to explore its effect on MG levels. In the patients of the responsive group, decreased serum levels of MG were observed early after initiation of the therapy, suggesting that early measurement of the serum MG may provide a useful indicator for monitoring the therapeutic response.

Introduction

Vitamin D administration has been used in the management of secondary hyperparathyroidism (2° HPT) due to impaired calcium and phosphorus metabolism and can allegedly be expected to prevent, to some extent, the progression of this pathological condition[1,6,7]. Unfortunately, however, it is not uncommon in routine clinical practice to find that 2° HPT of moderate or greater severity is resistant to treatment with these vitamin D preparations at appropriate oral therapeutic doses[4]. Slatopolsky *et al.*[8] reported that intravenous administration of 1,25-dihydroxy-cholecalciferol (1,25(OH)$_2$D$_3$) directly induced a marked decrease in parathyroid hormone (PTH), apparently not via the calcium-dependent feedback mechanism, in uraemic patients. On this basis, parenteral use of this active form of vitamin D attracted attention as a promising new therapeutic approach to 2° HPT. Oral active vitamin D pulse therapy has also been attempted, and there are increasing numbers of reports documenting its effectiveness[9].

On the other hand, studies have gradually clarified the metabolic pathways and physiological roles of guanidino compounds (GC) long known as uraemic toxins. Concerning methylguanidine (MG) among other GCs, its precursor has already been identified as creatinine (Cr)[3], and it is known that activated oxygen as well as PTH and vitamin D participate in the process leading to its formation.

In the present study, we gave oral $1,25(OH)_2D_3$ pulse therapy to patients with chronic renal failure for the treatment of 2° HPT in an attempt to assess the effectiveness of this therapeutic regimen and to explore its effect on MG levels.

Subjects

Twenty-one chronic renal failure patients with 2° HPT who were maintained on haemodialysis therapy at the Nihon University Itabashi Hospital were included in the study. The diagnostic criteria employed for 2° HPT were serum high-sensitivity parathyroid hormone (HS-PTH) concentrations of ≥ 10 ng/ml and the presence of bone pain and other putative symptoms of 2° HPT. The patients had a mean age of 52 ± 7 years (mean $\pm$ SEM) and a mean duration of haemodialysis therapy of 10.8 ± 0.9 (mean $\pm$ SEM).

Methods

Each patient was administered orally 6 µg of $1,25(OH)_2D_3$ once a week after completion of the first of the weekly dialysis sessions over a 12-week period. From the moment of inclusion in the study, medications involving any other active vitamin D preparation were withdrawn. During the study period, the serum HS-PTH level was determined at 4-week intervals. The serum MG was measured before each dialysis session during the first week of treatment and at 4-week intervals for the remaining treatment period.

Only those patients who showed more than a 20 per cent reduction in their serum HS-PTH under the influence of the study treatment were considered to be responders (the responsive group), and the others were regarded as resistant (the resistant group). Comparisons between both groups were made for the individual parameters.

The serum HS-PTH was determined by a double antibody radioimmunoassay method using a High-Sensitivity PTH Kit Yamasa (Yamasa Soy Co., Ltd.) capable of recognizing 44–46 fragments. The serum MG was measured by high performance liquid chromatography employing a JASCO G800 Series Autoanalyzer (Nihon Bunko Co.).

All data were expressed as the mean $\pm$ SEM and analysed for statistical significance by the paired t-test and unpaired t-test at the 5 per cent level ($P < 0.05$).

Results

Among the 21 patients treated, a decrease of more than 20 per cent in PTH was achieved by the end of the 12-week treatment period in 10 individuals (the responsive group), but not in the remaining 11 (the resistant group).

PTH

The pretreatment serum levels of PTH were significantly higher in the resistant than in the responsive group. They ranged from 10.1 to 68.8 ng/ml (mean, 29.8 ± 5.8 ng/ml) in the latter group against 40.3 to 135 ng/ml (mean, 75.3 ± 12.2 ng/ml) in the former. The time courses of the serum PTH level in the responsive group declined significantly from 29.8 ± 5.8 ng/ml at week 0 to 21.4 ± 3.7 ng/ml ($P < 0.01$) at week 4, 23.9 ± 4.8 ng/ml ($P < 0.01$) at week 8 and 22.7 ± 6.6 ng/ml ($P < 0.05$) at week 12 of treatment, whereas the corresponding values in the resistant group were 75.3 ± 12.2 ng/ml ($P < 0.01$) at week 8, 78.0 ± 11.4 ng/ml at week 4, 112.2 ± 14.7 ng/ml ($P < 0.01$) at week 8 and 104.7 ± 14.7 ng/ml ($P < 0.01$) at week 12, being significantly elevated at weeks 8 and 12.

Serum MG

The serum MG levels found after the initial dose of $1,25(OH)_2D_3$ and at 2 and 4 days thereafter were as follows: in the responsive group, all patients revealed a progressive reduction in serum MG with time, with the mean value being $62.0 \pm 11.0\ \mu g/dl$ initially, $60.5 \pm 11.5\ \mu g/dl$ ($P < 0.05$) at 2 days and $46.9 \pm 9.4\ \mu g/dl$ ($P < 0.05$) at 4 days. No significant changes were noted in the resistant group, the values at the corresponding time points being $53.7 \pm 7.0\ \mu g/dl$ and $50.2 \pm 7.0\ \mu g/dl$, respectively.

The serum MG levels observed after 4, 8 and 12 weeks of the study treatment are as follows: in the responsive group, the serum MG decreased significantly from an initial level of $62.0 \pm 11.0\ \mu g/dl$ to $36.4 \pm 4.1\ \mu g/dl$ ($P < 0.05$), $31.2 \pm 3.9\ \mu g/dl$ ($P < 0.05$) and $37.0 \pm 5.1\ \mu g/dl$ ($P < 0.05$) at weeks 4, 8 and 12, respectively. In the resistant group, in contrast, no significant changes in serum MG were noted during the study period, the values at the corresponding time points being $53.7 \pm 7.0\ \mu g/dl$, $55.2 \pm 7.8\ \mu g/dl$, $49.2 \pm 7.1\ \mu g/dl$ and $47.3 \pm 4.9\ \mu g/dl$, respectively.

Discussion

As indicated above, patients who responded to oral 1,25(OH)₂D₃ pulse therapy with at least a 20 per cent reduction in their serum PTH were considered as responders in the present study (a decrement of 20 per cent being the smallest difference achieving statistical significance). As was expected, the results obtained suggested that the therapy under investigation might be indicated in slight to moderate 2° HPT, but not in severe 2° HPT where surgical treatment is inevitable. The borderline serum PTH level has yet to be defined by additional studies.

MG is one of the GCs and investigations on it as a uraemic toxin have long been made. Jones *et al.*[2] reported that MG was formed from creatinine *in vivo*, and recent evidence indicates that reactions involved in the production of MG from creatinine take place in the absence of oxygen, with activated oxygen (especially hydroxyl radicals) playing an important role[5].

Concerning the relationships between MG and PTH and vitamin D, Ishizaki *et al.* reported the existence of a positive linear correlation between serum MG and serum PTH[2]. We noted decreased serum levels of MG at 1 week after parathyroidectomy in patients with 2° HPT undergoing oral 1,25-(OH)₂D₃ pulse therapy, and inferred that the active form of vitamin D administered might play a role in the observed fall in serum MG. Although a decrease in serum MG was found to occur concurrently with a fall in serum PTH in the responsive group in the present study, the decrease in serum MG was apparently less marked than that seen in parathyroidectomized patients. No reduction in serum MG was noted in the resistant group. Moreover, in the responsive group, the serum MG was already decreased very early after the initiation of oral 1,25-(OH)₂D₃ pulse therapy, suggesting the possibility that early measurement of MG could provide an indicator for assessing the effectiveness of the therapy.

At the present time, few reports are available on the involvement of vitamin D in the process of MG production, and further studies are needed to explain adequately the reasons why the administration of vitamin D in large doses results in a lowering of the serum MG and why vitamin D induces such a change in MG electively in those patients with 2° HPT who respond well to the therapy in question with a fall in serum PTH.

However, in so far as the present study is concerned, it seems probable that the vitamin D exerted some, as yet undefined, influence on the production of MG. Since the serum vitamin D concentration was not determined in any of the patients at the present study, it remains uncertain whether the observed rather poor overall therapeutic responsiveness to the drug regimen employed was due to an impaired intestinal absorption of the drug in the resistant group or was largely accounted for by 2° HPT which was sufficiently severe to defy even adequately raised blood levels of the vitamin. At all events, this issue remains to be clarified by future research.

Conclusion

Twenty-one haemodialysis patients with severe 2° HPT were treated with oral 1,25(OH)₂D₃ pulse therapy. The results obtained yielded the following conclusions:

1. The serum HPT level was significantly higher in patients who were resistant than in those who were responsive to the oral 1,25(OH)$_2$D$_3$ pulse therapy.

2. In the patients of the responsive group, decreased serum levels of MG were observed early after initiation of the therapy, suggesting that early measurement of the serum MG may provide a useful indicator for monitoring the therapeutic response.

References

1. Brickman, A.S., Coburn, J.W. & Norman, A.W. (1972): Action of 1,25(OH)$_2$D$_3$ a potent, kidney-produced metabolites of vitamin D in uremic man. *N. Engl. J. Med.* **289**, 891–899.

2. Ishizaki, M., Kitamura, K. & Kitamoto, Y. (1986): The correlation between serum parathyroid hormone and methylguanidine in hemodialysis patients. *Jpn. J. Nephrol.* **28**, 610.

3. Jones, J.D. & Burnett, P.C. (1972): Implication of creatinine and gut flora in the uremic syndrome: induction of creatininase in colon contents of the rat by dietary creatinine. *Clin. Chem.* **18**, 280–284.

4. Massry, S.G., Goldstein, D.A. & Malluche, H.H. (1980): Current status of the use of 1,25(OH)$_2$D$_3$ in the management of renal osteodystrophy. *Kidney Int.* **18**, 409–418.

5. Nagase, S., Aoyagi, K., Narita, M. & Tojo, S. (1986): Active oxygen in methylguanidine synthesis. *Nephron* **44**, 299–303.

6. Okada, K., Kuno, T., Yanai, M., Maeda, H., Takahashi, S. & Hatano, M. (1988): The effects of CaCO$_3$ on renal osteodystrophy. *Jpn. J. Nephrol.* **30**, 1053–1062 (in Japanese).

7. Okada, K., Takahashi, S. & Hatano, M. (1990): Administration of calcium carbonate with adequate doses of vitamin D metabolite to patients on hemodialysis improves mild secondary hyperparathyroidism. *Jpn. J. Nephrol.* **32**, 899–903.

8. Slatopolsky, E., Weort, C., Thielan, J., Horst, R., Harter, H. & Martin, K.J. (1984): Marked suppression of secondary hyperparathyroidism by intravenous administration of 1,25-dihydroxycholecalciferol in uremic patients. *J. Clin. Invest.* **74**, 2136–2143.

9. Tuskamoto, Y., Nomura, M. & Marumo, F. (1989): Pharmacological parathyroidectomy by oral 1,25(OH)$_2$D$_3$ pulse therapy. *Nephron* **51**, 130–131.

Guanidino Compounds in Biology and Medicine, eds. by P.P. De Deyn, B. Marescau, V. Stalon and
I.A. Qureshi. ©1992 John Libbey & Company Ltd., pp. 321–322.

Chapter 46

Dipyridamole decreases the urinary methylguanidine to creatinine ratio in proteinuric patients with preserved renal function

Yutaka KODA, Sachio TAKAHASHI, Masashi SUZUKI, Yoshihei HIRASAWA and
Moto-o NAKAJIMA[1]

*Kidney Centre, Shinrakuen Hospital, Niigata; and [1]Bioscience Research Laboratory, Kikkoman Corporation, Noda,
Japan*

Introduction

Some studies on experimental models on glomerular injury indicate oxygen radicals to be important mediators of tissue injury. It could thus be possible to obtain good clinical results by reducing oxidative stress in proteinuric patients by medication. Dipyridamole is an adenosine potentiator and adenosine inhibits the oxygen radical generation of some cells[2].

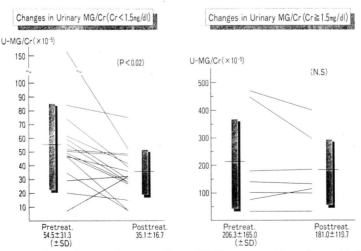

*Figs. 1 and 2. A significant decrease in mean urinary MG/Cr was noted in patients with S-Cr < 1.5 mg/dl
(Fig. 1, left) but not in patients with S-Cr > 1.5 mg/dl (Fig. 2, right).
MG: methylguanidine (μg/dl), Cr: creatinine (g/l).*

Methylguanidine (MG) has been determined in the serum of uraemic patients and found to be a toxin in the pathogenesis of uraemia. Recently, MG was synthesized from creatinine (Cr) by the action of oxygen radicals *in vitro*[1]. The ratio of MG to Cr (MG/Cr) can thus serve as a useful indicator of oxygen radical injury[1,4].

This study was conducted to determine whether dipyridamole improves oxidative stress in proteinuric patients with various renal diseases, using urinary MG/Cr as an indicator of oxidation.

Subjects and methods

Twenty three out-patients (eight with IgA nephropathy, three with membraneous nephropathy, four with chronic renal failure and eight with other miscellaneous renal diseases) with proteinuria were treated with dipyridamole at 300 mg/day for 60.3 ± 44.5 days. Patients mean age was 49.0 ± 14.5 years and mean serum Cr (S-Cr) before therapy, 1.8 ± 1.6 mg/dl (range 0.5–6.4). MG, Cr, protein excretion (UP) in spot urine and Cr in the serum were measured before and after therapy. MG was determined by enzymatic assay[3].

The results were expressed as mean $\pm$ SD. Statistical analysis was performed using Student's *t*-test for combined data and linear regression analysis. The null hypothesis was rejected at $P < 0.05$.

Results and discussion

In our patients, not only urinary MG (µg/dl) itself, but also urinary MG (µg/dl)/Cr(g/l) was found to be well correlated with S-Cr ($r = 0.8719$, $P < 0.01$). MG production from Cr may thus possibly be promoted especially in patients with advanced renal diseases, due to excessive oxygen radicals.

A significant decrease in mean urinary MG/Cr, from 54.5 ± 31.3 pretherapy to 35.1 ± 16.7 post therapy, was noted in 16 patients with S-Cr < 1.5 mg/dl but not in seven patients with S-Cr $\geq$ 1.5 mg/dl (Figs. 1 and 2). This is the key finding of this study. Oxygen radical injury can more easily be reduced in earlier stages of renal disease. Thus, treatment especially for oxidative injury should be instituted as soon as possible.

No change could be detected by UP/Cr or S-Cr during treatment but a significant positive correlation was found between per cent change of MG/Cr, and of UP/Cr excluding relapsed minimal change nephrotic syndrome. Although we could not confirm the prognostic implication of urinary MG/Cr in proteinuric patients, the present results suggest an important relationship between urinary MG/Cr and proteinuria. Attempts should be made to reduce urinary MG/Cr in proteinuric patients by any means.

IgA nephropathy is one of the commonest types of glomerulonephritis in Japan. MG/Cr decreased in a manner not statistically significant in six of eight IgA glomerulonephritis patients with normal renal function. This study was of rather short duration for confirming clinical results after treatment of chronic renal disease. Additional study should be made to assess the effectiveness of dipyridamole treatment.

In conclusion, dipyridamole appears to improve oxidative injury especially in patients with mild renal dysfunction. Dipyridamole should thus be used in the early stages of glomerular disease in view of possible oxygen radical injury.

References

1. Aoyagi. K., Nagase, S., Narita, M. & Tojo, S. (1987): Role of active oxygen on methylguanidine synthesis in isolated rat hepatocyte. *Kidney Int.* **32**, S229–S233.
2. Cronstein, B.N., Levin, R.I., Belanoff, J., Weissmann, G. & Hieschhozn, R. (1986): Adenosine: an endogenous inhibitor of neutrophil-mediated injury to endothelial cells. *J. Clin. Invest.* **78**, 760–770.
3. Nakajima, M., Nakamura, K., & Shirokane, Y. (1985): Enzymic determination of methylguanidine in urine. In *Guanidines*, eds. A. Mori, B.D. Cohen & A. Lowenthal, pp. 39–46. New York: Plenum Press.
4. Nakajima, M., Nakamura, K., Shirokane, Y. & Hirasawa, Y. (1989): Enzymic determination of methylguanidine in serum and plasma of hemodialysis patients as a marker of hydroxyl radical. In *Guanidines 2*, eds. A. Mori, B.D. Cohen & H. Koide, pp. 3–11. New York: Plenum Press.

Guanidino Compounds in Biology and Medicine, eds. by P.P. De Deyn, B. Marescau, V. Stalon and
I.A. Qureshi. ©1992 John Libbey & Company Ltd., pp. 323–325.

Chapter 47

Urinary guanidinoacetic acid excretion rate as a tool for clinical diagnosis of acute rejection in transplanted kidney

Yoshiharu TSUBAKIHARA, Waka SHIMOIDE, Makoto ARAI, Eisaku KITAMURA,
Noriyuki OKADA, Isao NAKANISHI, Nobutoshi IIDA, Kiichiro ITOH[1],
Shiro SAGAWA[1] and Akio ANDO[2]

*Department of Nephrology and Department of Urology[1], Osaka Prefectural Hospital, 3-1-56 Bandaihigashi,
Sumiyoshi-ku, Osaka 558; Faculty of Health and Sport Sciences, Osaka University[2], Osaka, Japan*

Summary

We evaluated the efficacy of urinary guanidinoacetic acid excretion rate (U-GAA) as a tool of differential diagnosis
between acute rejection (AR) and cyclosporine nephrotoxicity (CyA) in 5 patients with renal transplantion (TPX). AR
and CyA were diagnosed by renal biopsy. U-GAA/creatinine clearance (Ccr) as determination more than 2 months after
TPX with high performance liquid chromatographic system. Nine episodes of AR were diagnosed in four patients and
3 episodes of CyA in 2 cases. As a results, U-GAA/Ccr was significantly decreased in AR as compared with CyA.
U-GAA/Ccr was significantly correlated conversely with urinary FDP excretion which was speculated as an indicator
of AR. From these results, U-GAA/Ccr was considered as a good tool in differential diagnosis between AR and CyA.

Introduction

Urinary guanidinoacetic acid excretion rate (U-GAA) has been recognized as an indicator of metabolic function of the kidney, especially of the tubule, because GAA has been shown to be synthesized by tubular cells.

It is difficult to differentiate clinically between acute rejection (AR) and cyclosporine nephrotoxicity (CyA toxicity) in transplanted kidney. In this paper, we evaluated U-GAA determination as a tool for differential diagnosis between AR and CyA toxicity in seven patients suffering from AR and/or CyA toxicity with cadaveric renal transplantation (TPX).

Subjects and methods

Subjects were seven cases with cadaveric TPX suffering AR and/or CyA toxicity (Table 1). AR or CyA toxicity was diagnosed by renal biopsy when serum creatinine (Cr) significantly increased from basal level. U-GAA and U-FDP (fibrin degradation products) were determined every day more than 2 months after TPX. U-FDP (FDP-E) was determined with Latex Photometric immunoassay. U-GAA was measured with a HPLC system (Hitachi-L6200) using 9,10-phenanthrenequinone for the detection of the guanidino compounds[1].

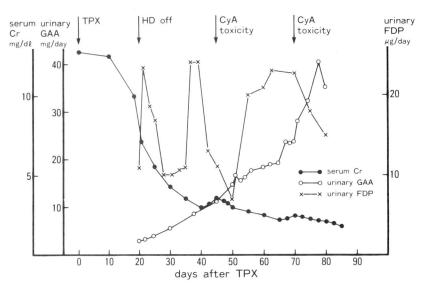

Fig. 1. Clinical course of case 1 suffering from cyclosporin (CyA) toxicity.

Results

Thirteen episodes of AR were diagnosed in six patients and four episodes of CyA toxicity in three cases by renal biopsy (Table 1). A typical patient suffering from CyA toxicity was Case 1 (Fig. 1). High serum concentrations (above 1300 ng/ml) of CyA after the administration were demonstrated at the periods of CyA toxicity, although the trough levels of CyA were within normal range. U-GAA gradually increased in spite of CyA toxicity. On the other hand, U-FDP fluctuated in the normal range. Similar changes were observed in cases 5 and 7.

Table 1. Patient profiles suffering from acute rejection and/or cyclosporin (CyA) toxicity with cadaveric renal transplantation

Case		Age	Sex	Disease	HD off (days)	Rejection (times)	CyA toxicity (times)
1	IN	23	F	CGN	20	0	2
2	MI	38	M	CGN	17	3	0
3	HF	46	M	CGN	15	2	0
4	EM	44	M	CGN	7	3	0
5	MM	49	F	CGN	7	1	1
6	EY	30	M	CGN	15	2	0
7	YT	31	F	CGN	16	2	1

HD off; days of suspension of haemodialysis therapy after transplantation.

Case 2 was a typical patient suffering from AR (Fig. 2). He experienced three episodes of AR during 3 months after TPX. U-GAA remarkably decreased and U-FDP significantly increased during all AR episodes, and they conversely changed during recovery phase. AR were well treated with solu-medrol pulse therapy and increments of immunosupressive therapy. In the other patients suffering from AR, similar changes were observed. In cases 5 and 7, AR appeared after decreasing the CyA dosage for the CyA toxicity. In these cases, U-GAA decreased remarkably during AR although it did not decrease during CyA toxicity periods. In all AR episodes, mean U-GAA (mean ± SD) before AR was 14.2 ± 10.4 mg/day, 4.0 ± 4.6 mg/day during AR and 22.0 ± 21.0 mg/day in the period of the recovery phase.

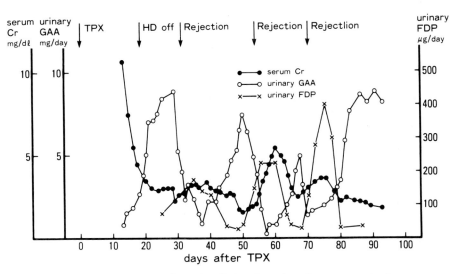

Fig. 2. Clinical course of case 2 suffering three times from acute rejections.

U-FDP conversely increased remarkably during AR, however it also fluctuated because of other factors.

Discussion

It has been difficult to differentiate AR and CyA toxicity, although many clinical and/or laboratory indexes such as urinary osmolarity, urinary Na/K ratio, eosinophilia and U-FDP were reported[2]. But these parameters have been recognized to be not specific for AR. Renal biopsy has been believed to be the most useful tool for the diagnosis in spite of its invasiveness.

As U-GAA has been recognized as an indicator of metabolic function of the kidney, it was expected to be a parameter of renal damage. Ishizaki[3] and Isoda[4] reported that U-GAA was remarkably reduced just before and during AR in the patients with living or cadaveric renal transplantation. From these results, they recommended U-GAA as a predictive tool for AR. We also reconfirmed these results and furthermore demonstrated that U-GAA was not influenced by CyA toxicity in this study. From this result, we recommend U-GAA as a differential diagnostic parameter between AR and CyA toxicity.

References

1. Ando, A., Kikuchi, T., Mikami, H., Orita, Y. & Abe, H. (1980): An automated analytical method for guanidino compounds. *Jpn. J. Clin. Chem.* **9**, 19–25.

2. Chatterjee, S.N. (1979): *Manual of renal transplantation.* ed. S.N. Chatterjee, pp. 135. New York: Springer-Verlag.

3. Ishizaki, M., Kitamura, H., Takahashi, H., Asana, H., Miura, K. & Okazaki, H. (1985): Evaluation of the efficacy of anti-rejection therapy using the quantitative analysis of guanidinoacetic acid (GAA) urinary excretion as a guide. In *Guanidines*, eds. A. Mori, B.D. Cohen & A. Lowenthal, pp. 353–363. New York, London: Plenum Press.

4. Isoda, K., Mitarai, T., Imamura, N., Honda, H., Nagasawa, R., Hirose, S., Sugimoto, K., Tanaka, R., Yokoyama, T. & Kashiwabara, H. (1989): The significance of serum and urinary guanidinoacetic acid level for the restoration of renal metabolic function in patients with kidney transplantation. In *Guanidines 2*, eds. A. Mori, B.D. Cohen & H. Koide, pp. 337–344. New York, London: Plenum Press.

Guanidino Compounds in Biology and Medicine, eds. by P.P. De Deyn, B. Marescau, V. Stalon and
I.A. Qureshi. ©1992 John Libbey & Company Ltd., pp. 327–328.

Chapter 48

The effects of stylene maleimide-superoxide dismutase(SMA-SOD) on guanidino compounds in rats with ischaemic acute renal failure

F. MURAKAMI, M. KITAURA, T. TSUJI, Y. KOSOGABE, K. KABUTAN, T. ITOH,
M. INOUE[1] and M. HIRAKAWA

*Department of Anaesthesiology and Resuscitology, Okayama University Medical School, 2-5-1 Shikata-Cho,
Okayama 700, Japan; [1]Department of Biochemistry, Kumamoto University Medical School, 2-2-1 Honjyo,
Kumamoto 860, Japan*

Introduction

Ischaemic acute renal failure (ARF) is thought to be caused by the oxygen free radicals generated by reperfusion following the renal ischaemia[1,4,5]. The protective role of free radical scavengers, such as superoxide dismutase (SOD), on ischaemic ARF is already reported[2,3,6]. However, SOD itself has a short half-life in the circulating blood. Therefore non-modified SOD has only a limited protective effect on reperfusion injuries. In this study, our purpose was to demonstrate the effect of stylene maleimide-SOD(SMA-SOD)[7], a long-acting SOD, on ischaemic ARF.

Materials and methods

Ischaemic ARF rats were produced by the mechanical clamping of bilateral renal vessels and ureters for 90 min. Ketamine was given at concentrations of 200–250 mg/kg to all rats for anaesthesia, before the operation. Rats were divided into three groups (n = 6). In group A, no drugs were administered to ischaemic ARF rats. In group B, SMA-SOD was administered to ischaemic ARF rats at the concentration of 10 mg/kg intravenously just before declamping. In group C, rats were sham-operated as a control. After 48 h, the concentrations of guanidino compounds in the serum, kidney and liver were analysed. Also blood serum urea nitrogen (BUN) levels were measured.

Results and discussion

BUN and serum creatinine levels were increased and methylguanidine (MG) was also detected in ischaemic ARF rats (group A). SMA-SOD administration lowered BUN and creatinine levels significantly ($P < 0.001$) and showed no detection of MG (group B). MG is focused as an index for oxidative stress, because MG is reported to be derived from creatinine reacting with active oxygen species. Our results suggest that SMA-SOD has a strong suppressive effect on oxidative stress,

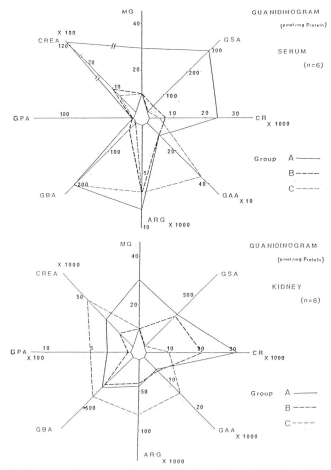

Fig. 1. Serum and kidney guanidinogram in three groups of rats. Groups A (ARF; no drugs), group B (ARF; SMA-SOD 10 mg/kg IV) and group C (sham-operated rats). Abbreviations:
MG: methylguanidine; GSA: guanidinosuccinic acid; CR: creatine; GAA: guanidinoacetic acid; ARG: arginine; GBA: guanidinobutyric acid; GPA: guanidinopropionic acid; CREA: creatinine.

because MG produced by renal ischaemia was suppressed by SMA-SOD administration. SMA-SOD would be a useful drug for the therapy of the ischaemic ARF.

References

1. Baker, G.L., Corry, R.J. & Autor, A.P. (1985): Oxygen free radical induced damage in kidneys subjected to warm ischemia and reperfusion. *Ann. Surg.* **202**, 628–641.
2. Bird, J.E., Milhoan, K., Wilson, C.B., Young, S.G., Mundy. C.A., Parthasarathy, S. & Blantz, R.C. (1988): Ischemic acute renal failure and antioxidant therapy in the rat. *J. Clin. Invest.* **81**, 1630–1638.
3. Bird, J.E., Evan, A.P., Peterson, O.W. & Blantz, R.C. (1989): Early events in ischemic renal failure in the rat: effects of antioxidant therapy. *Kidney Int.* **35**, 1282–1289.
4. Canavese, C., Stratta, P. & Vercellone, A. (1987): Oxygen free radicals in nephrology. *Int. J. Artif. Organs* **10**(6), 379–389.
5. Gamelin, L.M. & Zager, R.A. (1988): Evidence against oxidant injury as a critical mediator of postischemic acute renal failure. *Am. J. Physiol.* **255**, F450–F460.
6. Hansson, R., Johnson, O., Lundstam, S., Pettersson, S., Schersten, T. & Waldenström, J. (1983): Effects of free radical scavengers on renal circulation after ischemia in the rabbit. *Clin. Sci.* **65**, 605–610.
7. Oyanagui, Y. (1989): *SOD and active oxygen modulators, pharmacology and clinical trials*, pp. 213–215. Tokyo: Nihon Igakukan.

Guanidino Compounds in Biology and Medicine, eds. by P.P. De Deyn, B. Marescau, V. Stalon and I.A. Qureshi. ©1992 John Libbey & Company Ltd., pp. 329–331.

Chapter 49

The stability of creatol, an intermediate in the production of methylguanidine from creatinine and its analysis in physiological fluids

Ko NAKAMURA and Kazuharu IENAGA[1]

Institute of Bio-Active Science (IBAS), Nippon Zoki Pharm. Co. Ltd., Yashiro-cho, Kato-gun, Hyogo 673-14, Japan

Introduction

Our results[1,2] have already shown that creatinine (Cr) is oxidized in mammalian bodies into a uraemic toxin, methylguanidine (MG), via creatol (CTL)[2] and that an active oxygen species, probably a hydroxyl radical ($\cdot$OH), mediates the first step in this change. We now postulate that the Cr oxidative pathway producing MG is important in relation, not only to renal hypofunction, but also to oxygen stress in patient with CRF.

When we first isolated CTL, 5-hydroxycreatinine, from the urine of patients with CRF[2] it was not clear why this compound had not been detected previously. Three possibilities were considered: the content of CTL in mammals and/or the detectability of CTL by conventional analytical system(s) might be much lower than those of other intrinsic guanidino compounds; or the stability of CTL might cause some difficulty in the detection process. The answer to this question is now available: although CTL exists in mammals in a concentration at least equimolar to that of MG, its detectability is so poor that it cannot be detected or estimated with conventional fluorogenic agents[3]. Recently, we were able to increase the detectability of CTL 50-fold in order to overcome this problem (unpublished work). In the present article, we report the stability of CTL, a factor which could be important in obtaining true analytical data for MG as well as for CTL.

Materials and methods

Authentic analytically pure CTL was synthesized in our institute (IBAS). For HPLC analyses, the conventional guanidino compounds analyser[3] (JASCO, Japan) was modified to suit our purpose (unpublished work): an alkaline hydrolysis step was added between column separation and the reaction with a fluorogenic reagent, 9,10-phenanthrenequinone (PQ) (Fig. 1).

Sera were prepared at 4 °C from venous blood collected just before taking breakfast. The volume of urine, collected for 24 h, was recorded. Before analysis, serum and urine were deproteinized by TCA (final concentration: 10 per cent) and centrifuged (9000 rpm, 3 min). For stability tests, two kinds of

A) Conventional Method   B) Modified Method

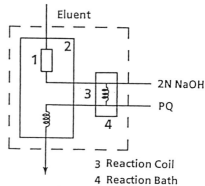

Eluent

1 Column
2 Heating Bath
3 Reaction Coil
4 Reaction Bath

Fig. 1. Flow diagram for creatol analysis with HPLC.
Analytical conditions; column: TSK gel SCX (Toso: 6 φ × 150 mm); two coils (0.5 φ mm × 5 m): in reaction bath (at 125 °C) and in column oven (at 65 °C); eluent; solutions A (0.4 M citrate/DMSO = 9/1: pH 5.20) and B (1 N-NaOH) for 20 and 10 min; sample injection: 11 min after start; flow rates for solutions A, B, PQ: 0.8, 0.5, 0.6 ml/min; fluorescence: 495 (em.) 340 nm (ex.).

stock solution were made: CTL was added to normal human sera (A-CTL), which contains undetectable CTL, and to buffer solutions with various pH values; the final CTL concentration in each case was 1 nmol/injection.

R = H :Cr
R = OH:CTL

Results and discussion

As shown in Fig. 2A, the stability tests on CTL in serum showed that CTL was not auto-degraded by hydrolysis below –20 °C but that it was so degraded at both 4 °C and 25 °C. This result should be noted, because CTL exists in the human serum at a concentration ten times higher than MG[1] (unpublished work): thus a 1 per cent degradation of CTL must cause a 10 per cent increase in MG. Therefore, we recommend that analytical samples should be stored below –20 °C, if possible at –80 °C where CTL is stable for more than 6 months. In the buffer solution also, CTL at ca pH 7 was stable below –20 °C (data not shown).

As shown already[1,2], CTL is hydrolysed into MG and glyoxylic acid. The stability of CTL added to buffer solutions at 37 °C showed a biphasal stability curve at various pH values (Fig. 2B): thus two peaks of CTL instability occurred, one under alkaline conditions (pH > 11) and the other under weakly acidic conditions (pH 5). Perhaps the neutral molecule (pH 9) and the di cation (pH 1) are stable, whereas the mono cation and the mono anion are unstable. Be that as it may, authentic samples of CTL should be stored at pH 1 (0.1 N HCl) and below –20 °C.

The fact that CTL is converted quantitatively into MG under alkaline conditions has been used to increase the detectability of CTL 50-fold (unpublished work): 125 °C proved to be the optimal reaction temperature for the conversion.

According to our analytical results on physiological fluids in mammals with and without renal failure, S-CTL is likely to be a good diagnostic index for judging the severity of renal failure (unpublished work) and especially for detecting incipient renal hypofunction. Furthermore the ratio S-CTL/S-Cr

(A) Human Serum   (B) Buffer Solutions

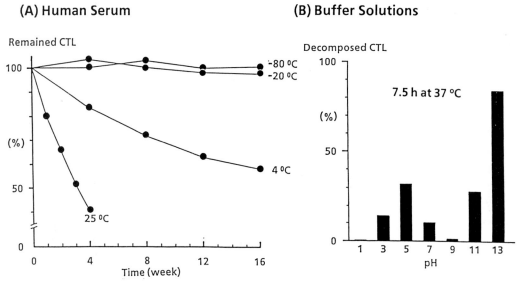

Fig. 2. Stabilities of creatol in human serum and buffer solutions.

might well be a useful index for oxygen stress (unpublished work). Analyses of CTL levels in other fluids, such as urine and CSF, are now in progress.

Acknowledgements

We thank Profs. H. Oura and T. Yokozawa, and Dr. D. J. Brown and Dr. K. Nakano MD for some advice.

References

1. Ienaga, K., Nakamura, K., Yamakawa, M., Toyomaki, Y., Matsuura, H., Yokozawa, T., Oura, H. & Nakano, K. (1991): The use of ^{13}C-labelling to prove that creatinine is oxidized by mammals into creatol and 5-hydroxy-1-methylhydantoin. *J. Chem. Soc. Chem. Commun.* 509–510.

2. Nakamura, K. & Ienaga, K. (1990): Creatol (5-hydroxycreatinine), a new toxin candidate in uremic patients. *Experientia* **46**, 470–472.

3. Yamamoto, Y., Manji, T., Saito, A., Maeda, K. & Ohta, K. (1979): Ion exchange chromatographic separation and fluorometric determination of guanidino compounds in physiologic fluids. *J. Chromatogr.* **162**, 327–340.

Guanidino Compounds in Biology and Medicine, eds. by P.P. De Deyn, B. Marescau, V. Stalon and
I.A. Qureshi. ©1992 John Libbey & Company Ltd., pp. 333–339.

Chapter 50

A new enzymatic assay of urinary guanidinoacetic acid

Yoshio SHIROKANE, Moto-o NAKAJIMA and Kiyoshi MIZUSAWA

Research and Development Division of Kikkoman Corporation, 399 Noda, Noda-shi, Chiba-ken 278, Japan

Summary

We described a new enzymatic determination of urinary guanidinoacetic acid (GAA) with three enzymes (GAA kinase, pyruvate kinase (PK) and lactate dehydrogenase (LDH)), which did not require a blank to correct for endogenous constituents. The present enzymatic method fulfilled the need for an accurate, specific and rapid assay of GAA, and was simple as compared to the HPLC method and our previous enzymatic method. There was a good correlation (r = 0.996) between the results obtained by the enzymatic and HPLC methods. GAA concentrations in 24-h urine samples were determined by the proposed method, and the urinary excretion of GAA was observed to decrease markedly in patients with renal failure.

Introduction

Guanidinoacetic acid (GAA) is produced from L-arginine (L-Arg) in the presence of glycine (Gly) by L-Arg:Gly amidinotransferase (transamidinase, EC 2.1.4.1) mainly in the kidney and is then transported to the liver, where it is methylated to yield creatine. Localization of this transamidinase in kidney has been demonstrated immunohistochemically in the proximal tubule[6], and the transamidinase activity was recently recognized only in the first and second portions of the proximal tubule[14]. Although GAA is physiologically excreted into urine, it significantly decreases in patients with renal dysfunction compared with healthy controls[8]. In recent years, the determination of GAA in urine has been reported to be useful as a sensitive index of renal tubular damage[4,5,7,13,14].

The urinary GAA has been determined by automated high performance liquid chromatography (HPLC)[1,2], but this method is time-consuming and inappropriate for use in a large number of analyses. Recently, in our laboratory, a new enzymatic method available for the determination of urinary GAA was first developed using guanidinoacetate (GAA) amidinohydrolase (EC 3.5.3.2) and urease (EC 3.5.1.5)[9,10]. However, this enzymatic method was found to be somewhat complicated because it required a sample blank to correct for endogenous colour-producing constituents in urine.

In order to establish a convenient determination of urinary GAA, we purified guanidinoacetate (GAA) kinase (EC 2.7.3.1) from a polychaete, *Perinereis* sp., because it was strictly specific to GAA[11], and developed a new enzymatic endpoint assay of urinary GAA using three enzymes (GAA kinase, pyruvate kinase (PK, EC 2.7.1.40) and lactate dehydrogenase (LDH, EC 1.1.1.27)). In our newly proposed two-reagent system, endogenous constituents (ADP and pyruvate) in urine are eliminated before determination, thus a quantity of GAA can be measured using the following reactions.

First step (elimination of ADP and pyruvate):

$$\text{ADP} + \text{phosphoenolpyruvate} \xrightarrow{\text{PK}} \text{ATP} + \text{pyruvate}$$

$$\text{Pyruvate} + \text{NADH} + \text{H}^+ \xrightarrow{\text{LDH}} \text{lactate} + \text{NAD}^+$$

Second step (determination of GAA):

$$\text{GAA} + \text{ATP} \xrightarrow{\text{GAA kinase}} \text{phosphoguanidinoacetate} + \text{ADP}$$

$$\text{ADP} + \text{phosphoenolpyruvate} \xrightarrow{\text{PK}} \text{ATP} + \text{pyruvate}$$

$$\text{Pyruvate} + \text{NADH} + \text{H}^+ \xrightarrow{\text{LDH}} \text{lactate} + \text{NAD}^+$$

The decrease of NADH in the second step, as measured at 340 nm, is proportional to the amount of GAA present.

Here, we describe a new enzymatic assay of urinary GAA and its application to urine samples.

Materials and methods

Urine specimens

Random urine specimens were obtained from healthy volunteers. Samples of 24-h urine were also collected from 34 outpatients (16 men and 18 women, ages 28–74) who went to Kikkoman General Hospital to take tests of renal function. These samples were stored at –20 °C until analysis.

Enzymes and chemicals

The purified GAA kinase was prepared by us from a polychaete, *Perinereis* sp.[11]. PK and LDH (both from rabbit muscle) were both purchased from Sigma Chemical Co. (St Louis, MO 63178, USA). GAA, ATP, NADH and phosphoenolpyruvate (PEP) were also obtained from Sigma Chemical Co. All other chemicals were of analytical grade.

Reagents

GAA standard solution was freshly prepared by dissolving 10.0 mg of GAA in 100 ml of distilled water. Reagent A was prepared to contain 10 kU of PK, 10 kU of LDH, 2.2 mmol of ATP, 0.84 mmol of PEP, 5.0 mmol of $MgSO_4$, 10 mmol of KCl and 0.2 mmol of NADH per litre in Tris-HCl buffer (100 mmol/l), pH 7.5. Reagent B was prepared to contain 150 kU of GAA kinase, 1.0 mmol of dithiothreitol and 5 per cent of glycerol per litre in Tris-HCl buffer (100 mmol/l), pH 7.5.

Procedure

GAA in urine was assayed by the procedure as described in Table 1 with use of microcuvettes. The absorbances were measured using a Hitachi Model 557 Double Wavelength Double Beam Spectrophotometer (Hitachi, Ltd, Tokyo, Japan). The amount of GAA in urine was calculated by the following formula.

$$\text{GAA (mg/dl)} = \frac{(\text{Es0} \times \text{F} - \text{Es1}) - (\text{Eb0} \times \text{F} - \text{Eb1})}{(\text{Est0} \times \text{F} - \text{Est1}) - (\text{Eb0} \times \text{F} - \text{Eb1})} \times 10$$

$$F = \frac{1.35}{1.45} \text{ (a factor for correction of the reaction volume)}$$

Table 1. Procedure for determination of GAA in urine

Microcuvette no.	I	II	III
Pipette into the microcuvettes:			
Urine	50 μl	–	–
Standard solution	–	50 μl	–
Distilled water	–	–	50 μl
Reagent A	1.3 ml	1.3 ml	1.3 ml
Mix and measure the absorbances at 340 nm against distilled water after incubation at 37 °C for 5 min.			
$A_{340\,nm}$	Es0	Est0	Eb0
Add to the microcuvettes:			
Reagent B	0.1 ml	0.1 ml	0.1 ml
Mix and measure the absorbances at 340 nm against distilled water after reaction at 37 °C for 10 min.			
$A_{340\,nm}$	Es1	Est1	Eb1

Results

Properties of GAA kinase

 GAA kinase was extracted and purified from a polychaete, *Perinereis* sp. by successive procedures involving ammonium sulphate fractionation and chromatographies on Butyl-Toyopearl 650C, Sephadex G-200 and DEAE-Sephacel[11]. The specific activity of the enzyme was 43–45 u/mg protein at 25 °C and adenosine-5′-triphosphatase activity was less than 0.002 per cent of GAA kinase activity. GAA kinase was strictly specific to GAA, and showed almost no susceptibility (less than 1 per cent of the rate obtained with GAA) toward other guanidino compounds in urine, such as creatinine, creatine, L-arginine, taurocyamine, 3-guanidinopropionate, 4-guanidinobutyrate, guanidine, guanidinosuccinate and methylguanidine. Some other properties were: K_m value for GAA, 4.1 mM; optimum pH, around 8.1; optimum temperature, 35 °C (pH 8.1); pH stability, 5.5–9.5; thermal stability, about 35 °C (pH 8.1).

These properties indicated that GAA kinase was appropriate for the determination of GAA in urine.

Assay conditions

Relatively large amounts of three coupling enzymes (GAA kinase, PK and LDH) were used for GAA assay system, so that the reactions in the two steps proceeded sufficiently rapidly. The GAA assay was carried out at pH 7.5 because all three enzymes were more active around pH 7–8. The optimal reaction times for the two steps were examined with use of GAA standard solution and/or urine samples. The reactions for the first and second steps went to completion within 5 min and 10 min, respectively. Thus, the final procedure for the determination of urinary GAA was established as outlined in Materials and methods.

Calibration curve

Figure 1 shows the calibration curve constructed from aqueous GAA standard solutions. Good linearity was obtained in a wide range of GAA concentrations (0–20 mg/dl).

Analytical recovery

The analytical recovery was investigated with normal urine samples of three different GAA values (2.47–8.93 mg/dl) by adding 3.0, 6.0 and 9.0 mg of GAA per decilitre. The average recovery of GAA added was 100.7 per cent (range 97.0–103.2 per cent).

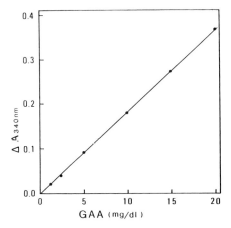

Fig. 1. Calibration curve obtained with various dilutions of GAA solutions.

Precision

To determine within-run and between-run precision (CVs), normal urine samples of four different GAA values (2.52–14.18 mg/dl) were assayed 10 times and six times, respectively. The within-run precision (CV) for urinary GAA was less than 3.6 per cent and the between-run precision (CV) for daily GAA analysis of urine was less than 4.8 per cent.

Effect of interfering substances

The effect of several substances, listed in Table 2, on the determination of urinary GAA was examined. Although GAA standard solution (10 mg/dl) was assayed with and without addition of possible interfering substances at the concentrations shown for each substance, these did not significantly affect the present method.

Table 2. Effect of substances added to GAA standard solution on the GAA assay

Substances	Concentration (mg/dl)	Apparent GAA (%)
None	–	100.0
Urea	3000	96.3
Creatinine	1000	99.3
Creatine	200	98.3
Glucose	5000	98.4
Albumin	5000	101.2
EDTA-Na$_2$	500	95.8
Citrate	500	98.8
Oxalate	100	97.7
Ascorbate	500	100.4
Glutathione	200	103.4

Correlation of results by the enzymatic and HPLC methods

The enzymatic method (x) was compared with an HPLC method[1] (y) with use of 39 urine samples for the determination of urinary GAA. A good correlation (r = 0.996) was obtained between the two methods (Fig. 2) and the regression equation for these data was y = 0.976 x + 0.236 mg/dl.

GAA excretion in 24-h urine samples

We determined the GAA concentrations in 24-h urine samples from 34 outpatients with suspected or

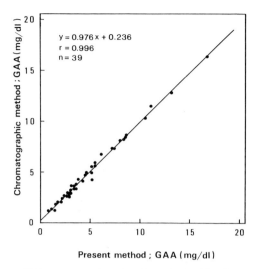

Fig. 2. Correlation between GAA concentrations in urines obtained by the present and chromatographic methods.

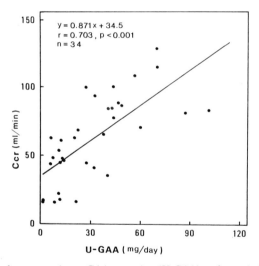

Fig. 3. Relationship between urinary GAA excretion (U-GAA) and creatinine clearance (Ccr).

proven renal insufficiency by the present method. Figures 3 and 4 show the relationship between the urinary GAA excretion (U-GAA) and the level of creatinine clearance (Ccr), blood urea nitrogen (BUN) or serum creatinine (S-Cr). The urinary GAA excretion indicated a significant positive correlation with Ccr ($r = 0.703$, $P < 0.001$), while there were significant negative correlations between the urinary GAA excretion and BUN ($r = -0.469$, $P < 0.01$), and between the urinary GAA excretion and S-Cr ($r = -0.515$, $P < 0.01$). However, no correlation was observed between the urinary GAA excretion and the urinary excretion of protein (mg/day) (data not shown). These results revealed that the urinary excretion of GAA was markedly lowered in proportion to the fall of Ccr below the normal level (70–130 ml/min), the rise of BUN over the normal level (7–20 mg/dl) or the rise of S-Cr over the normal level (0.5–1.4 mg/dl).

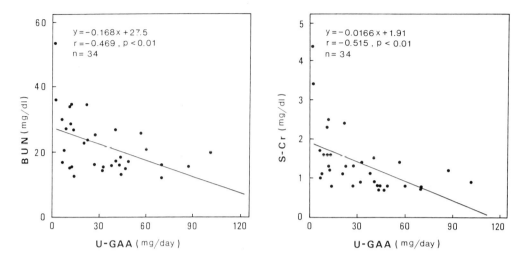

Fig. 4. Relationship between urinary GAA excretion (U-GAA) and blood urea nitrogen (BUN) or serum creatinine (S-Cr).

Discussion

We developed a new enzymatic endpoint assay including the two steps to obtain a simple and specific method of GAA analysis. Urinary GAA was quantitatively determined by the described method, because GAA kinase was strictly specific to GAA[11]. The present enzymatic method for urinary GAA was satisfactory, as shown by the linearity of the calibration curve and by the good analytical recovery of GAA. The precision of this method was similar to those of the HPLC method and our previous enzymatic method, and there was a good correlation (r = 0.996) between the results obtained by the present enzymatic and HPLC methods. Although we examined many urine samples from normal persons and outpatients with suspected or proven renal insufficiency, none was found to contain interfering substances on the GAA assay system as described in Table 1. The enzymatic method proposed here fulfilled the need for an accurate, specific and rapid assay of GAA, and was simple as compared to the HPLC method and our previous enzymatic method. In addition, the present enzymatic method can be adapted to automatic analysers presently used in many laboratories.

To confirm the relation between the decrease of GAA in urine and renal failure, we determined the GAA concentrations of 24-h urine samples from outpatients with suspected or proven renal insufficiency by the present method. Our results agreed with previous findings[3,7,13,15,16] which showed a correlation between renal dysfunction and the decrease of urinary GAA excretion and it is obvious that the urinary GAA excretion as well as the levels of Ccr, S-Cr and BUN are useful to detect renal dysfunction. In recent years measurement of β_2-microglobulin (β_2-MG) and/or N-acetyl-β-D-glucos-aminidase (NAG) have been widely used as indicators for detecting renal tubular damage. On the other hand, the decrease of urinary GAA excretion was observed without any alteration of urinary β_2-MG and NAG in patients with essential hypertension[12] and renal disorder secondary to medication with antibiotics[7], suggesting that the urinary excretion of GAA may be a more useful marker of renal tubular damage than that of β_2-MG or NAG. In conclusion, it is likely that the measurement of urinary GAA by the method proposed here is valuable and helpful for the early diagnosis of renal dysfunction, especially renal tubular metabolic dysfunction.

References

1. Hiraga, Y. & Kinoshita, T. (1981): Post-column derivatization of guanidino compounds in high performance liquid chromatography using ninhydrin. *J. Chromatogr.* **226**, 43–51.

2. Hung, Y., Kai, M., Nohta, H. & Ohkura, Y. (1984): High-performance liquid chromatographic analysis for guanidino compounds using benzoin as a fluorogenic reagent. *J. Chromatogr.* **305**, 281–294.

3. Ishizaki, M., Kitamura, H., Takahashi, H. *et al.* (1985): Evaluation of the efficacy of anti-rejection therapy using the quantitative analysis of guanidinoacetic acid (GAA) urinary excretion as a guide. In *Guanidines*, eds. A. Mori, B.D. Cohen & A. Lowenthal, pp. 353–363. New York: Plenum Press.

4. Isoda, K., Mitarai, T., Imamura, N. *et al.* (1989): The significance of serum and urinary guanidinoacetic acid level for the restoration of renal metabolic function patients with kidney transplantation. In *Guanidines 2*, eds. A. Mori, B.D. Cohen & H. Koide, pp. 337–344. New York: Plenum Press.

5. Kuwagaki, Y. & Sudo, J. (1989): Interrelation of urinary and plasma levels of guanidinoacetic acid with alteration in renal activity of glycine amidinotransferase in acute renal failure rats. *Chem. Pharm. Bull.* **37**, 781–784.

6. McGuire, D.M., Gross, M.D., Elde, R.P. & Van Pilsum, J.F. (1986): Localization of L-arginine-glycine amidinotransferase protein in rat tissues by immunofluorescence microscopy. *J. Histochem. Cytochem.* **34**, 429–435.

7. Nakayama, S., Kiyatake, I., Shirokane, Y. & Koide, H. (1989): Effect of antibiotic administration on urinary guanidinoacetic acid excretion in renal disease. In *Guanidines 2*, eds. A. Mori, B.D. Cohen & H. Koide, pp. 313–322. New York: Plenum Press.

8. Sasaki, M., Takahara, K. & Natelson, S. (1973): Urinary guanidinoacetate/guanidinosuccinate ratio: an indicator of kidney dysfunction. *Clin. Chem.* **19**, 315–321.

9. Shirokane, Y., Utsushikawa, M. & Nakajima, M. (1987): A new enzymic determination of guanidinoacetic acid in urine. *Clin. Chem.* **33**, 394–397.

10. Shirokane, Y., Nakajima, M. & Mizusawa, K. (1991): Easier enzymatic determination of guanidinoacetic acid in urine. *Clin. Chem.* **37**, 478–479.

11. Shirokane, Y., Nakajima, M. & Mizusawa, K. (1991): Purification and properties of guanidinoacetate kinase from a polychaete, *Perinereis* sp. *Agric. Biol. Chem.* **55**, 2235–2242.

12. Takano, Y., Gejyo, F., Shirokane, Y., Nakajima, M. & Arakawa, M. (1988): Urinary excretion of guanidinoacetic acid in essential hypertension. *Nephron* **48**, 167–168.

13. Takano, Y. (1989): Estimation of urinary excretion rate of guanidinoacetic acid in essential hypertension. *Jpn. J. Nephrol.* **31**, 1187–1196.

14. Takeda, M., Kiyatake, I., Yaguchi, Y. *et al.* (1990): Intrarenal biosynthesis of guanidinoacetic acid and its application to gentamicin nephrotoxicity. *Igaku No Ayumi.* **153**, 653–654.

15. Tsubakihara, Y., Iida, N., Yuasa, S. *et al.* (1985): Guanidinoacetic acid (GAA) in patients with chronic renal failure (CRF) and diabetes mellitus (DM). In *Guanidines*, eds. A. Mori, B.D. Cohen & A. Lowenthal, pp. 309–316. New York: Plenum Press.

16. Tsubakihara, Y., Yamato, E., Yokoyama, K. *et al.* (1989): Short-term protein load in assessment of guanidinoacetic acid synthesis in patients with chronic renal failure. In *Guanidines 2*, eds. A. Mori, B.D. Cohen & H. Koide, pp. 331–336. New York: Plenum Press.

Section VI
Guanidino compounds in inborn errors of metabolism

Guanidino Compounds in Biology and Medicine, eds. by P.P. De Deyn, B. Marescau, V. Stalon and
I.A. Qureshi. ©1992 John Libbey & Company Ltd., pp. 343–348.

Chapter 51

Argininaemia: clinical and biochemical aspects

J.P. COLOMBO

Department of Clinical Chemistry, Inselspital, University of Berne, CH 3010 Berne, Switzerland

Summary

Argininaemia is characterized by an increase of the amino acid arginine in plasma and cerebrospinal fluid. The accumulation of arginine in extracellular fluids is due to the deficiency of the liver type arginase. Human liver arginase cDNA clones have been isolated. The arginase locus has been mapped to band q23 of chromosome 6. The clinical course of classical cases leads to severe psychomotor retardation and spastic diplegia of the legs.

The increased plasma concentration of arginine leads to an augmented urinary excretion of this amino acid which induces a dibasic amino aciduria and cystinuria. A cystinuria pattern may be observed. The excretion of pyrimidines, particularly orotic acid, is elevated. Hyperammonaemia is not obligatory but may occur during the deranged metabolic episodes. Protein-restricted diet with an arginine free amino acid mixture eventually in combination with sodium benzoate can be used to treat the patients.

Introduction

Argininaemia is characterized by an increase of arginine in plasma. In our cases values up to 1500 μmol/l have been observed (normal range 91.6 ± 45 μmol/l). Argininaemia is an autosomal, recessively transmitted rare disorder. Up to now about 20 cases are known. The underlying cause is a deficiency of arginase, an enzyme of the urea cycle responsible for urea synthesis (Fig. 1). This enzyme splits arginine into ornithine and urea. The first two cases presenting with enzyme deficiency were described in 1969 by our group[27]. A severe neonatal form presenting with argininuria only during episodes with seizures observed earlier in Spain probably belongs to the same disease entity[22].

Clinical features

Our observations

The symptomatology of argininaemia is best illustrated with the clinical course of case number 5, first of our three affected patients[10] (Fig. 2). Basic developmental data: At 7 months sitting, 9 months first words, 13 months walking. At 2 years an episode of fever with seizures. The EEG presented sharp and slow-waves. At 3 years: uncoordinated 'tiptoe-gait', spasticity in the legs. At 4 years: head circumference too small, seizures, retroflection of head in sitting position, extreme spasticity in the legs. Hyperactive deep tendon reflexes, cloni at the feet, shortening of Achilles tendons. The child presented with a full picture of spastic diplegia. Moderate to severe psychomotor retardation was present. The patient was last seen by us at the age of 23 years (now 27). She is bound to a wheel-chair, unable to walk. She can however perform easy housework such as making coffee and laying the table.

343

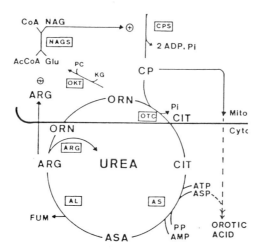

Fig. 1. Urea cycle.

OTC = ornithinetranscarbamylase; AS = argininosuccinate synthetase; AL = argininosuccinate lyase; Arg = arginase; OKT = ornithine-α-ketoglutarate transaminase; NAGS = N-acetylglutamate synthetase; CPS = carbamylphosphate synthetase; ORN = ornithine; CIT = citrulline; ASA = argininosuccinic acid; ARG = arginine; AcCoA = acetyl coenzyme A; Glu = glutamate; NAG = N-acetyl glutamate; PC = pyrrolinecarboxylic acid; KG = α-ketoglutarate; ASP = aspartate; FUM = fumorate; PP = pyrophosphate; Pi = phosphate.

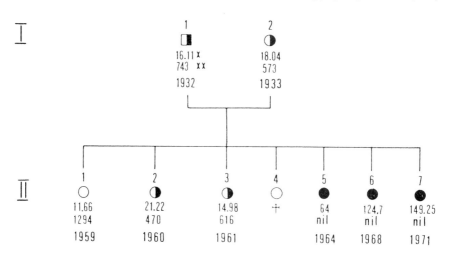

Fig. 2. Family with argininaemia. Arginine* concentration in plasma (controls: 9.16 ± 4.5 μmoles/100 ml) and red blood cell arginase activity** (controls: 793–1330 μmoles/h/g Hb).

She eats by herself, talks fairly well, has learned to write some letters and can knit. Patient 6 when last seen was 19 years old (now 23). She does not communicate, and is completely invalid in a wheel-chair. The youngest girl (patient 7) seen at the age of 16 years (now 20) is also in a wheel-chair, but can eat by herself, and talks and communicates rather well. She is however not able to do any housework.

Further symptoms in cases with argininaemia

In the cases described in the literature, further symptoms have been observed quite independent of the age of the patient. Those symptoms comprise: ataxia, drooling, swallowing difficulties, converging strabismus, spastic tetraplegia. Different techniques like, previously, pneumoencephalography, and more recently CAT-scan and magnetic resonance imaging disclose moderately large cerebral ventricles. Abnormal visual and acustic evoked potentials are also described[6]. Electromyogram and nerve conduction time were normal. One patient recently described from Portugal presented without spastic diplegia[30]. In contrast to the severe mental retardation, mild intellectual impairment has also been found. Of particular interest is a case of arginase deficiency who presented with a severe protein intolerance and died very early at the 49th day of life. Another interesting case features an 18-year-old girl who was hospitalized for many years in a psychiatric clinic due to complete disability and advanced neurological deterioration including diplegia. She was referred to a neurological unit because of severe starting seizures which were treated with valproate. During this treatment the girl developed hyperammonaemia with stupor. A deficiency of arginase was detected. The girl recovered from this episode after the treatment was stopped[9]. In some cases a hepatomegaly was detected, particularly during metabolic crisis with hyperammonaemia.

In summary: the major symptoms of argininaemia observed practically in all patients are progressive loss of mental and motor skill, and increasing spasticity of the lower extremities accompanied by hyperreflexia. These symptoms take a rapidly deteriorating course. The clinical picture of argininaemia is remarkably uniform and distinguishes these cases from other disorders of urea cycle enzyme deficiencies. Besides these typical symptoms, and particularly in early life, intermittent episodes of appetite loss, vomiting, irritability, lethargy, eventually progressing to coma and seizures are observed. These symptoms may also be found in other disorders of the urea cycle and are due to episodes of hyperammonaemia which however do not regularly occur in argininaemia.

Biochemical findings

In most cases, particularly after a protein load, a typical urinary amino acid pattern may emerge. Besides arginine, lysine, ornithine and cystine, sharing a common gastrointestinal and renal transport mechanism with arginine, are found in increased amounts, giving rise to a typical cystinuria pattern. In our patients, a diminished reabsorption capacity for lysine, ornithine and cystine was observed which was less pronounced than that found in cases with classical cystinuria (Table 1). We think that all retarded and neurologically handicapped children with 'cystinuria' should be screened for increased arginine concentrations in plasma.

Table 1. Percentage of tubular reabsorption of cystine, lysine, ornithine and arginine in patients 5 and 6, patients with 'classic' cystinuria (n = 10) and normal subjects

Patient	5	6	'Classic' cystinuria	Normal subjects
Cystine	68.4	80.4	7.4	97.4
Lysine	95.7	91.8	46.2	99.2
Ornithine	99.4	95.8	71.0	97.7
Arginine	99.4	97.9	42.5	99.8

An increased excretion of other amino acids has also been observed in some patients. Ten to fifteen times normal values of β-amino-isobutyric acid were found[7]. This may indicate an increased pyrimidine synthesis and breakdown with a consequent accumulation of β-amino-isobutyric acid. Excretion of homocitrulline and homoarginine suggests an enhanced synthesis of these amino acids in some patients[14]. Argininosuccinic acid has also been found in one patient[19].

Urinary uracyl excretion 10 times normal was reported in patients with arginase deficiency[21]. Uridine and orotic acid which are normally not detected can be excreted in large amounts and may be influenced by the protein intake, as has been observed in other patients with hyperammonaemia of various origin[2,29]. The augmented excretion of orotic acid is not quite understood. Orotate accumulation in argininaemia may be due to a stimulatory effect of arginine on N-acetylglutamate synthetase[2]. Ornithine depletion may also play a role[8,23]. Carbamyl phosphate would thus accumulate and be channelled into pyrimidine synthesis (Fig. 1). Orotic aciduria in argininaemia may well be the result of a combination of both these factors. An increased excretion of putrecine in argininaemia particularly under ornithine supplementation was also reported[15]. An acceleration of extramitochondrial ornithine metabolism due to an impaired mitochondrial uptake may cause the hyperexcretion of putrecine. Organic acid excretion was reported normal whereas a diminished excretion of total and free carnitine in the presence of normal plasma levels was observed[6].

In the *plasma and cerebrospinal fluid* (CSF) of all patients an increased concentration of arginine is present. The highest concentration in the CSF in our cases was 132 µmol/l (normal 14.2 ± 7.4). The concentrations in both fluids may be up to ten times normal. An increase of glutamine, glutamic acid and alanine as observed in other cases of urea cycle enzyme deficiencies is not present. This can be interpreted as a lack of permanent hyperammonaemia in these cases. Since accumulating nitrogen can, to a great extent, not be excreted as urea, ammonia accumulates in tissues leading to *hyperammonaemia* in these patients. It is however not persistent and is particularly difficult to detect in children with a self-imposed protein restriction[4,19]. Hyperammonaemia may be found in febrile illnesses, after an increase in daily protein intake or after loading tests with amino acids. The introduction of a low protein diet as a rule leads to a decrease of the hyperammonaemia. Ammonia levels in our patients ranged from 143 to 394 µmol/l on a normal protein intake and from 79 to 186 µmol/l on a low protein diet (normal < 80 µmol/l). *Guanidino compounds* are catabolites of arginine. Besides arginine, homoarginine, α-keto-δ-guanidinovaleric acid, argininic acid, and α-N-acetylarginine have been found in higher concentrations in plasma of argininaemic patients. This was also the case for the urinary excretion of α-keto-δ-guanidinovaleric acid and argininic acid[18]. This subject is treated in Chapter 54 by Marescau *et al.*

Enzyme studies

The underlying biochemical cause of argininaemia is a deficiency of arginase in the liver. Activities of arginase have been demonstrated in erythrocytes, leukocytes, platelets, kidney, skeletal and heart muscle, brain, intestine, pancreas, lungs, epidermis, placenta, salivary glands, testes, plasma and fibroblasts[11]. The metabolic function of extrahepatic arginase is still unclear. So far patients with argininaemia have been shown to lack arginase in the liver, erythrocytes and leukocytes and the stratum corneum of the skin. However the kidney activity was retained[26]. If liver biopsy is refused, lack of arginase can be tested by performing loading tests with arginine, as was done in our patients, and measuring its plasma disappearance rate. The rate is much slower than in control subjects[10].

Five forms of arginase were isolated from human tissues, two liver, and two kidney arginases and one salivary enzyme[31]. Arginase A1 from rat kidney showed complete immunological incompatibility with arginase A5 from liver. All main forms contain two similar subunits. The minor forms are hybrids being built of the corresponding two kinds of subunits. These multiple forms of arginase may be interpreted in the context of posttranslational changes and may not be related with the primary gene product. The diagnosis of argininemia can be made by examining the arginase activity in erythrocytes which contain the liver and the kidney form[31] and in liver tissue. 98 per cent of the arginase in erythrocytes is of the liver type[25]. Human liver arginase has a molecular weight of 107,000[5]. It is an oligomer and is localized in the cytosol of the liver cells. The subunit consists of 320 amino acid residues giving a molecular weight of 34,000 which could roughly correspond to a trimer. The gene for human liver arginase has been assigned to the chromosome 6 band q23[25]. The hepatic cytosolic arginase and the renal mitochondrial arginase are encoded by two separate genes with substantially different structures[12]. The arginase gene is 1.5 kb long and divided into eight exons. Recently a patient who is a compound heterozygote, inheriting one mutant allele with a 4-bp deletion for the father and one allele with the 1-bp deletion from the mother was described[13]. It seems to be the first evidence of a case of argininaemia caused by two different deletion mutations. Human liver arginase cDNA clones have been isolated[12]. Further aspects of the molecular biology of arginase are treated in Chapter 53 by S.D. Cederbaum. A normal activity of arginase was described in fibroblasts of argininaemic patients, so that these cells can not be used for diagnostic purposes[16]. Fibroblasts contained the kidney and the liver isoenzyme[17].

Treatment

Normalization of plasma arginine concentration may be achieved by the addition of arginine free amino acid mixture of essential amino acids to a diet of restricted natural protein (0.5–0.7 g/kg body weight per day). A supplement of sodium benzoate is open to discussion and may be given during periods of acute metabolic derangements with hyperammonaemia since this therapy is directed towards lowering plasma ammonia concentrations[3,24]. An arginine-restricted diet plus sodium benzoate resulted in a decrease in plasma and urinary arginine levels to near normal and orotate excretion normalized[24]. Other therapeutic measures such as attempts to introduce arginase into patients' erythrocytes or to reduce argininaemia by transfusion of normal erythrocytes were unsuccessful[1,19,20]. Gene replacement by intravenous injection of Shope papilloma virus, an arginase-inducing agent, was also unsuccessful[28]. New light into this type of therapy might be shed by isolation of the mutant genes and the construction of genomic libraries[12,13].

References

1. Adriaenssens, K., Karcher, D., Lowenthal, A. & Terheggen, H. (1976): Use of enzyme-loaded erythrocytes in *in-vitro* correction of arginase-deficient erythrocytes in familial hyperargininemia. *Clin. Chem.* **22**, 323–326.

2. Bachmann, C. & Colombo, J.P. (1980): Diagnostic value of orotic acid excretion in heritable disorders of the urea cycle and in hyperammonemias due to organic acidurias. *Eur. J. Pediatr.* **134**, 109–113.

3. Bachmann, C. (1990): Urea cycle disorders. In *Inborn metabolic diseases*, eds. J. Fernandes, J.-M. Saudubray & K. Tada, pp. 211–228. Berlin: Springer-Verlag.

4. Bernar, J., Hanson, R.A., Kern, R., Phoenix, B., Shaw, K.N.F. & Cederbaum, S.D. (1986): Arginase deficiency in a 12-year-old boy with mild impairment of intellectual function. *J. Pediatr.* **108**, 432–435.

5. Berüter, J., Colombo, J.P. & Bachmann, C. (1978): Purification and properties of arginase from human liver and erythrocytes. *Biochem. J.* **175**, 449–454.

6. Brockstedt, M., Smit, L.M.E., de Grauw, A.J.C., van der Klei-van Moorsel, J.M. & Jakobs, C. (1990): A new case of hyperargininaemia: neurological and biochemical findings prior to and during dietary treatment. *Eur. J. Pediatr.* **149**, 341–343.

7. Cederbaum, S.D., Shaw, K.N.F., Spector, E.B., Verity M.A., Snodgrass P.J. & Sugarmann G.I. (1979): Hyperargininemia with arginase deficiency. *Pediatr. Res.* **13**, 827–833.

8. Cederbaum, S.D., Moedjono S.J., Shaw, K.N.F., Carter, M., Naylor, E. & Walser, M. (1982): Treatment of hyperargininaemia due to arginase deficiency with a chemically defined diet. *J. Inherit. Metab. Dis.* **5**, 95–99.

9. Christmann, D., Hirsch, E., Mutschler, V., Collard, M., Marescaux, C. & Colombo, J.P. (1990): Argininémie congénitale diagnostiquée tardivement a l'occasion de la prescription de valproate de sodium. *Rev. Neurol.* **146**, 12, 764–766.

10. Colombo, J.P., Terheggen, H.G., Lowenthal, A., Van Sande, M. & Rogers, S. (1973): Argininaemia. In *Inborn errors of metabolism*, eds. F.A. Hommes & C.J. Van den Berg, pp. 239–254. London. New York: Academic Press.

11. Colombo, J.P. & Konorska, L. (1984): Arginase, In *Methods of enzymatic analysis*, 3rd edn., ed. H.U. Bergmeyer, pp. 285–294. Weinheim: Verlag Chemie.

12. Grody, W.W., Craig, A., Kern, R.M., Dizikes, G.J. Spector, E.B., *et al.* (1989): Differential expression of the two human arginase genes in hyperargininemia. *J. Clin. Invest.* **83**, 602–609.

13. Haraguchi, Y., Aparicio, R.J.M., Takiguchi, M., Akaboshi, I., Yoshino, M., Mori, M. & Matsuda, I. (1990): Molecular basis of argininemia. *J. Clin. Invest.* **86**, 347–350.

14. Kato, T., Sano, M., Mizutani, N. & Hayakawa, C. (1988): Homocitrullinuria and homoargininuria in hyperargininaemia. *J. Inherit. Metab. Dis.* **11**, 261–265.

15. Kato, T., Sano, M., Mizutani, N. & Hayakawa, C. (1987): Increased urinary excretion of putrescine in hyperargininaemia. *J. Inherit. Metab. Dis.* **10**, 391–396.

16. Konarska, L., Wiesmann, U. & Colombo, J.P. (1981): Arginase activity in human fibroblast cultures. *Clin. Chim. Acta* **115**, 85–92.

17. Konarska, L., Wiesmann, U., v. Fellenberg, R. & Colombo, J.P. (1983): Isoenzyme pattern and immunological properties of arginase in normal and hyperargininemia fibroblasts. *Enzyme* **29**, 44–53.

18. Marescau, B., De Deyn, P.P., Lowenthal, A., Qureshi, I.A., Antonozzi, I., Bachmann, C., Cederbaum, S.D., Cerone, R., Chamoles, N., Colombo, J.P. *et al.* (1990): Guanidino compound analysis as a complementary diagnostic parameter for hyperargininemia: follow-up of guanidino compound levels during therapy. *Pedriatr. Res.* **27**, 3, 297–303.

19. Michels, V. & Beaudet, A.I.. (1978): Arginase deficiency in multiple tissues in argininemia. *Clin. Genetics* **13**, 61–67.

20. Mizutani, N., Hayakawa, C., Maehara, M. & Watanabe, K. (1987): Enzyme replacement therapy in a patient with hyperargininemia. *Tohoku J. Exp. Med.* **151**, 301–307.

21. Naylor, E.W. & Cederbaum, S.D. (1981): Urinary pyrimidine excretion in arginase deficiency. *J. Inherit. Metab. Dis.* **4**, 207–210.

22. Peralta Serrano, A. (1965): Argininuria, convulsiones y oligofrenia: un nuevo error innato del metabolismo? *Rev. Clin. Esp.* **96**, 176–184.

23. Qureshi, I.A., Letarte, J., Ouellet, R., Larochelle, J. & Lemieux, B. (1983): A new French-Canadian family affected by hyperargininaemia. *J. Inherit. Metab. Dis.* **6**, 179–182.

24. Qureshi, I.A., Letarte, J., Quellet, R., Batshaw, M.L. & Brusilow, S. (1984): Treatment of hyperargininemia with sodium benzoate and arginine-restricted diet. *J. Pediatr.* **104**, 473–476.

25. Sparkes, R.S., Dizikes, G.J., Klisak, I., Grody, W.W., Mohandas, T. *et al.* (1986): The gene for human liver arginase (ARG 1) is assigned to chromosome band 6q23. *Am. J. Hum. Genet.* **39**, 186–193.

26. Spector, E.B., Rice, S.C.H & Cederbaum, S.D. (1983): Immunologic studies of arginase in tissues of normal human adult and arginase deficient patients. *Pediatr. Res.* **17**, 941–944.

27. Terheggen, H.G., Schwenk, A., Lowenthal, A., Van Sande, M. & Colombo, J.P. (1969): Argininemia with arginase deficiency. *Lancet* **ii**, 748.

28. Terheggen, H.G., Lowenthal, A., Lavinha, F., Colombo, J.P. & Rogers, S. (1975): Unsuccessful trial of gene replacement in arginase deficiency. *Z. Kinderheilk.* **119**, 1–3.

29. Van Gennip, A.H., van Bree-Blom, E.J., Grift, J., De Bree, P.K. & Wadman S.K. (1980): Urinary purines and pyrimidines in patients with hyperammonemia of various origins. *Clin. Chim. Acta* **104**, 227–239.

30. Vilarinho, L., Senra, V., Vilarinho, A., Barbosa, C., Parvy, P., Rabier, D. & Kamoun, P. (1990): A new case of argininaemia without spastic diplegia in a Portuguese male. *J. Inherit. Metab. Dis.* **13**, 751–753.

31. Zamecka, E. & Porembska, Z. (1988): Five forms of arginase in human tissues. *Biochem. Med. Metab. Biol.* **39**, 258–266.

Guanidino Compounds in Biology and Medicine, eds. by P.P. De Deyn, B. Marescau, V. Stalon and I.A. Qureshi. ©1992 John Libbey & Company Ltd., pp. 349–353.

Chapter 52

Structure and expression of liver-type arginase gene

Masaki TAKIGUCHI and Masataka MORI

Institute for Medical Genetics, Kumamoto University Medical School, Kuhonji 4-24-1, Kumamoto, Japan

Summary

Liver-type arginase, the last enzyme of the urea cycle, is expressed specifically in the liver of ureotelic animals. Its expression is developmentally and hormonally regulated in coordination with other urea cycle enzymes. A deficiency in this enzyme results in argininaemia, an inherited autosomal recessive disorder accompanied by hyperammonaemia. The rat and human arginase genes were cloned and their structures were determined. Both genes are about 12 kilobases long and are split into 8 exons. About 100 nucleotides of the immediately 5′-flanking region of the arginase gene are highly conserved between the rat and human genes. The promoter region of the rat gene was investigated with an *in vitro* transcription system using nuclear extracts prepared from rat tissues. Analysis of deletion mutants using liver extracts revealed a region required for efficient transcription. Two protected areas overlapping with this region were detected by DNase I footprinting. The more downstream footprint area was occupied competitively by two factors each related to CTF/NF-1 and Sp1. The other more upstream area was recognized by a factor related to a liver-enriched factor C/EBP, which was recently found to bind the regulatory regions of other urea cycle enzyme genes.

Introduction

Arginase catalyses the hydrolysis of arginine to urea and ornithine. Liver-type arginase is localized in the cytosol of the liver cells of ureotelic animals and catalyses the last step of the urea synthetic pathway. Arginase activities in the rat[15] and human[4,14] liver increase markedly in the perinatal period, in coordination with other urea cycle enzymes. The activities of arginase and of the other urea cycle enzymes are induced also in a coordinated manner by dietary protein[18,23] and hormones, such as glucagon and dexamethasone[2,17]. These inductions of the urea cycle enzyme activities, in most cases, are associated with increases of mRNA levels for the enzymes[19,21]. A deficiency in human liver arginase results in argininaemia, an autosomal recessive disorder accompanied by hyperammonaemia, motor difficulties, and mental retardation[1].

As the first step in studying the tissue-specific, developmental, and hormonal regulation of expression of the liver-type arginase gene and in determining the nature of mutation in argininaemia, we isolated cDNA clones for the rat[12,13] and human[5] enzyme and determined the entire nucleotide sequences. We found nucleotide and amino acid sequence heterogeneity of the human enzyme[6]. Sparkes *et al.*[24] also isolated cDNA clones for human liver-type arginase and assigned the human arginase gene to chromosome band 6q23. We attempted to construct an expression plasmid for the human enzyme and established a system for producing a large amount of active arginase protein[9]. We then isolated genomic clones for rat[22] and human[25] liver-type arginase and determined the entire organization of the genes. Based on this information, we analysed an argininaemic patient with no consanguinity and identified two discrete frame-shift deletions in the arginase gene[7].

We report here a comparison of the rat and human liver type arginase genes and their 5′-flanking sequences. We also report *in vitro* analysis of the promoter region of the rat arginase gene.

Results

Structure of the rat and human liver-type arginase genes

The genes for rat and human liver-type arginase were cloned and their structures were determined. Both genes are about 12 kilobases long and are split into 8 exons. The exon-intron organization of these two genes are very similar except for sizes of introns 1, 2 and 4. The locations of the 7 introns in the mRNA sequences are completely conserved between the rat and human genes. All of the splice donor and acceptor sites conform to the GT/AG rule.

Characterization of the 5′ ends of the genes

The cap sites of the rat and human genes were determined by nuclease S1 mapping and primer extension. The 5′ untranslated region of the human gene (58 nucleotides) is shorter than that of the rat gene (93 nucleotides). The situation of the cap site of the human gene is essentially identical with that of the rat gene (Fig. 1). Sequences of the 5′ end region of the rat and human arginase genes were compared. There are several highly conserved segments in the 5′-flanking regions between the rat and human genes up to position −600. The immediately 5′-flanking sequences of the rat and human genes up to position −95 (rat) and −105 (human) are 84 per cent identical without counting several small gaps (Fig. 2). The sequence TATAA (−27 to −23 for the rat gene and −28 to −24 for the human gene) identical with the canonical 'TATAA box' is situated at the ordinary location and is presumably functional.

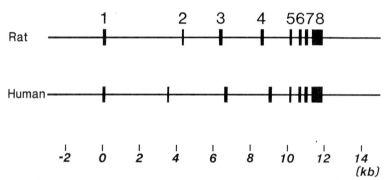

Fig. 1. Structures of the rat and human liver-type arginase genes. The structure of the rat gene is from Ohtake et al.[22] and that of the human gene is from Takiguchi et al.[25]. Boxes with numbers show exons.

In vitro analysis of the rat arginase promoter

In order to identify *cis*-acting DNA elements and *trans*-acting factors regulating the liver-specific, developmentally-regulated and coordinated expression of the arginase gene, we investigated its promoter region[26] using an *in vitro* transcription system originally developed by Gorski et al.[3]. This system employs nuclear extracts prepared from rat tissues and has been successfully used for analysis of a number of tissue-specific genes. A plasmid bearing the arginase promoter region from −193 to 286 was transcribed in nuclear extracts from rat liver and also from rat brain in which arginase activity was not detectable. The arginase promoter was transcribed several-fold more efficiently in liver nuclear extracts than in brain extracts. Therefore, the tissue-specificity of the arginase promoter activity *in vivo* could be reproduced qualitatively *in vitro*. A series of plasmids carrying various lengths of the 5′-flanking region of the gene was transcribed in liver nuclear extracts (Fig. 3). Deletion from −2.7 kb to −90 bp did not significantly affect transcription. A dramatic decrease of the transcription

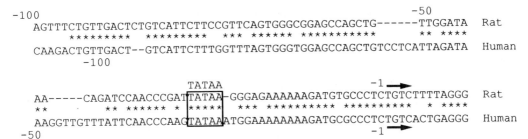

```
-100                                                        -50
   AGTTTCTGTTGACTCTGTCATTCTTCCGTTCAGTGGGCGGAGCCAGCTG------TTGGATA   Rat
         ********  ********   ***  ****** ***********  **  ****
   CAAGACTGTTGACT--GTCATTCTTTGGTTTAGTGGGTGGAGCCAGCTGTCCTCATTAGATA   Human
      -100
```

```
                 TATAA                        -1
   AA-----CAGATCCAACCCGATTATAA-GGGAGAAAAAAGATGTGCCCTCTGTCTTTTAGGG   Rat
   **      **  ****** *  *****   ***  ********** *********  *  ****
   AAGGTTGTTTATTCAACCCAAGTATAAATGGAAAAAAAAGATGCGCCCTCTGTCACTGAGGG   Human
      -50                                          -1
```

Fig. 2. The 5′ end regions of the rat and human arginase genes. Identical nucleotides are shown by asterisks. Arrows show transcription start sites.

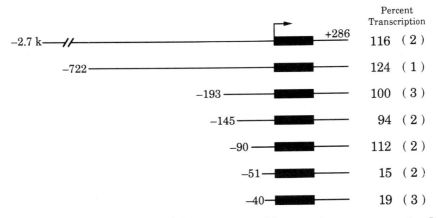

Fig. 3. In vitro transcription analysis of 5′ deletion mutants of the rat arginase promoter region. Boxes represent exon 1 of the arginase gene. Values on the right indicate transcription levels in percent relative to that obtained with the construct bearing the 5′-flanking region up to –193 bp, and show averages of the number of experiments in parentheses.

was brought about by deletion from –90 to –51 bp. Further deletion from –51 to –40 led to no significant change. Therefore, at least one positive *cis*-acting element exists within the region –90 to –51.

Factors binding to the arginase promoter region

To detect DNA-binding factors that may be involved in the transcriptional regulation, DNase I footprint analysis of the promoter region was carried out. With liver nuclear extracts, we detected three protected regions: region A from –95 to –82, region B from –70 to –39 and region C from +89 to +113 bp. Protected areas of region A and B with brain nuclear extracts were more restricted than those obtained with liver extracts, whereas footprint profiles of region C with liver extracts were indistinguishable from those with brain extracts. To further characterize the binding factors, footprint competition analysis was performed. Region B contains two overlapping sequence elements. One (–67 to –58 bp) coincides with the sequence for Sp1-binding sites[11] and the other (–60 to –55) is similar to that for CTF/NF-1-binding sites[10]. In fact, footprint competition analysis using synthetic oligonucleotides showed that the protection of region B can be attributed to two factors, each related to CTF/NF-1 and Sp1. These two factors bind to this region in a mutually exclusive manner and the CTF/NF-1-related factor is predominant between the two. It was also observed that a factor(s) binding to region C is also related to CTF/NF-1. The footprint of region A is rather liver-specific. The sequence at positions –94 to –87 bp is similar to the C/EBP-binding site sequence[16]. Actually, the protection

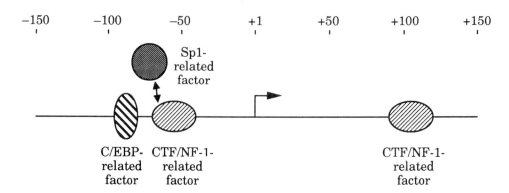

Fig. 4. Schematic representation of the rat liver-type arginase promoter. The horizontal arrow indicates the transcription start site. Ovals show binding factors. The dual arrow represents mutually exclusive binding. From Takiguchi and Mori[26].

of region A was partially removed by the C/EBP-binding oligonucleotides. Thus, a factor(s) binding to region A is related to C/EBP. This was further confirmed by gel shift competition analysis.

Discussion

Factors binding to the promoter region of the rat liver-type arginase gene are shown schematically in Fig. 4. Region A binds a factor(s) related to C/EBP, a liver-enriched transcription factor, and region B binds a factor(s) related to CTF/NF-1. Region B can also bind a Sp1-related factor(s) when the CTN/NF-1 related factor is titrated out. In addition to these regions, region C also binds a CTF/NF-1-related factor(s), although functional analysis of this region remains to be carried out. Since footprint profiles of region B as well as region A exhibit liver-specific features, both factors interacting with these regions may contribute to determination of liver-specificity of the arginase promoter. This should be examined in a future study. Recently, Howell *et al.*[8] reported that C/EBP binds to the promoter region of the gene for rat carbamyl phosphate synthetase, the first enzyme of the urea cycle, and that the C/EBP-binding site is essential for efficient transcription of this gene. On the other hand, we showed that a factor(s) binding to two sites in the enhancer region of the gene for rat ornithine transcarbamylase, the second enzyme of the cycle, is also related to C/EBP[20]. Thus, including the arginase promoter in the present study, C/EBP or its relate factor(s) binds to regulatory regions of three urea cycle enzyme genes thus far analysed in five genes. Therefore, C/EBP or its related factor(s) may play a role in the coordinated regulation of urea cycle enzyme genes.

References

1. Brusilow, S.W. & Horwich, A.L. (1989): Urea cycle enzymes. In *The metabolic basis of inherited disease*, ed. C.R. Scriver, *et al.*, pp. 629–663. New York: McGraw-Hill.

2. Gebhardt, R. & Mecke, D. (1979): Permissive effects of dexamethasone on glucagon induction of urea cycle enzymes in perfused primary monolayer cultures of rat hepatocytes. *Eur. J. Biochem.* **97**, 29–35.

3. Gorski, K., Carneiro, M. & Schibler, U. (1986): Tissue-specific *in vitro* transcription from mouse albumin promoter. *Cell* **47**, 767–776.

4. Guha, S.K. & Mukherjee, K.L. (1974): Urea biosynthesis in normal human fetuses. *Biochim. Biophys. Acta* **372**, 285–290.

5. Haraguchi, Y., Takiguchi, M., Amaya, Y., Kawamoto, S., Matsuda, I. & Mori, M. (1987): Molecular cloning and nucleotide sequence of cDNA for human liver arginase. *Proc. Natl. Acad. Sci. USA* **84**, 412–415.

6. Haraguchi, Y., Takiguchi, M., Matsuda, I. & Mori, M. (1988): Sequence heterogeneity of human liver arginase cDNA and restriction fragment length polymorphism of the gene locus. *Jpn. J. Hum. Genet.* **33**, 305–313.

7. Haraguchi, Y., Aparicio R.J.M., Takiguchi, M., Akaboshi, I., Yoshino, M., Mori, M. & Matsuda, I. (1990): Molecular basis of argininemia *J. Clin. Invest.* **86**, 347–350.

8. Howell, B.W., Lagace, M. & Shore, G.C. (1989): Activity of the carbamyl phosphate synthetase I promoter in liver nuclear extracts is dependent on a *cis*-acting C/EBP recognition element. *Mol. Cell. Biol.* **9**, 2928–2933.

9. Ikemoto, M., Tabata, M., Miyake, T., Kano, T., Mori, M., Totani, M. & Murachi, T. (1990): Expression of human liver arginase in *Escherichia coli. Biochem. J.* **270**, 697–703.

10. Jones, K.A., Kadonaga, J.T., Rosenfeld, P.J., Kelly, T.J. & Tjian, R. (1987): A cellular DNA-binding protein that activates eukaryotic transcription and DNA replication. *Cell* **48**, 79–89.

11. Kadonaga, J.T., Jones, K.A. & Tjian, R. (1986): Promoter-specific activation of RNA polymerase II transcription by Sp1. *Trends Biochem. Sci.* **11**, 20–23.

12. Kawamoto, S., Amaya, Y., Oda, T., Kuzumi, T., Saheki, T., Kimura, S. & Mori, M. (1986): Cloning and expression in *Escherichia coli* of cDNA for arginase of rat liver. *Biochem. Biophys. Res. Commun.* **136**, 955–961.

13. Kawamoto, S., Amaya, Y., Murakami, K., Tokunaga, F., Iwanaga, S., Kobayashi, K., Saheki, T., Kimura, S. & Mori, M. (1987): Complete nucleotide sequence of cDNA and deduced amino acid sequence of rat liver arginase. *J. Biol. Chem.* **262**, 6280–6283.

14. Kemman, A.L. & Cohen, P.P. (1974): Ammonia detoxication in liver from humans. *Proc. Soc. Exp. Biol. Med.* **106**, 170–173.

15. Lamers, W.H., Mooren, P.G., De Graaf, A. & Charles, R. (1985): Perinatal development of the liver in rat and spiny mouse. *Eur. J. Biochem.* **146**, 475–480.

16. Landschulz, W.H., Johnson, P.F., Adashi, E.Y., Graves, B.J. & McKnight, S.L. (1988): Isolation of a recombinant copy of the gene encoding C/EBP. *Genes Dev.* **2**, 786–800.

17. Lin, R.C., Snodgrass, P.J. & Rabier, D. (1982): Induction of urea cycle enzymes by glucagon and dexamethasone in monolayer cultures of adult rat hepatocytes. *J. Biol. Chem.* **257**, 5061–5067.

18. Mori, M., Miura, S., Tatibana, M. & Cohen, P.P. (1981): Cell-free translation of carbamyl phosphate synthetase I and ornithine transcarbamylase messenger RNAs of rat liver. *J. Biol. Chem.* **256**, 4127–4132.

19. Morris, S.M., Jr., Moncman, C.L., Rand, K.D., Dizikes, G. J., Cederbaum, S.D. & O'Brien, W.E. (1987): Regulation of mRNA levels for five urea cycle enzymes in rat liver by diet, cyclic AMP, and glucocorticoids. *Arch. Biochem. Biophys.* **256**, 343–353.

20. Murakami, T., Nishiyori, A., Takiguchi, M. & Mori, M. (1990): Promoter and 11-kilobase upstream enhancer elements responsible for hepatoma cell-specific expression of the rat ornithine transcarbamylase gene. *Mol. Cell. Biol.* **10**, 1180–1191.

21. Neves, V.L. & Morris, S.M. Jr. (1988): Regulation of messenger ribonucleic acid levels for five urea cycle enzymes in cultured rat hepatocytes. *Mol. Endo.* **2**, 444–451.

22. Ohtake, A., Takiguchi, M., Shigeto, Y., Amaya, Y., Kawamoto, S. & Mori, M. (1988): Structural organization of the gene for rat liver-type arginase. *J. Biol. Chem.* **263**, 2245–2249.

23. Schimke, R.T. (1962): Adaptive characteristics of urea cycle enzymes in the rat. *J. Biol. Chem.* **237**, 459–468.

24. Sparkes, R.S., Dizikes, G.J., Klisak, I., Grody, W.W., Mohandas, T., Heinzmann, C., Zollman, S., Lusis, A.J. & Cederbaum, S.D. (1986): The gene for human liver arginase (*ARG1*) is assigned to chromosome band 6q23. *Am. J. Hum. Genet.* **39**, 186–193.

25. Takiguchi, M., Haraguchi, Y. & Mori, M. (1988): Human liver-type arginase gene: structure of the gene and analysis of the promoter region. *Nucl. Acids Res.* **16**, 8789–8802.

26. Takiguchi, M. & Mori, M. (1991): *In vitro* analysis of the rat liver-type arginase promoter. *J. Biol. Chem.* **266**, 9186–9193.

353

Guanidino Compounds in Biology and Medicine, eds. by P.P. De Deyn, B. Marescau, V. Stalon and I.A. Qureshi. ©1992 John Libbey & Company Ltd., pp. 355–362.

Chapter 53

The arginases in human arginase deficiency

Deborah KLEIN, Wayne W. GRODY and Stephen D. CEDERBAUM

Departments of Psychiatry, Pediatrics and Pathology and the Mental Retardation Research Center, UCLA School of Medicine, Los Angeles, CA 90024, USA

Summary

The data presented demonstrate that two isozymes of arginase exist, encoded in different genes, and the differences between them are highlighted. These differences include their presence in different cellular compartments, their susceptibility to different forms of regulation and their differences in apparent physiological function. In inherited deficiency of the primarily catabolic liver arginase, AI, in liver, arginine accumulates and AII is 'induced' in direct proportion to this accumulation. Antibody to AII is now available to define the nature of this induction. Finally, the demonstrated defects in AI leading to its deficiency are reviewed and evidence supporting its absence of induction in hyperargininaemia is presented.

Introduction

Urea formed from arginine by hydrolysis in the liver cytosol is the primary end product of ammonia metabolism in man, who may excrete 30 g or more per day. Because of this, it is not surprising that arginine and its hydrolytic catalyst arginase, are thought of primarily in this catabolic role[8]. This vision does not, however, give adequate consideration to the multiplicity of metabolic fates of both arginine and ornithine, the precursor and second product of the arginase reaction respectively, or of their importance in biology. Some of those numerous roles are discussed extensively elsewhere in this book. The possibility that more than one arginase isozyme existed in mammals was first entertained 20 years ago and was supported by biochemical and immunological studies[6,7,16,29]. Knox and his co-workers first suggested that a second isozyme, designated AII or A4, might be involved in proline biosynthesis[21,29], and Roberts and his colleagues believed that this arginase might be involved in γ-aminobutyric acid (GABA) synthesis as well[30]. These roles were supported by the demonstration that GABA was a feedback inhibitor of ornithine aminotransferase and that proline was an effective feedback inhibitor of AII, but not AI[1]. This is shown in Fig. 1.

The existence and independent genetic determination of AII was clarified and proven by studies of patients with liver arginase (AI or A1) deficiency and following the cloning of the AI gene by Dr. Mori and his collaborators[14,18] and ourselves[4,5]. The first indication that a second form of arginase might exist, derived from clinical investigations in which virtually all patients studied by Dr. Colombo and his collaborators[27,28] and by ourselves[2,3] showed substantial, persistent and indeed increasing ureagenesis with increased protein intake. The increased plasma urea levels with increasing protein intake occurring in two siblings is shown in Table 1. Urinary urea excretion for one of the siblings rose from 5.3 g/g creatinine on the 1 g protein/kg body weight/day diet to 10.2 on the 3 g diet; if urea cycle function were normal, a tripling would have been expected.

Table 1. Urea cycle and related amino acids in the plasma of two patients with arginase deficiency[a]

	Protein intake[b]					Normal Values
	Patient MU			Patient RU		
	1.0	2.0	3.0	1.0	2.5	
Arginine	637	591	677	786	913	21.2–151.0
Citrulline	52.5	37.7	45.7	39.4	57.6	12.0–55.4
Ornithine	46.1	38.6	40.1	74.9	95.3	29.5–126.0
Lysine	127	86.2	80.0	86.9	194	82.8–237.0
Glutamine	392	440	451	556	607	415.0–694.0
Other amino acids	normal	some low	some low	normal	normal	–
Urea (mg/dl)	1.35	14.7	18.4	15.5	20.3	20.0–30.0

[a]Plasma concentrations in µmol/l. All specimens were collected 3 h after a meal containing 1/3 of the prescribed diet. Fasting samples were not distinguishable from the postprandial ones. [b]Protein intake in g/kg body weight/day. Reprinted from Cederbaum et al.[3] with permission.

This inference was confirmed when kidney arginase levels proved to be elevated, rather than reduced, in liver arginase deficiency[25]. Immunological characterization of the specimen proved that all residual enzyme activity failed to be precipitated by anti-AI antibody whereas 50 per cent of the activity in normal kidney was precipitated by this same antibody[25]. This is shown in Fig. 2. Subsequently, we demonstrated no cross-hybridization between a cDNA probe for AI and mRNA derived from a cell line (HEK) that apparently produces AII alone[5]. Moreover, neither Mori's group nor our own could

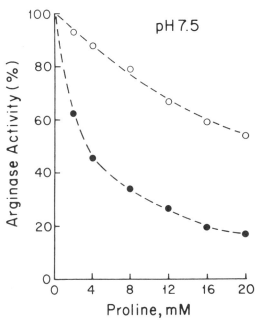

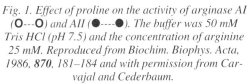

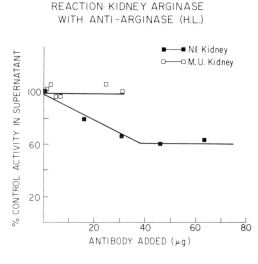

Fig. 1. Effect of proline on the activity of arginase AI (O---O) and AII (●----●). The buffer was 50 mM Tris HCl (pH 7.5) and the concentration of arginine 25 mM. Reproduced from Biochim. Biophys. Acta, 1986, **870**, 181–184 and with permission from Carvajal and Cederbaum.

Fig. 2. Reaction of normal kidney arginase and patient (MU) kidney arginase with antihuman liver arginase antibody. Extracts from normal kidney (■, 736 mU) and from MU kidney (□, 75 mU) were incubated with increasing amounts of antihuman liver arginase antibody (0.65 and 0.33 µg, respectively). Reprinted from Spector et al.[25], with permission.

produce any evidence for either a second transcribed locus or a pseudogene for AI[4].

Table 2. Comparative properties of arginase AI and AII

	AI	AII
Activity	high	low
Tissue specificity	restricted	general
Cellular location	cytosol	mitochondrial matrix
pI	9.3	6.8
Subunit mol. wt.	35,000d	40,000d
Immunoreactivity (anti-liver arginase)	+	−
Immunoreactivity (anti-mammary gland arginase)	−	+
Proline inhibition	±	++
Isoleucine inhibition	++	±
Activity in hyperarginaemia	deficient	enhanced
Hydrocortisone induction	+	−
Arginine induction	−	+

The comparative properties of AI and AII are listed in Table 2. AI is cytoplasmic, varies widely in expression from tissue to tissue, but is highest in liver, is regulated by protein intake and hormones, is highly basic and is subject to feedback inhibition by the branched-chain amino acids. AII is mitochondrial, is more widely and uniformly expressed, is apparently regulated by arginine levels, is neutral, and is subject to feedback inhibition by proline. Both are expressed in utero and share a similar K_m for arginine, K_i for lysine and ornithine, a divalent cation dependency for both stability and activity, preferring manganese, and have an identical and very high heat stability[8]. We have inferred that while they differ substantially from one another in primary structure, they must share critical sequences to confer these similarities in properties.

The behaviour of AI and AII in hyperargininaemia due to AI deficiency

AI deficiency leads to greatly increased levels of arginine in the blood and to a progressive neurodegenerative disorder[8,27,28]. The clinical condition is presumed to be the result of the toxicity

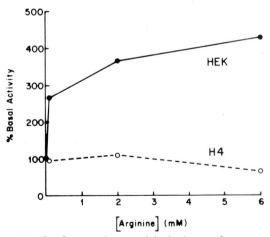

*Fig. 3. Effect of increased arginine levels on arginase activity in tissue culture extracts of HEK cells (●) and H4 cells (O). Enzyme activity was measured 5 days after adding varying amounts of arginine to the culture media. Reproduced from J. Clin. Inves., 1989, **83**, 602–609 by copyright permission of the Am. Soc. Clin. Invest.*

of arginine or some other compounds derived from it (perhaps other guanidino compounds or nitric oxide) to the cerebral cortex, in particular[20].

The first hint that AII might be 'induced' by very high arginine levels was obtained from our first kidney biopsy from an AI deficient patient, in which the AII level was found to be about fivefold higher than in normal controls[25]. This finding has recently been confirmed and extended. Grody *et al.* demonstrated apparent fourfold induction of AII in the kidney obtained at autopsy from an AI deficient patient, and could replicate this finding in tissue culture (Fig. 3)[9]. The residual activity was characterized by both the immunoprecipitation profile and by kinetic criteria, both of which were typical of AII. The latter is shown in Fig. 4.

More recently we have studied two more patients, one who died acutely with massive hyperargininaemia and the other who died with more moderate elevations. AII[10] was induced 25-fold in the kidney of the first and 10-fold in the second, again with enzymatic properties completely consistent with AII. Arginine levels were more highly elevated than in the two previous patients, suggesting a direct relationship between the augmentation of AII activity and substrate levels. Residual activity in the liver of these patients, judged to be AII by immunological and kinetic criteria, did not appear to be 'induced' by the hyperargininaemia (data not shown). This latter conclusion must be qualified by the uncertainty of the true AII levels in liver when estimated against the background of the very high AI present. If true however, it suggests that the regulation of AII in liver may differ considerably from that in kidney, adding still further to the complexity of arginase regulation in man.

Recently we have obtained an anti-arginase antibody that appears to be specific for AII, so that approximate quantitation of AII protein antigen should now be possible. The regulation of AI by arginine can be inferred from some of these same studies. A minimal deviation rat liver hepatoma cell line (H4) failed to alter arginase activity in response to variation of arginine in the growth medium (Fig. 3)[9]. In a single patient who has cross-reacting material (CRM)-positive and mRNA-positive AI deficiency, normal levels of both were found in both kidney and liver, by Western and Northern blotting respectively (data not shown). These data are consistent with those obtained by Schimke nearly 30 years ago *in vivo* and *in vitro*[23,24] and others[22].

Taken together, these data support the notion of substrate regulation of AII, at least in some tissues, but not of AI. AI, on the other hand, is regulated by corticosteroids, both in tissue culture and *in vivo*, and by diet *in vivo*, along with other urea cycle enzymes[12,13,17]. We have failed to show regulation of AII by testosterone in kidney cells in culture, but others have observed this in the living animal. The physiological significance remains to be demonstrated.

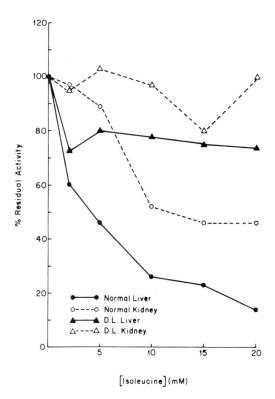

Fig. 4. Effect of isoleucine on activity of arginase in tissue extracts from normals and patient DL: normal liver (●), normal kidney (O), DL liver (Δ), DL kidney (Δ). Each point represents the average of triplicate determinations from one experiment. Each tissue was tested at least twice. Reproduced from J. Clin. Invest ., 1989, 83, 602–609 by copyright permission of the Am. Soc. Clin. Invest.

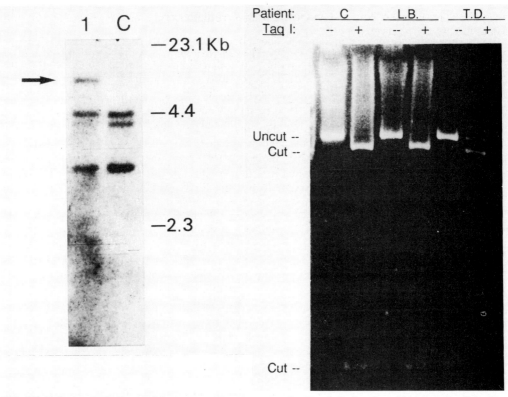

Fig. 5a (Top left). Southern blot analysis of the AI gene cut with Taq I in a patient (1) and control (C) probed with a near-full-length AI cDNA. The loss of the Taq I site, in heterozygosity, is apparent from the larger-than-expected band and the light band at about 4.2 kb.

Fig. 5b (Top right). Analysis by PCR or exon 8 in patient TD, a second patient LB and a control C. After amplification the DNA fragment was run without (–) and with (+) Taq I digestion for each. The control and patient LB digest completely into two smaller fragments, whereas only 50 per cent digestion occurs in TD.

Fig. 5c (Bottom left). Direct sequence analysis of the PCR-amplified region around the exon 8 Taq I site in arginase-deficient patient TD. The sequencing ladder reveals a heterozygous cytosine-to-guanine substitution just outside the TCGA Taq I site in patient TD. The boxes delineate the nucleotide that comprises the Taq I recognition site. Reprinted from Klein et al.[19] with permistsion.

Genetic defects in AI deficiency

To our knowledge the precise mutation has been determined in only four arginase-deficient alleles and the significance of one of these is in doubt. Haraguchi, working with Dr. Mori and his group, has determined that each allele of a single Japanese patient had short deletions of 1 and 4 bases respectively[15], leading to the CRM-negative status in erythrocytes that we found[11]. We have scanned the arginase CRM status in the erythrocytes of 15 patients from around the world and found that all but two were devoid of residual antigen[11]. No gross deletions were found in this sample of genomic DNAs, but three were found to have suffered the loss of a Taq I restriction endonuclease recognition site. In one patient whose parents are consanguineous, studied previously by Drs. Colombo, Lowenthal, Marescau and their collaborators[27,28], the codon for an arginine residue in exon 8, near the 3′ terminus of the enzyme has been mutated to a stop codon in both alleles[19]. Another patient from Australia appeared to have the same Taq I site change, but this time in heterozygosity[11]. To our surprise this mutation, changing a threonine to a serine at position 290 in the protein lay, 1 base outside of the canonical Taq I site (and not in a CpG dinucleotide) and demonstrated that adjacent bases could influence susceptibility to cutting by restriction endonucleases. This is illustrated in Fig. 5, a composite of the Southern blot, a series of polymerase chain reactions of the involved area and the gene sequence in that area. Less clear is whether the semiconservative substitution of a serine for a threonine is sufficient to cause the loss of enzymatic activity or whether another mutation not yet ascertained is responsible for this.

A third patient from a consanguineous family was found to have a Taq I site loss at the 5′ end of the gene, a change seen only in heterozygosity in the father. Examination of the exon sequences and those at the exon intron junction published by Takiguchi et al.[26] failed to reveal any other Taq I sites, the loss of which might result in a functional change in the gene or gene product. Thus this mutation either causes enzyme loss in some manner not apparent to us or it represents a private polymorphism in the inbred isolate from which the patient came.

Discussion and future studies

Two major questions in the biology and pathobiology of the arginases are the nature of the function, structure and evolution of the two (or more) arginase gene loci and their products and the means of exploiting this fortuitous evolutionary event for the treatment of human arginase AI deficiency. We have tantalizing clues that the mitochondrially-located AII may be 'induced' and assume some of the burden for arginine hydrolysis imposed by the loss of function of AI. We have no idea if a larger and more widespread gratuitous induction of AII (could we achieve it) would be sufficient to cure the hyperargininaemia or if this augmented activity could be tolerated without depleting the arginine necessary for vital function in one or another compartment of the body.

A second question derives from the same observation concerning the apparent role of AII in the mitigation of the hyperammonaemia and hyperargininaemia in AI deficiency. Can we develop a gene replacement strategy that is viable and involves an approach other than getting AI expressed in hepatocytes? Would such an approach work and would it be safe?

The availability of an antibody to AII has allowed our study of the arginases to cross an important Rubicon. The answers to the numerous unknowns that fall within the two larger questions depend on the cloning, sequencing and mapping of the AII cDNA and gene. We are well on our way now to accomplishing this goal and being able to compare AII to AI by the many criteria outlined by both Mori et al. and ourselves. As we use these immunological and molecular data we will gain insight into the role of arginase also in the metabolism of other guanidino compounds and of nitric oxide.

Acknowledgements

Supported in part by the Mental Retardation Research Program at UCLA, USPHS Grants: HD-06576, HD-04612 and the March of Dimes Birth Defects Foundation Grants 6-592 and 5-733 (a Basil O'Connor Award through funds received from the Lifespring Foundation).

References

1. Carvajal, N. & Cederbaum, S.D. (1986): Kinetic of inhibition of rat liver and kidney arginases by proline and branched-chain amino acids. *Biochem. Biophys. Acta* **870**, 181–184.

2. Cederbaum, S.D., Shaw, K.N.F. & Valente, M. (1977): Hyperargininemia. *J. Pediatr.* **90**, 569–573.

3. Cederbaum, S.D., Shaw, K.N.F., Spector, E.B., Verity, M.A., Snodgrass, P.J. & Sugarman, G.I. (1979): Hyperargininemia with arginase deficiency. *Pediatr. Res.* **13**, 828–833.

4. Dizikes, G.J., Spector, E.B. & Cederbaum, S.D. (1986): Cloning of rat liver arginase cDNA and the elucidatin of the regulation of arginase gene expression in H4 rat hepatoma cells. *Somat. Cell Mol. Genet.* **12**, 375–384.

5. Dizikes, G. J., Grody, W.W., Kern, R.M. & Cederbaum, S.D. (1986): Isolation of human arginase cDNA and absence of homology between the two liver arginase genes. *Biochem. Biophys. Res. Commun.* **141**, 53–59.

6. Gasiorowska, I., Porembska, Z., Jachimowicz, J. & Mochnacka, I. (1970): Isozymes of arginase in rat tissues. *Acta Biochem. Polon.* **17**, 19–30.

7. Glass, R.D. & Knox, W.E. (1973): Arginase isozymes of rat mammary gland, liver, and other tissues. *J. Biol. Chem.* **248**, 5785–5789.

8. Grody, W.W., Dizikes, G.J. & Cederbaum, S.D. (1987): Human arginase isozymes. In *Isozymes: current topics in biological and medical research* eds M.C. Rattazzi, J.G. Scandalios & G.S. Whitt, pp. 181–214. New York: Alan R. Liss Inc.

9. Grody, W.W., Argyle, C., Kern, R.M., Dizikes, G.J., Spector, E.B., Strickland, A.D., Klein, D. & Cederbaum, S.D. (1989): Differential expression of the two human arginases genes in hyperargininemia: enzymatic pathologic, and molecular analysis. *J. Clin. Invest.* **83**, 602–609.

10. Grody, W.W., Kern, R.M., Klein, D., Dodson, A.E., Wissman, P.B., Barsky, S. & Cederbaum, S.D. (1990): Delayed catastrophic sequelae with massive induction of a kidney arginase isozyme in liver arginase deficiency. *Am. J. Hum. Genet.* **47**, A157.

11. Grody, W.W., Dodson, A., Klein, D., Kern, R.M., Bassand, P. & Cederbaum, S.D. (1989): Molecular genetic study of human arginase deficiency. *Am. J. Hum. Genet.* **45**, A191.

12. Haggerty, D.F., Spector, E.B., Lynch, M., Kern, R., Frank, L.B. & Cederbaum, S.D. (1982): Regulation by glucocorticoids of arginase and argininosuccinate synthetase in cultured rat hepatoma cells. *J. Biol. Chem.* **257**, 2246–2253.

13. Haggerty, D.F., Spector, E.B., Lynch, M., Kern, R., Frank, L.B. & Cederbaum, S.D. (1983): Regulation by expression of genes for enzymes of the mammalian urea cycle in permanent cell-culture lines of hepatic and non-hepatic origin. *Mol. Cell. Biochem.* **53/54**, 57–76.

14. Haraguchi, Y., Takiguchi, M., Amaya, Y., Kawamoto, S., Matsuda, I. & Mori, M. (1987): Molecular cloning and cDNA sequence of cDNA for human liver arginase. *Proc. Natl. Acad. Sci. USA* **84**, 412–415.

15. Haraguchi, Y., Aparicio, J.M., Takiguchi, M., Akaboshi, I., Yoshiro, M., Mori, M. & Matsuda, I. (1990): Molecular basis of argininemia: identification of two discrete frame-shift deletions in the liver type arginase gene. *J. Clin. Invest.* **86**, 347–350.

16. Herzfeld, A., Rosenoer, V.M. & Raper, S.M. (1976): Glutamate dehydrogenase, alanine aminotransferase, thymidine kinase, and arginase in fetal and adult human and rat liver. *Pediatr. Res.* **10**, 960–964.

17. Husson, A., Bouazza, M., Buquet, C. & Vaillant, R. (1984): Precocious induction of arginase in primary cultures of fetal rat hepatocytes. *In Vitro* **20**, 314–320.

18. Kawamoto, S., Amaya, Y., Murakami, K., Tokunaga, F., Iwanaga, S., Kobayashi, K., Saheki, T., Kimura, S. & Mori, M. (1987): Complete nucleotide sequence of cDNA and deduced amino acid sequence of rat liver arginase. *J. Biol. Chem.* **262**, 628–631.

19. Klein, D., Dodson, A.E., Tabor, D.E., Cederbaum, S.D. & Grody, W.W. (1991): Effect of an adjacent base on deletion of a point mutation by restriction enzyme digestion. *Som. Cell Mol. Genet.* **17**, 369–375.

20. Marescau, B., Lowenthal, A., De Deyn, P.D. *et al.* (1990): Guandinino compound analysis as a complementary diagnostic parameter for hyperargininemia. Follow-up of guanidino compound level during therapy. *Pediatr. Res.* **27**, 297–303.

21. Mezl, V.A. & Knox, W.E. (1977): Metabolism of arginine in lactating rat mammary gland. *Biochem. J.* **166**, 105–113.

22. Morris, S.M., Jr., Moncman, C.L., Rand, K.D., Dizikes, G.J., Cederbaum, S.D. & O'Brien, W.D. (1987): Effects of diet, cyclic AMP, and glucocorticoids on mRNA levels for five urea cycle enzymes in rat liver. *Arch. Biochem. Biophys.* **256**, 343–353.

23. Schimke, R.T. (1963): Studies on factors affecting the levels of urea cycle enzymes in rat liver. *J. Biol. Chem.* **238**, 1012–1018.

24. Schimke, R.T. (1964): Enzymes of arginine metabolism in cell culture: studies on enzyme induction and repression. *Natl. Cancer Inst. Monogr.* **13**, 197–217.

25. Spector, E.B., Rice, S.C.H. & Cederbaum, S.D. (1983): Immunologic studies of arginase in tissues of normal human adult and arginase-deficient patients. *Pediatr. Res.* **17**, 941–944.

26. Takiguchi, M., Haraguchi, Y. & Mori, M. (1988): Human liver-type arginase gene: structure of the gene and analysis of the promotor. *Nucl. Acids Res.* **16**, 8789–8802.

27. Terheggen, H.G., Schwenk, A.I, Lowenthal, A., van Sande, M. & Colombo, J.P. (1970): Hyperargininämie mit arginasedefekt: eine neue familiäre Stoffwechselstörung. I. Klinische Befunde. *Z. Kinderheilk* **107**, 298–312.

28. Terheggen, H.G., Schwenk, A.l, Lowenthal, A., van Sande, M. & Colombo, J.P. (1970): Hyperargininämie mit arginasedefekt: eine neue familiäre Stoffwechselstörung. II.Biochemische Untersuchungen. *Z. Kinderheilk* **107**, 313–323.

29. Yip, M.C.M. & Knox, W.E. (1972): Function of arginase in lactating mammary gland. *Biochem. J.* **128**, 893–899.

30. Yoneda, Y., Roberts, E. & Dietz, G.W., Jr. (1982): A new synaptosomal biosynthetic pathway of glutamate and GABA from ornithine and its negative feedback inhibition by GABA. *J. Neurochem.* **38**, 1686–1694.

Guanidino Compounds in Biology and Medicine, eds. by P.P. De Deyn, B. Marescau, V. Stalon and
I.A. Qureshi. ©1992 John Libbey & Company Ltd., pp. 363–371.

Chapter 54

Guanidino compounds in hyperargininaemia

B. MARESCAU, P. P. DE DEYN, I.A. QURESHI, I. ANTONOZZI, C. BACHMANN,
S. D. CEDERBAUM, R. CERONE, N. CHAMOLES, J.P. COLOMBO, M. DURAN,
A. FUCHSHUBER, E. HIRSCH, K. HYLAND, R. GATTI, C. JAKOBS, S.S. KANG,
M. LAMBERT, C. MARESCAUX, N. MIZUTANI, I. POSSEMIERS, I. REZVANI,
B. ROTH, L.M.E. SMIT, S.E. SNYDERMAN, H.G. TERHEGGEN,
L. VILARINHO and M. YOSHINO

*Department of Medicine-UIA, Laboratory of Neurochemistry-BBS, University of Antwerp, UIA T 504,
Universiteitsplein 1, 2610 Antwerp, Belgium*

Summary

The aim of this collaborative study was to demonstrate the importance of analytical studies of guanidino compounds in hyperargininaemia. Our study shows that determination of arginine in urine alone can yield false-negative diagnoses of hyperargininaemia. However, hyperargininaemic patients with normal urinary arginine levels have always increased excretion levels of α-keto-δ-guanidinovaleric acid and argininic acid. The serum arginine levels are always increased in the patients studied. The increase of the serum α-keto-δ-guanidinovaleric acid and argininic acid levels is at least 10-fold. Thus we can conclude that guanidino compound analysis can be used as a complementary biochemical diagnostic parameter for hyperargininaemia.

After institution of protein restriction together with a supplementation of essential amino acids with or without sodium benzoate, the serum arginine levels of hyperargininaemic patients can be brought to levels found in serum of individuals with normal liver arginase activity. The residual arginase activity in the patients probably determines the serum levels of the direct catabolites of arginine. Quantitative determination of guanidino compounds in hyperargininaemic patients shows that the levels of arginine, homoarginine, α-keto-δ-guanidinovaleric acid, argininic acid and α-*N*-acetylarginine are highly increased. The secondary catabolic pathway that is most activated in hyperargininaemic patients is the one with the formation of α-keto-δ-guanidinovaleric acid and argininic acid. Our study has also shown that the biosynthesis of guanidinosuccinic acid in hyperargininaemic patients is decreased and a metabolic relationship between urea and guanidinosuccinic acid was further demonstrated. The results of this collaborative analytical study clearly demonstrate that the determination of guanidino compounds in hyperargininaemia have diagnostic, therapeutic, pathophysiological and metabolic-biochemical implications.

Introduction

The aim of this collaborative study was to demonstrate the importance of analytical studies of guanidino compounds in hyperargininaemia. In order to achieve this goal it was of paramount importance to create a collaborative team consisting of patients, parents, nurses, paediatricians and researchers. This autosomal, recessively transmitted disorder characterized by a deficiency of arginase leading to an accumulation of arginine levels and specific neurological symptoms was described first clinically and biochemically by our group[18–20]. The clinical and biochemical data of the published and some unpublished patients up to 1989 have been reviewed by De Deyn[3]. In the following sections we will give some examples showing the implications of analytical studies of guanidino compounds in these patients. The results were obtained by liquid cation-exchange chromatography and the fluorescence ninhydrin detection method as described earlier[9,11].

Diagnostic implications

We started this collaborative study to investigate whether the pathobiochemistry of the guanidino compounds could provide complementary diagnostic parameters for hyperargininaemia. Early diagnosis is very important to prevent neurological complications[17]. At present hyperargininaemia is biochemically diagnosed by increased arginine levels in serum. The diagnosis can be confirmed by measuring the arginase activity in erythrocytes. Determination of arginine in urine alone can yield a false-negative diagnosis[1]. Up to now (1992) we analysed guanidino compounds in 21 serum and 29 urine samples of different hyperargininaemic patients. The following guanidino compounds were determined: α-keto-δ-guanidinovaleric acid, guanidinosuccinic acid, creatine, guanidinoacetic acid, α-N-acetylarginine, argininic acid, β-guanidinopropionic acid, creatinine, γ-guanidinobutyric acid, arginine, homoarginine, guanidine and methylguanidine. For the purpose of the study, one would ideally consider samples of untreated patients. However, it is medically and ethically unacceptable to interrupt the therapies. Therefore, samples of patients under no treatment (n = 4), protein restriction (n = 12), or combined therapy of protein restriction with supplementation of essential amino acids with (n = 8) or without (n = 6) sodium benzoate were included in this study.

From the 13 guanidino compounds quantitated in serum of hyperargininaemic patients, the levels of β-guanidinopropionic acid, guanidine and methylguanidine were below detection limit in all patients and comparable to controls. The serum levels of creatine and creatinine were within the normal range for all the hyperargininaemic patients. Serum arginine, homoarginine, α-keto-δ-guanidinovaleric acid, argininic acid and α-N-acetylarginine levels of all hyperargininaemic patients were above normal range. The serum guanidinoacetic acid levels of some hyperargininaemic patients fell within the normal range, for other patients the levels were higher than normal. The γ-guanidinobutyric acid levels of most of the studied hyperargininaemic patients were comparable to controls. The serum guanidinosuccinic acid levels of most of the hyperargininaemic patients were below control levels. Also in patients with other urea cycle diseases there is a significant decrease of the guanidinosuccinic acid levels, however in hyperargininaemic patients this decrease is much more pronounced[10]. Figure 1 gives the serum levels of α-keto-δ-guanidinovaleric acid and argininic acid, the two guanidino compounds which are most increased, together with the serum arginine levels. The considerable patient-to-patient variations in the levels are probably a consequence of differences in therapy[8].

The urinary excretion of guanidine, methylguanidine and creatine were within the normal range for all studied hyperargininemic patients. The excretion of guanidinopropionic acid from most of the 29 patients was also comparable to controls, but eight were abnormal: seven had excretion values about four times and one about 15 times higher than the upper limit of the controls. The urinary excretion of α-keto-δ-guanidinovaleric acid and argininic acid of all the studied hyperargininaemic patients was higher than the normal range. The urinary excretion of α-N-acetylarginine and γ-guanidinobutyric acid of most of the patients was higher than normal, but the majority had arginine, homoarginine and guanidinoacetic acid excretion within the normal range. As for the serum values, the urinary excretion of guanidinosuccinic acid of all the studied hyperargininaemic patients was lower than the control levels. Figure 2 shows the urinary excretion levels of arginine, α-keto-δ-guanidinovaleric acid and argininic acid. The considerable patient-to-patient variations in the urinary excretion of the guanidino compounds are probably, as in serum, also a consequence of therapy.

It must be stressed that in this collaborative study half of the hyperargininaemic patients were under a combined therapy of protein restriction together with essential amino acid supplementation with or without sodium benzoate. Still higher guanidino compound levels are to be expected in the biological

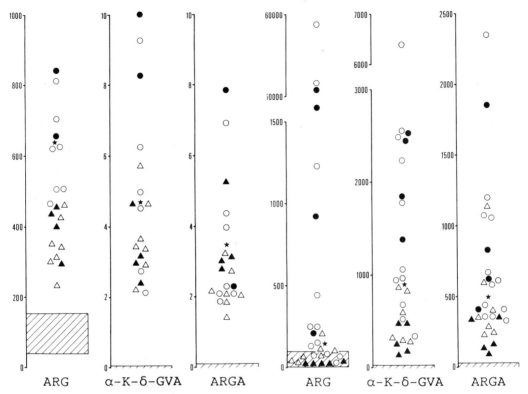

Fig. 1. Serum guanidino compound levels (μM) of ARG, arginine; α-K- δ-GVA, α-keto-δ-guanidino-valeric acid; and ARGA, argininic acid, in different hyperargininaemic patients. Free diet (n = 2) ●; protein restriction (n = 7) □; protein restriction + essential amino acids (n = 4) ▲: protein restriction + essential amino acids + sodium benzoate (n = 7) Δ; protein restriction + essential amino acids + Orn and Lys (n = 1) ✶. Shaded areas = range of normal values.

Fig. 2. Urinary excretion levels (μ moles/g crea-tinine) of ARG, arginine; α-K-δ-GVA, α-keto-γ-gua-nidinovaleric acid; and ARGA, argininic acid, in different hyperargininaemic patients. Free diet (n = 4) ●; protein restriction (n = 12)□; protein restric-tion + essential amino acids (n = 5) ▲; protein re-striction + essential amino acids + sodium benzoate (n = 7) Δ; protein restriction + essential amino acids + Orn and Lys (n = 1) ✶. Shaded areas = range of normal values.

fluids of untreated patients. Untreated and protein-restricted hyperargininaemic patients can be considered as one group because hyperargininaemic patients, as with the other urea cycle patients, mostly impose a restricted protein diet upon themselves[1]. Considering this, our study shows that determination of arginine in urine alone can yield false-negative diagnoses. We have also seen that in urine of hyperargininaemic patients with normal urinary arginine levels, the excretion levels of α-keto-δ-guanidinovaleric acid and argininic acid are increased. The serum arginine levels are always increased in the studied patients. The increase of the serum α-keto-δ-guanidinovaleric acid and argininic acid levels is at least 10-fold. Thus we can conclude that guanidino compound analysis can be used as a complementary biochemical diagnostic parameter for hyperargininaemia.

Therapeutic implications

The guanidino compounds are candidate neurotoxins in hyperargininaemia. Spasticity and epilepsy are frequently seen in these patients. The epileptogenic character of most of the studied guanidino compounds has been shown experimentally[12]. Recently more progress has been made in the possible

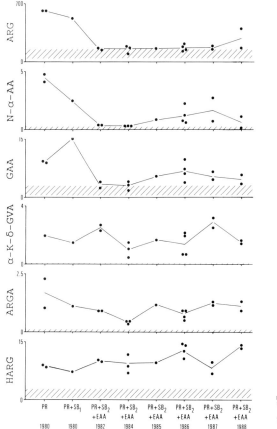

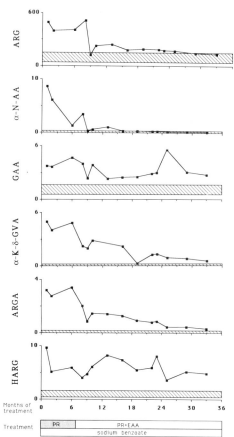

Fig. 3 Follow-up of the serum guanidino compound levels (μM) during 9 years of therapy in a hyperargininaemic girl. For abbreviations of guanidino compounds see Fig. 1. GAA = guanidinoacetic acid; N-α-AA = α-N-acetylarginine; HARG = homoarginine; PR = protein restriction; PR + SB₁ = protein restriction + sodium benzoate 250 mg/kg/day; PR + SB₂ + EAA = protein restriction + sodium benzoate 350 mg/kg/day + essential amino acids. Shaded areas = range of normal values.

Fig. 4. Follow-up of the serum guanidino compound levels (μM) during 3 years of therapy in a hyperargininaemic girl. For abbreviations of guanidino compounds see Fig. 1 and 3.

knowledge of the mechanistic action of the studied guanidino compounds as neurotoxins[4-6]. As patients with other urea cycle disorders, hyperargininaemic patients impose protein restriction upon themselves[1]. Protein restriction was also the first approach in therapy of hyperargininaemia[1,19]. More promising results were obtained by Snyderman *et al.*[17], who prevented neurological complications by treating a boy from the day of birth with protein restriction and supplementation of essential amino acids. The administration of sodium benzoate has also been found to be beneficial in ammonia detoxification[14].

Since the guanidino compounds are candidate neurotoxins we followed up the serum levels during therapeutic interventions in different hyperargininaemic patients. Figure 3 gives the follow-up of the serum guanidino compound levels during 9 years of therapy in a hyperargininaemic girl. The first clinical and biochemical description of this girl was published in 1981[15]. Only through restriction of the daily arginine intake (protein restriction and supplementation of essential amino acids without arginine) could the serum arginine levels be lowered to the upper limit of the control range. During the first years of a combined therapy of protein restriction together with a supplementation of essential amino acids with sodium benzoate, the serum guanidinoacetic acid and α-*N*-acetylarginine levels were brought into the upper control levels. However, a longer follow-up shows that these guanidino compound levels could not be normalized. The decrease and the normalization of the serum arginine levels did certainly not normalize the serum α-keto-δ-guanidinovaleric acid, argininic acid and homoarginine levels, which remained considerably increased. The follow-up of the serum guanidino compound levels during almost 3 years of therapy in another hyperargininaemic girl is shown in Fig. 4. Clinical and biochemical description of this patient was done by Lambert *et al.*[7]. Also in this patient, after the introduction of protein restriction and supplementation of essential amino acids without arginine, the argininaemia could be normalized. A normalization of the α-*N*-acetylarginine levels was obtained as well. The levels of α-keto-δ-guanidinovaleric acid and argininic acid tend to the upper normal levels. The levels of guanidinoacetic acid and homoarginine still remain abnormal during the therapy.

Figure 3 and 4 clearly show that the argininaemia can be 'normalized' or lowered to levels found in serum of individuals with normal liver arginase activity. The residual arginase activity in the patients probably determines the serum levels of the direct catabolites of arginine. The normalization of the serum arginine levels can only be obtained by a combined therapy of protein restriction together with a supplementation of essential amino acids with or without sodium benzoate. Prevention and/or decrease of neurological complications obtained by institution of the former therapeutic regimen[7,16,17], coincides with a substantial lowering of arginine together with the guanidino compounds and further supports the hypothesis that guanidino compounds are neurotoxins in hyperargininaemia.

Pathophysiological, metabolic-biochemical implications

Arginine and the guanidino compounds have been determined in the biological fluids and blood cells of hyperargininaemic patients, however no information about the guanidino compound levels in vital organs has been published up to now. Table 1 gives the serum guanidino compound levels in 11 different hyperargininaemic patients under free or protein-restricted diet.

Table 1. Serum guanidino compound levels (µM) in 11 different hyperargininaemic patients under free or protein restricted diet

Guanidino compound	Controls n = 107	FFL	LC	CP	AR	FB	PR	RM	RL	FF	CC	O
α-K-δ-GVA	< 0.050 – 0.200	8.28	ND	4.57	4.83	6.25	4.96	9.29	2.09	3.15	12.6	2.70
GSA	0.100 – 0.500	0.110	0.025	< 0.025	< 0.025	< 0.025	0.08	< 0.025	< 0.025	0.04	ND	0.03
CT	10 – 200	143	198	139	93.6	144	138	109	129	105	107	110
GAA	0.400 – 3.00	2.98	8.83	3.06	4.69	4.51	2.59	3.08	3.78	2.32	3.57	4.68
α-N-AA	< 0.025 – 0.300	0.98	4.74	2.54	1.28	10.3	3.04	0.89	ND	2.11	2.74	1.33
ArgA	< 0.025 – 0.200	7.88	2.27	2.08	3.39	4.39	3.96	6.93	2.12	2.23	2.30	1.73
β-GPA	< 0.025 – 0.085	< 0.025	< 0.025	< 0.025	< 0.025	< 0.025	< 0.025	< 0.025	< 0.025	< 0.013	< 0.013	< 0.013
CTN	10 – 100	43.8	69.1	23.9	42.6	60	19.3	27.9	29.3	36.2	26.4	22.6
γ-GBA	< 0.025 – 0.080	0.71	ND	0.22	< 0.025	< 0.025	< 0.025	< 0.025	< 0.025	0.07	0.13	0.16
Arg	40 – 155	837	620	464	411	814	625	506	703	417	649	508
Harg	0.500 – 2.80	8	8.45	6.40	5.88	8.62	4.42	4.43	9.81	7.66	4.17	5.00
G	< 0.200 – 0.400	< 0.200	< 0.200	2.82	< 0.200	< 0.200	< 0.200	< 0.200	< 0.200	< 0.060	0.14	0.26
MG	< 0.100	< 0.100	< 0.100	< 0.100	< 0.100	< 0.100	< 0.100	< 0.100	< 0.100	< 0.020	< 0.020	< 0.020

Abbreviations
α-K-δ-GVA = α-keto-δ-guanidino acid; GSA = guanidinosuccinic acid; CT = creatine; GAA = guanidinoacetic acid; α-N-AA = α-*N*-acetylarginine; ArgA = argininic acid; β-GPA = β-guanidinopropionic acid; CTN = creatinine; γ-GBA = γ-guanidinobutyric acid; Arg = arginine; Harg = homoarginine; G = guanidine; MG = methylguanidine; ND = not determined.

Table 2. Levels of guanidino compounds in liver (nmoles/g tissue) of a hyperargininaemic patient

Guanidino compounds	Control autopsy	Autopsy	Biopsy
α-Keto-δ-guanidinovaleric acid	< 0.100	28.3	30.5
Guanidinosuccinic acid	15.7 – 127	0.72	ND
Creatine	1956 ± 776	1498	1750
Guanidinoacetic acid	2.88 ± 1.49	21.3	22.1
α-N-acetylarginine	< 0.090	21.6	23.3
Argininic acid	< 0.080 – 1.95	70.2	56.5
β-Guanidinopropionic acid	< 0.080 – 1.11	< 0.080	ND
Creatinine	171 ± 50.3	172	ND
γ-Guanidinobutyric acid	ND	1.90	ND
Arginine	23.3 ± 10.4	979	1491
Homoarginine	< 0.190 – 1.83	27.5	28.0
Guanidine	< 0.290 – 6.42	4.62	ND
Methylguanidine	0.120 – 0.710	< 0.080	ND

ND = not determined.

Table 2 shows the levels of guanidino compounds in the liver of a hyperargininaemic boy of 4 years who died of circulatory arrest. The autopsy was performed a few hours after death. Similar concentration levels in autopsy and biopsy material were obtained. The arginine and homoarginine levels were about 50 times higher in the patient than in controls. The increase of the α-N-acetylarginine and α-keto-δ-guanidinovaleric acid levels was at least 250 to 300 times. Argininic acid, another main catabolite of arginine, was increased 50 times, guanidinoacetic acid only eight times. The levels of the other guanidino compounds were comparable to controls except guanidinosuccinic acid which was clearly decreased (the same decrease is also seen in serum and urine of hyperargininaemic patients).

Table 3 gives the levels of guanidino compounds in kidney autopsy material of the same hyperargininaemic patient. In the patient's kidney, which possesses another arginase type than liver, the arginine and homoarginine levels were about four times higher than those found in controls. The main catabolites of arginine (α-keto-δ-guanidinovaleric acid, argininic acid and α-N-acetylarginine) were approximately 25 to 50 times higher. The other guanidino compound levels were comparable to controls except guanidinosuccinic acid which was decreased.

Table 3. Levels of guanidino compounds in kidney autopsy material (nmoles/g tissue) of a hyperargininaemic patient

Guanidino compounds	Control	Hyperargininaemia
α-Keto-δ-guanidinovaleric acid	< 0.100 – 2.57	16.5
Guanidinosuccinic acid	3.86 – 46.7	< 0.050
Creatine	586 – 4287	1489
Guanidinoacetic acid	92.9 ± 46.7	62.9
α-N-acetylarginine	< 0.090	4.74
Argininic acid	0.540 – 5.84	54.0
β-Guanidinopropionic acid	< 0.080 – 2.43	2.81
Creatinine	749 ± 306	190
γ-Guanidinobutyric acid	ND	1.12
Arginine	466 ± 247	1188
Homoarginine	< 0.190 – 3.87	7.58
Guanidine	< 0.290 – 7.99	65
Methylguanidine	< 0.120 – 2.21	< 0.12

ND = not determined.

Given the neurological symptoms seen in hyperargininaemic patients we determined the guanidino compound levels in different brain regions. Table 4 shows the average guanidino compound levels in six cerebral cortex regions of the hyperargininaemic patient studied. Although the arginine levels of the patient were comparable to control levels, the α-keto-δ-guanidinovaleric acid levels were at least 25 times and the levels of homoarginine, argininic acid and guanidinoacetic acid four to seven times higher than in controls.

Table 4. Levels of guanidino compounds in cerebral cortex (nmoles/g tissue) of a hyperargininaemic patient

Guanidino compounds	Control	Hyperargininaemia
α-Keto-δ-guanidinovaleric acid	< 0.100	1.24 ± 0.706
Guanidinosuccinic acid	< 0.050 – 0.370	< 0.050
Creatine	7631 ± 678	4788 ± 1247
Guanidinoacetic acid	2.31 ± 0.29	10.3 ± 2.34
α-N-acetylarginine	< 0.090	< 0.090
Argininic acid	< 0.080 – 0.120	4.10 ± 1.15
β-Guanidinopropionic acid	< 0.080	< 0.080
Creatinine	287 ± 34.3	304 ± 50.3
γ-Guanidinobutyric acid	1.56 ± 0.25	0.99 ± 0.145
Arginine	393 ± 75.7	435 ± 34.2
Homoarginine	1.32 ± 0.12	5.20 ± 1.30
Guanidine	< 0.290	1.92
Methylguanidine	< 0.120 – 0.300	1.02

Quantitative determination of guanidino compounds (Tables 1, 2, 3 and 4) in hyperargininaemic patients shows that some guanidino compounds are highly increased. As a consequence of the arginase deficiency the arginine and homoarginine levels are increased. The secondary catabolic pathway that is most activated in hyperargininaemic patients is the one with the formation of α-keto-δ-guanidino-valeric acid. This formation is probably catalysed by a transaminase. The increased activity of the hydrogenation of α-keto-δ-guanidinovaleric acid to argininic acid is also remarkable. The increase

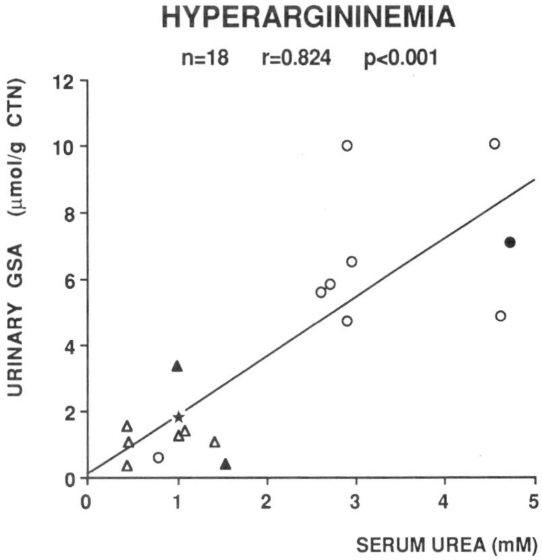

HYPERARGININEMIA

n=18   r=0.824   p<0.001

Fig. 5. Relation between the serum urea levels of 18 hyperargininaemic patients and their corresponding urinary excretion levels of guanidinosuccinic acid.
Range of normal urinary excretion levels of guanidinosuccinic acid: 15 to 160 μmoles/g creatinine. Free diet (n = 1) ●; Protein restriction (n = 8) □; protein restriction + essential amino acids (n = 2) ▲; protein restriction + essential amino acids + sodium benzoate (n = 6) Δ; protein restriction + essential amino acids + Orn and Lys (n = 1) ★.

of the transamidination activity involved in the biosynthesis of guanidinoacetic acid, γ-guanidinobutyric acid and β-guanidinopropionic acid in hyperargininaemic patients is less pronounced. The biosynthesis of guanidinosuccinic acid is related to urea[13]. Hyperargininaemic patients have a decreased urea cycle activity as a consequence of the arginase deficiency. This is reflected in the decreased serum urea levels[10]. Table 1 clearly shows that the guanidinosuccinic acid levels are decreased in hyperargininaemic patients. Autopsy as well as biopsy material of the studied hyperargininaemic patient is characterized by decreased guanidinosuccinic acid levels. The metabolic relationship between urea and guanidinosuccinic acid was further demonstrated in our collaborative study. Figure 5 gives the relation between the serum urea levels of 18 different hyperargininaemic patients and the corresponding urinary excretion levels of guanidinosuccinic acid. A significant positive linear correlation (r = 0.824; $P < 0.001$) was found: patients with higher serum urea levels having higher urinary guanidinosuccinic acid levels. The correlation between the urinary excretion levels of guanidinosuccinic acid and the corresponding serum arginine levels is smaller and less significant (in another hypothesis it was suggested that guanidinosuccinic acid could be formed by transamidination of arginine to aspartic acid[2]). Figure 5 also shows that hyperargininaemic patients under a combined therapy of protein restriction together with a supplementation of essential amino acids with or without sodium benzoate have a lower urea biosynthesis activity than untreated and protein-restricted patients. Indeed the 'waste' nitrogen formation in these patients is also lower as a consequence of therapy. These patients have also the lowest urinary guanidinosuccinic acid levels.

To conclude, the results of this collaborative analytical study clearly demonstrate that the pathobiochemistry of the guanidino compounds in hyperargininaemia has diagnostic implications. Since guanidino compounds are candidate neurotoxins in hyperargininaemia, performing guanidino compound determinations for each patient seems to be indicated, especially with each change in therapeutic regimen attempting to lower these toxins as much as possible. The pathobiochemistry in hyperargininaemia also gave us more insight into the metabolism of arginine and the guanidino compounds.

Appendix

Institutional affiliations of each author:

B. Marescau, P.P. De Deyn and I. Possemiers; University of Antwerp (Antwerp). I.A. Qureshi and M. Lambert; Hôpital Sainte Justine (Montréal). I. Antonozzi; Istituto di Neuropsychiatria Infantile (Roma). C. Bachmann; Centre Hospitalier Universitaire Vaudois (Lausanne). S.D. Cederbaum; University of California (Los Angeles). R. Cerone and R. Gatti; Istituto G. Gaslini (Genova). N. Chamoles; Laboratory of Neurochemistry, Uriarte 2383 (Buenos Aires). J.P. Colombo; Inselspital Bern (Bern). M. Duran; University of Utrecht (Utrecht). A. Fuchshuber and B. Roth; University of Cologne (Cologne). E. Hirsch and C. Marescaux; Hôpital Civil (Strasbourg). K. Hyland; Institute of Child Health (London). C. Jakobs and L.M.E. Smit; Free University Hospital (Amsterdam). S.S. Kang; Rush Presbytarian-St. Luke's Medical Center (Chicago). N. Mizutani; University of Nagoya (Nagoya). I. Rezvani, St Christopher's Hospital for Children (Philadelphia). S.E. Snyderman; New York University Medical Center (New York). H.G. Terheggen; Maizweg 16 (Cologne). L. Vilarinho; Instituto de Genetica Medica (Porto). M. Yoshino; Kurume University (Kurume).

Acknowledgements

This work was supported by the Born-Bunge Foundation, the Universitaire Instelling Antwerpen, the MRC Canada MT 9124, the United Fund of Belgium and the NFWO grants N° 3.0044.92 and NDE 58.

References

1. Cederbaum, S.D., Shaw, K.N.F. & Valente, M. (1977): Hyperargininemia. *J. Pediatr.* **90**, 569–573.

2. Cohen, B. (1970): Guanidinosuccinic acid in uremia. *Arch. Intern. Med.* **126**, 846–850.

3. De Deyn, P.P. (1989): Analytical studies and pathophysiological importance of guanidino compounds in uremia and hyperargininemia. Thesis submitted to obtain the degree of 'Geaggregeerde van het Hoger Onderwijs', UIA, Belgium: University of Antwerp.

4. De Deyn, P.P. & Macdonald, R.L. (1990): Guanidino compounds that are increased in uremia inhibit GABA- and glycine-responses on mouse neurons in cell culture. *Ann. Neurol.* **28**, 627–633.

5. De Deyn, P.P., Marescau, B. & Macdonald, R.L. (1991): Guanidino compounds that are increased in hyperargininemia inhibit GABA- and glycine-responses on mouse neurons in cell culture. *Epilepsy Res.* **8**, 134–141.

6. D'Hooghe, R., Manil, J., Colin, F. & De Deyn, P.P. (1991): Guanidinosuccinic acid inhibits excitatory synaptic transmission in CA1 region of rat hippocampal slices. *Ann. Neurol.* **30**, 622–623.

7. Lambert, M.A., Marescau, B., Desjardins, M., Laberge, M., Dhondt, J.L., Dallaire, L., De Deyn, P.P. & Quershi, I.A. (1991): Hyperargininemia: intellectual and motor improvement related to changes in biochemical data. *J. Pediatr.* **118**, 420–424.

8. Marescau, B., De Deyn, P.P., Löwenthal, A., Qureshi, I.A., Antonozzi, I., Bachmann, C., Cederbaum, S.D., Cerone, R., Chamoles, N., Colombo, J.P., Hyland, K., Gatti, R., Kang, S.S., Letarte, J., Lambert, M., Mizutani, N., Possemiers, I., Rezvani, I., Snyderman, S.E., Terheggen, H.G. & Yoshino, M. (1990): Guanidino compound analysis as a complementary diagnostic parameter for hyperargininemia: follow-up of guanidino compound levels during therapy. *Ped. Res.* **27**, 297–303.

9. Marescau, B., De Deyn, P.P., Van Gorp, L. & Löwenthal, A. (1986): Purification procedure for some urinary guanidino compounds. *J. Chromat.* **377**, 334–338.

10. Marescau, B., Qureshi, I.A., De Deyn, P.P., Letarte, J., Chamoles, N., Yoshino, M., De Potter, W.P. & Löwenthal, A. (1989): Serum guanidinosuccinic acid levels in urea cycle diseases. In *Guanidines 2*, eds. A. Mori, B.D. Cohen & H. Koide, pp. 245–249. New York: Plenum Press.

11. Marescau, B., Qureshi, I.A., De Deyn, P.P., Letarte, J., Ryba, R. & Löwenthal, A. (1985): Guanidino compounds in plasma, urine and cerebrospinal fluid of hyperargininemic patients during therapy. *Clin. Chim. Acta* **146**, 21–27.

12. Mori, A. (1987): Biochemistry and neurotoxicity of guanidino compounds. History and recent advances. *Pavlov J. Biol. Sci.* **22**, 85–94.

13. Natelson, S. & Sherwin, J.E. (1979): Proposed mechanism for urea nitrogen reutilisation: relationship between urea and proposed guanidine cycles. *Clin. Chem.* **25**, 1343–1344.

14. Qureshi, I.A., Letarte, J., Ouellet, R., Batshaw, M.L. & Brusilow, S. (1984): Treatment of hyperargininemia with sodium benzoate and arginine restricted diet. *J. Pediatr.* **104**, 473–476.

15. Qureshi, I.A., Letarte, J., Ouellet, R., Lelièvre, M. & Laberge, C. (1981): Ammonia metabolism in a family affected by hyperargininemia. *Diabet. Met.* **7**, 5–11.

16. Snyderman, S.E., Sansaricq, C., Chen, W.J., Norton, P.M. & Phansalkar, S.V. (1977): Argininemia. *J. Pediatr.* **90**, 563–568.

17. Snyderman, S.E., Sansaricq, C., Norton, P.M. & Goldstein, F. (1979): Argininemia treated from birth. *J. Pediatr.* **95**, 61–63.

18. Terheggen, H.G., Schwenk, A., Löwenthal, A., van Sande, M. & Colombo, J.P. (1969): Argininaemia with arginase deficiency. *Lancet* **ii**, 748–749.

19. Terheggen, H.G., Schwenk, A., Löwenthal, A., van Sande, M. & Colombo, J.P. (1970): Hyperargininämie mit Arginasedefekt: eine neue familiäre Stoffwechselstörung. I. Klinische Befunde. *Z. Kinderheilk.* **107**, 298–312.

20. Terheggen, H.G., Schwenk, A., Löwenthal, A., van Sande, M. & Colombo, J.P. (1970): Hyperargininämie mit Arginasedefekt: eine neue familiäre Stoffwechselstörung. II. Biochemische Untersuchungen. *Z. Kinderheilk* **107**, 313–323.

Guanidino Compounds in Biology and Medicine, eds. by P.P. De Deyn, B. Marescau, V. Stalon and
I.A. Qureshi. ©1992 John Libbey & Company Ltd., pp. 373–378.

Chapter 55

Guanidino compounds in HHH syndrome during sodium benzoate therapy

I.A. QURESHI[1], B. MARESCAU[2], M. LAMBERT[1], J.F. LEMAY[1] and P.P. DE DEYN[2]

[1]*Research Centre, Hôpital Sainte-Justine and Département de Pédiatrie, Université de Montréal, Montreal, Quebec,
Canada H3T 1C5;* [2]*Laboratory of Neurochemistry, BBF University of Antwerp, UIA 2610 Antwerp, Belgium*

Summary

Guanidino compounds were determined in serum and 24 h urine samples of patients suffering from hyperornithinaemia, hyperammonaemia and homocitrullinuria (HHH) syndrome treated with sodium benzoate. The fasting guanidinosuccinic acid levels in five out of six patients with HHH syndrome and urinary guanidinosuccinic acid excretion in four out of six patients was significantly lower than normal, thus indicating an impairment of the guanidinosuccinic acid synthesis as seen in other urea cycle disorders. Serum urea concentrations had a general tendency to be lower amongst children.

Serum guanidinoacetic acid concentration in only one out of six patients, and urinary guanidinoacetic acid in only two out of six patients was significantly lower than normal. Each one of the patients also showed lower than normal excretion of arginine and/or creatine. This indicated the possibility of only an insignificant effect of either a metabolic arginine deficiency and/or glycine depletion, due to sodium benzoate therapy, on creatine synthesis in HHH syndrome.

Introduction

Ornithine, in the mammalian liver, is mainly produced by the action of the cytosolic arginase (EC 3.5.3.1) on dietary arginine[22]. A second ornithine producing reaction is that of arginine-glycine transamidinase (EC 2.1.4.1), which in humans is thought to be located in the cytosol[14], as against the mitochondrial matrix in the chicken[6] or on the inner mitochondrial membrane in the rat kidney[12]. Ornithine enters the mitochondrial space through carrier-mediated transport[7], where it acts as a substrate for ornithine transcarbamylase (EC 2.1.3.3), or alternately for ornithine-δ-amino-transferase (EC 2.6.1.13).

There are now two human inborn errors of metabolism, in which hyperornithinaemia is known to occur[22]. The first one is the gyrate atrophy of the choroid and retina, which is caused by a mutation of the ornithine-δ-aminotransferase gene. It does not result in elevated plasma ammonia levels. Over 100 cases of this disease are known, mostly in Finland. The second hyperornithinaemic syndrome is associated with hyperammonaemia and homocitrullinuria and is known as the HHH syndrome. First described in the year 1969[20], it is postulated to be caused by a defect in the ornithine translocator protein[5]. About 30 cases of the HHH syndrome have been described in the literature[22]. This syndrome may be more common amongst people of French Canadian origin. We have studied a series of patients with HHH syndrome, all of French Canadian ancestry, which were diagnosed on the basis of hyperornithinaemia and hyperammonaemia. These patients have been treated with protein restriction and sodium benzoate therapy. The therapeutic action of sodium benzoate against hyperammonaemia is primarily based on its conjugation with glycine[2]. This amino acid is not only highly ammon-iagenic[19], but its metabolic synthesis may also be ammonia depleting[1]. Consequently, its elimination as hippurate reduces ammoniagenicity and the overload on an impaired urea cycle[18].

We decided to test a hypothesis in our patients during sodium benzoate therapy, postulating: (1) that an increased utilization and a possible depletion of glycine could impair the activity of arginine-gly-cine transamidinase, and thus effect the synthesis of guanidinoacetic acid and creatine, and (2) that this effect could be additive to a possible inhibitory effect of ornithine or a deficiency of metabolic arginine seen in HHH syndrome. An earlier work by Dionisi Vici et al.[4], which was based on two HHH patients, had indicated that there may be an impaired synthesis of creatine, and that this may be caused by a feedback inhibition of the arginine-glycine transamidinase by an ornithine excess. However, a measurement of guanidinoacetic acid was not done in the above mentioned study, and these patients were not reported to be under sodium benzoate therapy. Ornithine inhibition of guanidinoacetic acid synthesis has earlier been demonstrated by Sipila[21] in an experimental rat model. The author postulated that the effect may be similar amongst patients with gyrate atrophy of the choroid and retina, caused by hyperornithinaemia as a result of the mutations in the ornithine-δ-ami-notransferase gene. Our results do not show evidence of a significant effect of either a metabolic arginine deficiency in HHH syndrome and/or glycine depletion due to sodium benzoate therapy on the synthesis of guanidinoacetic acid and creatine in the patients studied.

Materials and methods

Patients and their treatment protocol

Six patients, male and female of various ages, have been diagnosed at the Service de Génétique Médicale, Hôpital Sainte-Justine, Montreal on the basis of their plasma ornithine and ammonia levels. The diagnosis of HHH syndrome was confirmed by a measurement of ^{14}C-ornithine/^{3}H-leucine incorporation into protein in fibroblasts. The biochemical data at the time of diagnosis and the neurological, ophthalmological and psychological evaluation of the six patients have already been presented[10], and are detailed in a separate publication[11]. Table 1 summarizes the identification, age, biochemical data and treatment at the time of the present evaluation. All patients were in stable metabolic condition. For the purpose of the present evaluation and the need for proper comparisons to normal values of guanidino compounds in various age groups, the six patients were divided into two groups: children (5.1 to 9.6 years) and adults (17.5 to 23.5 years).

Table 1. Treatment and biochemical data of patients diagnosed for HHH syndrome

Group of patients	Case ID	Age at evaluation	Treatment at the time of evaluation			Biochemical data at the time of the evaluation		
		(years)	Sodium benzoate mg/kg/day	Dietary protein g/day	Dietary protein g/kg/day	Ornithine[a] AC	Ammonia[b] AC	Ammonia[b] 1H PC
Children	(BL)	5.1	210	15	0.88	246	48	53
	(MA)	6.5	275	18	0.95	267	36	37
	(CL)	9.6	205	32	1.28	423	24	26
Adults	(EP)	17.5	170	35	0.65	334	62	49
	(MP)	19.0	175	40	0.69	426	69	49
	(DG)	23.5	100	50	1.00	510	42	38

Normal values: [a]25–85 µmol/l; [b]17–54 µmol/l.

Sodium benzoate therapy was administered orally with the total intake divided into three doses. Plasma samples were obtained before breakfast, while 24 h urine samples were collected for 24 h following the blood sampling. All samples were immediately frozen at $-70\,°C$ until analysis. Plasma amino acid determinations were performed on a Beckman system 7300 (Beckman Instruments Co. Inc., Palo Alto, California), as recommended by the manufacturer. Blood ammonia was measured by the cation exchange resin technique of Hutchison & Labby[8].

Measurement of guanidino compounds

The concentrations of arginine and various guanidino compounds were determined using a Biotronic LC 6001 amino acid analyser adapted for guanidino compound determination. The guanidino compounds were separated over a cation exchange column using sodium citrate buffers and were detected with the fluorescence ninhydrin method as has been reported in detail earlier[16].

Serum urea nitrogen was determined with diacetylmonoxime, as described by Ceriotti[3]. The interference of the guanidino compounds on the urea determination was carefully controlled[14].

The results of individual guanidino compounds analysed from the plasma and urine samples of each patient were compared against the statistical range (mean $\pm$ 2 SD) values obtained from normal children and adults. The criteria for the selection of normal control groups have been described in an earlier publication[11].

Results and discussion

The results of the fasting plasma values of urea, guanidinosuccinic acid, guanidinoacetic acid, creatine, creatinine, arginine and homoarginine from the children and adults with HHH syndrome are presented in Tables 2 and 3 respectively. The same parameters in respect of the 24 h urinary excretion/g of creatinine can be seen in Tables 4 and 5, relating to the groups of three children and three adults respectively.

Table 2. Fasting serum guanidino compound levels in childgren diagnosed for HHH syndrome during sodium benzoate therapy

	BL	MA	CL	Normal values[c]
Guanidino compounds[a]				
Guanidinosuccinic acid	0.03*	0.06*	0.12*	0.13 – 0.48
Guanidinoacetic acid	1.09	1.36	0.90	0.56 – 1.88
Creatine	50.8	32.1	55.3	24.2 – 109.4
Creatinine	25.0	26.6	42.5	11.1 – 49.6
Arginine	72.3	107.0	162.0	43.0 – 150.3
Homoarginine	0.75	1.04	1.33	0.49 – 1.54
Urea[b]	1.83*	1.32*	3.97	1.92 – 7.11

[a]μmol/l; [b]mmol/l; [c]n = 74, range; *below the normal range.

Measurements of other guanidino compounds, e.g. α-keto-δ-guanidinovaleric acid, α-N-acetylarginine, argininic acid, β-guanidinopropionic acid, γ-guanidinobutyric acid, guanidino and methylguanidino, were also made in the serum and urinary samples from all the patients. However, the data are not presented, since these did not represent any abnormality in respect to the hypothesis being studied.

The levels of serum guanidinosuccinic acid are significantly lower than normal in all three children (Table 2) and in two out of three adult patients (Table 3). A similar situation is seen in Tables 4 and 5, where two out of three children and adults show a significantly decreased secretion of guanidinosuccinic acid in the 24 h urine. This confirms an earlier study[15], in which only two of these six patients (MP and CL) were available at the time, and were shown to have serum guanidinosuccinic acid values lower than normal. The HHH syndrome is thus similar to hyperargininaemia[13,15] and other urea cycle disorders[15] in showing a significantly lower synthesis of guanidinosuccinic acid. This compound is hypothesized to originate as an inert overflow product through the guanidine cycle[17]. The guanidine cycle is an alternate metabolic cycle proposed for the conversion of urea nitrogen to creatine. Serum urea values in the present study were lower than control in two out of three children (Table 2), but were not affected in the adult patients (Table 3).

Table 3. Fasting serum guanidino compound levels in adult patients diagnosed for HHH syndrome during sodium benzoate

	EP	MP	DG	Normal range[c]
Guanidino compounds[a]				
Guanidinosuccinic acid	0.05*	0.06*	0.12	0.07 – 0.45
Guanidinoacetic acid	1.64	1.92	0.89*	1.58 – 3.64
Creatine	26.3	24.5	9.77	5.4 – 54.7
Creatinine	50.0	50.0	58.3	45.5 – 116.2
Arginine	83.9	65.9	77.1	65.7 – 157.2
Homoarginine	1.12	1.25	1.23	0.71 – 3.24
Urea[b]	3.30	2.98	5.46	2.22 – 6.61

[a]μmol/litre; [b]mmol/litre; [c]n = 66, range; *below the normal range.

The guanidinoacetic acid and creatine levels in the serum and urine are variable. Only one adult patient shows a significant decrease in the serum level of guanidinoacetic acid (Table 3), while two children show a decreased excretion of urinary guanidinoacetic acid (Table 4). Serum values of creatine are within normal limits in all children or adults. However, one of the children (BL) shows a diminished excretion of urinary creatine (Table 4) which is consistent with the lower guanidinoacetic acid seen in the same child. The other child (CL) who had shown a decreased excretion of guanidinoacetic acid, has a rather high level of urinary creatine. Considering that metabolic arginine is an essential substrate for the activity of arginine-glycine transamidinase and that sodium benzoate therapy may cause a deficiency of glycine[9] and/or arginine synthesis through the urea cycle, we expected to see some correlation between the levels of guanidinoacetic acid/creatine and arginine. However, the results do not indicate any such correlation. Serum arginine values are normal in all the patients (Tables 2 and 3). Urinary arginine excretion is significantly decreased in only one of the children (CL) which (incidentally) also shows a diminished urinary guanidinoacetic acid (Table 4). Serum and urinary homoarginine levels do not show any abnormality, indicating a normal functioning of arginase.

Table 4. Guanidino compounds in 24 h urinary samples from children diagnosed for HHH syndrome during sodium benzoate therapy

Guanidino compounds[a]	BL	MA	CL	Normal values[b]
Guanidinosuccinic acid	9.6*	26.4	7.7*	25 – 100
Guanidinoacetic acid	54.7*	519	168*	250 – 850
Creatine	163*	5307	2426	400 – 3200
Arginine	30.3	71.2	7.7*	10 – 60
Homoarginine	< DL	39.7	< DL	< DL – 20

[a]μmol/g creatinine; [b]n = 30, range; *below the normal range; < DL = below the detection limit.

Our results, therefore, do not corroborate the study of Dionisi Vici et al.[4], in which two children suffering from HHH syndrome were shown to have a low urinary excretion of creatine, which could be normalized by arginine or citrulline administration. We also do not see any potentiating effect of sodium benzoate therapy. Unfortunately, due to ethical considerations, we could not stop the benzoate therapy to see the effect on the guanidino compounds with or without therapy. The metabolism of guanidinoacetic acid in the HHH syndrome may also not be similar to that seen in the gyrate atrophy patients, as has been speculated by Sipila[21] in an experimental study.

Table 5. Guanidino compounds in 24 h urinary samples from adult patients diagnosed for HHH syndrome during sodium benzoate therapy

Guanidino compounds[a]	EP	MP	DG	Normal range[b]
Guanidinosuccinic acid	8.9*	7.0*	11.3	9 – 55
Guanidinoacetic acid	56.6	59.2	69.1	15 – 500
Creatine	146.0	82.1	81.6	30 – 3250
Arginine	13.4	12.5	10.5	5 – 100
Homoarginine	< DL	< DL	< DL	< DL – 15

[a]μmol/g creatinine; [b]n = 30, range; *below the normal range.

Acknowledgements

This work was supported by the Medical Research Council of Canada (grant #MT-9124), the Born-Bunge Foundation, the United Fund of Belgium and NFWO grants N° 3.0044.92 and NDE 58. Technical assistance by Ms. Elaine Larouche and secretarial work by Mrs. Raffaela Ballarano is appreciated and acknowledged.

References

1. Brusilow, S., Tinker, J. & Batshaw, M.L. (1980): Amino acid acylation: a mechanism of nitrogen excretion in inborn errors of urea synthesis. *Science* **207**, 659–661.

2. Brusilow, S.V., Valle, D.L. & Batshaw, M.L. (1979): New pathways of nitrogen excretion in inborn errors of urea synthesis. *Lancet* **ii**, 452–454.

3. Ceriotti, G. (1971): Ultramicro determination of plasma urea by reaction with diacetylmonoxime-antipyrine without deproteinization. *Clin. Chem.* **17**, 400–402.

4. Dionisi Vici, C., Bachmann, C., Gambarara, M., Colombo, J.P. & Sabetta, G. (1987): Hyper-ornithinemia-hyperammonemia-homocitrullinuria syndrome: low creatine excretion and effect of citrulline, arginine or ornithine supplement. *Pediatr. Res.* **22**, 364–367.

5. Fell, V., Pollitt, R.J., Sampson, G.A. & Wright, T. (1974): Ornithinemia, hyperammonemia and homocitrullinuria. A disease associated with mental retardation and possibly caused by defective mitochondrial transport. *Am. J. Dis. Child* **127**, 752–756.

6. Grazi, E., Magri, E. & Balboni, G. (1975): On the control of arginine metabolism in chicken kidney and liver. *Eur. J. Biochem.* **60**, 431–436.

7. Hommes, F.A., Eller, A.G., Evans, B.A. & Carter, A.L. (1984): Reconstitution of ornithine transport in liposomes with lubrol extracts of mitochondria. *FEBS Lett.* **170**, 131–134.

8. Hutchison, J.H. & Labby, P.H. (1962): New method for the micro determination of blood ammonia by use of cation exchange resin. *J. Lab. Clin. Med.* **60**, 170–178.

9. Jackson, A.A., Badaloo, A.V., Forrester, T., Hibbert, J.M. & Persand, C. (1987): Urinary excretion of 5-oxyproline (pyroglutamic aciduria) as an index of glycine insufficiency in normal man. *Br. J. Nutr.* **58**, 207–214.

10. Lemay, J.F., Arbour, J.F., Dubé, J., Flessas, J., Laberge, M., Lafleur, L., Orquin, J., Qureshi, I.A., Vanasse, M., Déry, R., Mitchell, G.A. & Lambert, M.A. (1990): HHH syndrome: neurologic, ophthalmologic and psychological evaluation of 6 patients. Programme of the V International Congress of Inborn Errors of Metabolism. Asilomar, CA. Abstract, #OC1.1.

11. Lemay, J.F., Lambert, M.A., Mitchell, G.A., Vanasse, M., Valle, D., Arbour, J.F., Dubé, J., Flessas, J., Laberge, M., Lafleur, L., Orquin, J., Qureshi, I.A. & Déry, R.: HHH syndrome: neurologic, ophthalmologic and psychological evaluation of 6 patients. *J. Pediatr.* (in press).

12. Magri, E., Balboni, G. & Grazi, E. (1975): On the biosynthesis of creatine: intramitochondrial localisation of transamidinase from rat kidney. *FEBS Lett.* **55**, 91–93.

13. Marescau, B., De Deyn, P.P., Löwenthal, A., Qureshi, I.A., Antonozzi, I., Bachmann, C., Cederbaum, S.D., Cerone, R., Chamoles, N., Colombo, J.P., Hyland, K., Gatti, R., Kang, S.S., Letarte, J., Lambert, M., Mizutani, N., Possemiers, I., Rezvani, I., Snyderman, S.E., Terheggen, H.G. & Yoshino, M. (1990): Guanidino compound analysis as a complementary diagnostic parameter for hyperargininemia: follow-up of guanidino compound levels during therapy. *Pediatr. Res.* **27**, 297–303.

14. Marescau, B., De Deyn, P.P., Qureshi, I.A., De Broe, M.E., Antonozzi, I., Cederbaum, S.D., Cerone, B., Chamoles, N., Gratti, R., Kang, S.S., Lambert, M., Possemiers, I., Snyderman, S.E. & Yoshino, M. : The pathobiochemistry of uremia and hyperargininemia further demonstrates a metabolic relationship between urea and guanidinosuccinic acid. *Metabolism* (in press).

15. Marescau, B., Qureshi, I.A., De Deyn, P.P., Letarte, J., Chamoles, N., Yoshino, M., De Potter, W.P. & Löwenthal, A. (1989): Serum guanidinosuccinic level in urea cycle diseases. In *Guanidines 2*, eds. A. Mori, B.D. Cohen & H. Koide, pp. 245–249. New York: Plenum Press.

16. Marescau, B., Qureshi, I.A., De Deyn, P.P., Letarte, J., Ryba, R. & Löwenthal, A. (1985): Guanidino compounds in plasma, urine and cerebrospinal fluid hyperargininemic patients during therapy. *Clin. Chim. Acta* **146**, 21–27.

17. Natelson, S. & Sherwin, J.E. (1979): Proposed mechanism for urea nitrogen reutilization: relationship between urea and proposed guanidine cycles. *Clin. Chem.* **25**, 1343–1344.

18. Qureshi, I.A., Rouleau, T., Letarte, J. & Ouellet, R. (1986): Significance of transported glycine in the *in vivo* conjugation of benzoate in spf-mutant mice with ornithine transcarbamylase deficiency. *Biochem. Int.* **12**, 839–846.

19. Rudman, D., Galambos, J.T., Smith, R.B., Solam, A.A. & Warren, W.D. (1973): Comparison of the effect of various amino acids upon the blood ammonia concentration of patients with liver disease. *Am. J. Clin. Nutr.* **26**, 916–925.

20. Shih, V.E., Efron, M.L. & Moser, H.W. (1969): Hyperornithinemia, hyperammonemia and homocitrullinuria. A new disorder of amino acid metabolism associated with myoclonic seizures and mental retardation. *Am. J. Dis. Child.* **117**, 83–92.

21. Sipila, I. (1980): Inhibition of arginine-glycine amidinotransferase by ornithine. A possible mechanism for the muscular and chorio-retinal atrophies in gyrate atrophy of the choroid and retina with hyperornithinemia. *Biochem. Biophys. Acta* **613**, 79–84.

22. Valle, D. & Simell, O. (1989): The hyperornithinemias. In *The metabolic basis of inherited disease*, 6th edn., eds. C.R. Scriver, A.L. Beaudet, W.S. Sly & D.Valle, pp. 599–627. New York: McGraw Hill.

Guanidino Compounds in Biology and Medicine, eds. by P.P. De Deyn, B. Marescau, V. Stalon and I.A. Qureshi. ©1992 John Libbey & Company Ltd., pp. 379–383.

Chapter 56

Low guanidinoacetic acid and creatine concentrations in gyrate atrophy of the choroid and retina (GA)

Ilkka SIPILÄ[1], David VALLE[2] and Saul BRUSILOW[2]

[1]*Children's Hospital, University of Helsinki, SF-00290 Helsinki, Finland;* [2]*The Johns Hopkins Hospital, Department of Pediatrics, Baltimore, Maryland 21205, USA*

Summary

In order to elucidate the pathogenesis of gyrate atrophy of the choroid and retina (GA), we measured the concentrations of the guanidino compounds creatine (Crea) and guanidinoacetic acid (GAA) in 15 GA patients and 36 controls. Plasma GAA concentration and urinary excretion of GAA was clearly lower in the patients without any overlap with the controls. Creatine concentration in plasma and urinary excretion of creatine was lower in patients with a considerable overlap with the controls, but tissue concentration of creatine in erythrocytes and muscle was clearly lower in the patients without overlap with the controls. We believe that the synthesis of GAA is inhibited by the high ornithine concentrations in GA patients leading to deficient formation of creatine in tissues like skeletal muscle, retina and choroid. This inadequacy of creatine for energy storing in phosphocreatine may lead to lack of energy and further to tissue atrophy.

Introduction

Gyrate atrophy of the choroid and retina (GA) is a rare, slowly progressive chorioretinal degeneration, which is inherited as an autosomal recessive trait. Although the biochemical phenotype is a systemic accumulation of ornithine to levels 10- to 20-fold normal in all bodily fluids, the clinical phenotype is mainly limited to the eye[4,6,11]. Affected individuals develop myopia in early childhood, followed by constriction of visual fields and loss of night vision in adolescence with progressive tunnel vision eventually leading to blindness in the third to the sixth decade of life. Posterior subcapsular cataracts are noted in the second to third decade. Aside from a mild proximal muscle weakness noted in a minority of adult patients[7] there are no other consistent clinical manifestations. The primary biochemical defect is a near complete deficiency of ornithine amino-transferase (OAT), a mitochondrial matrix enzyme which catalyses the first step in ornithine catabolism. The OAT cDNA and structural gene have been cloned and sequenced and a variety of mutations causing GA have been delineated. Despite this progress in understanding the molecular basis for GA, our understanding of the pathophysiology of the retinal degeneration is incomplete.

One proposed mechanism for the pathophysiology of GA is that creatine biosynthesis is inhibited leading to deficiency of creatine and creatine phosphate[7]. Glycine transamidinase (GTA), the first enzyme in creatine biosynthesis, catalyses the transfer of the guanidino group of arginine to glycine forming GAA and ornithine[13]. GAA is then methylated to form creatine in a reaction catalysed by guanidinoacetate methyltransferase. Available data indicate that transamidinase is only expressed in the kidney and pancreas, and guanidinoacetate methyltransferase is expressed in the liver. On the other hand, 98 per cent of creatine is found in skeletal muscle and most of the remainder in the central nervous system. Thus creatine formed centrally is transferred to peripheral tissues via the plasma.

Previous *in vitro* studies by one of us (IS)[8] and others[5] have shown that the activity of GTA is competitively inhibited by ornithine at concentrations within the range of those observed in GA patients (K_i = 0.253 mM; plasma ornithine values in GA range from 450 to 1200 μM). Sipilä *et al.* suggested that plasma creatine and creatinine concentration may be low in GA patients[8]. They also showed that urinary guanidinoacetate excretion following an arginine load (1.1 mmol per kg, iv) was reduced in the GA-patients as compared to normals[9]. We now report that by use of a specific and sensitive high performance liquid chromatography (HPLC) method for assay of GAA and creatine in biological fluids we found a striking reduction in GAA and creatine in plasma, CSF, erythrocyte and muscle of GA-patients. Our results indicate that creatine biosynthesis is impaired in GA and that deficient levels of tissue creatine may play a role in the pathophysiology of GA.

Materials and methods

Patients and controls

Blood samples were obtained after an overnight fast from 15 GA patients. Simultaneous blood and CSF samples were drawn from adult patients undergoing neurological evaluation of headache for which no specific diagnosis was found. CSF was obtained from eight Finnish GA patients as a part of their initial metabolic evaluation. Muscle samples were obtained by biopsy of the left vastus lateralis muscle with a Tru-Cut® needle.

The diagnosis of GA in all patients was made by demonstration of hyperornithinaemia, typical ophthalmological findings, deficiency of fibroblast OAT activity and direct molecular analysis of the OAT gene. At the time of the guanidino compound measurements all patients were on a free diet without medication.

Samples

Plasma and CSF samples were deproteinized using micropartition vials (Amicon, USA). The deproteinized samples were acidified with HCl to a final concentration of 0.1 N. Packed erythrocytes were washed once with phosphate buffered saline at 4 °C, haemolysed by adding 10 volumes of distilled water, and thereafter treated as the plasma samples. Urine samples were collected at 4 °C and an aliquot was deproteinized by filtering in a micropartition vial and acidified with HCl to a final concentration of 0.1 N.

Analytical

Guanidino compounds (guanidinosuccinate, GAA, creatine, guanidinoethylsulphonate, and arginine) were separated by HPLC followed by postcolumn derivatization with an alkaline ninhydrin reaction and fluorometric detection. We used reversed phase with sulphosalicylic acid as the ion-pairing agent. After an isocratic run of 25 min, a gradient from 0 to 30 per cent methanol was developed in 5 min and maintained for 30 min to strip the column for subsequent run.

Results

We found good separation of guanidinosuccinic acid, GAA and creatine, the main guanidino compound of interest in the scope of this study. The detection limit for GAA was 0.1 μmol/l and for creatine 0.01 μmol/l with linear increase in fluorescence up to concentrations of 20 μM for GAA and 500 μM for Crea. In the urine samples of both patients and controls, we found several unidentified fluorescent compounds.

In Table 1 we show the concentrations of guanidino compounds in GA patients as compared to controls. GAA was measurable in the plasma and urine of GA patients with a mean value in plasma which was 20 per cent that of controls. The amount excreted was even more strikingly reduced with a mean value that was 5 per cent of the controls. These results which show a pronounced reduction of the concentration of the product of the GTA reaction are consistent with the *in vitro* observations of inhibition of this enzyme by ornithine. A peak of GAA was consistently observed but was below the limit of integration.

Table 1. Guanidino compounds in plasma, CSF fluid, erythrocytes, muscle and 24 h urine samples of GA patients and controls

Sample source	N	GAA mean μM	SD	Creatine mean μM	SD	Range
Plasma						
GA	15	0.40	0.17	6.3	3.3	1.6 – 14.9
Controls	36	1.99	0.68	35.9	*(26.9)	7.8 – 56.9
CSF-fluid						
GA	8			18.1	6.9	5.2 – 26.2
Controls	19			57.6	*(32.4)	22.2 – 163.2
Erythrocytes						
GA	4			188.3	36.7	
Controls	8			369.5	62.5	
Muscle						
GA	13			5.9	2.3	2.1 – 10.3
Controls	3			31.4	9.5	23.5 – 42.0
24-h urine			Range			
GA	8	16.9	0 – 65	96.9	*(128.8)	1 – 355
Controls	9	321.1	116 – 913	355.8	*(455.8)	51 – 1441

* Observations do not follow normal distribution.

Creatine was detectable in all samples and again was much lower in the GA patients as compared to the controls. The mean values in the plasma and urine of GA patients were 17 and 27 per cent of control, respectively. Tissue values for creatine concentrations were similarly reduced in GA patients ranging from 19 per cent (muscle) to 51 per cent (erythrocytes) of control values. The tissue values in GA patients never overlapped with the values in controls. These results are consistent with deficient tissue creatine content in GA. It is interesting that the creatine concentration levels in CSF of the patients are 2–3 times higher than the creatine levels found in serum of the studied patients. Unfortunately we do not have simultaneous plasma and CSF samples in any of the GA patients. All 19 controls for whom we have CSF data had plasma values obtained simultaneously. In these individuals, the mean ratio of creatine concentrations in these two compartments was 1.08 with a range of 0.39 to 2.59 with no significant correlation between plasma and CSF values.

In CSF we found occasionally a low peak probably representing guanidinoethylsulphonate in both the patients and controls. Only one subject, a GA patient, had a clear peak for guanidinoethylsulphonate, which was 9.3 μmol/l. There was no significant sex or age difference for the guanidino compounds in plasma, CSF, erythrocytes or muscle. The urinary excretion of all guanidino compounds were lower in the two children in the patients group.

Discussion

The metabolic pathway for the synthesis of creatine has been known for almost 50 years[13]. The guanidino group of arginine is transferred to glycine forming GAA in the first and rate-limiting reaction of the pathway catalysed by glycine transamidinase. GAA is then methylated forming creatine in a reaction catalysed by guanidinoacetate methyltransferase. Studies of creatinine excretion indicate that a 70 kg man requires approximately 2 g of creatine a day from dietary or endogenous sources[2].

Our results show a strikingly reduced level of creatine and, most importantly, its precursor GAA in a GA patient. These results strongly support the hypothesis that the high levels of ornithine characteristic of gyrate atrophy inhibit creatine biosynthesis at the level of glycine transamidinase. The levels of skeletal muscle, the tissue which contains approximately 98 per cent of the total body creatine pool, are reduced to 24 per cent of controls. These low muscle levels may account for the tubular aggregates, and type 2 fibre atrophy characteristic of most GA patients as well as the mild proximal weakness noted in some. In support of this latter idea, Sipilä *et al.* have shown that the histological abnormalities can be reversed by creatine supplementation (2 g/day for 1 year)[10].

The role of creatine in the CNS and especially in the retina is not known. However, the CNS is the major site of non-skeletal muscle creatine. Thus it is intriguing that our data on CSF creatine levels suggests that the reduction of creatine in the CNS of GA patients is similar to that of muscle. Recently it has been shown that mitochondrial creatine kinase can be used as a marker for the differentiation of photoreceptor cells in chicken retina[14]. The appearance of Mi-CK at the time of hatching in autophagous birds gives more evidence for the importance of the creatine–phosphocreatine shuttle in visual function.

Interestingly the plasma concentration of GAA was low in the patients but the creatine concentration overlapped with the controls. On the other hand, tissue creatine was clearly low in the patients. This may reflect the possible transportation of GAA to the tissues and local methylation, which may be more important for the retina than for the muscle. There is no evidence of the expression of guanidinoacetate methyltransferase in the retina, but various other methyltransferases could play a role in this methylation. In experimental therapy of GA patients with creatine, muscle atrophy was entirely cured while the chorioretinal degeneration continued. This may suggest that there is poor entry of creatine into the retina[12].

It is conceivable that the pathophysiology of GA involves depletion of creatine in the photoreceptor cell with resulting disruption of its energy metabolism and disturbance of its normal rhythmical disc shedding. Such an alteration could over time have adverse effects on the photoreceptor cells and the pigment epithelial cells which must phagocytose and degrade the shed outer segment discs[3,1]. Additional studies on the role of creatine in the mammalian retina and particularly the photoreceptor cell are warranted.

Acknowledgements

We thank Dr. Hannu Somer for the plasma and CSF samples. The study was supported by NIH.

References

1. Besharse, J.C. (1982): The daily light-dark cycle and rhythmic metabolism in the photoreceptor-pigment epithelial complex. In *Progress in retinal research*, eds. N.N. Osborne & G.J. Chader, pp. 81–124. Oxford: Pergamon Press.

2. Heymsfield, S.B. (1983): Measurement of muscle mass in humans: validity of the 24-hour urinary creatine method. *Am. J. Clin. Nutr.* **37**, 478–494.

3. Hollyfield, J.G., Besharse, J.C. & Rayborn, M.E. (1977): Turnover of rod photoreceptor outer segments I. Membrane addition and loss in relationship to temperature. *J. Cell Biol.* **75**, 490–506.

4. Jacobsohn, E. (1888): Ein Fall von Retinitis pigmentosa atypica. *Klin. Monatbl. Augenheilk.* **26**, 202–206.

5.   Ratner, S. & Rochovansky, O. (1956): Biosynthesis of guanidinoacetic acid. I. Purification and properties of transamidinase. *Arch. Biochem. Biophys.* **63**, 277–295.

6.   Simell, O. & Takki, K. (1973): Raised plasma ornithine and gyrate atrophy of the choroid and retina. *Lancet* **i,** 1031–1033.

7.   Sipilä, I., Simell, O., Rapola, J., Sainio, K. & Tuuteri, L. (1979): Gyrate atrophy of the choroid and retina: tubular aggregates and type 2 fiber atrophy in muscle. *Neurology* **29**, 996–1005.

8.   Sipilä, I. (1980): Inhibition of arginine-glycine aminotransferase by ornithine. A possible mechanism for the muscular and chorioretinal atrophies in gyrate atrophy of the choroid and retina with hyperornithinemia. *Biochim. Biophys. Acta.* **613**, 79–84.

9.   Sipilä, I., Simell, O. & Arjomaa, P. (1980): Gyrate atrophy of the choroid and retina with hyperornithinemia. Deficient formation of guanidinoacetic acid from arginine. *J. Clin. Invest.* **66**, 684–687.

10.   Sipilä, I., Rapola, J., Simell, O. & Vannas, A. (1981): Supplementary creatine as a treatment for gyrate atrophy of the choroid and retina. *N.E.J.M.* **304**, 867–870.

11.   Valle, D. & Simell, O. (1989): The hyperornithinemias. In *The metabolic basis of inherited disease*, 6th edn., ed. C.R. Scriver, A.L. Beaudet, W.S. Sly & D. Valle, pp. 599–627. New York: McGraw-Hill.

12.   Vannas-Sulonen, K., Sipilä, I., Vannas, A., Simell, O. & Rapola, J. (1985): A five-year follow-up of creatine supplementation. *Ophthalmology* **92**, 1719–1727.

13.   Walker, J.B. (1979): Creatine: biosynthesis, regulation, and function. In *Advances in enzymology*, ed. A. Meister, pp. 177–242. New York: John Wiley & Sons.

14.   Wegmann, G., Huber, R., Zanolla, E., Eppenberger, H.M. & Walliman, T. (1991): Differential expression and localization of brain-type and mitochondrial creatine kinase isoenzymes during development of the chicken retina: Mi-CK as a marker for differentiation of photoreceptor cells. *Differentiation* **46**, 77–87.

Section VII

Guanidino compounds in renal insufficiency

Guanidino Compounds in Biology and Medicine, eds. by P.P. De Deyn, B. Marescau, V. Stalon and I.A. Qureshi. ©1992 John Libbey & Company Ltd., pp. 387–393.

Chapter 57

Effect of guanidino compounds on membrane fluidity of rat synaptosomes

Midori HIRAMATSU, Sachiko OHBA[1], Rei EDAMATSU[1], Dai KADOWAKI[1] and Akitane MORI[1]

Yamagata Technopolis Foundation, Yamagata and [1]Department of Neurochemistry, Institute for Neurobiology, Okayama University Medical School, 2-5-1 Shikata-cho, Okayama 700, Japan

Summary

Methylguanidine, α-guanidinoglutaric acid, guanidinosuccinic acid, guanidine, homoarginine, guanidinopropionic acid and γ-guanidinobutyric acid, in descending order, decreased the synaptosomal membrane fluidity of rat cerebral cortex while guanidinoethane sulphonic acid increased the fluidity. Taurine decreased membrane fluidity and the mixture of taurine and guanidinoethane sulphonic acid resulted in an intermediate effect tending towards the control state. Pentylenetetrazol and bicucullin, glutamic acid and aspartic acid decreased the fluidity while IP3 increased it. Potassium chloride and EGTA also lowered the fluidity but magnesium chloride, calcium chloride and flunarizine had no effect. Hypoxanthine-xanthine oxidase system and the Fenton reagents also decreased the membrane fluidity. These results could suggest that the guanidino compounds-induced seizure mechanism might be related with the decreased membrane fluidity.

Introduction

Guanidinoethane sulphonic acid[20], guanidinoacetic acid[14], N-acetylarginine[26], γ-guanidinobutyric acid[15], methylguanidine[18], α-guanidinoglutaric acid[30] and homoarginine[32], which are detected in the mammalian brain, induced epileptic discharges and/or caused convulsions after intracisteral injection or local administration into rodents. Levels of creatinine and guanidinoacetic acid in the cerebrospinal fluid of rabbits were increased at pre-convulsive stage induced by pentyle-netetrazol[10]. N-Acetylarginine, homoarginine and guanidine levels were elevated in the brain at pre-convulsive running phase induced by ultra sounds in audiogenic rats[31]. α-Guanidinoglutaric acid level was dramatically raised in the cobalt-induced epileptic foci of cats[21]. Though the induction of seizures by guanidino compounds has been suggested to be related to neurotransmitters[23,30] or reactive oxygen species[7], the mechanisms are still unclear.

There have been reports suggesting that the abnormalities in epilepsy are related to membrane fluidity. A decrease in the fluidity of epileptogenic freeze-lesioned synaptic membranes was found in the cat[25]. Anticonvulsant drugs, including diazepam[19], valproic acid[28] and phenobarbital[3], increased the fluidity of the synaptosomal membrane in the hippocampus and whole brain. Neuronal membrane may be related to neurotransmission[17], lipid peroxidation[2] or a high cholesterol level[17]. In the present study, the effect of guanidino compounds on membrane fluidity in rat cortical synaptosomes was further examined to clarify the seizure mechanism induced by these compounds.

Materials and methods

Chemicals

16-Doxylstearic acid (16-DS), pentylenetetrazol and bicucullin were purchased from SIGMA Chemical Co. (St Louis, MO). Xanthine oxidase (cow milk) was sourced from Boehringer Mannheim Ingelheim (Germany). All other chemicals and reagents were of the highest grade available from commercial suppliers.

Animals

Male Sprague-Dawley rats of 7 weeks old were obtained from Charles River Japan Inc. (Japan). Animals were decapitated and the cerebral cortex was rapidly removed on an ice plate and used for the preparation of synaptosomal fraction.

Preparation of crude synaptosomal fraction

Tissues were weighed and added to 20 volumes of 0.32 M sucrose buffer solution containing 1 mM EDTA and 10 mM Tris-HCl (pH 7.4). The samples were homogenized and centrifuged at $1300 \times g$ for 3 min. The supernatant was obtained and centrifuged at $17,000 \times g$ for 10 min. The pellet containing the synaptosomal fraction was washed with 0.32 M sucrose buffer and recentrifuged at $17,000 \times g$ for 10 min. The final pellet was suspended in 0.32 M sucrose buffer and stored at $-80\,°C$. The fraction consisted of about 80 per cent synaptosomes as determined by electron microscopy.

Membrane fluidity analysis

16-DS was dissolved in ethanol. Eight μg of label per mg protein was dried with nitrogen. Preliminary experiments showed that 8 μg label/mg protein was suitable. One hundred μl of synaptosomal fraction was added and mixed well for 2 min and the samples were then put into flat cells (JEOL Ltd.). Each parameter of the spin labels was assayed using an electron spin resonance (ESR) spectrometer (JEOL JES- FE1×G). Conditions of ESR spectrometry were as follows:

magnetic field, 327 ± 5 mT and ± 10 mT; modulation, 0.2 mT; response, 0.3 s; amplitude, 3.2×10^3; sweep time, 2 and 4 min; and temperature, 37 °C. The motion parameter, corresponding to the rotational correlation time in hydrophobic areas, was used for the 16-DS labelled samples as described by Eletr and Inesi[6].

Protein assay

Protein assay was carried out using a protein assay reagent (Pierce, Rockford, IL).

Statistical analysis

Statistical analysis was performed using the Student's t-test.

Results

The concentration of guanidino compounds used for the study of membrane fluidity was in the range of 0.025 to 25 mM. Methylguanidine, guanidinopropionic acid, guanidinosuccinic acid, α-guanidinoglutaric acid, guanidine, homoarginine and δ-guanidinobutyric acid increased the order parameter, that is, decreased membrane fluidity in the core of synaptosomal membranes in the range of 0.025 to 25 mM (Figs. 1 and 2). Guanidinopropionic acid decreased the membrane fluidity at the concentration of 0.025 and 0.25 mM but had no effect at 2.5 and 25 mM (Fig. 1). On the other hand, guanidinoethane sulphonic acid in the range of 0.025 to 2.5 mM increased the membrane fluidity while taurine, which is an antagonist of guanidinoethane sulphonic acid, decreased the fluidity in the range of 0.025 to 25 mM. The combined application of taurine and guanidinoethane sulphonic acid kept the level of membrane fluidity somewhat between that of taurine and guanidinoethane sulphonic acid alone (Fig. 3).

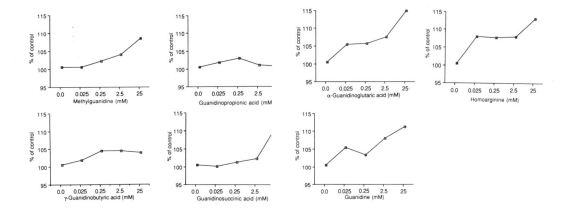

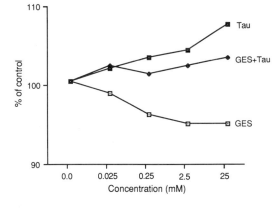

Fig. 1. (Top left) Effect of guanidino compounds on the motion parameter of rat cerebral cortical synaptosomes–1.

Fig. 2. (Top right) Effect of guanidino compounds on the motion parameter of rat cerebral cortical synaptosomes–2.

Fig. 3. (Bottom left) Effect of guanidinoethane sulphonic acid (GES) and taurine (TAU) on the motion parameter.

Convulsants, pentylenetetrazol and bicucullin, linearly decreased the membrane fluidity between 0.0025 and 2.5 mM. Excitatory amino acid neurotransmitters, glutamic acid and aspartic acid, also decreased the fluidity between 0.0025 and 2.5 mM. The second messenger, inositol-1,4,5-triphosphate (IP_3), increased the fluidity at 0.025, 0.25 and 2.5 mM (Fig. 4).

Potassium chloride decreased the membrane fluidity in the range of 0.0125 to 12.5 mM. At 2.5 mM each, both magnesium chloride and calcium chloride showed a substantial decreasing effect on membrane fluidity. Flunarizine showed no effect on membrane fluidity at 0.025, 0.25 and 2.5 mM, but it dramatically increased the fluidity at 25 mM. EGTA decreased the fluidity steadily from

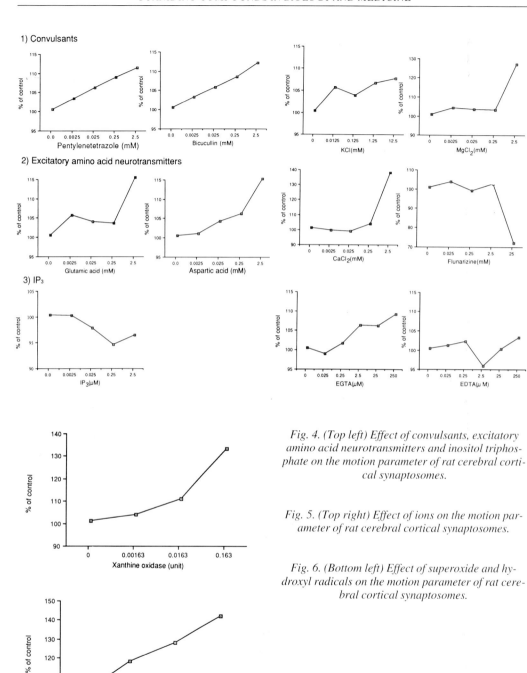

Fig. 4. (Top left) Effect of convulsants, excitatory amino acid neurotransmitters and inositol triphosphate on the motion parameter of rat cerebral cortical synaptosomes.

Fig. 5. (Top right) Effect of ions on the motion parameter of rat cerebral cortical synaptosomes.

Fig. 6. (Bottom left) Effect of superoxide and hydroxyl radicals on the motion parameter of rat cerebral cortical synaptosomes.

0.25–250 µM while at the concentration of 2.5 µM, EDTA increased the fluidity (Fig. 5). The system of hypoxanthine and xanthine oxidase and the Fenton reagents, which generate superoxide and hydroxyl radicals respectively, both decreased the membrane fluidity dose-dependently (Fig. 6).

Discussion

Peroxidation induces biological membrane rigidity[1,4,5,29]. Superoxide and hydroxyl radicals generated by the system of hypoxanthine and xanthine oxidase and the Fenton reagents were detected using ESR spectrometry with dimethyl-1-pyrroline-1-oxide (DMPO) spin trap and the reactivity of guanidino compounds with hydroxyl radicals was reported[11]. The addition of hypoxanthine and xanthine oxidase system or the Fenton reagents to rat cerebral cortical synaptosomes decreased the membrane fluidity and the results ascertained the above phenomena, that is, peroxidation induces biological membrane rigidity.

It was found previously that each solution of α-guanidinoglutaric acid, homoarginine, guanidine, arginine, methylguanidine, guanidinosuccinic acid, guanidinoethane sulphonic acid and guanidinopropionic acid was detected to generate carbon-centred radicals and hydroxyl radicals using ESR spin trapping technique (unpublished data). Levels of methylguanidine and guanidinoacetic acid, N-acetylarginine and arginine were increased with the Fenton reagents in the water solution[11]. Methylguanidine was synthesized from creatinine through creatol catalysed by hydroxyl radicals[24]. Methylguanidine and guanidinoacetic acid were also elevated in the brain after iron solution injection into the rat sensory motor cortex and these phenomena are suggested to be due to the oxygen and other free radicals generated in the cerebral cortex after the iron injection[7]. From these reports and the results of our study, it is implied that the decreased membrane fluidity after the addition of guanidino compounds may be due to the generated reactive oxygen species.

Aside from guanidino compounds used in the study which have been known to decrease membrane fluidity in rat cortical synaptosomes to induce convulsions or seizures, it is interesting to note that pentylenetetrazol and bicucullin likewise decreased the synaptosomal membrane fluidity. In addition, glutamic acid and aspartic acid, which are excitatory neurotransmitters, induce convulsions[8] and are regarded to play a role in seizure mechanism[9,16,27], decreased the membrane fluidity. IP$_3$ is an intracellular second messenger which mobilizes intracellular calcium stores in response to cell stimulation by neurotransmitters[22]. Increased membrane fluidity by IP$_3$ suggests an increase in the calcium ion permeability in the membranes. Chloride, calcium or magnesium ions did not seem to affect the synaptosomal membrane fluidity, while potassium decreased the membrane fluidity and this result may relate to seizure induction. However, EGTA, which is a chelating agent of calcium ion, decreased the membrane fluidity. This fact, as well as the increase of membrane fluidity by 2.5 mM EDTA are hard to explain.

Guanidinoethane sulphonic acid and taurine compete against each other. Taurine antagonizes the guanidinoethane sulphonic acid-induced epileptic-like discharges[20]; whereas the brain taurine level is decreased by the addition of guanidinoethane sulphonic acid[12,13]. In the present study, our results showed increased membrane fluidity in rat synaptosomes by guanidinoethane sulphonic acid and decreased fluidity by taurine. The reason for these phenomena is not known and it may be due to the structural formula of guanidinoethane sulphonic acid which includes sulphur.

Anticonvulsants, phenobarbital, valproate and diazepam have been reported to increase the brain membrane fluidity[3,19,28]. Considering these phenomena with our results, it is suggested that guanidino compound-induced seizure mechanisms are involved in decreased membrane fluidity and reactive oxygen species in brain synaptosomes.

Acknowledgements

This study was partly supported by Special Coordination Funds of the Science and Technology Agency of the Japanese Government.

References

1. Barrow, D.A. & Lentz, B. (1981): A model for the effect of lipid oxidation on diphenylhexatriene fluorescence in phospholipid vesicles. *Biochim. Biophys. Acta* **645**, 17–23.

2. Bruch, R.C. & Thayer, W.S. (1983): Differential effect of lipid peroxidation on membrane fluidity as determined by electron spin resonance probes. *Biochim. Biophys. Acta* **733**, 216–222.

3. Deliconstantinos, G. (1983): Phenobarbital modulates the (Na^+, K^+)-stimulated ATPase and Ca^{2+}-stimulated ATPase activities by increasing the bilayer fluidity of dog brain synaptosomal plasma membranes. *Neurochem. Res.* **8**, 1143–1152.

4. Dobretsov, T.A., Borschevska, T.A., Petrov, V.A. & Vladimirov, Yu. A. (1977): The increase of phospholipid bilayer rigidity after lipid peroxidation. *FEBS Lett.* **84**, 125–128.

5. Eichenberger, K., Böhni, P., Winterhalter, K.H., Kawato, S. & Richter, C. (1982): Microsomal lipid peroxidation causes an increase in the order of the membrane lipid domain. *FEBS Lett.* **142**, 59–62.

6. Eletr, S. & Inesi, G. (1972): Phase changes in the lipid moieties of sarcoplasmic reticulum membranes induced by temperature and protein conformational changes. *Biochim. Biophys. Acta* **290**, 178–185.

7. Fukushima, M. (1987): Guanidino compounds in iron-induced epileptogenic foci of rats. *Okayama-Igakkai-Zasshi* **99**, 787–806.

8. Hayashi, T. (1954): Effects of sodium glutamate on the nervous system. *J. Med.* **3**, 183–192.

9. Higashihara, Y., Hiramatsu, M. & Mori, A. (1990): Aspartic acid release from cerebral cortical slices of El mice with high seizure susceptibility. *Res. Commun. Chem. Pathol.* **70**, 155–171.

10. Hiramatsu, M., Edamatsu, R., Fujikawa, N., Shirasu, A., Yamamoto, N.M., Suzuki, A.S. & Mori, A. (1988): Measurement during convulsions of guanidino compound levels in cerebrospinal fluid collected with a catheter inserted into the cisterna magna of rabbits. *Brain Res.* **455**, 38–42.

11. Hiramatsu, M., Edamatsu, R., Kohno, M. & Mori, A. (1989): The reactivity of guanidino compunds with hydroxyl radicals. In *Guanidines 2*, eds. A. Mori, B. Cohen, & H. Koide, pp. 97–105. New York: Plenum Press.

12. Hiramatsu, M., Niiya-Nishihara & Mori, A. (1982): Effect of taurocyamine on taurine and other amino acids in the brain, liver and muscle of mice. *Neurosciences* **8**, 289–294.

13. Huxtable, R.J. & Lippincott, S.E. (1981): Comparative metabolism and taurine-depleting effects of guanidinoethanesulfonate in cats, mice and guinea pigs. *Arch. Biochem. Biophysics* **210**, 698–709.

14. Jinnai, D., Mori, A., Mukawa, J., Ohkusu, H., Hosotani, M., Mizuno, A. & Tye, L.C. (1969): Biological and physical studies on guanidino compounds induced convulsion. *Jpn. J. Brain Physiol.* **106**, 3668–3673.

15. Jinnai, D., Sawai, A. & Mori, A. (1966): γ-Guanidinobutyric acid as a convulsive substance. *Nature* **212**, 617.

16. Koyama, I. (1972): Amino acid in the cobalt-induced epileptogenic and non-epileptogenic cat's cortex. *Can. J. Physiol.* **50**, 740–752.

17. Maguire, P.A. & Druse, M.J. (1989): The influence of cholesterol on synaptic fluidity, dopamine D_1 binding and dopamine-stimulated adenylate cyclase. *Brain Res. Bull.* **23**, 69–74.

18. Matsumoto, M., Kobayashi, K., Kishikawa, H. & Mori, A. (1976): Convulsive activity of methylguanidine in cats and rabbits. *IRCS Med. Sci.* **4**, 65.

19. Menini, T., Ceci, A., Caccia, S., Garattini, S., Masturzo, P. & Salmona, M. (1984): Diazepam increases membrane fluidity of rat hippocampus synaptosomes. *FEBS Lett.* **173**, 255–258.

20. Mizuno, A., Mukawa, J., Kobayashi, K. & Mori, A. (1975): Convulsive activity of taurocyamine in cats and rabbits. *IRCS* **3**, 385.

21. Mori, A., Akagi, M., Katayama, Y. & Watanabe, Y. (1980): α-Guanidinoglutaric acid in cobalt-induced epileptogenic cerebral cortex of cats. *J. Neurochem.* **35**, 603–605.

22. Nahorski, S.R., Kendall, D.A. & Batty, I. (1986): Receptors and phosphoinositide metabolism in the central nervous system. *Biochem. Pharm.* **35**, 2447–2453.

23. Nakae, I. (1991): Synthesis of *N,N'*-dibenzoylguanidine and its convulsive action . *Neurosciences* **17**, 205–217.

24. Nakamura, K., Ienaga, K., Yokozawa, T., Fujitsukla, N. & Oura, H. (1991): Production of methylguanidine from creatinine via creatol by active oxygen species: analyses of the catabolism *in vitro*. *Nephron* **58**, 42–46.

25. Nelson, L. & Delgado-Escueta, A.V. (1986): Fluidity of normal and epileptogenic freeze-lesioned synaptic membranes in the cat. *Brain Res.* **386**, 32–40.

26. Ohkusu, H. (1970): Isolation of *N*-acetyl-L-arginine from calf brain and convulsive seizure induced by this substance. *Osaka-Igakkai- Zasshi* **21**, 49–50.

27. Osaki, Y., Hirarnatsu, M. & Mori, A. (1990): Uptake and K⁺-stimulated release of glutamate from cerebral cortical and hippocampal slices in El mice, and dose effect of glutamate. *Neurosciences* **16**, 561–567.

28. Perlman, B.J. & Goldstein, D.B. (1984): Membrane-disordering potency and anticonvulsant action of valproic acid and other short-chain fatty acids. *Mol. Pharmacol.* **26**, 83–89.

29. Rice-Evans, C. & Hochstein, P. (1981): Alterations in erythrocyte membrane fluidity by phenylhydrazine-induced peroxidation of lipids. *Biochem. Biophys. Res. Commun.* **100**, 1537–1542.

30. Shiraga, H., Hiramatsu, M. & Mori, A. (1986): Convulsive activity of α-guanidinoglutaric acid and the possible involvement of 5-hydroxytryptamine in the α-guanidinoglutaric acid induced seizure mechanism. *J. Neurochem.* **47**, 1832–1836.

31. Wiechert, P., Marescau, B., De Deyn, P. & Lowenthal, A. (1987): Guanidino compounds in serum and brain of audiogenic sensitive rats during the preconvulsive running phase of cerebral seizures. *Neurosciences* **13**, 35–39.

32. Yokoi, I., Toma, J. & Mori, A. (1984): The effect of homoarginine on the EEG of rats. *Neurochem. Pathol.* **2**, 295–300.

Guanidino Compounds in Biology and Medicine, eds. by P.P. De Deyn, B. Marescau, V. Stalon and I.A. Qureshi. ©1992 John Libbey & Company Ltd., pp. 395–402.

Chapter 58

Guanidinosuccinic acid affects excitatory postsynaptic response in rat hippocampal slices

R. D'HOOGE[3], J. MANIL[1], F. COLIN[2], M. DILTOER[1], G. NAGELS[1], M. VERVAECK[1] and P.P. DE DEYN[3]

[1]*Laboratory of Physiology and Pathophysiology at Free University of Brussels (VUB);*
[2]*Laboratory of Electrophysiology at Free University of Brussels (ULB); and*
[3]*Laboratory of Neurochemistry, Born-Bunge Foundation, University of Antwerp (UIA), Belgium*

Summary

Since guanidinosuccinic acid (GSA) is suggested to play a role in uraemic encephalopathy, we set out to investigate the effects of GSA on excitatory synaptic neurotransmission and on potentiation of this transmission in rat hippocampal slices. In addition, we investigated the influence of the allosteric *N*-methyl-D-aspartate (NMDA) receptor agonist D-serine on the GSA-induced effects. A high-frequency stimulation paradigm was employed. Every 4.1 s, a bipolar nickel-chrome electrode stimulated the Schaffer collaterals in 300 μm thick parasagittal hippocampal slices, with a maximal intensity of 2 mA. Glass micropipettes recorded the resulting field potentials in CA1 region of the slices. Multiple tetanic bursts (20 pulses at 100 Hz) were delivered at fixed intervals. Under the influence of 1000, 100, and 50 μM GSA in the perfusion fluid, excitatory postsynaptic potentials surface remained the same or decreased in all slices tested, in contrast to the usual increase observed under standard conditions; 25 or 10 μM GSA did not produce such effects. The inhibitory effect of GSA (100 μM) on potentiation of synaptic efficacy was completely antagonized by the coapplication of 100 μM D-serine. Application of 100 μM D-serine alone did not result in any significant distortion of the standard pattern. Although our results could indicate an involvement of NMDA receptors, the exact mechanism of GSA's interference with excitatory synaptic transmission in hippocampus remains to be investigated further.

Introduction

Protein and amino acid metabolism results in the production of nitrogen compounds such as guanidines which depend upon kidney function for their excretion and which accumulate in the body of uraemic patients. Renal failure is typically associated with brain dysfunction[7,8,17,18], and guanidino compounds have been suggested to play a role in this uraemia-associated encephalopathy[3,4,7]. Indeed, De Deyn *et al.*[4] reported that guanidinosuccinic acid (GSA), guanidine, methylguanidine, creatinine and argininic acid are increased in serum and in cerebrospinal fluid of uraemic patients with neurological symptomatology. Impairment in memory storage and retrieval is one of the most striking deficits in cognitive functioning in uraemia[8,17,18], and Ginn *et al.*[7] found short-term memory performance to be significantly correlated with serum creatinine levels in uraemic patients.

Therefore, since cognitive functioning is apparently impaired in renal failure patients, we set out to investigate the effect of GSA on the evoked excitatory postsynaptic response at the Schaffer collateral-pyramidal cell synapse in the CA1 region of rat hippocampus. We explored the possibility

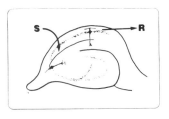

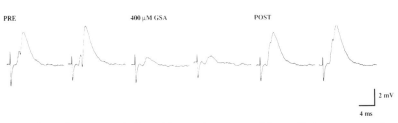

Fig. 1. Depiction of a rat hippocampal slice with an enlargement of the studied synapse (top figure). The stimulation (S) was applied to the Schaffer collaterals, the resulting field potentials were recorded (R) on CA1 region pyramidal cells. The effect of GSA on this excitatory postsynaptic response was studied (bottom graph, an example). A concentration of e.g. 400 μM GSA induces a dramatic decrease in the amplitude and the surface of the evoked field potential. Six consecutive averaged responses are shown: two before injection (PRE), two during (400 μM GSA), and two after injection (POST).

to use this neural system as a model system in the study of uraemic encephalopathy since it is widely accepted that the hippocampal neural circuit plays an important role in cognitive functioning (e.g. ref.[16]). Also, we tried to infer the possible synaptic mechanism of the GSA-induced effect. A preliminary observation on the effect of GSA upon CA1 synaptic transmission has already been reported by D'Hooge *et al.*[5].

Materials and methods

Adult albino rats were decapitated under ether anaesthesia and under hypothermia. Transverse slices (300 μm thick) were cut from the hippocampus and maintained for at least 30 min at 37 °C in a standard artificial cerebrospinal fluid (ACSF) solution containing (in mM): NaCl 122, KCl 2.5, $CaCl_2$ 2.65, $MgSO_4$ 3.38, $NaHCO_3$ 18 and glucose 20, which was continuously gassed with 95 per cent O_2/5 per cent CO_2 at 35 °C. Single slices were placed at the gas–liquid interface in a modified Oslo chamber. The chamber was continuously perfused with ACSF solution at a rate of 1.5 m/s, a temperature of 37 °C and at a pressure maintained at 40 mmHg under the ambient atmospheric pressure in order to avoid the formation of bubbles under the slice. GSA and D-serine were dissolved in ACSF solution and were very slowly injected at a controlled rate in the perfusing flow. GSA was purchased from Sigma Chemical Company (St Louis, MO, USA), all other chemicals used were obtained from Merck (Dalmstadt, Germany) and were of analytical grade.

Field potentials, population excitatory postsynaptic potentials (EPSPs) and population spikes (PSs), were recorded from the CA1 region with micropipettes filled with 3 M NaCl (see also Fig. 1, top figure) and amplified by a conventional AC coupled high impedance amplifier (band pass from 1 Hz to 10 kHz). Electrical current pulses, generated by a biphasic isolated constant current stimulator, were applied to the Schaffer collaterals, every 5 s, with a bipolar isolated 30 μm nickel-chrome wire at an intensity level subliminal to the PS generation. A personal computer (PC) controlled the whole experimental process. Four consecutive responses were averaged on line and stored on hard disk as averaged response (AR) of 76.8 ms duration. Each experiment began with the recording of four baseline ARs. Afterwards, a variable number of runs were registered. Each run started with the

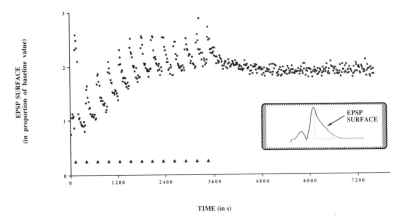

TIME (in s)

Fig. 2. A control experiment demonstrating potentiation of the CA1 excitatory postsynaptic response in the high-frequency stimulation paradigm. EPSP surface was measured between time-limits determined after visual inspection of the whole record: the first limit was chosen after the PS and the second one at the end of the EPSP (see insert). Post-tetanic potentiation (PTP) appeared as a sudden but transient increase in the late EPSP surface following each tetanus (indicated by black triangles at the bottom of the figure), and a subsequent decrease with time constant of about 10 s. Long-term potentiation (LTP) appeared as a steady rise of the baseline surface which was sustained during many hours after cessation of tetanic stimulation. Abscissa depicts the time in s, ordinate depicts the relative EPSP surface in proportions of the value during the four baseline averaged responses.

recording of two ARs followed by a short conditioning tetanus (20 pulses at 100 Hz). The conditioning tetanus was followed by 14 control ARs. Thus, one single run lasted 320 s (4 × 5 × (2 + 14) s). Usually, three to four runs were recorded before drug administration at a time when long-term potentiation (LTP) of the responses was clearly initiated but well before the maximal effect was obtained. Special analysis software was developed for the measurement of EPSP surface. The EPSP tail was preferred to the initial slope, which is often contaminated by the end of the stimulus artefact or by early PS components (see also Fig. 2, insert). This surface was measured between time-limits determined after visual inspection of the whole record. The first limit was chosen after the PS and the second one at the end of the EPSP. Zero line was calculated by averaging either the very first points before the stimulus or the very last points of the averaged response, choosing whatever reference appeared to be the most reliable. Relative EPSP surface was expressed in proportions of the four baseline ARs.

Results

Concentrations of 25, 50, 100, 400 and 1000 μM GSA were applied to hippocampal slice during field potential recording in CA1 region following high-frequency Schaffer collateral stimulation. Concentrations of 50 μM and more were able to depress the CA1 field potential (see Fig. 1, for an example). EPSP surface and amplitude, and concomitant PS were affected by GSA bath application.

In most slices submitted to our stimulation paradigm (n > 100) tetanic stimulation resulted in two distinct forms of potentiation of synaptic efficacy (see Fig. 2). Post-tetanic potentiation (PTP) appeared as a sudden but transient increase in the late EPSP surface and a subsequent decrease with time constant of about 10 s. LTP appeared as a steady rise of the baseline surface which was sustained during many hours as demonstrated in Fig. 2. Usually, this rise approached a saturation level near the end of the experiment. An interruption of the EPSP surface increase or a reversal of the slope were taken as evidence of inhibition.

The effect of GSA upon CA1-evoked responses appeared to be concentration-dependent (see Fig. 3). Whereas 100 μM inhibited only the PTP peaks and the rise in baseline surface, 1000 μM abolished

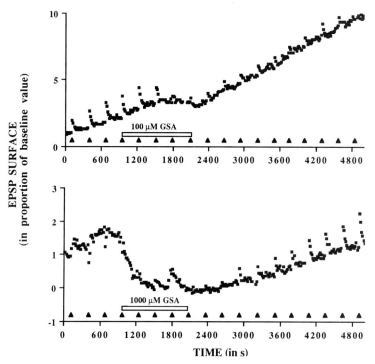

Fig. 3. Qualitative difference in the effect of 100 and 1000 μM GSA (two typical experiments). Whereas 100 μM inhibited only the PTP peaks and the rise in baseline surface, 1000 μM abolished the evoked response altogether. GSA application is indicated by a white rectangle at the bottom of the figure. This figure is constructed like Fig. 2.

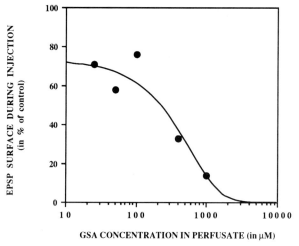

GSA CONCENTRATION IN PERFUSATE (in μM)

Fig. 4. Concentration-dependent decrease of CA1 postsynaptic response following GSA bath application. Between 25 and 1000 μM GSA the EPSP surface during injection, in per cent of the relative value of control experiments (ordinate), decreased according to the GSA concentration in the perfusate during the injection, in μM GSA (abscissa). This figure is derived from figures like Fig. 3; each black dot represents the mean relative EPSP surface of the last responses of the four runs during GSA injection in three to five experiments.

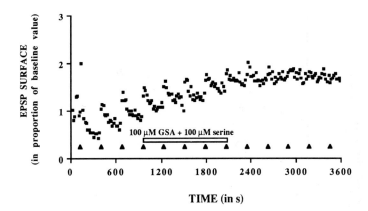

*Fig. 5. Effect of GSA and D-serine coapplication. Coapplication of 100 µM GSA and 100 µM D-serine is indi-
cated by a white rectangle at the bottom of the figure. This equimolar application of GSA and D-serine had
no apparent effect upon EPSP surface (of three experiments one is shown). This figure is constructed like
Fig. 2.*

the evoked response altogether. In some slices, the dramatic effect at the highest concentrations was
preceded by a short excitatory phase (multiple PSs). The effect of GSA was reversible, even at the
highest concentrations used. In all five experiments, LTP development was markedly inhibited during
injection of 100 µM GSA in the perfusion fluid, and PTP was also influenced by the injection. After
the injection, during the wash-out period, PTP and LTP resumed.

The relative EPSP surface values at the end of the four injection runs were compared to the values
obtained in control experiments (Fig. 4). In the concentration range of 25 to 1000 µM GSA, a
concentration-dependent decrease in relative EPSP surface was observed during GSA injection.
During GSA injection, EPSP surface decreased on the average to 71 per cent of the relative value
observed in control experiments in the case of 25 µM (due to a decrease in rising slope) and to 14 per
cent in the case of 1000 µM (due to inhibition of the response beneath baseline value, even complete
inhibition of the response).

Injection of 100 µM D-serine in the perfusion fluid had no apparent effect upon the CA1-evoked
responses (data not shown, n = 3). Coapplication of 100 µM D-serine together with 100 µM GSA
abolished the inhibitory effect of GSA upon LTP development in our high-frequency stimulation
paradigm (see Fig. 5). However, when higher concentrations of GSA were used, the coapplication of
100 µM D-serine did not inhibit the GSA effect (data not shown).

Discussion

The effect of GSA upon the evoked excitatory postsynaptic response in CA1 region of rat hippocampus was investigated in this study. It was found that this postsynaptic response in CA1 region is
mediated by α-amino-3-hydroxy-5-methyl-4-isoxazole proprionic acid (AMPA) receptors, whereas
the induction of the increase in the size of the EPSP is dependent upon *N*-methyl-D-aspartate (NMDA)
receptor activation (see ref. [2]).

It was shown that GSA affects the AMPA receptor-mediated response as well as the NMDA
receptor-mediated LTP induction. Apparently, the EPSP surface, and thus the AMPA receptor
function, was decreased by GSA in a concentration-dependent fashion following bath application of
this guanidino compound. However, the mechanism of action of this inhibitory effect of GSA remains
conjectural, and specific receptor or ion channel blockade, enzyme inhibition as well as specific
actions (*eg* on cell membrane fluidity) could be proposed. D'Hooge *et al.*[5] proposed the blocking of

excitatory amino acid receptor-associated ion channels as a possible mode of action of GSA at the receptor level. Receptor binding studies should be able to provide some answers to these questions (see *eg* the contribution of Reynolds and Rothermund, Chapter 64).

The lower concentrations of GSA that were applied in this study could have exerted their effect at the NMDA receptor exclusively since baseline EPSP surface was not (or was only slightly) affected at these concentrations, only the increase in EPSP size. The inhibition of this effect of GSA by D-serine suggests at least two possible sites of action of GSA at the NMDA receptor: at the NMDA receptor-associated ion channel or at the NMDA receptor glycine binding site.

A strychnine-insensitive high-affinity glycine binding site, sensitive to D-serine and D > L-alanine as well, is thought to be combined with the NMDA receptor and to bring about an allosteric activation of the receptor[11,19]. Binding of D-serine to the NMDA receptor glycine binding site could 'unplug' the receptor channel since it is suggested that this binding site is responsible for the modulation of the opening of the NMDA receptor-associated ion channel[14]. The integrity of this ion channel is thought to play an essential role in the enhancement of synaptic efficacy in the hippocampus[2]. Interestingly, as the morphology of the NMDA receptor has often been compared with that of the GABA-benzodiazepine receptor complex (*eg*[6,19]), it could be hypothesized that GSA blocks the NMDA receptor- associated ion channel in a way similar to the blocking of the GABA receptor ion channel postulated by De Deyn & Macdonald[3].

On the other hand, the nullification of the inhibitory effect of GSA by D-serine could also indicate that GSA has some affinity for the NMDA receptor glycine binding site itself. Johnson & Ascher[10] discovered that glycine potentiates NMDA responses of mouse neurons in cell culture. The activation of NMDA receptors expressed in *Xenopus* oocytes also required glycine[13] and in slices of adult rat neocortex, glycine enhanced NMDA receptor-mediated EPSPs[20]. Thus, the action of GSA could share some properties with the action of competitive antagonists of the NMDA receptor-associated glycine binding site, which are typified by kynurenate, an endogenous tryptophan metabolite[12]. Izumi *et al.*[9] found that these glycine antagonists blocked the induction of LTP without affecting baseline synaptic transmission. Bashir *et al.*[1] suggested that the action of 7-chlorokynurenate was at the glycine binding site because its effect on NMDA receptor-mediated synaptic transmission and on LTP induction was reversed by D-serine. In their experiments, both PTP and LTP were inhibited by 7-chlorokynurenate.

Of the guanidino compounds accumulated in uraemia, GSA was shown to be the most potent inhibitor of iontophoretically induced responses to GABA and glycine on mouse neurons in cell culture[3], and it was suggested that this inhibition is due to the blocking of the chloride channels of the inhibitory GABA and glycine receptors. This action of GSA on inhibitory amino acid transmission reported by De Deyn & Macdonald[3] together with GSAs ability to induce behavioural convulsions (see Pei *et al.*, Chapter 67) and the proposed role of GSA as an endogenous convulsant (see De Deyn *et al.*, Chapter 60) are not in accordance with the inhibitory effect of GSA on excitatory postsynaptic response reported here. However, two points must be taken into consideration when one is to use the CA1 synapse in the study of the action of a compound upon central nervous system function in order to explain the underlying mechanism of behavioural phenomena. Firstly, the observed effect of GSA on CA1 evoked response does not mean that GSA has no effect at all on inhibitory GABAergic function in CA1 because the latter function is at least in part mediated by feedback circuits (see ref.[15]). These inhibitory circuits are stimulated by the Schaffer collateral-commissural nerve terminals as well and by feeding back they depress excitatory synaptic action in CA1. Thus, when the excitatory processes are inhibited by GSA to begin with, feedback circuits are no longer activated either, and so nothing can be inferred from our findings about the concomitant effects of GSA upon GABA receptor-mediated function. Secondly, the effect of a compound on an isolated neural system *in vitro* does not necessarily represent the action of this compound upon the action of the central nervous system as a whole and *in vivo*.

Although the exact mechanism of action of the observed effect of GSA upon excitatory postsynaptic response in CA1 region of rat hippocampal slices remains as yet hypothetical, the involvement of excitatory amino acid receptors, including the widely acclaimed NMDA receptor, seems plausible. But as GSA and related guanidino compounds have been implicated in disorders like uraemic encephalopathy, these kind of results could aid in the elucidation of the complicated pathophysiology of these disorders.

Acknowledgements

The authors very much acknowledge the expert technical assistance of E. Cotman, R. Peters and P. Reygaert. Financial support was received from the Bom-Bunge Foundation, NFWO grants 3.0044.92 and NDE.58, and FGWO grant 3.9001.88.

References

1. Bashir, Z.I., Tam. B. & Collingridge, G.L. (1990): Activation of the glycine site in the NDMA receptor is necessary for the induction of LTP. *Neurosci. Lett.* **108**, 261–266.

2. Collingridge, G.L. & Singer, W. (1990): Excitatory amino acid receptors and synaptic plasticity. *Trends Pharmacol. Sci.* **11**, 290–296.

3. De Deyn, P.P. & Macdonald, R.L. (1990): Guanidino compounds that are increased in uremia inhibit GABA- and glycine-responses on mouse in cell culture. *Ann. Neurol.* **28**, 627–633.

4. De Deyn, P.P., Marescau, B., Cuykens, J.J., Van Gorp, L., Löwenthal, A. & De Potter, W.P. (1987): Guanidino compounds in serum and cerebrospinal fluid of non-dialysed patients with renal insufficiency. *Clin. Chim. Acta* **167**, 81–88.

5. D'Hooge, R., Manil, J., Colin, F. & De Deyn, P.P. (1991): Guanidinosuccinic acid inhibits excitatory synaptic transmission in CA1 region of rat hippocampal slices. *Ann. Neurol.* **30**, 622–623.

6. Foster, A.C. & Kemp, J.A. (1989): Glycine maintains excitement. *Nature* **338**, 377–378.

7. Ginn, H.E., Teschan, P.E., Walker, P.J., Bourne, J.R., Fristoe, M., Ward, J.W., McLain, W., Johnston, H.B. & Hamel, B. (1975): Neurotoxicity in uremia. *Kidney Int.* **7**, 357–360.

8. Glaser, G.H. (1974): Brain dysfunction in uremia. In *Brain dysfunction in metabolic disorders*, Vol. 53, ed. F. Plum, Res. Publ. Assoc. Nerv. Ment. Dis., pp. 173–199. New York: Raven Press.

9. Izumi, Y., Clifford, D.B. & Zorumski, C.F. (1990): Glycine antagonists block the induction of long-term potentiation in CA1 of rat hippocampal slices. *Neurosci. Lett.* **112**, 251–256.

10. Johnson, J.W. & Ascher, P. (1987): Glycine potentiates the NMDA response in cultured mouse brain neurons. *Nature* **325**, 529–531.

11. Johnson, J.W. & Ascher, P. (1988): The NMDA receptor and its channel. Modulation by magnesium and by glycine. In *Excitatory amino acids in health and disease*, ed. D. Lodge, pp. 143–163. London: John Wiley & Sons.

12. Kessler, M., Terramani, T., Lynch, G. & Baudry, M. (1989): A glycine site associated with *N*-methyl-D-aspartic acid receptors: characterization and identification of a new class of antagonists. *J. Neurochem.* **52**, 1319–1328.

13. Kleckner, N.W. & Dingledine, R. (1988): Requirement for glycine in activation of NMDA-receptors expressed in *Xenopus* oocytes. *Science* **241**, 835–837.

14. Lamdani-Itkin, H., Kloog, Y. & Sokolovsky, M. (1990): Modulation of glutamate-induced uncompetitive blocker binding to the NMDA receptor by temperature and by glycine. *Biochemistry* **29**, 3987–3993.

15. Lopes da Silva, F.H., Witter, M.P., Boeijinga, P.H. & Lohman, A.H.M. (1990): Anatomic organization and physiology of the limbic cortex. *Physiol. Rev.* **70**, 453–511.

16. Morris, R.G.M., Kandel, E.R. & Squire, L.R. (1988): The neuroscience of learning and memory: cells, neural circuits and behaviour. *Trends Neurosci.* **11**, 125–127.

17. Osberg, J.W., Meares, G.J., McKee, D.C. & Burnett, G.B. (1982): Intellectual functioning in renal and chronic dialysis. *J. Chron. Dis.* **35**, 445–457.

18. Souheaver, G.T., Ryan, J.J. & Dewolfe, A.S. (1982): Neuropsychological patterns in uremia. *J. Clin. Psychol.* **38**, 490–496.

19.   Thomson, A.M. (1989): Glycine modulation of the NMDA receptor/channel complex. *Trends Neurosci.* **10**, 349–353.

20.   Thomson, A.M., Walker, V.E. & Flynn, D.M. (1989): Glycine enhances NMDA-receptor mediated synaptic potentials in neocortical slices. *Nature* **338**, 422–424.

Guanidino Compounds in Biology and Medicine, eds. by P.P. De Deyn, B. Marescau, V. Stalon and
I.A. Qureshi. ©1992 John Libbey & Company Ltd., pp. 403–408.

Chapter 59

The effect of hyperbaric oxygen on the guanidino compounds in rat brain

T. ITOH[1], K. YUFU[2], R. EDAMATSU[1], H. YATSUZUKA[2] and A. MORI[1]

[1]*Department of Neurochemistry, Institute for Neurobiology;* [2]*Department of Anesthesiology and Resuscitology,
Okayama University Medical School, 2-5-1 Shikata-cho Okayama, 700 Japan*

Summary

In this study, we determined the effect of hyperbaric oxygen (HBO) on guanidino compounds in rat cerebral cortex in
relation to reactive oxygen species and antioxidants. Rats were exposed to HBO at 3 atmospheres absolute of 100 per
cent oxygen for 110 min. No rat had seizures induced by hyperoxia in this condition. The active oxygen species,
thiobarbituric acid reactive substances (TBARS), guanidino compounds and amino acid contents, and superoxide
dismutase (SOD)-like activity were measured. The HBO exposure resulted in the increase of carbon centred radical,
TBARS, SOD-like activity and arginine levels while there was decrease in ammonia content. There was no significant
change in methylguanidine level. These data suggest that the exposure to the HBO was an oxidative stress for the rat
brain. However, the enhancement of the protective mechanism against the oxidative stress seemed to prevent the
seizures.

Introduction

Hyperbaric oxygen (HBO)-induced seizures in small animals are useful models for epilepsy
research[18]. The seizures usually occur within a few minutes after the initiation of HBO
exposure. The amino acid metabolism[3,6], active oxygen and lipid peroxidation[5,11,14,20,21] and
antioxidants[9,19] in relation to HBO have been some of the major interests.

As the animals usually have convulsions at pressures greater than 3 atmosphere absolute (ATA), most
of the former experiments were done at pressures from 4 ATA to 6 ATA, or much higher.

In the clinical HBO therapy, however, seizures or any other central nervous system (CNS) symptom
must be carefully prevented. Therefore, therapeutic HBO exposures are carried out at pressures lower
than 3 ATA[7].

In our present study, we determined the effects of lower pressure of HBO on guanidines and amino
acid metabolism, lipid peroxidation and antioxidants in the rat brain. Our purpose was to demonstrate
the mechanisms of the beneficial effects of clinical HBO as well as its side effects based on oxidative
stress.

Materials and methods

Animals and preparation of samples

We used 7-week-old male Sprague-Dawley rats. Every rat was fed *ad libitum* prior to HBO, whereas they were not fed during the HBO procedure. Control rats were kept in room air and had 2 h starvation time to match with the experimental group. Immediately after HBO, they were sacrificed by either method described below, and we prepared the samples as follows.

Sample preparations for guanidino compounds, superoxide dismutase (SOD)-like activity, free radical concentration and thiobarbituric acid reactive substances (TBARS) assays

Animals were anaesthetized with light ether and cannulated in their aorta. Cold saline was perfused rapidly before the removal of the brain. We dissected the brain into seven regions on ice according to the method of Glowinski & Iversen[8]. The cerebral cortex was used for these measurements. Each cortex was homogenized in ice cold 0.1 M phosphate buffer solution adjusted to pH 7.8. We prepared mitochondrial and cytosolic fractions by the following procedure[4]. The crude homogenate was centrifuged at $1650 \times g$ for 15 min. The supernatant was centrifuged at $11,000 \times g$ for 30 min. We obtained mitochondrial fraction by resuspending the pellet in the phosphate buffer. The cytosolic fraction was obtained by centrifuging the supernatant at $105,000 \times g$ for 60 min. Samples were kept at $-80\,°C$ until measurement procedures.

Sample preparation for amino acid, urea and ammonia analysis

Animals were decapitated and the head was immediately immersed in liquid nitrogen for 7 s. After this the cerebral cortex was dissected by the same method as above. We homogenized each cortex in 10 volumes of 1 per cent 2,4,6-trinitrophenol solution for deproteinization. The homogenate was centrifuged at $1650 \times g$ and $4\,°C$ for 15 min. The excess 2,4,6-trinitrophenol was removed from the supernatant with a column of Dowex 2- X8(Cl⁻). The colourless eluate was dried up *in vacuo*. Residue was dissolved in hydrochloric acid (pH 2.2) and samples were kept at $-20\,°C$ until measurement procedures.

Guanidino compounds assay

We deproteinized 1.5 ml of cytosolic fraction with 150 µl of 30 per cent trichloroacetic acid solution. Then we measured guanidino compounds in filtrated supernatant fluorometrically using an automated guanidino compounds analyser system (JASCO, G-520, Tokyo). We used 9,10-phenanthrenequinone as fluorogenic compound.

SOD-like activity assay and free radical assay

We assayed SOD-like activity using the electron spin resonance (ESR) spectrometry method[10]. Fifty µl of 2 mM hypoxanthine solution, 35 µl of 10.98 mM diethylenetriaminepentaacetic acid solution, 50 µl of sample, 15 µl of 5,5-dimethyl-1-pyrroline-*N*-oxide (DMPO), 50 µl of xanthine oxidase (XOD) (0.326 u/ml), was added and vortexed in a test tube and transferred to a quartz cell. The decrease in ESR signal intensity of O^{2-}-DMPO spin adduct was measured using ESR spectrometer(JEOL-FE1 XG, Tokyo) 60 s after the addition of XOD. We used human Cu, Zn-SOD as a standard. For the free radical assay, 200 µl of the sample and 20 µl of DMPO were vortexed in a test tube and the mixture was immediately transferred to the quartz cell. We recorded the ESR signal 60 s after addition of DMPO. ESR parameters were: field 335 ± 5 mT; modulation 0.2 mT, response 0.1 s; sweeptime 10 mT/30s. We made all the ESR measurements at room temperature.

Amino acid, urea and ammonia assay

We determined amino acid, urea and ammonia contents using an automated amino acids analyser (IRICA 5500, Tokyo).

Lipid peroxide assay

We evaluated lipid peroxidation by measuring TBARS value[15]. We used 1,1,3,3-tetramethoxypropane as a standard. A hundred µl of the sample homogenate, 200 µl of 8.1 per cent SDS solution, 1.5 ml of 20 per cent sodium acetate buffer (pH 3.5), 1.5 ml of 0.8 per cent thiobarbituric acid solution, and 0.7 ml of 0.1 mM sodium phosphate buffer were added and vortexed. The mixture was boiled for 60 min. After boiling, 1.0 ml of distilled water and 5 ml of 1-butanol was added. Then we shook the mixture for 2 min. After centrifugation at $1650 \times g$ for 10 min, the 1-butanol layer was removed. The fluorescence intensity was measured using a fluorescence spectrophotometer (Hitachi 650–10S, Tokyo). The excitation wave length was 515 nm while the emission wavelength was 555 nm.

Protein assay

We determined protein content using a BCA® (bicinchoninic acid) protein assay kit manufactured by Pierce Co., USA.

Hyperbaric chamber operation

We used a hyperbaric oxygen chamber specially designed for animal experiments (Tabai PHC-special, Tokyo). Medical pure oxygen gas was flushed into the chamber before pressurization at a flow rate of approximately 15 l/min. Then we gradually increased the oxygen pressure up to 3 atmospheres absolute (ATA) for 5 min. During the maintenance period, we maintained continuous gas flow at a rate of 10 l/min. Total duration of HBO exposure to rats were 120 min. We placed soda lime in the chamber to adsorb carbon dioxide. Temperature, oxygen concentration, carbon dioxide concentration, and pressure in the chamber were continuously monitored. Temperature was $24 \pm 1°C$, oxygen concentration was 95–99 per cent, carbon dioxide concentration was less than 0.5 per cent and inner pressure was 3.0 ± 0.1 ATA. We observed the behaviour of the rats carefully. No rats had seizures, tremor or any other convulsive symptoms.

Data analysis

We used Student's *t*-test in the analysis of significance.

Results

Guanidino compounds

Table 1 shows the content of guanidino compounds. Hyperbaric oxygen increased guanidinoacetic acid and arginine content, while there was no significant change in creatinine, homoarginine and methylguanidine.

Table 1. Effect of hyperbaric oxygen on guanidino compounds in rat cerebral cortex

	Guanidino compounds (nmol/g tissue)	
	Control	HBO
Guanidinoacetic acid	7.68 ± 0.61	$9.37 \pm 0.32*$
Arginine	539.5 ± 13.1	$728.2 \pm 33.7**$
Creatinine	97.6 ± 6.92	93.1 ± 4.38
Homoarginine	2.15 ± 0.21	2.25 ± 0.13
Methylguanidine	0.900 ± 0.15	1.01 ± 0.21

Each value represents mean $\pm$ SEM of seven rats. $*P < 0.05$, $**P < 0.005$ *vs* control.

SOD-like activity

Table 2 shows the SOD-like activity in the samples. Hyperbaric oxygen increased SOD-like activity in both fractions of the rat cerebral cortex.

405

Table 2. Effect of hyperbaric oxygen on SOD-like activity in rat cerebral cortex

	SOD-like activity (unit/mg protein)	
	Control	HBO
Mitochondrial fraction	0.690 ± 0.049	1.08 ± 0.107**
Cytosolic fraction	10.73 ± 0.473	12.65 ± 0.600*

Each value represents mean ± SEM of six or seven rats.
*$P < 0.05$, **$P < 0.025$ *vs* control.

Free radicals

We identified and quantified the carbon-centred radical, hydroxyl radical and hydrogen radical. Table 3 shows the results. The carbon-centred radical level increased by HBO, whereas the hydroxyl radical level decreased. There was no significant change in the hydrogen radical level.

Table 3. Effect of hyperbaric oxygen on free radical content in rat cerebral cortex

	Free radicals ($\times 10^{15}$ spins/ml)	
	Control	HBO
Carbon-centred radical	1.272 ± 0.024	1.437 ± 0.009*
Hydroxyl radical	0.130 ± 0.022	0.062 ± 0.009**
Hydrogen radical	0.359 ± 0.033	0.3914 ± 0.019

DMPO adduct of free radicals in crude homogenate was measured. Each value represents eight or nine rats.
*$P < 0.05$, **$P < 0.01$ *vs* control.

Amino acids

Table 4 shows amino acids, urea and ammonia contents. Taurine, phosphoethanolamine and glutamine contents increased. There was no change in aspartic acid, glutamic acid, citrulline and ornithine contents. Urea level increased while ammonia level decreased.

Table 4. Effect of hyperbaric oxygen on amino acids, urea and ammonia contents

	Amino acid, urea and ammonia (μmol/g tissue)	
	Control	HBO
Taurine	5.445 ± 0.398	6.232 ± 0.925*
Phosphoethanolamine	1.978 ± 0.130	2.300 ± 0.230***
Aspartic acid	2.60 ± 0.171	2.43 ± 0.229
Asparagine	0.180 ± 0.025	0.162 ± 0.017
Glutamic acid	11.13 ± 0.408	11.61 ± 0.886
Glutamine	3.415 ± 0.188	4.864 ± 0.976**
Citrulline	0.018 ± 0.002	0.020 ± 0.007
γ-Aminobutyric acid	2.10 ± 0.119	2.05 ± 0.306
Ornithine	0.015 ± 0.002	0.017 ± 0.005
Urea	6.771 ± 0.269	9.39 ± 0.510***
Ammonia	1.821 ± 0.024	1.618 ± 0.080*

Each value represents mean ± SEM of eight or nine rats.
*$P < 0.05$, **$P < 0.01$, ***$P < 0.005$ *vs* control.

Lipid peroxides

TBARS value in the HBO group was 2.30 ± 0.13 (nmol/mg protein) while in the control group, it was 1.91 ± 0.13 (nmol/mg protein). HBO increased TBARS value in rat cerebral cortex ($P < 0.05$).

Protein assay

We observed the change in protein content per g tissue weight. In the HBO group protein content was 82.5 ± 2.01 (mg protein/g tissue), whereas it was 99.6 ± 3.81 (mg protein/g tissue) in the control group ($P < 0.005$).

Discussion

In our experiments, the major findings were: (1) increase in the TBARS level, SOD-like activity, arginine (Arg) and taurine levels, (2) decrease in hydroxyl radical and ammonia content, and (3) no change in γ-aminobutyric acid (GABA) and methylguanidine (MG) levels. No rats had seizures during HBO. In the former reports, HBO-induced convulsion was mostly explained by the decrease of GABA content, ammonia accumulation or peroxidation of lipids in the brain[5,14]. Because of the increase in TBARS level and carbon-centred radical content, we considered that our condition of HBO exposure was sufficiently oxidative to cause peroxidation in rat tissue. Nevertheless, there was no change in GABA content. We also observed a decrease in the ammonia content and an increase in the urea content. Glutamine level was also increased. These data suggest that ammonia metabolism was enhanced in our system, and that accumulation of ammonia did not occur. Together with the increase in SOD-like activity in both mitochondrial and cytosolic fractions, these findings seem to favour the protection mechanism against the HBO stress. The increase of Arg might be due to the suppression of arginase by active oxygen species, and the effect of the increase in Arg content is not clear. According to the report of Krichevskaia[12], Arg might have a protective effect on oxidative stress.

Another interest in guanidines was the change in the MG content, since MG is a potent convulsant[13]. Recently, MG is thought to be generated by the reaction of creatinine and active oxygen species[2,16]. For this reason, MG is proposed to be one of the indices for oxidative stress[17]. Though we expected some change in MG content in the brain, this was not the case. Maybe our condition for HBO exposure was too mild to cause enhancement of MG generation. The mechanism and the role of the increase in taurine content are unclear. Taurine might have a protective role in seizures, but its effect on HBO-induced seizure is questionable[1]. In conclusion, we observed the alteration in both guanidino compounds and amino acid metabolism caused by HBO exposure in rats. The defence mechanism was enhanced and it seemed to prevent seizures at a clinically relevant pressure of HBO.

Acknowledgements

The authors express thanks to Dr H. Hashimoto for his kind advice throughout our work, to Mr. H. Okamoto and Mr. S. Tamaru for their kind assistance in amino acid analysis and to Dr. S. Kira for his kind suggestions on preparing the manuscripts.

References

1. Adembri, G., Bartolini, A., Bartolini, R., Giotti, A. & Zilleti, L. (1974): Anticonvulsive action of homotaurine and taurine. *Br. J. Pharmacol.* **52**, 439–440.

2. Aoyagi, K., Nagase, S., Narita, M. & Tojo, S. (1989): Active oxygen in methylguanidine synthesis by isolated rat hepatocytes. In *Guanidines 2*, eds. A. Mori, B.D. Cohen, & H. Koide, pp. 79–85. New York: Plenum Press.

3. Banister, E.W., Bhakthan, N.M.G. & Singh, A.K. (1976): Lithium protection against oxygen toxicity in rats: ammonia and amino acid metabolism. *J. Physiol.* **260**, 587–596.

4. Danh, H.C., Benedetti, M.S. & Dostert, P. (1983): Differential changes in superoxide dismutase activity in brain and liver of old rats and mice. *J. Neurochem.* **40**, 1003–1007.

5. Dirks, R.C. & Faiman, M. (1982): Free radical formation and lipid peroxidation in rat and mouse cerebral cortex slices exposed to high oxygen pressure. *Brain Res.* **248**, 355–360.

6. Evans, G., Nelson, D.R. & Huggins, A.K. (1978): Effects of acute hyperbaric exposure on the concentration of alanine and other metabolites in mouse tissues. *Biochem. Soc. Trans.* **6**, 1022–1025

7. Fischer, B., Jain, K.K., Braun, E. & Lehrl, S. (1988): Oxygen toxicity. In *Handbook of hyperbaric oxygen therapy*, pp. 35–46. Heidelberg: Springer-Verlag.

8. Glowinski, J. & Iversen, L. (1966): Regional studies of catecholamines in the rat brain. *J. Neurochem.* **13**, 655–669.

9. Harabin, A.L., Braisted, J.C. & Flynn, E.T. (1990): Response of antioxidant enzymes to intermittent and continuous hyperbaric oxygen. *J. Appl. Physiol.* **69**, 328–335.

10. Hiramatsu, M. & Kohno, M. (1987): Determination of superoxide dismutase activity by electron spin resonance spectrometry using the spin trap method. *JEOL News*, **23A**, 7–9.

11. Jerrett, S.A., Jefferson, D. & Mengel, C.E. (1973): Seizures, H_2O_2 formation and lipid peroxides in brain during exposure to oxygen under high pressure. *Aerospace Med.* **44**, 40–44.

12. Krichevskaia, A.A., Shugalei, V.S., Ananian, A.A. & Zigova, I.G. (1978): Protective effect of arginine in hyperoxia. Activity of cerebral glutaminase and glutamate decarboxylase. (translation) *Vopr. Med. Khim.* **24**, 42–46.

13. Matsumoto, M., Kobayashi, K., Kishikawa, H. & Mori, A. (1976): Convulsive activity of methylguanidine in cats and rabbits. *ICRS Med. Sci.* **4**, 65.

14. Noda, Y., McGeer, P.L. & McGeer, E.G. (1983): Lipid peroxide distribution in brain and the effect of hyperbaric oxygen. *J. Neurochem.* **40**, 1329–1332.

15. Ohkawa, H., Ohishi, N. & Yagi, K. (1979): Assay of lipid peroxides in animal tissues by thiobarbituric acid reaction. *Anal. Biochem.* **95**, 351–358.

16. Sakamoto, M., Aoyagi, K., Nagase, S., Ohba, S., Miyazaki, M., Narita, M. & Tojo, S. (1989): Effect of active oxygen on guanidine synthesis *in vitro*. In *Guanidines 2*, eds. A. Mori, B.D. Cohen, & H. Koide, pp. 87–95. New York: Plenum Press.

17. Takemura, K., Aoyagi, K., Nagase, S., Sakamoto, M., Ishikawa, T., Kawamura, E. & Narita, M. (1989): Methylguanidine as an index of active oxygen species in hyperbaric oxygen therapy (translation). In *Abstracts of the twelfth Japanese conference on guanidino compounds*, 42–43.

18. Wood, J.D. (1972): Systemic oxygen derangements, 3. Hyperbaric oxygen. In *Experimental models of epilepsy*, eds. D.P. Purpura, J.K. Penry, D.B. Tower, D.M. Woodbury & R.D. Walter, pp. 461–476. New York: Raven Press.

19. Yusa, T., Crapo, J.D. & Freeman, B.A. (1984): Liposome-mediated augmentation of brain SOD and catalase inhibits CNS O_2 toxicity. *J. Appl. Physiol.* **57**, 1674–1681.

20. Yusa, T., Beckman, J.S., Crapo, J.D. & Freeman, B.A. (1987): Hyperoxia increases H_2O_2 production by brain *in vivo*. *J. Appl. Physiol.* **63**, 353–358.

21. Zirkle, L.G., Mengel, C.E., Horton, B.D. & Duffy, E.J. (1965): Studies of oxygen toxicity in the central nervous system. *Aerospace Med.* **36**(11), 1027–1032.

Guanidino Compounds in Biology and Medicine, eds. by P.P. De Deyn, B. Marescau, V. Stalon and
I.A. Qureshi. ©1992 John Libbey & Company Ltd., pp. 409–417.

Chapter 60

Possible mechanisms of action of guanidino compounds as candidate convulsants in uraemia and hyperargininaemia

P.P. DE DEYN[1], R. D'HOOGE[1], Y.Q. PEI[1], B. MARESCAU[1] and
R.L. MACDONALD[2]

[1]*Laboratory of Neurochemistry, Born-Bunge Foundation, UIA, Universiteitsplein 1, 2610 Wilrijk and Department of
Neurology, AZ Middelheim, Lindendreef 1, 2020 Antwerp, Belgium; and* [2]*Department of Neurology, University of
Michigan Medical Centre, 1103 E. Huron Street Ann Arbor, Michigan 48104-1687, USA*

Summary

The mechanisms through which guanidino compounds induce epilepsy in several experimental species and might contribute to the neurological complications in uraemia and hyperargininaemia have hitherto remained unknown. The effects of guanidino compounds, increased in cerebrospinal fluid of uraemic and hyperargininaemic patients, on postsynaptic responses to iontophoretically applied GABA and glycine, were assessed on mouse spinal cord neurons in primary dissociated cell culture. Guanidine, methylguanidine, creatinine and guanidinosuccinic acid (increased in uraemia) and arginine, homoarginine, α-keto-δ-guanidinovaleric acid and argininic acid (increased in hyperargininaemia) inhibited GABA and glycine responses in a concentration-dependent manner. Guanidinosuccinic acid displayed significant effects at concentrations similar to those found in cerebrospinal fluid. The guanidino compounds were equally potent in decreasing GABA and glycine responses and the benzodiazepine receptor antagonist CGS 9896 did not antagonize the guanidino compound-induced inhibition of GABA responses. These results suggest that the studied guanidino compounds inhibit responses to the inhibitory neurotransmitters by blocking the chloride channel.

Moreover, we further behaviourally assessed the epileptogenic properties of the uraemic guanidino compounds after intracerebroventricular administration in mice. A dose-dependent epileptogenic effect was observed. Interestingly, the same potency order was obtained in the behavioural experiments as in the electrophysiological ones (in order of decreasing potency: guanidinosuccinic acid, methylguanidine, guanidine, creatinine). The combined electrophysiological and behavioural findings support the hypothesis that inhibition of inhibitory neurotransmission could underlie the epileptogenicity of these guanidino compounds. The presented findings might be pathophysiologically important with regard to epileptogenesis in uraemia and hyperargininaemia.

Introduction

Hyperargininaemia, an inborn error of the urea cycle, is characterized by a neurological syndrome consisting of spasticity, epilepsy and mental retardation. The first clinical and biochemical descriptions of this inborn error of metabolism date from 1969 and 1970[20–22] and the syndrome has recently been reviewed by De Deyn[2]. Hyperargininaemic patients share their epileptic symptomatology as a common clinical feature with uraemic subjects. In addition, renally insufficient patients may suffer from encephalopathy and polyneuropathy. The basis for the majority of the neurological complications, including the epileptic symptomatology, seen in uraemia and hyperargininaemia is uncertain and the pathophysiology still remains to be elucidated. Our previously published analytical results and those presented elsewhere in this book (see Chapter 54), obtained by performing cation-exchange chromatography with fluorimetric detection, demonstrated the increase of several guanidino compounds in biological fluids (including cerebrospinal fluid) and different brain regions from uraemic and hyperargininaemic patients[2,5–7,9,13,14]. Most guanidino compounds have been shown to have convulsant effects[16]. The mechanisms through which guanidino compounds induce epilepsy and encephalopathy in several experimental models and possibly in uraemia and hyperargininaemia, remained unknown.

The reviewed electrophysiological and presented behavioural studies were performed in an attempt to determine the mechanisms through which the tested guanidino compounds produce their epileptogenic effects in experimental conditions and might induce epileptic symptomatology in hyperargininaemia and uraemia. We determined the effects of the guanidino compounds that accumulate in uraemia and hyperargininaemia on postsynaptic responses to GABA and glycine on mouse neurons in cell culture. In addition, the convulsive action of the guanidino compounds which were increased in uraemia were behaviourally assessed after intracerebroventricular administration in adult mice.

Materials and methods

Influence of GCs on GABA and glycine responses evoked on primary dissociated spinal cord neurons

Cultures of spinal cord neurons were prepared from dissected spinal cords and attached dorsal root ganglia from 12- to 14-day-old foetal mice as described previously[19]. Intracellular recordings were made from the somata of mouse spinal cord neurons in primary dissociated cell culture using glass micropipettes filled with 3M KCl. All recordings were made in Dulbecco's phosphate-buffered saline (DPBS) with elevated magnesium ion concentration. GABA and Gly responses were evoked by iontophoresis. The use of 3M KCl-filled micropipettes resulted in elevation of the intracellular chloride ion concentration and a shift in the chloride equilibrium potential from about –65 mV to about –20 mV. Under these conditions, an increase in chloride conductance resulted in an outward chloride current giving depolarizing responses[1,17]. The studied drugs were applied by miniperfusion. During study of the effect of co-application of guanidino compounds or guanidino compounds and the benzodiazepine receptor antagonist CGS 9896, the drugs were applied through one perfusion micropipette to avoid flow artefacts.

Solutions of the drugs were always made on the day of the experiment. CGS 9896 was dissolved in dimethylsulphoxide while the guanidino compounds were dissolved in DPBS to form stock solutions. Aliquots were removed and diluted in bathing medium to obtain the applied concentrations. At all applied concentrations, mean values and standard deviations were calculated for the effects on GABA and Gly responses. All effects are expressed as percent change from control GABA-response. The statistical significance of differences between GABA and Gly responses with and without drug application was calculated using the two-tailed Student's t-test; a P value of less than 0.05 was considered statistically significant. A detailed description of the methodology can be found elsewhere[3].

Behavioural assessment of the convulsive activity of uraemic guanidino compounds in mice after intracerebroventricular administration

For intracerebroventricular administration, the guanidino compounds were dissolved in saline and delivered with a Hamilton microsyringe into the left lateral brain ventricle. Dose-dependent effects of the different guanidino compounds were evaluated in groups of 5 to 10 animals. Administered dosages for creatinine ranged from 20 to 300 µmol/kg, for guanidine from 4 to 15 µmol/kg, for guanidinosuccinic acid from 0.5 to 4.0 µmol/kg and for methylguanidine from 0.5 to 7.0 µmol/kg.

Immediately after administration of the guanidino compounds, the mice were placed in individual cylindrical plastic cages for observation of their behaviour according to Irwin[11]. The character, onset time latency, evolution and duration of presumed epileptic activity was noted within a 1 h observation period. Systematic observation was assured by the use of a scoring procedure. A score was assigned to indicate the severity of the convulsive activity of the animal: 0 = normal behaviour; scores 1 and 2 were indicative of rather focal and short-lasting phenomena, while scores 3 to 5 represented generalized seizures. 3 = continuous, generalized epileptiform activity lasting ≥ 5 s: jumping without resting, running fits and automatisms; 4 = generalized clonic attack: vigorous and uncoordinated clonic movements of the body and limbs with falling and loss of righting reflex; 5 = tonic extension. Convulsive dose in 50 per cent of the animals (CD50) and their 95 per cent confidence intervals for each of the four GCs were determined by probit analysis or moving average interpolation.

Results

Direct effects of guanidino compounds that are increased in uraemia on GABA and Gly responses[4]

Application of recording solution did not significantly alter GABA responses. Guanidine, methyl-guanidine, creatinine and guanidinosuccinic acid rapidly and reversibly reduced GABA responses in a concentration-dependent manner as demonstrated in Fig. 1. Significant decreases of 23.9 ± 8.45 per cent ($P < 0.01$)(n = 10) and 34.7 ± 14.5 per cent ($P < 0.001$)(n = 16) of GABA responses were obtained at 10 mM for guanidine and methylguanidine respectively. Creatinine was devoid of any significant effect on GABA responses at concentrations up to 10 mM. At 20 mM however, a significant reduction of GABA responses by 23.5 ± 7.55 per cent ($P < 0.01$)(n = 9) was obtained. The weak inhibition of GABA responses of approximately 4 per cent (3.72 ± 4.76 per cent; n = 6) by guanidinosuccinic acid applied at 10 µM was not statistically significant. However, a significant 12.1 ± 3.69 per cent guanidinosuccinic acid-induced decrease ($P < 0.01$)(n = 6) of GABA responses was obtained at 100 µM and an almost complete inhibition of GABA responses was obtained at 10 mM. Guanidinosuc-cinic acid was approximately ten times more potent than guanidine and methylguanidine in inhibiting GABA responses.

Table 1. Effects of the tested 'uraemic' and 'hyperargininaemic' guanidino compounds on glycine responses on mouse spinal cord neurons

	Concentration	No of cells	Gly responses % decrease (mean ± SD)
'Uraemic' guanidino compounds			
Creatinine	10 mM	5	1.86 ± 2.68
	20 mM	6	31.0 ± 7.77*
Guanidine	10 mM	5	28.1 ± 6.31
Guanidinosuccinic acid	100 µM	4	14.8 ± 6.00*
	10 mM	9	96.6 ± 6.45*
Methylguanidine	10 mM	5	28.1 ± 8.52*
'Hperargininaemic' guanidino compounds			
Arginine	10 mM	7	54.8 ± 7.12*
Argininic acid	10 mM	6	92.5 ± 15.0*
Homoarginine	10 mM	4	78.9 ± 13.4*
α-Keto-δ-guanidinovaleric acid	10 mM	8	72.8 ± 32.5*

*$P < 0.01$ compared to controls.

Co-application of methylguanidine and guanidine resulted in an additive inhibition of GABA

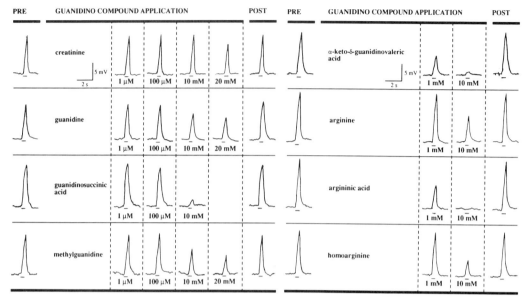

Fig. 1. Reversible and concentration-dependent inhibition of GABA responses on mouse neurons in primary dissociated cell culture by creatinine (top series), guanidine (upper middle series), guanidinosuccinic acid (lower middle series) and methylguanidine (bottom series). PRE shows stable GABA-responses before drug application. The four middle responses illustrate the effect of the guanidino compound. GABA responses returned to control values (POST) within 2 min following removal of the guanidino compound-containing micropipette. Iontophoretic application of GABA is indicated with a dash.

Fig. 2. Reversible and concentration-dependent inhibition of GABA responses on mouse neurons in primary dissociated cell culture by α-keto-δ-guanidinovaleric acid (top series), arginine (upper middle series), argininic acid (lower middle series) and homoarginine (bottom series). PRE shows stable GABA responses before drug application. The two middle responses illustrate the effect of the superfused guanidino compound. GABA responses returned to control values (POST) within 2 min following removal of the guanidino compound-containing micropipette. Iontophoretic application of GABA is indicated with a dash.

responses. Co-application of methylguanidine (10 mM) and guanidine (10 mM) resulted in a significant larger inhibition of GABA responses than when either of these compounds were applied separately at 10 mM. Whereas methylguanidine (10 mM) reduced GABA responses 25.7 ± 7.37 per cent (n = 3) and guanidine (10 mM) reduced GABA responses 23.3 ± 4.1 per cent (n = 3), their co-application resulted in a 39.7 ± 7.93 per cent (n = 3) decrease of GABA responses. Guanidine (10 mM), methylguanidine (10 mM), creatinine (20 mM) and guanidinosuccinic acid (100 μM and 10 mM) rapidly and reversibly decreased Gly responses in a highly statistically significant manner (Table 1). The guanidino compounds were equally potent in decreasing Gly and GABA responses.

Direct effects of guanidino compounds that are increased in hyperargininaemia on GABA and Gly responses[8]

The guanidino compounds arginine, homoarginine, α-keto-δ-guanidinovaleric acid and argininic acid that were found to be increased in the cerebrospinal fluid and brain tissue of hyperargininaemic patients, were subjected to the same experimental paradigms as those applied for the catabolites accumulating in uraemia. Again, the tested compounds inhibited GABA and Gly responses in a reversible and dose-dependent manner. Argininic acid was the most potent drug in inhibiting GABA and Gly responses, followed in decreasing potency by α-keto-δ-guanidinovaleric acid, homoarginine and arginine. While arginine was devoid of any significant effect on GABA responses at 1 mM, it significantly decreased GABA responses to 49.3 ± 7.02 per cent of control values at 10 mM (n = 5). Homoarginine decreased GABA responses to 38.7 ± 12.9 per cent ($P < 0.01$, n = 6) at 10 mM. α-keto-δ-guanidinovaleric acid significantly decreased GABA responses at all concentrations tested (100 μM, 1 mM and 10 mM). A small decrease of 3.64 ± 2.94 per cent ($P < 0.01$, n = 6) was obtained at 100 μM while a decrease of 86.4 ± 8.24 per cent ($P < 0.01$, n = 6) was induced at 10 mM. Argininic acid was applied at the same concentrations. This guanidino compound reduced GABA responses with 4.2 ± 2.30 per cent ($P < 0.01$, n = 6) at 100 μM and with 100 ± 0.00 per cent ($P < 0.01$, n = 5) at 10 mM. Arginine (10 mM) and argininic acid (10 mM) rapidly and reversibly decreased Gly responses (Table 1). The 'hyperargininaemic' guanidino compounds were equally effective in decreasing Gly and GABA responses.

Table 2. The pure benzodiazepine receptor antagonist CGS 9896 did not antagonize the effects of the 'uraemic' and 'hyperargininaemic' guanidino compounds on GABA-responses on mouse spinal cord neurons

	Number of cells	GABA responses % decrease (mean ± SD)
'Uremic' guanidino compounds		
Creatinine (20 mM)	3	21.0 ± 5.81
Creatinine (20 mM) + CGS 9896 (1 μM)	3	21.1 ± 7.39
Guanidine	3	24.9 ± 0.75
Guanidine (10 mM) + CGS 9896 (1 μM)	3	24.3 ± 9.10
Guanidinosuccinic acid (10 mM)	3	94.7 ± 9.12
Guanidinosuccinic acid (10 mM) + CGS 9896 (1 μM)	3	93.3 ± 11.5
Methylguanidine (10 mM)	5	28.5 ± 3.69
Methylguanidine (10 mM) + CGS 9896 (1 μM)	5	32.2 ± 6.74
'Hyperargininaemic' guanidino compounds		
α-Keto-δ-guanidinovaleric acid (10 mM)	3	81.5 ± 11.6
α-Keto-δ-guanidinovaleric acid (10 mM) + CGS 9896 (1 μM)	3	81.5 ± 10.0
Arginine (10 mM)	3	54.5 ± 10.9
Arginine (10 mM) + CGS 9896 (1 μM)	3	52.0 ± 10.6
Argininic acid (10 mM)	3	100 ± 0.00
Argininic acid (10 mM) + CGS 9896 (1 μM)	3	100 ± 0.00
Homoarginine (10 mM)	3	65.6 ± 10.9
Homoarginine (10 mM) + CGS 9896 (1 μM)	3	61.0 ± 9.5

No statistical differences were found for the paired samples of (a) application of guanidino compound alone and (b) co-application of guanidino compound and CGS 9896.

Effects of CGS 9896, a pure benzodiazepine receptor antagonist on the guanidino compound-induced inhibition of GABA responses

In order to evaluate whether the guanidino compounds induced their inhibition of GABA through interaction with the benzodiazepine receptor, we studied the effect of co-application of the guanidino compounds with the pyrazoloquinoline CGS 9896. In a concentration of 1 μM, this pure benzodiazepine receptor antagonist[3] did not significantly antagonize the guanidine, methylguanidine, creatinine and guanidinosuccinic acid induced inhibition of GABA responses (Table 2). Nor did CGS 9896 (1 μM) influence the inhibition of GABA responses induced by arginine (10 mM), homoarginine (10 mM), α-keto-δ-guanidinovaleric acid (10 mM) and argininic acid (10 mM) (Table 2).

Convulsive effects of the 'uraemic' guanidino compounds creatinine, guanidine, guanidinosuccinic acid and methylguanidine after intracerebroventricular administration in mice

While sham icv injections of saline in 25 animals produced generalized convulsions in only one

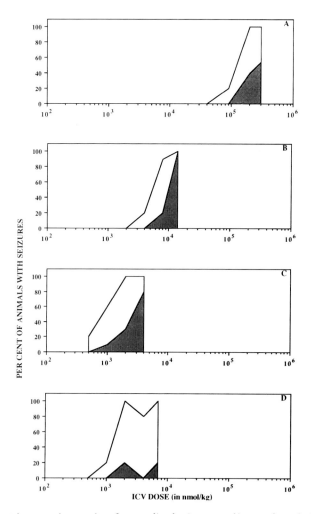

Fig. 3. Dose-dependent increase in severity of generalized seizures and/or number of mice presenting with seizures after intracerebroventricular injection of one of the four 'uraemic' guanidino compounds: creatinine (graph A), guanidine (graph B), guanidinosuccinic acid (graph C) and methylguanidine (graph D). The abscissa depicts dose of compound injected in μmol/kg, the ordinate depicts proportion of animals presenting with generalized convulsions in per cent of total number of 5 or 10 animals per dose. White areas correspond to generalized clonic convulsions (scores 3 and 4), and dark shaded areas with score 5 or tonic extension.

animal, all four tested guanidino compounds induced generalized convulsions (score ≥ 3) dose-dependently. Usually, seizures appeared 20 to 30 s after injection of the guanidino compounds. Low doses caused only local spasmodic movements and twitches in head, body and limbs of the animals. Higher doses induced generalized convulsions. An initial excitatory phase consisted of fast circling in the direction opposite to the injection side, running fits and jumping up and down the walls of the observatory cages and, dependent on the dose injected, this phase was followed by severe clonic seizures with loss of righting reflex, increased salivation, micturation and defecation. Several clonic attacks occurred which sometimes led to tonic extension and death by asphyxia. Also, the animals often died from prolonged clonic convulsions.

Figure 3 illustrates the dose-dependent increase in severity of generalized convulsions and/or number of mice presenting with generalized seizures after intracerebroventricular injection of the four individually studied 'uraemic' guanidino compounds. CD50 for induction of generalized convulsions (score ≥ 3; white area in Fig. 3) were inferred from the behavioural observations. The CD50 values for induction of generalized convulsions were (in $\mu mol/kg$): 101 (77–133) (n = 33) for creatinine; 5.0 (3.8–6.6) (n = 25) for guanidine; 0.8 (0.5–1.2) (n = 30) for guanidinosuccinic acid and 1.0 (0.6–1.5) (n = 25) for methylguanidine. Guanidinosuccinic acid was the most potent compound in the induction of generalized convulsions, followed in decreasing potency by methylguanidine, guanidine and creatinine. Guanidinosuccinic acid induced tonic convulsions more readily than methylguanidine.

Discussion

The tested guanidino compounds, that were all previously found to be increased in biological fluids (including cerebrospinal fluid) and brain of uraemic and hyperargininaemic patients, dose-dependently decreased GABA-and Gly-responses on mouse neurons in primary dissociated cell culture. Guanidinosuccinic acid, argininic acid and α-keto-δ-guanidinovaleric acid were most potent in the applied experimental paradigms. The observed inhibition of GABA and Gly responses by the tested guanidino compounds, was not mediated through an interaction with the benzodiazepine receptor. Indeed, CGS 9896, a pyrazoloquinoline and a pure antagonist at the benzodiazepine receptor, did not antagonize the guanidino compound-induced inhibition of GABA responses. Since GABA and Gly exert their inhibitory effects by activation of chloride conductance through interaction with different receptors and since guanidino compounds did inhibit GABA and Gly responses in a manner not significantly different, our observations suggest that the tested guanidino compounds, shown not to be benzodiazepine receptor ligands, inhibit postsynaptic effects of GABA and Gly by blocking chloride channels.

All four of the tested 'uraemic' guanidino compounds displayed the ability to induce full-blown clonic-tonic convulsions after intracerebroventricular administration in a dose-related manner. This paper is the first to demonstrate the behavioural epileptogenicity of guanidinosuccinic acid. Creatinine was clearly the least epileptogenic of the four guanidino compounds tested. Guanidine was consistently less epileptogenic than guanidinosuccinic acid or methylguanidine but much more than creatinine. The ranking of the compounds according to their epileptogenicity paralleled with their potency order in reducing GABA and Gly responses.

The observed inhibitory effects of the tested guanidino compounds on inhibitory neurotransmission could, in agreement with the so-called 'GABA hypothesis' of epilepsy[12,15,18], explain their epileptogenicity in mice and other animal models and might have pathophysiological implications with regard to epileptogenesis in uraemia and hyperargininaemia. Also, the epileptogenic potency order after intracerebral administration of the 'uraemic' guanidino compounds paralleled the potency order of these compounds in their GABA antagonism, indicating that inhibition of inhibitory amino acid neurotransmission might underlie their ability to induce behavioural convulsions. Moreover, guanidinosuccinic acid displayed significant effects at concentrations similar to those found in cerebrospinal fluid and brain tissue of uraemic patients[2]. The other guanidino compounds that reduced GABA and Gly responses only at concentrations higher than those hitherto found in cerebrospinal fluid and brain of uraemic and hyperargininaemic patients, might have additive effects, perhaps in combination with still other toxins. The additive effects of guanidine and methylguanidine have been reported here.

Some of our data suggest an involvement of still other neurotransmitter systems than the GABA and glycinergic ones in the behavioural effects of the tested guanidino compounds. Indeed, guanidinosuccinic acid was found to be ten times more potent than methylguanidine in the inhibition of GABA responses while it was only 1.25 times more potent than methylguanidine in behavioural epileptogenesis. It can be concluded that the presented results further establish the guanidino compounds as experimental convulsants and candidate convulsants that might contribute to the epileptic symptomatology in uraemia and hyperargininaemia. While an inhibitory effect on GABA and glycinergic neurotransmission is demonstrated, our combined electrophysiological and behavioural findings further suggest the involvement of other neurotransmitter systems (see also Chapter 58 by D'Hooge *et al.* and [10]).

Acknowledgements

This work was supported by the Born-Bunge Foundation, the Universitaire Instelling Antwerpen, the United Fund of Belgium and the NFWO grants N° 3.0044.92 and NDE 58.

References

1. Curtis, D.R., Hösli, L., Johnston, G.A.R. & Johnston I.H. (1986): The hyperpolarization of spinal motoneurons by glycine and related amino acids. *Exp. Brain Res.* **5**, 235–258.

2. De Deyn, P.P. (1989): Analytical studies and pathophysiological importance of guanidino compounds in uremia and hyperargininemia. Thesis submitted to obtain the degree of 'Geaggregeerde van het hoger onderwijs', Belgium: University of Antwerp.

3. De Deyn, P.P. & Macdonald, R.L. (1987): CGS 9896 and ZK 91296, but not CGS 8216 and Ro 15-1788, are pure benzodiazepine receptor antagonist on mouse neurons in cell culture. *J. Pharmacol. Exp. Ther.* **242**, 48–55.

4. De Deyn, P.P. & Macdonald, R.L. (1990): Guanidino compounds that are increased in cerebrospinal fluid and brain of uremic patients inhibit GABA and glycine responses on mouse neurons in cell culture. *Ann. Neurol.* **28**, 627–633.

5. De Deyn, P.P., Marescau, B., Cuykens, J.J., Van Gorp, L. & Lowenthal, A. (1987): Guanidino compounds in serum and cerebrospinal fluid of non-dialysed patients with renal insufficiency. *Clin. Chim. Acta* **167**, 81–88.

6. De Deyn, P.P., Marescau, B., Lornoy, W., Becaus, I. & Lowenthal, A. (1986): Guanidino compounds in uraemic dialysed patients. *Clin. Chim. Acta* **157**, 143–150.

7. De Deyn, P.P., Marescau, B., Lornoy, W., Becaus, I., Van Leuven, I. & Lowenthal, A. (1987): Serum guanidino compound levels and the influence of a single hemodialysis in uremic patients undergoing maintenance hemodialysis. *Nephron* **45**, 291–295.

8. De Deyn, P.P., Marescau, B. & Macdonald, R.L. (1991): Guanidino compounds that are increased in hyperargininemia inhibt GABA and glycine responses on mouse neurons in cell culture. *Epi. Res.* **8**, 134–141.

9. De Deyn, P.P., Marescau, B., Swartz, R.D., Hogaerth, R., Possemiers, I. & Lowenthal, A. (1990): Serum guanidino compound levels and clearances in uremic patients treated with continuous ambulatory peritoneal dialysis. *Nephron* **54**, 307–312.

10. D'Hooge, R., Manil, J., Colin, F. & De Deyn, P.P. (1991): Guanidinosuccinic acid inhibits excitatory synaptic transmission in Ca1 region of rat hippocampal slices. *Ann. Neurol.* **30**, 622–623.

11. Irwin, S. (1968): Comprehensive observational assessment: Ia. A systematic, quantitative procedure for assessing the behavioral and physiologic state of the mouse. *Psychopharmacologia* **13**, 222–257.

12. Lloyd, K.G., Munari, C. & Bossi, L. (1981): Biochemical evidence for the alterations of GABA-mediated synaptic transmission in pathological brain tissue (stereo EEG or morphological definition) from epileptic patients. In *Neurotransmitters, seizures and epilepsy*, eds. P.L. Morsell, K.G. Lloyd & W. Löscher, pp. 325–338. New York: Raven Press.

13. Marescau, B., De Deyn, P.P., Lowenthal, A., Qureshi, I.A., Antonozzi, I., Bachman, C., Cederbaum, C., Cerone, R., Chamoles, N., Colombo, J.P., Hyland, K., Gatti, R., Kang, S.S., Letarte, J., Lambert, M., Mizutani, N., Possemiers, I., Rezvani, I., Snyderman, S.E., Terheggen, H.G. & Yoshino, M. (1990): Guanidino compound analysis as a complementary diagnostic parameter for hyperargininemia: follow-up of guanidino compounds levels during therapy. *Ped. Res.* **27**, 297–303.

14. Marescau, B., Qureshi, I.A., De Deyn, P.P., Letarte, J., Ryba, R. & Lowenthal, A. (1985): Guanidino compounds in plasma, urine and cerebrospinal fluid of hyperargininemic patients during therapy. *Clin. Chim. Acta* **146**, 21–27.

15. Meldrum, B. (1979): Convulsant drugs, anticonvulsants and GABA-mediated neuronal inhibition. In *GABA-neurotransmitters*, eds. P. Krogsgaard-Larsen, J. Scheel-Kruger & H. Kofod, pp. 390–405. Copenhagen: Munksgaard.

16. Mori, A. (1987): Biochemistry and neurotoxicity of guanidino compounds. History and recent advances. *Pavlov. J. Biol. Sci.* **22**, 85–94.

17. Nowak, L.M., Young, A.B. & Macdonald, R.L. (1982): GABA and bicuculline action on mouse spinal cord and cortical neurons in cell culture. *Brain Res.* **244**, 155–164.

18. Olsen, R.W. (1981): The GABA postsynaptic membrane receptor-ionophore complex: site of action of convulsant and anticonvulsant drugs. *Mol. Cell Biochem.* **39**, 261–279.

19. Ransom, B.R., Neale, E., Henkart, M., Bullock, P.N. & Nelson, P.G. (1977): Mouse spinal cord in cell culture. I. Morphology and intrinsic neuronal electrophysiologic properties. *J. Neurophysiol.* **40**, 1132–1150.

20. Terheggen, H.G., Schwenk, A., Lowenthal, A., van Sande, M. & Colombo, J.P. (1969): Argininaemia with arginase deficiency. *Lancet* **ii**, 748–749.

21. Terheggen, H.G., Schwenk, A., Lowenthal, A., van Sande, M. & Colombo, J.P. (1970): Hyperargininämie mit Arginasedefekt: eine neue familiäre Stoffwechselstörung. I. Klinische Befunde. *Z. Kinderheilk* **107**, 298–312.

22. Terheggen, H.G., Schwenk, A., Lowenthal, A., van Sande, M., & Colombo, J.P. (1970): Hyperargininämie mit Arginasedefekt: eine neue familiäre Stoffwechselstörung. II. Biochemische Untersuchungen. *Z. Kinderheilk* **107**, 313–323.

Guanidino Compounds in Biology and Medicine, eds. by P.P. De Deyn, B. Marescau, V. Stalon and
I.A. Qureshi. ©1992 John Libbey & Company Ltd., pp. 419–424.

Chapter 61

Increase of guanidinoacetic acid and methylguanidine in the brain following amygdala kindling in rats

Yoshio HIRAYASU, Kiyoshi MORIMOTO[1], Saburo OTSUKI[1] and Akitane MORI[2]

Division of Neurosciences and Neurology, University of British Columbia, Vancouver, Canada;
Department of [1]Neuropsychiatry and [2]Neurochemistry, Institute for Neurobiology, Okayama University Medical
School, Okayama 700, Japan

Summary

Some guanidino compounds are known to be endogenous chemoconvulsants. In this study, we investigated the regional and time-dependent changes of these compounds in the rat brain following amygdala (AM) kindling and compared these results with those found after electric convulsive shock (ECS) seizures using high performance liquid chromatography. Twenty-eight days after the last AM-kindled seizure, guanidinoacetic acid and methylguanidine levels significantly increased bilaterally in the AM compared with those of control rats, which had been implanted with an electrode but were not stimulated. Both compounds, however, tended to decrease in the bilateral AM after ECS seizure. In addition, we measured these compounds following the induction of one afterdischarge (AD) in the AM. These compounds increased significantly 7 days after AD induction in the stimulated AM. These findings suggest that the increase of these compounds is coincident with AD generation in the AM, and specific to AM-kindled seizure.

Introduction

It has been well documented that guanidino compounds are intrinsic chemoconvulsants. Intracisternal injection of γ-guanidinobutyric acid[7], guanidinoethanesulphonic acid (GES)[12], guanidinoacetic acid (GAA), creatine, creatinine, creatine phosphate[8], *N*-acetylarginine[17], and methylguanidine (MG)[9] were all reported to induce epileptic discharges and convulsions in rodents. Recently, a change in the level of guanidino compounds was reported to occur following seizure in several experimental epileptic models. In the focal cobalt model of cats, α-guanidinoglutaric acid has been shown to increase not only in the ipsilateral but also in the contralateral cerebral cortex 24 h after unilateral cobalt placement[15]. The levels of *N*-acetylarginine, creatinine and guanidine were found to increase during the preconvulsive running phase in audiogenic sensitive rats[19]. Also in rats, a two to threefold increase of GAA and MG was reported to occur in the cerebral cortex and hippocampus (HP) 2 months after the injection of ferric chloride into the motor cortex inducing epileptogenic focus[1]. At the onset of pentylenetetrazol-induced convulsion in rabbits, the levels of GAA and creatinine in the cerebrospinal fluid (CSF) were increased, returning to normal 2 h after convulsion[5]. All of these findings suggest that guanidino compounds are participating in the expression of experimental epileptic seizures. In human epilepsy, high levels of GES have been found[10,16] in the CSF; a finding that was not reproduced in non-epileptic subjects[16]. These findings suggest that accumulation of GES may play a role in the seizure mechanism.

In the present study, we investigated the regional and long-lasting changes of guanidino compounds in the brain of amygdala (AM)-kindled rats to determine the role of those compounds in AM kindled seizures compared with electric convulsive shock (ECS) seizures. Kindling is the process whereby repeated electrical stimulation in the limbic brain regions or related forebrain sites results in progressive and permanent intensification of electrical discharges culminating in a generalized seizure[3,4]. In spite of many investigations, the pathophysiological mechanisms of kindling remain unclear. This study attempted to understand the role of guanidino compounds on the excitatory mechanisms of AM kindling.

Materials and methods

AM kindling

Male Sprague-Dawley rats weighing 300–350 g at the time of surgery were housed with free access to food and water under a 12-h light/12-h dark cycle. A tripolar electrode (Diamel-insulated Nichrome wire, 0.18 mm in diameter) was stereotaxically placed in the left AM under pentobarbital anaesthesia (50 mg/kg, ip). All coordinates were with the incisor bar 5 mm above the interaural line, the AM being at 0 mm anterior and 5 mm lateral to bregma, and 8.0 mm below the dura. In addition, one screw electrode which served as the reference electrode was placed in the frontal bone. Kindling began on the 7th postoperative day. The stimulus was delivered twice daily to the left AM in a 1 s train of 60 Hz sine waves at a current intensity of 200 μA. Seizure development was assessed using Racine's classification[18]. After stage 4 seizure, the kindling stimulus was administered once daily until five consecutive stage 5 seizures occurred. Control animals were implanted with the electrode but were not stimulated. AM-kindled rats were decapitated 24 h (n = 8) and 7 days (n = 8) after the last kindled seizure, along with the controls (n = 7). The brains were immediately divided into seven regions (cerebral cortex, striatum, thalamus, left AM, right AM, left HP, and right HP) following the method of Glowinsky & Iversen[2].

Lasting changes after kindled seizure

To investigate the long-lasting change in levels of guanidino compounds, both kindled rats (n = 11) and controls (n = 9) were decapitated 28 days after the last kindled seizure. These brains were divided into four regions (left AM, right AM, left HP and right HP).

Changes following one afterdischarge (AD)

Sixteen animals experienced one AD following stimulation of the left AM at a current of 200 μA. They were decapitated 24 h (n = 8), and 7 days (n = 8) after stimulation, along with the controls (n = 8). Three regions (left AM, right AM and left HP) were immediately removed.

ECS seizure

Seven animals with a tripolar electrode in the left AM were given ECS via a corneal electrode: 0.2 s, 60 Hz, 75 mA, once daily for 5 days. All rats receiving ECS stimulation displayed clonic-tonic seizures, lasting for 10–15 s. These animals were decapitated 24 h after the last ECS seizure, along with the controls which underwent the same operation without ECS (n = 8). Four regions (left AM, right AM, left HP and right HP) were immediately removed.

Analysis of guanidino compound levels

Brain samples were homogenized in ice-cold 1 per cent picric acid and centrifuged at 3000 rpm for 10 min. The supernatant was passed through a column of Dowex 2 × 8 [Cl⁻ form] (1.0 × 2 cm), and the colourless eluate was dried *in vacuo* and dissolved with 1 ml of diluted HCl (pH 2.2). Two hundred μl of these samples were applied to a high-performance liquid chromatograph (Jasco full automatic guanidino compounds analysing system), and the phenanthrenequinone reaction method was used for detection[14,20]. GAA, creatinine, arginine and MG were determined in rat brain. Conditions of the guanidino compound analysis were as follows: column, guanidino pack (6.0×50 mm, Jasco); column temperature, 70 °C; flow rate of eluent, 1.0 ml/min; flow rate of 0.5 per cent (w/v) 9,10-phenanthrene-quinone in dimethylformamide, 0.5 ml/min; flow rate of 2 M sodium hydroxide, 0.5 ml/min; reaction coil, 0.5 mm id $\times$ 5 m; detector, fluorometer (Jasco FP-110, Ex 365 nm, Em 495 nm); eluent, (1) 0.2 M sodium citrate buffer, pH 3.00 (10.3 min), (2) 0.2 M sodium citrate buffer, pH 3.50 (4.0 min), (3) 0.2 M sodium citrate buffer, pH 5.25 (4.2 min), (4) 0.2 M sodium citrate buffer, pH 10.00 (9.4 min), (5) 1 M sodium hydroxide (5.2 min); fluorescent reagent, 2 M sodium hydroxide and 0.05 per cent (w/v) 9,10-phenanthrenequinone in dimethylformamide. Other details were obtained utilizing the method of Jinnai *et al.*[8] and Mori *et al.*[14].

Results

AM kindling

In the stimulated left AM, the level of GAA and MG increased significantly 24 h after kindled seizure. In the right AM and both HP, the levels of GAA and MG 24 h after kindled seizure were also higher than in the control. These increases of GAA and MG in the limbic area lasted 7 days after kindled seizure (Table 1). In the left AM, GAA and MG increased 7 days after kindled seizure when compared with the control. In the cerebral cortex, increased level of GAA and MG were found at 24 h but not 7 days after kindled seizure. Creatinine and arginine showed no significant change.

Table 1. Guanidinoacetic acid and methylguanidine levels in the regions of rat brain 24 h and 7 days after kindled seizure (Kin 24 h and 7 d)

Tissue	Kin 24 h (n = 8)	Kin 7 d (n = 8)	Control (n = 7)
Guanidinoacetic acid (nmol/g tissue wet weight)			
Cortex	4.48 ± 0.58^a	4.67 ± 1.18	3.10 ± 0.65
Striatum	5.05 ± 1.78	5.45 ± 2.35	4.48 ± 0.57
L.AM	9.03 ± 0.90^a	10.75 ± 1.50^b	6.46 ± 1.04
R.AM	12.92 ± 1.89^c	10.41 ± 3.54^b	4.90 ± 0.61
L.HP	15.66 ± 1.56^c	15.66 ± 1.56^c	10.98 ± 1.09
R.HP	13.09 ± 1.69^c	15.21 ± 2.42^c	6.73 ± 0.60
Thalamus	7.53 ± 2.20	6.08 ± 1.37	5.46 ± 1.62
Methylguanidine (nmol/g tissue wet weight)			
Cortex	0.173 ± 0.035^a	0.173 ± 0.052	0.128 ± 0.012
Striatum	0.485 ± 0.101	0.672 ± 0.270	0.483 ± 0.063
L.AM	1.114 ± 0.397^a	1.226 ± 0.117^c	0.732 ± 0.134
R.AM	1.332 ± 0.261^c	1.099 ± 0.207^c	0.464 ± 0.134
L.HP	1.206 ± 0.151^c	1.476 ± 0.313^c	0.704 ± 0.078
R.HP	1.270 ± 0.242^c	1.428 ± 0.260^c	0.666 ± 0.175
Thalamus	0.468 ± 0.140	0.360 ± 0.124	0.353 ± 0.078

Values are expressed as mean $\pm$ SD.
$^aP < 0.05$; $^bP < 0.01$; $^cP < 0.001$ as compared to control (one way ANOVA, *t*-test). AM, amygdala; HP, hippocampus; L, left; R, right.

Lasting change after kindled seizure

Twenty-eight days after kindled seizure the level of MG was significantly elevated in both AM (Table 2), and both HP. GAA was significantly increased only in the AM (left and right). This finding suggests, therefore, that the increase of both GAA and MG in the AM lasts for at least 4 weeks.

Table 2. Guanidinoacetic acid and methylguanidine levels in the left and right amygdala (L.AM and R.AM) 28 days after kindled seizure

Tissue	Kin 28 d (n = 9)	Control (n = 11)
Guanidinoacetic acid (nmol/g tissue wet weight)		
L.AM	13.08 ± 2.62*	6.82 ± 1.28
R.AM	10.75 ± 2.67*	5.48 ± 0.80
Methylguanidine (nmol/g tissue wet weight)		
L.AM	0.921 ± 0.292*	0.428 ± 0.127
R.AM	0.793 ± 0.190*	0.372 ± 0.122

Values are expressed as mean ± SD. $*P < 0.001$ as compared to control (Student's t-test).

Changes following one AD

The levels of GAA and MG were significantly increased in the stimulated left AM 24 h and 7 days after stimulation. However, no significant change was observed in the non-stimulated right AM and ipsilateral HP (Table 3).

Table 3. Guanidinoacetic acid and methylguanidine levels in the left amygdala, right amygdala and left hippocampus (L.AM, R.AM and L.HP) after one kindling stimulation (AD 24 h and 7 d)

Tissue	AD 24 h (n = 8)	AD 7 d (n = 8)	Control (n = 8)
Guanidinoacetic acid (nmol/g tissue wet weight)			
L.AM	14.95 ± 4.16[b]	18.03 ± 2.12[b]	8.97 ± 1.39
R.AM	11.77 ± 2.21	12.16 ± 3.65	10.46 ± 2.96
L.HP	8.01 ± 2.77	12.88 ± 3.81	8.32 ± 3.45
Methylguanidine (nmol/g tissue wet weight)			
L.AM	1.460 ± 0.540[a]	1.522 ± 0.328[b]	0.759 ± 0.341
R.AM	1.228 ± 0.103	1.375 ± 0.394	1.008 ± 0.269
L.HP	0.585 ± 0.224	0.862 ± 0.280	0.617 ± 0.396

Values are expressed as mean ± SD.
[a]$P < 0.05$; [b]$P < 0.01$ as compared to control (one way ANOVA, t-test).

Electric convulsive shock (ECS)

After five ECS induced tonic-clonic seizures, GAA and MG tended to decrease in both AM and HP 24 h after the last ECS seizure (Table 4).

Table 4. Guanidinoacetic acid and methylguanidine levels in the bilateral amygdala and hippocampus (AM and HP) 24 h after electric convulsive shock (ECS 24 h) seizure

Tissue	ECS 24 h (n = 7)	Control (n = 8)
Guanidinoacetic acid (nmol/g tissue wet weight)		
L.AM	39.6 ± 1.02*	8.96 ± 1.39
R.AM	7.22 ± 0.75	10.50 ± 2.96
L.HP	8.70 ± 2.06	8.32 ± 3.45
R.HP	6.56 ± 2.12*	11.62 ± 1.62
Methylguanidine (nmol/g tissue wet weight)		
L.AM	0.623 ± 0.128	0.759 ± 0.341
R.AM	0.772 ± 0.190	1.008 ± 0.269
L.HP	0.526 ± 0.215	0.617 ± 0.396
R.HP	0.572 ± 0.332	0.761 ± 0.118

Values are expressed as mean ± SD. $*P < 0.001$ as compared to control (Student's t-test).

Discussion

It is known that 5 mg (42.7 μmol)/kg of GAA and 1.5 mg (20.5 μmol)/kg of MG induces severe tonic-clonic convulsions after intracisternal administration in rabbit[8,9]. These findings suggest that both GAA and MG are endogenous convulsants. In some experimental epileptic models, altered levels of guanidino compounds have been reported to follow convulsive seizures. However, the cellular localization of these compounds and the mechanism by which seizures are induced by guanidino compounds remains unknown. Recently, it has been reported that some guanidino compounds directly effect the neurotransmitter, for instance, MG inhibits acetylcholinesterase[11] as well as Na,K-ATPase activity[10].

Other findings suggest that MG generates hydroxyl ($\cdot$OH) and another unidentified radical ($\cdot$X)[6]. These facts suggest that a part of the neurotoxicity of guanidino compounds may be based in the neurotransmitter system and peroxidation of polyunsaturated fatty acids, which are important components of neuronal membrane, by such peroxide radicals generated from guanidino compounds. The action of these compounds at a cellular and molecular level in the brain is unknown. However, it has been reported that 2–10 mM of guanidine produced a near complete inhibition of the fast voltage-dependent K current on presynaptic membrane at the mouse neuromuscular junction. This effect is expected to allow an increase in Ca^{2+} influx and, thereby, in transmitter release upon nerve stimulation[13].

As potential underlying excitatory mechanisms of kindling, several putative neurotransmitter systems and neuropeptides have been investigated. In the present study, we investigated the regional and time-dependent change of guanidino compounds in AM-kindled rats. We found that the increase of GAA and MG persists for 28 days after the last kindled seizure. Since ECS seizure was not followed by such change, these increases of GAA and MG may be specific to AM kindling leading to the kindled state. However, further studies are required, because other areas of brain kindling are not tested similarly. In addition, we found that a single AD induced in the AM results in an increase of GAA and MG lasting for 7 days. In AM kindling, AD is initially localized to the kindling site with subsequent gradual propagation to distant brain sites, becoming bilateral and eventually culminating in a kindled generalized seizure. In the present study the increase of these compounds was unilaterally localized to the stimulated left AM following one AD but became bilateral upon completion of AM kindling. Therefore, the increase of GAA and MG in the limbic area is coincident with AD generation and dissemination induced by AM kindling. Another recent study showed significant increases of GAA in the stimulated AM 3 months after the last kindled seizure while no increase of MG was found (unpublished data by Morimoto *et al.*). These findings suggest that GAA and MG are likely to participate in excitatory mechanisms of epileptogenesis in the kindled site induced by AM kindling.

The difference in the results between the AM-kindled seizure and ECS seizure is difficult to interpret, but suggests that participation of these compounds is specific to kindled seizure.

Acknowledgements

We acknowledge the help of the staff of the Department of Neurochemistry, Institute for Neurobiology, Okayama University for technical advice.

References

1. Fukushima, M. (1987): Guanidino compounds in iron-induced epileptogenic foci of rats. *Okayama-Iggakai-Zasshi* **99**, 787–806. (Japanese).

2. Glowinski, J. & Iversen, L.L. (1966): Regional studies of catecholamines in the rat brain–I. The disposition of [³H]norepinephrine, [³H]dopamine and [³H]dopa in various regions of the brain. *J. Neurochem.* **13**, 655–669.

3. Goddard, G.V. (1967): Development of epileptic seizures through brain stimulation at low intensity. *Nature* **214**, 1020–1021.

4. Goddard, G.V., McIntyre, D.C. & Leech, C.K. (1969): A permanent change in brain function resulting from daily electrical stimulation. *Exp. Neurol.* **25**, 295–330.

5. Hiramatsu, M., Edamatsu, R., Fujikawa, N., Shirasu, A., Yamamoto, M., Suzuki, S. & Mori, A. (1988): Measurement during convulsions of guanidino compound levels in cerebrospinal fluid collected with a catheter inserted into the cisterna magna of rabbits. *Brain Res.* **455**, 38–42.

6. Hiramatsu, M., Edamatsu, R., Kohno, M. & Mori, A. (1989): The reactivity of guanidino compounds with hydroxyl radicals. In *Guanidines 2*, eds. A. Mori, B.D. Cohen & H. Koide, pp. 97–105. New York, London: Plenum Press.

7. Jinnai, D., Sawai, A. & Mori, A. (1966): γ-Guanidinobutyric acid as a convulsive substance. *Nature* **212**, 617.

8. Jinnai, D., Mori, A., Mukawa, J., Ohkusu, H., Hoshitani, M., Mizuno, A. & Tye, L.C. (1969): Biological and physiological studies of guanidino compound-induced convulsion. *Jpn. J. Brain Physiol.* **106**, 3668–3673.

9. Matsumoto, M., Kobayashi, K., Kishikawa, H. & Mori, A. (1976): Convulsive activity of methylguanidine in cats and rabbits. *IRCS Med. Sci.* **4**, 65.

10. Matsumoto, M. & Mori, A. (1976): Effects of guanidino compounds on rabbit brain microsomal Na-K-ATPase activity. *J. Neurochem.* **27**, 635–636.

11. Matsumoto, M., Fujiwara, M., Mori, A. & Robin, Y. (1977): Effet des dérivés guanidiques sur la cholinestérase et sur l'acétylcholinestérase du cerveau de lapin. *Comp. Rend. Soc. Biol.* **171**, 1226–1229.

12. Mizuno, A., Mukawa, J., Kobayashi, K. & Mori, A. (1975): Convulsive activity of taurocyamine in cats and rabbits. *IRCS Med. Sci.* **3**, 385.

13. Molgò, J. & Mallart, A. (1988): The mode of action of guanidine on mouse motor nerve terminal. *Neurosci. Lett.* **89**, 161–164.

14. Mori, A., Katayama, W., Higashidate, S. & Kimura, S. (1979): Fluorometrical analysis of guanidino compounds in mouse brain. *J. Neurochem.* **32**, 643–644.

15. Mori, A., Akagi, M., Katayama, Y. & Watanabe, Y. (1980): α-Guanidinoglutaric acid in cobalt-induced epileptogenic cerebral cortex of cats. *J. Neurochem.* **35**, 603–605.

16. Mori, A., Watanabe, Y. & Fujimoto, N. (1982): Fluorometrical analysis of guanidino compounds in human cerebrospinal fluid. *J. Neurochem.* **38**, 448–450.

17. Ohkusu, H. (1970): Isolation of α-N-acetyl-L-arginine from calf brain and convulsive seizure induced by this substance. *Osaka-Iggakai-Zasshi* **21**, 49–55.

18. Racine, R.J. (1975): Modification of seizure activity by electrical stimulation. Cortical areas. *Electroenceph. Clin. Neurophysiol.* **38**, 1–12.

19. Wiechert, P., Marescau, B., De Deyn, P. & Lowenthal, A. (1987): Guanidino compounds in serum and brain of audiogenic sensitive rats during the preconvulsive running phase of cerebral seizures. *Neurosciences* **13**, 35–40.

20. Yamamoto, Y., Saito, A., Manji, H., Nishi, H., Ito, K., Maeda, K., Ohta, K. & Kobayashi, K. (1978): A new automated analytical method for guanidino compounds and their cerebrospinal fluid levels in uremia. *Trans. Am. Soc. Artif. Intern. Organ.* **24**, 61–68.

Guanidino Compounds in Biology and Medicine, eds. by P.P. De Deyn, B. Marescau, V. Stalon and I.A. Qureshi. ©1992 John Libbey & Company Ltd., pp. 425–432.

Chapter 62

Antioxidant action of dipyridamole

Midori HIRAMATSU[1], Daiichi KADOWAKI, Rei EDAMATSU, Jiankang LIU, Xiaoyan WANG, Sachiko OHBA and Akitane MORI

[1]*Yamagata Technopolis Foundation, Yamagata and Department of Neurochemistry, Institute for Neurobiology, Okayama University Medical School, 2-5-1 Shikata-cho, Okayama 700, Japan*

Summary

Dipyridamole scavenged hydroxyl and DPPH radicals dose-dependently and quenched superoxide slightly. It also scavenged carbon centred radicals generated by ascorbic acid and $FeCl_2$ in the brain, kidney and liver homogenates and inhibited the TBARS formation in the same samples. The formation of methylguanidine which was synthesized from creatinine with $FeCl_3$ and H_2O_2 dose-dependently in the brain, kidney and liver homogenates was decreased with the administration of dipyridamole in these tissues. Methylguanidine and guanidinoacetic acid levels were elevated in the ipsilateral cortex 30 min after the iron injection into the rat sensory motor cortex while pretreatment with dipyridamole lowered the guanidinoacetic acid and methylguanidine levels in the cortex of the same animal models. These results indicate that the ability of dipyridamole to decrease the formation of methylguanidine *in vitro* and *in vivo* is due to the effect of its hydroxyl radical scavenging action.

Introduction

There have been some reports on oxygen free radicals and renal failure, which suggest that oxygen free radicals may play a role in the induction of renal failure[2,7,17–19]. Dipyridamole, which has anti-aggregative activity, has been used for the treatment of nephritis. However, a study on the scavenging activity of dipyridamole on oxygen free radicals and other organic free radicals has yet to be reported. In the present study, the effect of dipyridamole on hydroxyl radicals, superoxide, 1,1-diphenyl-2-picrylhydrazyl (DPPH) radicals, carbon centred radicals, thiobarbituric reactive substances (TBARS) formation and methylguanidine synthesis was examined using electron spin resonance (ESR) spectrometry, spectrofluorometry and high performance liquid chromatography to clarify its pharmacological action in *in vitro* and *in vivo* experiments.

Materials and methods

Source of chemicals

Dipyridamole, 2,6-bis(diethanol amino)-4,8-dipiperidinopyrimido-[5,4-d] dipyrimidine was kindly supplied by Nippon Boehringer Ingelheim Co., Ltd. (Kawanishi, Japan). DPPH was obtained from Sigma Chemical Co. (St Louis, MO, USA). Xanthine oxidase (XOD, cow milk) was from Boehringer Mannheim Ingelheim Co. Ltd. (Germany) and 5,5-dimethyl-1-pyrroline-1-*N*-oxide (DMPO) was from Daiichi Pure Chemicals Co., Ltd. (Tokyo, Japan). All other chemicals and reagents were of the highest grade available from commercial suppliers.

Animals

Male Sprague-Dawley rats (250–300 g) were obtained from Charles River Japan Inc. (Japan). Animals were decapitated after which the brain, kidney and liver were rapidly removed and used for the assay of carbon centred radicals, TBARS and methylguanidine.

Iron-induced epileptic foci

Rats were anaesthetized with ether and placed in a stereotaxic apparatus. A burr hole was made in the left calvarium 1 mm posterior and 1 mm lateral to the bregma. A 25-gauge needle attached to a microsyringe firmly held in a stereotaxic micromanipulator, was inserted into the cortex to 2.5 mm below the exposed dura. A freshly prepared 100 mM ferric chloride aqueous solution (5 μl) was injected over a period of 5 min. For controls, 5 μl saline with the same pH as the ferric chloride aqueous solution was injected[27]. Two dipyridamole doses, 2 and 20 mg/kg, were intraperitoneally injected into rats twice a day for 2 weeks. The control group (saline-injected animals) was also given an ip injection of *vehicle*. The animals were killed by decapitation 30 min following the cortical injection. After removing the whole brain, the cortex, hippocampus, hypothalamus, striatum, pons-medulla-oblongata and cerebellum were dissected on an ice plate and kept at –80 °C until guanidino compound analysis.

Dipyridamole

Dipyridamole was dissolved in acidic distilled water for *in vitro* experiment and dissolved in a mixed solution of tartaric acid and polyethyleneglycol 600 for *in vivo* experiment.

Free radical analysis

DPPH radicals

One hundred μl of 30 μ M DPPH ethanol solution and 100 μM of dipyridamole solution were mixed for 10 s and transferred to a special flat cell (JEOL Ltd., Tokyo) for the estimation of the amount of DPPH radicals by electron spin resonance (ESR) spectrometer (JEOL JES-FE1XG, Tokyo, Japan) 60 s after adding DPPH solution. The signal intensities were evaluated by the peak height of the second of the five peak signals of DPPH radicals[4].

Superoxide analysis

Fifty μl of 2 mM hypoxanthine, 35 μl of 10.98 mM diethylenetriaminepentaacetic acid (DETAPAC), 50 μl of dipyridamole solution, 15 μl of DMPO and 50 μl of XOD were placed in a test tube and mixed. The mixture was then transferred to the ESR spectrometry cell and the levels of DMPO spin adducts of superoxide (DMPO-O_2^-) were quantified 60 s after addition of XOD. The signal intensities were evaluated by the peak height of the first of the 12 signals of DMPO-O_2^- spin adducts[4,10].

Hydroxyl radical analysis

75 μl of 1 mM $FeSO_4$ and DETAPAC solution, 75 μl of 1 mM hydrogen peroxide, 50 μl of dipyridamole solution and 20 μl of DMPO were put into a test tube and mixed. The concentration of DMPO spin adducts of hydroxyl radicals (DMPO-OH) formed was estimated exactly 60 s after the addition of DMPO. The signal intensities were evaluated by the peak height of the second signal of the quartet of DMPO-OH spin adducts.

Analysis of the carbon centred radicals

The cerebral cortex was homogenized in 20 vol (w/v) of 30 mM Tris-HCl buffer (pH 7.4). To 0.5 ml homogenate, 0.4 ml of dipyridamole solution, and 0.1 ml of activating solution which consisted of 2 mM Fe^{2+} and 2 mM ascorbic acid were added, and the solution was then incubated at 37 °C for 15 min. Two hundred μl of the incubated solution and 20 μl of DMPO were put into a test tube and mixed. The amount of carbon centred radicals (·C) as DMPO spin adducts (DMPO-C) formed was measured exactly 1 min later[4].

Conditions of ESR spectrometry

The conditions of the ESR spectrometer to estimate the DPPH radicals were as follows: magnetic field, 335 ± 10 mT; response, 0.3 s; sweep time, 0.5 min; amplitude, 3.2×1000. To estimate the superoxide, the magnetic field, the response, the sweep time and the amplitude were, respectively, 335 ± 5 mT; 0.1 s; 2 min; and 6.3×100. To estimate the hydroxyl radicals and carbon centred radicals, they were respectively 335 ± 5 mT; 0.03 s; 0.5 min; and 2.5×1000. Manganese oxide was used as the internal standard. Spin number was calculated using the ratio of signal height intensities of 2,2,6,6-tetramethyl-4-hydroxyl-piperidine-oxyl with known spin quantities.

TBARS assay

The analysis of TBARS was carried out as follows[16]: briefly, to 0.2 ml incubated solution or to 0.2 ml of brain tissue homogenate with sodium chloride, 0.2 ml of 8.1 per cent sodium dodecyl sulphate, 1.5 ml of 20 per cent sodium acetate buffer (pH 3.5), 1.5 ml of sodium thiobarbituric acid and 0.6 ml distilled water were added. After mixing, the solution was put into 100 °C water bath for 60 min. The TBARS were extracted with a mixture of *n*-butanol:pyridine (15:1, v/v), and the fluorescence was measured (Ex 515 nm, Em 532 nm) using spectrofluorometry.

Determination of guanidino compound levels

The tissue was homogenized with ice-cold 1 per cent picric acid and centrifuged at 10,000 rpm for 10 min. The supernatants were mixed with Dowex 1–8 [Cl⁻] vigorously and the solution was passed through filter paper. The colourless eluates were dried in vacuo and the resulting samples were dissolved in diluted hydrochloric acid (pH 2.2). Guanidino compounds were analysed using a high performance liquid chromatograph (Jasco G-520). The phenanthrenequinone reaction method was used for detection. The details of the method have been described elsewhere[3,11].

Statistical analysis

Results were analysed with Student's *t*-test.

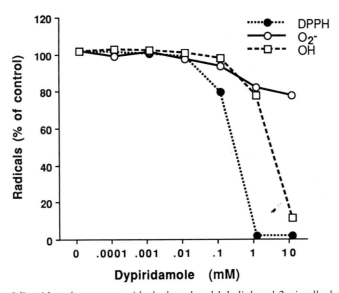

Fig. 1. Effects of dipyridamole on superoxide, hydroxyl and 1,1-diphenyl-2-picrylhydrazyl radicals.

427

Results

Dipyridamole dose-dependently scavenged superoxide (3.27×10^{16} spins/ml) in the range of 0.1~10 mM, hydroxyl radicals (9.14×10^{16} spins/ml) in the range of 0.1–10 mM and DPPH radicals (7.06×10^{15} spins/ml) in the range of 0.01–10 mM. Its quenching actions at maximal concentration on DPPH radicals, hydroxyl radicals and superoxide were about 100 per cent, 90 per cent and 25 per cent, respectively (Fig. 1). Carbon centred radicals were increased after the addition of ascorbic acid and $FeCl_2$ in the rat homogenates of brain, kidney and liver but dipyridamole decreased the generation of these radicals in the brain, kidney and liver at 0.1 and 1 mM, 0.01 and 0.1 mM, and 0.01–1 mM, respectively. However, it increased the radicals in the kidney at a high concentration of 10 mM (Fig. 2). In the same samples, dipyridamole inhibited the TBARS formation in the brain, kidney and liver dose-dependently (Fig 3).

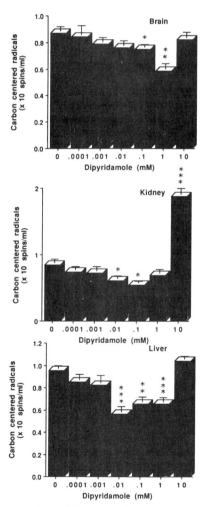

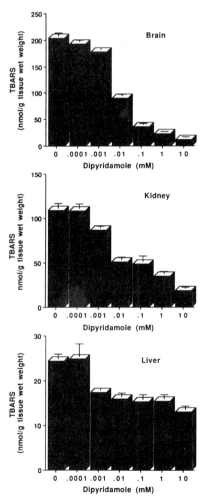

Fig. 2. Effects of dipyridamole on carbon centered radicals generated by $FeCl_2$ and ascorbic acid in rat homogenates. Each bar represents the mean ± SEM of 5–6 determinations.
*P < 0.05, **P < 0.01, ***P < 0.001 vs control.

Fig. 3. Effects of dipyridamole on thiobarbituric acid reactive substances (TBARS) formed by $FeCl_2$ and ascorbic acid in rat organ homogenates. Each bar represents the mean ± SEM of 3–6 determinations.

428

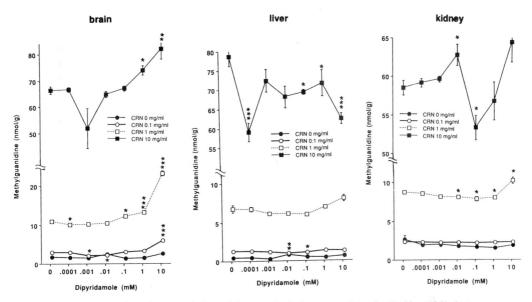

Fig. 4. Effects of dipyridamole on methylguanidine synthesis from creatinine by FeCl₃ and H₂O₂ in rat organ homogenates. Each bar represents the mean ± SEM of 2–6 determinations.
*P < 0.05, **P < 0.01, ***P < 0.001 vs control.*

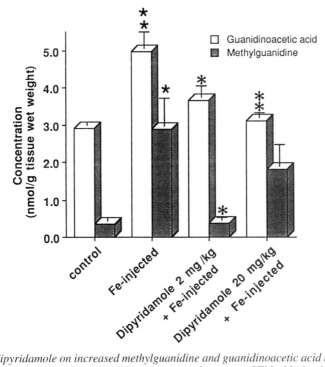

Fig. 5. Effects of dipyridamole on increased methylguanidine and guanidinoacetic acid levels in iron-induced epileptic focus of rats. Each bar represents the mean ± SEM of 8–6 animals.
*P < 0.05, **P <0.005 vs control; *P < 0.05, ** P < 0.01 vs Fe-injected.*

Methylguanidine was synthesized in the brain, liver and kidney homogenates after the addition of creatinine with $FeCl_3$ and H_2O_2 at increasing dosages of 0–10 mM. Otherwise, the addition of dipyridamole inhibited the formation of methylguanidine at low doses in the brain, liver and kidney homogenates and accelerated the formation at high doses in the brain and kidney homogenates (Fig. 4). Thirty min after the injection of iron solution into the rat cortex, methylguanidine and guanidinoacetic acid levels were increased in the ipsilateral cortex. The chronic administration of dipyridamole inhibited the increase in methylguanidine and guanidinoacetic acid levels at the dose of 2 mg/kg; and it further decreased the guanidinoacetic acid level at the dose of 20 mg/kg (Fig. 5).

Discussion

Dimethylthiourea, which is a scavenger of hydroxyl radicals, SOD and allopurinol have protected renal function after recirculation in ischaemia[2]. In addition, hydroxyl radicals have been suggested to relate to the induction of gentamicin-induced acute renal failure in rats[23]. In our study, dipyridamole was found to have a scavenging action on hydroxyl and DPPH radicals and carbon centred radicals generated by ascorbic acid and $FeCl_2$ in brain, liver and kidney homogenates. Methylguanidine, which is a uraemic toxin, has been reported to be synthesized via creatol from creatinine with hydroxyl radicals[15]. The increase of methylguanidine level in the rat brain after the iron injection into the rat cortex has been related to the generation of hydroxyl radicals[1].

Iron-induced epileptic foci are considered an excellent animal model for traumatic epilepsy. The mechanism for induction of epileptic seizures is regarded as follows; the hydroxyl radicals generated and the free radicals subsequently induced by iron injected solution attack double bonds of unsaturated fatty acids in lipids in the cell membranes. Lipid peroxides are formed, membrane functional activities are lessened and epileptic discharges are then induced[5,6,21,24,26]. Pretreatment with Japanese herbs, Sho-saiko-to-go-keishi-ka-shakuyaku-to (TJ-960) and epigallocatechin, which have a quenching action on hydroxyl radicals, superoxide and DPPH radicals[4,22] inhibited the appearance of epileptic discharges induced by iron injection into the rat cortex[13,14]. TJ-960 also suppressed the increase in TBARS level in the rat cortex after iron injection into this brain part[24]. In addition, the traditional Chinese medicine, Guilingji, which has a quenching action on hydroxyl radicals, superoxide and DPPH radicals[9], and the hydroxyl radical scavenger, L-ascorbic acid 2-[3,4-dihydro-2,5,7,8-tetra-methyl-2-(4,8,12-trimethyltridecyl)-2H-1-benzopyran-oyl-hydrogen phosphate, have also inhibited the increase of TBARS in rat brain induced by the iron injection[8,12]. Vitamin E and selenium depressed the epileptic discharges induced by iron injection[20,25]. These facts show that the formation of iron-induced epileptic foci is due to the oxygen free radicals and other organic free radicals generated by the iron injection. The inhibition of dipyridamole on the increase in methylguanidine and guanidinoacetic acid levels may be due to its quenching effect on hydroxyl radicals.

The effective dose of dipyridamole in our experiment was 1–100 μM and its effective blood concentration was about 2 μM (1 μg/ml). These results suggest that the effect of dipyridamole on nephritis may be due to its scavenging action on hydroxyl and other free radicals and inhibition of lipid peroxidation.

Acknowledgements

This research was partly supported by Special Coordination Funds of the Science and Technology Agency of the Japanese Government.

References

1. Fukushima, M. (1987): Guanidino compounds in iron-induced epileptogenic foci of rats. *Okayarna Igakkai Zasshi* **99**, 787–806.

2. Hansson, R., Johnsson, S., Johnsson, O., Pettersson, S., Scherstein, T. & Waldenstrom, J. (1986): Kidney protection by pretreatment with free radical scavengers and allopurinol: renal function at recirculation after warm ischemia in rabbits. *Clin. Sci.* **71**, 245–251.

3. Hiramatsu, M., Edamatsu, R., Fujikawa, N., Shirasu, A., Yamamoto, M., Suzuki, S. & Mori, A. (1988): Measurement during convulsions of guanidino compound levels in cerebrospinal fluid collected with a catheter inserted into the cisterna magna of rabbits. *Brain Res.* **455**, 38–42.

4. Hiramatsu, M., Edamatsu, R., Kohno, M. & Mori, A. (1988): Scavenging of free radicals by TJ-960. In *Recent advances in the pharmacology of KAMPO (Japanese Herbal) medicines*, eds. E. Hosoya & Y. Yamamura, pp. 120. Amsterdam: Excerpta Medica.

5. Hiramatsu, M., Khochi, Y. & Mori, A. (1983): Active oxygen and free radical reaction in Fe^{3+}-induced epileptogenic focus of rat. *Folia Psychiat. Neurol. Jpn.* **37**, 295–296.

6. Hiramatsu, M., Mori, A. & Kohno, M. (1984): Formation of peroxyl radical after $FeCl_3$ injection into rat isocortex. *Neurosciences* **10**, 281–284.

7. Johnson, R.J., Ochi, R.F., Adler, D., Baker, P., Sparks, L. & Couser, W.G. (1987): Participation of the myeloperoxidase H_2O_2-halide system in immune complex nephritis. *Kidney Int.* **32**, 342–349.

8. Liu, J., Edamatsu, R., Kabuto, H. & Mori, A. (1990): Antioxidant action of Guilingji in the brain of rats with $FeCl_3$-induced epilepsy. *Free Rad. Biol. Med.* **9**, 451–454.

9. Liu, J., Edamatsu, R. & Mori, A. (1990): Scavenging effect of Guilingji on free radicals. *Neurosciences* **16**, 623–630.

10. Mitsuta, K., Mizuta, Y., Kohno, M., Hiramatsu, M. & Mori, A. (1990): The application of ESR spin trapping technique to the evaluation of SOD-like activity of biological substances. *Bull. Chem. Soc. Jpn.* **63**, 187–191.

11. Mori, A., Akagi, M., Katayama, Y. & Watanabe, Y. (1980): α-Guanidinoglutaric acid in cobalt-induced epileptogenic cerebral cortex of cats. *J. Neurochem.* **35**, 603–605.

12. Mori, A., Edamatsu, R., Kohno, M. & Ohmori, S. (1989): A new hydroxyl radical scavenger: EPC-K₁. *Neurosciences* **15**, 371–376.

13. Mori, A. & Hiramatsu, M. (1983): Inhibitory effect of Sho-saiko-to-go-keishi-ka-shakuyaku-to on iron-induced epileptic seizures, a model of traumatic epilepsy. *Kanpoigaku* **7**, 12–16 (in Japanese).

14. Mori, A., Hiramatsu, M., Yokoi, I. & Edamatsu, R. (1990): Biochemical pathogenesis of post-traumatic epilepsy. *Pav. J. Biol. Soc.* **25**, 54–62.

15. Nakamura, K., Ienaga, K., Yokozawa, T., Fujitsukla, N. & Oura, H. (1991): Production of methylguanidine from creatinine via creatol by active oxygen species: analyses of the catabolism *in vitro*. *Nephron* **58**, 42–46.

16. Ohishi, S. (1978): Method for determination of lipid peroxidation. *Saishin Igaku* **33**, 660–663.

17. Peller, M.S., Hoidal, J.R. & Ferris, T.F. (1984): Oxygen free radicals in ischemic acute renal failure in the rat. *J. Clin. Invest.* **74**, 1156–1164.

18. Rehan, A., Johnson, K.J., Kunkel, R.G. & Wiggins, R.C. (1985): Role of oxygen radicals in phorbol myristate acetate-induced glomerular injury. *Kidney Int.* **27**, 503–511.

19. Richard, M.J., Arnaud, J., Jurkovitz, C., Hachache, T., Meftani, H., Laporte, F., Foret, M., Favier, A. & Cordonnier, D. (1991): Trace elements and lipid peroxidation abnormalities in patients with chronic renal failure. *Nephron* **57**, 10–15.

20. Rubin, J.J. & Willmore, L.J. (1980): Prevention of iron-induced epileptiform discharges in rats by treatment with antiperoxidants. *Exp. Neurol.* **67**, 472–480.

21. Triggs, W.J. & Willmore, L.J. (1984): *In vivo* lipid peroxidation in rat brain following intracortical Fe^{2+} injection. *J. Neurochem.* **42**, 976–980.

22. Uchida, S., Edamatsu, R., Hiramatsu, M., Mori, A., Nonaka, G., Nishioka, I., Niwa, M. & Ozaki, M. (1987): Condensed tannins scavenge active oxygen free radicals. *Med. Sci. Res.* **15**, 831–832.

23. Walker, P.D. & Shah, S.V. (1988): Evidence suggesting a role for hydroxyl radical in gentamicin-induced acute renal failure in rats. *J. Clin. Invest.* **81**, 334–341.

24. Willmore, L.J., Hiramatsu, M., Kochi, H. & Mori, A. (1983): Formation of superoxide radicals after $FeCl_3$ injection into rat isocortex. *Brain Res.* **277**, 393–396.

25. Willmore, L.J. & Rubin, J.J. (1981): Antiperoxidant pretreatment and ironinduced epileptiform discharges in the rat: EEG and histopathologic studies. *Neurology* **31**, 63–69.

26.   Willmore, L.J. & Rubin, J.J. (1984): The effect of tocopherol and dimethyl sulfoxide on focal edema and lipid peroxidation induced by isocortical injection of ferrous chloride. *Brain Res.* **296**, 389–392.

27.   Willmore, L.J., Sypert, G.W. & Munson, J.B. (1978): Chronic focal epileptiform discharges induced by injection of iron into rat and cat cortex. *Science* **200**, 1501–1503.

Guanidino Compounds in Biology and Medicine, eds. by P.P. De Deyn, B. Marescau, V. Stalon and
I.A. Qureshi. ©1992 John Libbey & Company Ltd., pp. 433–439.

Chapter 63

Effects of guanidinoethanesulphonic acid and 2-oxothiazolidine-4-carboxylic acid on electrocorticograms and contents of amino acids in rat and mouse brain

Isao YOKOI, Hideaki KABUTO, Tateki HATANO and Akitane MORI

Department of Neurochemistry, Institute for Neurobiology, Okayama University Medical School, Okayama 700, Japan

Summary

In this study, the effects of 2-oxothiazolidine-4-carboxylic acid (OTC) on guanidinoethanesulphonic acid (GES)-induced changes in ECoG and the levels of amino acids in the brain were examined. While spike discharges induced by GES (1 µmol) applied onto the rat sensorimotor cortex were completely suppressed by the subsequent application of taurine (1 µmol), OTC (up to 10 mmol/kg, ip) had no effect on the discharge frequency or pattern. Taurine was specifically decreased by GES administration (po) for 15 days in all brain regions tested, but did not increase following OTC (5 mmol/kg, ip).

On the other hand, OTC decreased glutamic acid, aspartic acid, glycine, alanine and arginine levels in cerebral cortex as well as in the cerebellum, but lowered GABA only in the striatum. These data indicate that OTC failed to influence the course of recovery from neurochemical and neurophysiological abnormalities induced by GES.

Introduction

Taurine (Tau) is contained in high concentrations in the central nervous system and, among others, may serve as a neurotransmitter or neuromodulator[3,9]. Tau is also known to have anticonvulsant effects on experimental epilepsy induced by cobalt[17], ouabain[7], pentylenetetrazol[8] and penicillin G[12]. On the other hand, *N*-amidino-taurine, *ie* guanidinoethanesulphonic acid (GES), competitively inhibits Tau uptake and thereby decreases the content of Tau in various organs[5,6]. GES induces convulsions and epileptic discharges in electrocorticograms (ECoG) when administered into cerebral ventricles or when applied to the pia mater of sensorimotor cortex[10,11,16,18]; supplemental Tau inhibits epileptic phenomena induced by GES[11,16]. It has been reported that the intraperitoneal administration of 5 mmol/kg of 2-oxothiazolidine-4-carboxylic acid (OTC) increased cysteine (Cys), a precursor of Tau, and caused a threefold increase in liver Tau content[15]. In the present study, we examined the effects of OTC on GES induced changes in the ECoG, and the levels of amino acids in the brain.

Materials and methods

Animals

Male Sprague Dawley rats weighing 250–350 g were used for electrophysiological studies, and ddY mice weighing 20–30 g were used for neurochemical studies. Rats and mice were maintained on a laboratory diet (protein content 24 per cent), MF (Oriental Yeast Co. Ltd., Tokyo, Japan), and water *ad libitum*.

Biochemical study

Mice were divided into four groups, and treated in the manner outlined in Fig. 1. Mice were killed by microwave irradiation to the head 150 min after the injection, followed by removal of the cerebral cortex, cerebellum and striatum. The analytical method used for amino acid determinations by HPLC was described previously[14]. Statistical analysis was performed using analysis of variance.

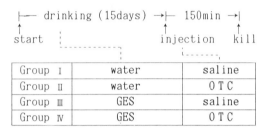

Fig. 1. Experimental procedure.

Electrophysiological study

Rats were immobilized with succinylcholine chloride under artificial ventilation with tracheal intubation. Four electrodes were placed epidurally at the sites described previously[16]. ECoG were recorded with a model EEG-5210 electroencephalograph (Nihon Koden, Japan) from four unipolar leads. For the topical application of GES, a trephine hole, 3 mm in diameter, was made over the left sensorimotor cortex, and the dura mater was removed. Each animal was allowed to recover for at least 2 h after the preparation and 10 μl of 100 mM GES solution was applied on the exposed cortex through the hole. After the completion of spike discharges induced by GES, 5 mmol/kg of OTC was injected ip.

Results

Sporadic spike discharges began 2–5 min after GES application onto the pia mater, and thereafter the frequency increased to 5–10 spikes/min. Spike discharges lasted until the end of recording for 3 h (Fig. 2).

While spike discharges induced by GES were completely suppressed by a supplement of Tau (1 μmol) on the pia mater, OTC administered intraperitoneally in doses up to 10 mmol/kg did not have any effect.

GES administration caused a selective decrease of Tau in all brain regions studied, on the other hand Tau was not increased by OTC injections (Figs. 3, 4 and 5).

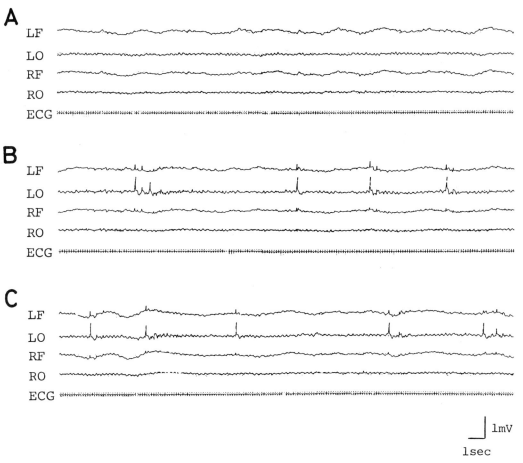

Fig. 2. Effects of guanidinoethanesulphonic acid (GES) and 2- oxothiazolidine-4-carboxylic acid (OTC) on electrocorticograms (ECoG) of rat unipolar recording.
A: control, before administration of GES. B: ECoG recorded 20 min after the administration of GES (1 μmol) onto the sensorimotor cortex. Sporadic spike discharges are observed. C: 60 min after the injection of OTC (10 mmol/kg, ip), i.e. 80 min after the GES administration. Sporadic spike discharges continue.
LF: left frontal cortex, LO: left occipital cortex, RF: right frontal cortex, RO: right occipital cortex, and ECG: electrocardiogram.

On the other hand, glutamic acid (Glu), aspartic acid (Asp), glycine (Gly), alanine (Ala) and arginine (Arg) levels in cerebral cortex were decreased by OTC administration (Fig. 3). In cerebellum, Asp, Gly, Ala and Arg were decreased (Fig. 4), and in striatum only GABA was decreased (Fig. 5) by OTC administration.

Though OTC showed a significant effect on some amino acid levels, no significant interaction between the effects of GES and OTC was noted.

Discussion

Tau is known to be synthesized from Cys[4]. However, Cys is known to be toxic when it is administered to animals in order to increase Tau, coenzyme A or glutathione content[1,13]. OTC was synthesized for the purpose of serving *in vivo* as a nontoxic Cys precusor[19,20]. OTC is effectively transported into

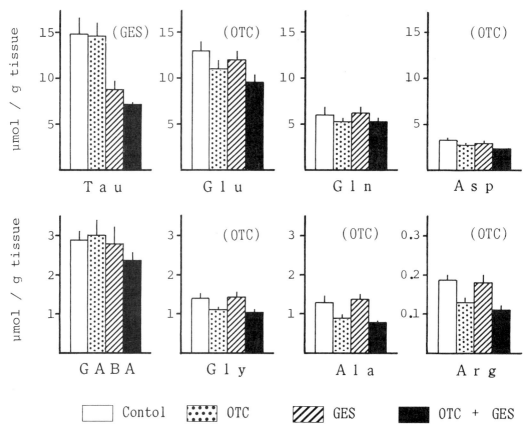

Fig. 3. Effect of guanidinoethanesulphonic acid (GES) and 2-oxothiazolidine-4-carboxylic acid (OTC) on amino acid levels in cerebral cortex of ddY mouse.

GES (1 per cent) was added in drinking water for 15 days, and OTC was injected ip (5 mmol/kg). Amino acids were analysed 150 min after the injection of OTC. Each bar represents amino acid levels (μmol/g tissue, n = 6: mean ± SEM). The treatments which caused a significant main effect (P < 0.05) by analysis of variance are shown between parentheses.

Tau: taurine; Glu: glutamic acid; Gln: glutamine; Asp: aspartic acid; GABA: 4-aminobutanoic acid; Ala: alanine; and Arg: arginine.

cells and is enzymatically hydrolysed by 5-oxoprolinase to Cys[2] which might then be metabolized to coenzyme A, Tau or glutathione. Taguchi et al.[15] have reported that OTC, corresponding to nearly 10 per cent of the dose administered, was metabolized to Tau and excreted in the urine. Though OTC is transported into the brain and known to increase Cys content, levels of brain glutathione do not change after OTC or Cys administration[2]. On the other hand, GES decreases the content of Tau in various organs[5,6] and induces spike discharges which are suppressed by Tau[11,16].

In this study, OTC did not increase or, after GES, normalize Tau levels in three brain regions of mice. Though the main effect of GES on Tau and the main effect of OTC on various amino acids were significant by analysis of variance, there was no significant interaction between OTC and GES. Cys in the cerebellum has been reported to increase by 250 per cent, 2 h after the OTC injection (10 mmol/kg)[2]. As only 5 mmol/kg of OTC was injected in this study, it might be one of the reasons why OTC failed to increase Tau level even in the cerebellum.

In the electrophysiological study, OTC (5–10 mmol/kg) had no effect on spike discharges induced

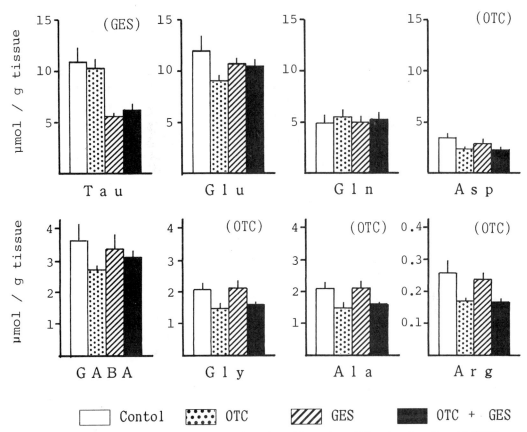

Fig. 4. Effect of guanidinoethanesulphonic acid (GES) and 2-oxothiazolidine-4-carboxylic acid (OTC) on amino acid levels in cerebellum of ddY mouse.
GES (1 per cent) was added in drinking water for 15 days, and OTC was injected ip (5 mmol/kg). Amino acids were analysed 150 min after the injection of OTC. Each bar represents amino acid levels (µmol/g tissue, n = 6: mean ± SEM).
A significant main effect by analysis of variance is indicated in parentheses. Tau: taurine; Glu: glutamic acid; Gln: glutamine; Asp: aspartic acid; GABA: 4-aminobutanoic acid; Ala: alanine; and Arg: arginine.

by GES. Doses of OTC higher than 10 mmol/kg were not tested in this study since it induces spike discharges in rats 20 to 60 min after administration of 15 mmol/kg. Though the neurotransmission role or the neuromodulatory effects by Tau may participate in the induction mechanism of GES-induced seizure activities[16], OTC dose not play a role in this mechanism. These data indicate that OTC fails to increase Tau in mice brain or to aid recovery from neurochemical and neurophysiological abnormalities induced by GES.

Acknowledgements

The authors are indebted to Professor Nico M. van Gelder, University of Montreal, Canada, for useful and invaluable discussion in this study.

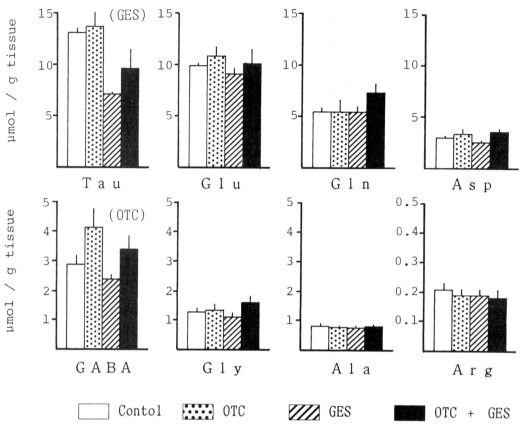

Fig. 5. *Effect of guanidinoethanesulphonic acid (GES) and 2-oxothiazolidine-4-carboxylic acid (OTC) on amino acid levels in striatum of ddY mouse.*

GES (1 per cent) was added in drinking water for 15 days, and OTC was injected ip (5 mmol/kg). Amino acids were analysed 150 min after the injection of OTC. Each bar represents amino acid levels (μmol/g tissue, n = 6: mean ± SEM). The treatments which had a significant main effect are shown between parentheses. Tau: taurine; Glu: glutamic acid; Gln: glutamine; Asp: aspartic acid; GABA: 4-aminobutanoic acid; Ala: alanine; and Arg: arginine.

References

1. Anderson, M.E. & Meister, A. (1987): Intracellular delivery of cysteine. *Methods Enzymol.* **143**, 313–325.

2. Anderson, M.E. & Meister, A (1989): Marked increase of cysteine levels in many regions of the brain after administration of 2-oxothiazolidine-4-carboxylate. *FASEB J.* **3**, 1632–1636.

3. Barbeau, A., Inoue, N., Tsukada, Y. & Butterworth, R.F. (1975): The neuropharmacology of taurine. *Life Sci.* **17**, 669–678.

4. Hays, K.C. & Sturman, J.A. (1981): Taurine metabolism. *Annu. Rev. Nutr.* **1**, 401–425.

5. Huxtable, R.J., Bonhaus, D., Nakagawa, K., Laird, H.F. & Pasantes-Morales, H. (1985): Taurine and the action of guanidinoethane sulfonate. In *Guanidine*, eds. A. Mori, B.D. Cohen & A. Lowenthal, pp. 213–225. New York: Plenum Press.

6. Huxtable, R.J., Lehmann, A., Sandberg, M. & Shindo, S. (1989): Guanidinoethane sulfonate and the investigation of taurine and other neuroactive amino acids. In *Guanidines 2*, eds. A. Mori, B.D. Cohen & H. Koide, pp. 189–197. New York: Plenum Press.

7. Izumi, K., Donaldson, J., Minnich, J. & Barbeau, A. (1973): Ouabain-induced seizures in rats: suppressive effects of taurine and GABA. *Can. J. Physiol. Pharmacol.* **51**, 885–889.

8. Izumi, K., Igusu, H. & Fukuda, T. (1974): Suppression of seizures by taurine – specific or non specific. *Brain Res.* **76**, 171–173.

9. Kuriyama, K. (1980): Does taurine have a function? Taurine as a neuromodulator. *Fed. Proc.* **39**, 2680–2684.

10. Mizuno, A., Mukawa, J., Kobayashi, K. & Mori, A. (1975): Convulsive activity of taurocyamine in cats and rabbits. *IRCS Med. Sci.* **3**, 385.

11. Mori, A., Katayama, Y., Yokoi, I. & Matsumoto, M. (1981): Inhibition of taurocyamine (guanidinotaurine)-induced seizures by taurine: In *The effects of taurine on excitable tissues*, eds. S.W. Schaffer, S.I. Baskin & J.J. Kocsis, pp. 41–48. New York: Spectrum Publications.

12. Mutani, R., Bergamini, L., Gariello, R. & Delsedime, M. (1974): Effect of taurine on cortical acute epileptic foci. *Brain Res.* **70**, 170–173.

13. Nishiuch, Y., Sasaki, M., Nakayasu, M. & Oikawa, A. (1976): Cytotoxicity of cysteine in culture media. *In Vitro* **12**, 635–638.

14. Shimada, M., Kabuto, H. & Yokoi, I. (1987): The effect of α-phosphono-ω-aminocarboxylic acids on seizures and brain amino acid levels in El mice. *Res. Commun. Chem. Pathol. Pharmacol.* **57**, 359–373.

15. Taguchi, T., Akagi, R. & Ubuka, T. (1990): Tissue contents and urinary excretion of taurine after administration of L-cysteine and 2-oxothiazolidine-4-carboxylate to rats. *Acta Med. Okayama* **44**, 123–128.

16. Toda, H., Shimizu, Y., Yokoi, I., Kabuto, H., Akiyama, K. & Mori, A. (1989): Effects of taurine and anticonvulsants on spike discharges induced by guanidinoethanesulfonate in rat. *Neurosciences* **15**, 193–195.

17. Van Gelder, M.N. (1972): Antagonism by taurine of cobalt induced epilepsy in cat and mouse. *Brain Res.* **47**, 157–165.

18. Watanabe, Y., Watanabe, S., Yokoi, I. & Mori, A. (1991): Effect of guanidinoethanesulfonic acid on brain monoamines in the mouse. *Neurochem. Res.* **16**, 1149–1154.

19. Williamson, J.M. & Meister, A. (1981): Stimulation of hepatic glutathione formation by administration of L-2-oxothiazolidine-4-carboxylate, a 5-oxo-L-prolinase substrate. *Proc. Natl. Acad. Sci. USA* **78**, 936–939.

20. Williamson, J.M. & Meister, A. (1982): New substrates of 5-oxo-L-prolinase. *J. Biol. Chem.* **257**, 13971–13976.

Guanidino Compounds in Biology and Medicine, eds. by P.P. De Deyn, B. Marescau, V. Stalon and
I.A. Qureshi. ©1992 John Libbey & Company Ltd., pp. 441–448.

Chapter 64

Multiple modes of NMDA
receptor regulation by guanidines

Ian J. REYNOLDS[1] and Kristi ROTHERMUND

*Department of Pharmacology, University of Pittsburgh, E1354 Biomedical Science Tower,
Pittsburgh PA 15261, USA*

Summary

We have investigated the interaction of a number of guanidino compounds with the *N*-methyl-D-aspartate (NMDA)
receptor complex. We have previously shown the bisguanidine arcaine is a potent antagonist of the action of polyamines
at the NMDA receptor. We have also shown that 1,10-bis(guanidino)decane is an effective NMDA antagonist, but
probably mimics the action of Zn^{2+} rather than blocking the polyamine site. In this study we have examined the effects
of guanidine, methylguanidine, guanidinosuccinic acid and creatinine on the NMDA receptor using [^{3}H]dizocilpine
binding and intracellular Ca^{2+} measurements. The monoguanidines at concentrations between 1–100 mM inhibited [^{3}H]
dizocilpine binding and increased its dissociation in a similar manner to Mg^{2+}. These drugs also inhibited NMDA and
glycine-stimulated increases in intracellular Ca^{2+}. Thus, although several different guanidino compounds interact with
the NMDA receptor complex they do not share a common mechanism of action despite ostensible structural similarities.

Introduction

The family of ionotropic glutamate receptors includes three principal types, identified by their
preferential agonists: *N*-methyl-D-aspartate (NMDA), kainate and α-amino-3-hydroxyl-5-
methyl-4-isoxazolepropionic acid (AMPA). The widespread distribution of these receptors in
the central nervous system implies a fundamental physiological role. However, greater interest has
been focused on these receptors for their potential pathophysiological involvement in disorders like
cerebral ischaemia, traumatic head injury, epilepsy, and possibly some neurodegenerative disorders[6].
As most of these diseases putatively involve over-activation of glutamate receptors in general, and
NMDA receptors in particular, there has been a great emphasis on understanding the factors which
regulate the activation of these receptors, and on developing effective antagonists of the action of
glutamate as potential therapeutic tools[6,7,26].

The NMDA receptor has attracted particular attention in the study of the pathophysiology of glutamate
receptors because it appears to be a key trigger in eliciting delayed neuronal death in cell culture
models of cerebral ischaemia[6]. This action of NMDA critically depends on the influx of Ca^{2+} from
the extracellular environment through the NMDA receptor-associated inonophore[5]. A number of
recent studies have also demonstrated that the activation of the NMDA receptor can be modulated by
a wide array of ligands normally found in the extracellular milieu. Thus, Mg^{2+} blocks NMDA receptor
activity by binding to a site located within the ion channel[13]. This action of Mg^{2+} can be reversed by
cell depolarization, which effectively makes the NMDA receptor a coincidence detector. The action
of NMDA and glutamate is also significantly potentiated by, and may require, glycine in low
concentrations[12]. Glycine binds to a site on the NMDA receptor complex that is distinct from that

which recognizes NMDA and glutamate, and is also different from the previously described strychnine-sensitive site. An additional site on the NMDA receptor complex recognizes Zn^{2+} in low micromolar concentrations[16,27]. Zn^{2+} non-competitively reduces the effects of NMDA and glycine. Aizenman and colleagues[2] have described a site probably on the extracellular face of the receptor that is sensitive to sulphhydryl redox reagents. Oxidation of this site diminishes NMDA receptor activity, while reduction potentiates responses. The details of the physiological role of this site have not yet been established, although it is interesting to note that oxygen radicals effectively oxidize these sulphhydryls[1]. A site that recognizes polyamines, including spermine and spermidine, was recently identified[17]. This will be discussed further below. Finally, of pharmacological interest, is the observation that drugs like phencyclidine and dizocilpine (MK801) are potent and specific NMDA antagonists that bind to a site located within the ion channel in a use-dependent fashion[4,30].

Studying the interactions between these modulators has proved to be challenging. In our laboratory we have taken advantage of the specificity and potency of [³H]dizocilpine to develop a radioreceptor binding assay that is sensitive to all of the previously described modulators[23]. These studies have revealed two principal mechanisms by which modulators can alter [³H]dizocilpine binding. As the binding site is located within the ion channel, the rate of ligand binding is controlled by the presence of agonists that activate the receptor. Thus, NMDA and glycine increase the rate of binding, while antagonists for these sites slow binding. By measuring binding under non-equilibrium conditions it is possible to monitor agonist effects at these sites. Drug effects can also be detected using direct measurements of association and dissociation. In addition to the NMDA and glycine sites, Zn^{2+} appears to operate predominantly by altering channel activation, and clearly shows the rate of binding[22,28]. The second mechanism for controlling [³H]dizocilpine binding site. This effect is very apparent with Mg^{2+}, for example, which is an effective negative allosteric modulator[22]. It is important to note that while the effects observed in the binding assay of ligands which change the rate of binding usually correlate well with functional activity, this has not always been the case with allosteric modulators.

The effects of the polyamines spermine and spermidine were initially discovered by Ransom & Stec using the [³H]dizocilpine binding assay[17]. The observation that spermidine dramatically increased [³H]dizocilpine binding initially suggested that polyamines might represent a novel class of agent that might potentiate NMDA responses in an analogous fashion to glycine. Similarly, it might be anticipated that a competitive polyamine antagonist might inhibit [³H]dizocilpine and NMDA responses. Several competitive polyamine antagonists have described. Of these, the most potent is arcaine (1,4-bis[guanidino]butane)[18]. Studies with arcaine have revealed the complexity of the polyamine site. It appears that polyamines exert at least two effects on the NMDA receptor[19]. The high affinity effect is the prominent increase in [³H]dizocilpine that has been widely observed, and is sensitive to arcaine. In addition, a lower affinity effect is apparent: this appears to resemble the action of Mg^{2+} and is insensitive to arcaine. While it is difficult to exclude an effect of polyamines on the rate of [³H]dizocilpine binding, and thereby the activation of the receptor, it is clear that polyamines change the equilibrium affinity of [³H]dizocilpine binding[19,29]. This indicates that polyamines fall into the class of allosteric modulators rather than channel activators. It is perhaps not surprising, then, that neither spermine, spermidine nor arcaine produce effects on functional NMDA responses that are pharmacologically consistent with a receptor activation role for polyamines (unpublished observations).

In an effort to develop more potent polyamine antagonists we recently tested a number of guanidine analogues of arcaine. These studies showed that 1,10-bis(guanidino)decane (BG10) was also an effective inhibitor of [³H]dizocilpine binding and displayed a similar potency to arcaine[21]. However, somewhat surprisingly given the similarity in structure, the pharmacological profile of BG10 was distinct from arcaine. The most obvious difference was that BG10 effectively inhibited NMDA-induced intracellular Ca^{2+} ($[Ca^{2+}]_i$) responses in cultured neurons at concentrations similar to those that inhibited [³H]dizocilpine binding, whereas, as previously noted, arcaine was ineffective. BG10 was also an effective NMDA antagonist *in vivo*. A closer look at the mechanism of action of BG10 suggested that this guanidine more closely resembled Zn^{2+} than arcaine with respect to its interaction with the NMDA receptor complex. In particular, BG10 showed the dissociation rate. This property shared by Zn^{2+}, is consistent with closing the NMDA receptor-associated ionophore and trapping the ligand, and can fully account for its functional activity. Whether the actions of BG10 are actually mediated by the Zn^{2+} site, or are the result of binding to a separate site with a similar action, remains to be determined.

Recent studies have implicated a number of guanidine in the central nervous system pathophysiology of uraemia. These guanidines, including guanidine, methylguanidine, guanidinosuccinic acid (GSA) and creatinine, are significantly elevated in the cerebrospinal fluid (CSF) of patients with uraemia[9]. In the light of our prior studies of the interaction of guanidine with the NMDA receptor, and in view of the suggestion that guanidines may interact with several different ion channels, we have investigated the interaction of several monoguanidines with the NMDA receptor complex.

Materials and methods

Receptor binding assays

Binding assays were performed using well-washed rat brain membranes prepared as previously described[24]. Assays contained 0.5 nM [^{3}H]dizocilpine (22.5 mCi/mmole, Du Pont/NEN, Boston, MA), 0.1 mg membrane protein, glutamate, glycine, spermidine and drugs as appropriate in a final volume of 1.0 ml 10 mM HEPES/NaOH, pH 7.4. After incubation at room temperature (19–20 °C) for 2 h, which represents non-equilibrium conditions, assays were terminated by vacuum filtration over glass fibre filters and radioactivity determined by liquid scintillation counting. Non-specific binding was defined by 30 µM dizocilpine. Drugs were obtained from Sigma Chemical Company (St Louis, MO). For dissociation assays ligand and tissue were incubated for 2 h prior to 120-fold dilution into HEPES buffer containing 100 µM glutamate, 30 µM glycine and drugs as appropriate. This mixture was allowed to incubate for a further 60 min, following which assays were terminated by filtration. For these assays data are expressed as a per cent of the level of binding present before dissociation occurs. Typically, binding is reduced to 70 per cent of control levels after 60 min. Drugs that produce lower levels of binding at this time do so by increasing the dissociation of the ligand.

Intracellular Ca^{2+} recordings

$[Ca^{2+}]_i$ recordings were made as previously described[25]. Primary culture of rat forebrain neurons (20–35 days *in vitro*) grown on glass coverslips were loaded with 5 µM fura-2 for 60 min at 37 °C. After rinsing coverslips were mounted on a recording chamber that allows rapid drug application. The recording solution contained (mM): NaCl 137, KCl 5, $MgSO_4$ 0.9, $CaCl_2$ 1.4, $NaHCO_3$ 3, Na_2HPO_4 0.6, KH_2PO_4 0.4, glucose 5.6, and HEPES 20, pH adjusted to 7.4 with NaOH. Mg^{2+} was omitted for recording NMDA responses. When 50 mM K^+ was used for the stimulus NaCl was iso-osmotically substituted by KCl. Drugs were applied by superfusion. Typically, two control agonist responses were obtained. Guanidines were then applied for 200 s and then agonists re-applied in the presence of the guanidine. The drug response was then expressed as a per cent of the control response.

Results

Guanidine, methylguanidine, GSA and creatinine inhibited non-equilibrium [^{3}H]dizocilpine binding at concentrations of 1–100 mM (Fig. 1). In general the inhibition profile was similar in the presence of low or high concentrations of glutamate and glycine or spermidine, although with 0.1 µM glutamate and 0.03 µM glycine GSA increased [^{3}H]dizocilpine binding. The approximate order of potency was guanidine = methylguanidine > creatinine > GSA, with IC_{50} values falling between 1 and 10 mM. Several compounds decreased [^{3}H]dizocilpine binding below the level of non-specific binding

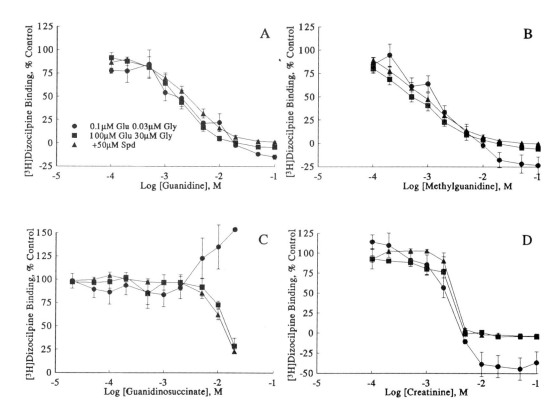

Fig. 1. Guanidines inhibit non-equilibrium [³H]dizocilpine binding to the NMDA receptor. Inhibition curves were constructed using (A) guanidine, (B) methylguanidine, (C) GSA and (D) creatinine in the presence of 0.1 μM glutamate and 0.03 μM glycine (●), 100 μM glutamate and 30 μM glycine (■), or 100 μM glutamate 30 μM glycine and 50 μM spermidine (▲). Results represent the mean (± SEM) of three to four separate experiments performed in duplicate.

defined by 30 μM dizocilpine. We have previously observed this phenomenon with Mg^{2+} (data not shown).

To further investigate the mechanism of action of these monoguanidines we monitored their effects on the dissociation of [³H]dizocilpine. Each of these compounds increased the dissociation of [³H] dizocilpine, as indicated by a reduction in the level of binding remaining after 60 min (Fig. 2). Creatinine was clearly less potent than the other compounds tested. However, in general the drugs increased the dissociation of [³H]dizocilpine at similar concentrations to those required to inhibit non-equilibrium binding. Shown for comparison is Mg^{2+}, which was somewhat more potent.

To test the functional consequences of this action of monoguanidines we examined their ability to inhibit $[Ca^{2+}]_i$ responses produced by the addition of NMDA (30 μM) together with glycine (1 μM) in single rat forebrain neurons (Fig. 3). NMDA/glycine produces very reproducible responses in these cells. Guanidine (5 mM), methylguanidine (5 mM) and GSA (1 mM) all significantly inhibited NMDA/glycine-induced increases in $[Ca^{2+}]_i$ (Table 1), and these effects could be rapidly reversed by superfusion with drug-free buffer. GSA also produced small increases in baseline $[Ca^{2+}]_i$ (Fig. 3C). As a simple test for specificity we also monitored the effects of these drugs on increases in $[Ca^{2+}]_i$ produced by 50 mM KCl, which predominantly activates voltage-sensitive Ca^{2+} channels. Guanidine produced a small but significant inhibition of KCl-induced $[Ca^{2+}]_i$ changes, while methylguanidine and GSA were ineffective.

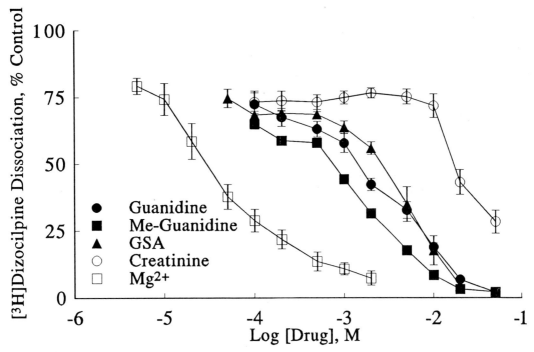

Fig. 2. Guanidines increase the dissociation of [³H]dizocilpine from the NMDA receptor. After equilibration of membranes and [³H]dizocilpine for 2 h the mixture was diluted 120-fold into buffer containing 100 μM glutamate 30 μM glycine and the drugs at the concentration indicated. The incubation was then allowed to proceed for 60 min prior to filtration. Control binding represents the level of binding prior to any dissociation. The downward slope indicates that the drugs increased the dissociation of [³H]dizocilpine. The data represents the mean (± SEM) of three experiments performed in duplicate. The Mg²⁺ data were adapted from ref[20].

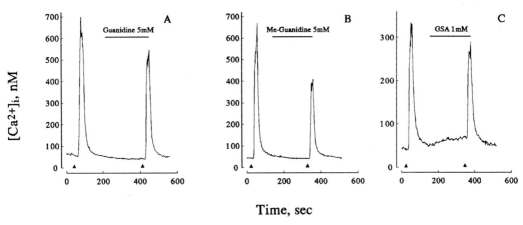

Fig. 3. Guanidines inhibit responses to 30 μM NMDA with 1 μM glycine. Responses were obtained in the absence of drug and then after a 200s incubation with (A) 5 mM guanidine, (B) 5 mM methylguanidine or (C) 1 mM GSA as indicated by the horizontal bars. Agonists were added at the arrowheads. These data represent traces from different cells that were repeated in five to six additional cells with essentially similar results.

Table 1. Effects of monoguanidine on $[Ca^{2+}]_i$ changes in single rat forebrain neurons. Responses were obtained by the addition of 30 μM NMDA with 1 μM glycine, or with 50 mM KCl as shown in Fig. 3. Control responses are expressed as the mean change in $[Ca^{2+}]_i$, while the responses in the presence of guanidines represent a per cent of the control value

Control	Increase in $[Ca^{2+}]_i$	
	NMDA/Gly	KCl
Control	477 ± 80 nM	261 ± 27 nM
+ Guanidine (5 mM)	$74.8 \pm 5.5\%$**	$89.1 \pm 2.3\%$**
+ Me-Guanidine (5 mM)	$50.9 \pm 7.1\%$**	$94.9 \pm 6.3\%$
+ GSA (1 mM)	$87.9 \pm 4.1\%$**	$92.9 \pm 4.7\%$

Data represent the mean ($\pm$ SEM) of 5–7 cells.
**Significantly different from control ($P < 0.05$, t-test).

Discussion

These studies have shown that guanidine, methylguanidine, GSA and creatinine can interact with the NMDA receptor complex *in vitro*, as indicated by their ability to inhibit [³H]dizocilpine binding, accelerate the dissociation of [³H]dizocilpine, and inhibit NMDA-induced increases in $[Ca^{2+}]_i$ in cultured central neurons.

The experiments performed allow some insights into the likely mechanism of action of these drugs. By comparing the affinity of the monoguanidines in the presence of different concentrations of the various agonists it is possible to monitor actions at the NMDA, glycine and polyamine sites. With the concentrations of agonists used the affinity of these drugs as antagonists should shift by 10–1000-fold. However, as virtually no difference was seen in the presence of either low or high levels of agonist it is unlikely that these drugs bind to any of the three agonists' sites. This is notable with respect to the polyamine site as arcaine is an effective and potent polyamine antagonist[18]. The action of these drugs on the NMDA receptor complex is associated with an increase in the dissociation of [³H] dizocilpine. This is a prominent effect of the monoguanidines tested, and occurs at the same concentrations that inhibit binding. Moreover, we have previously reported a similar profile of activity for Mg^{2+} [22]. Thus, it seems likely that the monoguanidines tested mimic the actions of Mg^{2+} on the NMDA receptor. This would also be consistent with the ability of guanidine, methylguanidine and GSA to inhibit NMDA-induced $[Ca^{2+}]_i$ changes.

Several previous studies have reported channel blocking activity of guanidines. For example De Deyn and colleagues[8] have shown that these monoguanidines can block the $GABA_A$-receptor associated ion channel in similar concentrations to those used in the present study. Previous studies have also shown that methylguanidine can block the nicotinic acetylcholine receptor channel[10]. Interestingly, several other drugs that interact with the nicotinic receptor channel also block NMDA receptors, including phencyclidine, mecamylamine and nicotine[3,14,15]. This may represent evidence for a structural similarity between the two receptor types.

GSA showed some minor differences in its overall profile of activity. Thus, in the presence of low levels of glutamate and glycine GSA increased [³H]dizocilpine binding. This may be indicative of a low efficacy agonist action of GSA that would effectively disappear when these sites were fully occupied by glutamate and glycine. Structurally this would not be surprising as the NMDA recognition site recognizes dicarboxylic acids. A small agonist effect of GSA could also account for its ability to increase baseline $[Ca^{2+}]_i$ in neurons (Fig. 3C). However, this may also be attributable to depolarization resulting from loss of GABA receptor inhibition, as GSA also blocks $GABA_A$ receptors at the concentrations used in this study[8].

Elevated CSF concentrations of various guanidines are believed to contribute to the range of central nervous system complications associated with uraemia[8]. The spectrum of neurological dysfunctions extends from minor sensory abnormalities through seizures to coma[11]. It is clear that modulation of NMDA receptor function can produce significant neurological effects, and channel blocking drugs like phencyclidine, ketamine and dizocilpine produce hallucinations and a dissociative anaesthetic state, for example. The extent to which the NMDA receptor–guanidine interaction described in this study contributes to the neurological syndrome associated with uraemia will depend on whether these agents are present in appropriate concentrations in CSF. Measurements of guanidines in CSF during uraemia have found low micromolar concentrations, which are somewhat lower than the effective concentrations described herein[9]. Thus, it seems unlikely that profound NMDA receptor blockade occurs during uraemia. However, it is possible that in combination several guanidines might produce an effect of pathophysiological significance.

In conclusion, a number of mono- and bis-guanidines interact with the NMDA receptor. However, despite ostensible structural similarities there does not appear to be a common mechanism whereby the NMDA receptor complex recognizes guanidines. Indeed, the compounds tested to date bind to at least three different sites including the polyamine site, the Zn^{2+} site and the Mg^{2+} site on the NMDA receptor complex.

References

1. Aizenman, E., Harnett, K.A. & Reynolds, I.J. (1990): Oxygen free radicals regulate *N*-methyl-D-aspartate receptor function via a redox modulatory site. *Neuron* **5**, 841–846.

2. Aizenman, E., Lipton, S.A. & Loring, R.H. (1989): Selective modulation of NMDA responses by reduction and oxidation. *Neuron* **2**, 1257–1263.

3. Aizenman, E., Tang, L-H., & Reynolds, I.J. (1991): Effects of nicotine agonists on the NMDA receptor. *Brain Res.* **551**, 355–357.

4. Anis, N.A., Berry, S.C. Berry, N.R. Burton & Lodge, D. (1983): The dissociative anaesthetics, ketamine and phencyclidine, selectively reduce excitation of central mammalian neurones by *N*-methyl-aspartate. *Br. J. Pharmacol.* **79**, 565–575.

5. Choi, D.W. (1987): Ionic dependence of glutamate neurotoxicity. *J. Neurosci.* **7**, 369–379.

6. Choi, D.W. (1988): Glutamate neurotoxicity and diseases of the nervous system. *Neuron* **1**, 623–634.

7. Collingridge, G.L. & Lester, R.A.J. (1989): Excitatory amino acid receptors in the vertebrate central nervous system. *Pharmacol. Rev.* **41**, 143–210.

8. De Deyn, P.P. & Macdonald, R.L. (1990): Guanidino compounds that are increased in cerebrospinal fluid and brain of uremic patients inhibit GABA and glycine responses on mouse neurons in cell culture. *Ann. Neurol.* **28**, 627–633.

9. De Deyn, P.P., Marescau, B., Cuykens, J.J., Van Gorp, L., Lowenthal, A. & De Potter, W.P. (1987): Guanidino compounds in serum and cerebrospinal fluid of non-dialyzed patients with renal insufficiency. *Clin. Chim. Acta* **167**, 81–88.

10. Farley, J.M., Yeh, J.Z., Watanabe, S. & Narahashi, T. (1981): Endplate channel block by guanidine derivatives. *J. Gen. Physiol.* **77**, 273–293.

11. Fraser, C.L. & Arieff, A.I. (1988): Nervous system complications in uremia. *Ann. Intern. Med.* **109**, 143–153.

12. Johnson, J.W. & Ascher, P. (1987): Glycine potentiates the NMDA response in cultured mouse brain neurons. *Nature* **325**, 529–531.

13. Nowak, L., Bregestovski, P., Ascher, P., Herbert, A. & Prochiantz, A. (1984): Magnesium gates glutamate-activated channels in mouse central neurons. *Nature* **307**, 462–465.

14. O'Dell, T.J. & Christensen, B.N. (1988): Mecamylamine is a selective non-competitive antagonist of *N*-methyl-D-aspartate and aspartate-induced currents in horizontal cells dissociated from catfish retina. *Neurosci. Lett.* **94**, 93–98.

15. Oswald, R.E., Heidmann, T. & Changeux, J.P. (1983): Multiple affinity states for non-competitive blockers revealed by [³H]phencyclidine binding to acetylcholine receptor rich membrane fragments from *Torpedo marmorata*. *Biochemistry* **22**, 3128–3136.

16.   Peters, S., Koh, J. & Choi, D.W. (1987): Zinc selectively blocks the action of *N*-methyl-D-aspartate on cortical neurons. *Science* **236**, 589–593.

17.   Ransom, R.W. & Stec, N.L. (1988): Cooperative modulation of [^{3}H]MK-801 binding to the *N*-methyl-D-aspartate receptor-ion channel complex by L-glutamate, glycine and polyamines. *J. Neurochem.* **51**, 830–836.

18.   Reynolds, I.J. (1990): Arcaine is a competitive antagonist of the polyamine site on the NMDA receptor. *Eur. J. Pharmacol.* **177**, 215–216.

19.   Reynolds, I.J. (1990): Arcaine uncovers dual interaction of polyamines with the *N*-methyl-D-aspartate receptor. *J. Pharmacol. Exp. Ther.* **255**, 1001–1007.

20.   Reynolds, I.J. (1991): The spider toxin argiotoxin $_{636}$, binds to a Mg^{2+} site on the *N*-methyl-D-aspartate receptor complex. *Br. J. Pharmacol.* **103**, 1373–1376.

21.   Reynolds, I.J., Baron, B.M. & Edwards, M.L. (1991): 1,10- bis(guanidino)decane inhibits *N*-methyl-D-aspartate responses *in vitro* and *in vivo*. *J. Pharmacol. Exp. Ther.* **259**, 626–632.

22.   Reynolds, I.J. & Miller, R.J. (1988): Multiple sites for the regulation of the *N*-methyl-D-aspartate receptor. *Mol. Pharmacol.* **33**, 581–584.

23.   Reynolds, I.J. & Miller, R.J. (1990): Allosteric modulation of *N*-methyl-D-aspartate receptors. *Adv. Pharmacol.* **21**, 101–126.

24.   Reynolds, I.J. & Palmer, A.M. (1991): Regional variations in [^{3}H]MK801 binding to rat brain NMDA receptors. *J. Neurochem.* **56**, 1731–1740.

25.   Reynolds, I.J. Rush, E.A. & Aizenman, E. (1990): Reduction of NMDA receptors with dithiothreitol increases [^{3}H] MK801 binding and NMDA-induced Ca^{2+} fluxes. *Br. J. Pharmacol.* **101**, 178–182.

26.   Rothman, S.M. & Olney, J.W. (1987): Excitotoxicity and the NMDA receptor. *Trends Neurosci.* **10**, 299–302.

27.   Westbrook, G.L. & Mayer, M.L. (1987): Micromolar concentrations of Zn^{2+} antagonize NMDA and GABA responses of hippocampal neurons. *Nature* **328**, 640–643.

28.   Williams, K., Dawson, V.L., Romano, C., Dichter, M.A. & Molinoff, P.B. (1990): Characterization of polyamines having agonist, antagonist and inverse agonist effects at the polyamine recognition site of the NMDA receptor. *Neuron* **5**, 199–208.

29.   Williams, K., Romano, C. & Molinoff, P.B. (1989): Effects of polyamines on the binding of [^{3}H]MK801 to the *N*-methyl-D-aspartate receptor: pharmacological evidence for the existence of a polyamine recognition site. *Mol. Pharmacol.* **36**, 575–581.

30.   Wong, E.H.F., Kemp, J.A., Priestley, T., Knight, A.R., Woodruff, G.N. & Iversen, L.L. (1986): The anticonvulsant MK 801 is a potent *N*-methyl-D-aspartate antagonist. *Proc. Natl. Acad. Sci. USA.* **83**, 7104–7108.

Guanidino Compounds in Biology and Medicine, eds. by P.P. De Deyn, B. Marescau, V. Stalon and
I.A. Qureshi. ©1992 John Libbey & Company Ltd., pp. 449–451.

Chapter 65

Effects of guanidino compounds on monoamineoxidase and catechol-O-methyltransferase activity

Yutaka NISHIJIMA, Katsuhisa HUKUYAMA, Hideaki KABUTO, Isao YOKOI and
Akitane MORI

*Department of Neurochemistry, Institute for Neurobiology, Okayama University Medical School, 2-5-1 Shikata-cho,
Okayama 700, Japan*

Introduction

It is well known that many guanidino compounds induce convulsive seizures[5] and some of them alter the monoaminergic neurotransmission to induce convulsions. For example, dopamine (DA) in the rat striatum increased during 5-guanidinovaleric acid (GVA)-induced convulsions[2], and 3,4- dihydroxyphenylacetic acid (DOPAC) content in the rat brain increased markedly during 2-guanidinoethanol (GEt)-induced convulsions[6]. Furthermore, DA and norepinephrine decreased during 2-guanidinoglutaric acid (GGA)-induced convulsions[3]. These findings suggest that some guanidino compounds might influence catabolic enzymes of catecholamines.

In this study, we observed the effects of guanidino compounds on monoamineoxidase (MAO) and catechol-O-methyltransferase (COMT) activities.

Materials and methods

Sprague-Dawley rats weighing 250–350 g were decapitated, and the cerebral cortex was removed on ice. It was homogenized with phosphate buffer (100 mM, pH 7.4) for MAO experiments or with KCl (150 mM) for COMT experiments, and then centrifuged. The supernatant was used as a crude source of enzyme. Final concentration of 2-N-acetylarginine, arginine, creatine, guanidine, guanidinoacetic acid, 4-guanidinobutyric acid, guanidinoethanesulphonic acid, GEt, 3-guanidinopropionic acid, guanidinosuccinic acid, homoarginine and 4-hydroxyarginine in the enzyme solution was 5 mM, and final concentration of creatinine (CRN), GGA, GVA and methylguanidine (MG) varied from 47 μM to 5 mM for the kinetic studies.

The MAO activity was determined using a colorimetric assay kit, MAO-B Test Wako 273–31101 (Wako Pure Chemical Ind. Ltd., Osaka), with some modification by us for kinetic studies. The method of the determination of COMT activity was described previously[1]. In this method, the amount of homovanillic acid (HVA) formed from DOPAC by COMT was measured using HPLC with electro-chemical detection after incubation of enzyme solution with S-adenosylmethionine and DOPAC. Paired *t*-test was used to evaluate the results, and kinetic parameters of enzyme reaction were analysed by nonlinear regression[4]. Protein content in the crude enzyme solution was determined according to Lowry.

Results and discussion

CRN, GVA and MG inhibited MAO activity to 89.3, 91.1 and 96.1 per cent respectively. The inhibition mechanisms by CRN, MG and GVA were all competitive, and K_i values were 14.0, 9.5 and 30.4 mM, respectively (Fig. 1). Though GGA significantly increased MAO activity to 119 per cent, 12 other guanidino compounds had no effect on MAO activity. The strongest inhibitor for MAO examined in this study seemed to be GVA. However GVA was not such a strong MAO inhibitor because the K_i value of 3,4-dihydroxyphenylalanine which is a well known MAO inhibitor was 1. 74 mM when acting as competitive inhibitor in this experiment.

As we used excess substrate for the determination of the effect of guanidino compounds on the MAO activity, it is difficult to say that the guanidino compounds tested, except CRN, MG, GVA and GGA, did not inhibit MAO in a noncompetitive, uncompetitive or linear mixed manner. As DOPAC content in the rat brain increased markedly while only small increases in the HVA content were observed during GEt-induced convulsions, COMT activity has been thought to be inhibited by GEt[6]. In this study, none of the guanidino compounds tested at 5 mM inhibited COMT activity (data not shown).

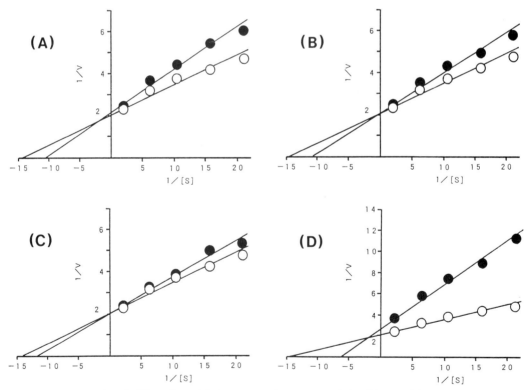

Fig. 1. Effect of guanidinio compounds on monoaminieoxidase activity.
In each double reciprocal plot: O, control: ●, 5 mM of 5-guanidinovaleric acid (GVA), creatinine (CRN), methylguanidine (MG) or 3,4-dihydroxyphenylalanine (L-DOPA) added. (A), GVA (K_i = 9.5 mM); (B), CRN (K_i = 14.0 mM); (C), MG (K_i = 30.4 mM); and (D), L-DOPA (K_i = 1.74 mM). K_i values are calculated from six independent assays.

However, we also used excess substrate in COMT studies, so it is difficult to say whether these guanidino compounds inhibit COMT activity in a competitive manner or not. Further studies at stoichiometric amounts of substrate are needed.

References

1. Akiyama, K., Shimizu, Y., Yokoi, I., Kabuto, H., Mori, A. & Ozaki, M. (1989): Effects of EGC and EGCG on COMT activity in mice brain. *Neurosciences* **15**, 262–264.

2. Kabuto, H., Iwaya, K., Yokoi, I. & Mori, A. (1992): Effects of 5-guanidinovaleric acid on monoamine release in rat striatum. In *Guanidino compounds in biology and medicine,* eds. P.P. De Deyn, B. Marescau, V. Stalon & I.A. Qureshi, pp. 453–456. London: John Libbey.

3. Shiraga, H., Hiramatsu, M. & Mori, A. (1989): Involvement of catecholamines on the seizure mechanism induced by α-guanidinoglutaric acid in rats. In *Guanidines 2* eds. A. Mori, B.D. Cohen & H. Koide, pp. 213–222. New York: Plenum Press.

4. Watanabe, Y., Yokoi, I. & Mori, A. (1989): Biosynthesis of 2-guanidinoethanol. In *Guanidines 2* eds. A. Mori, B.D. Cohen & H. Koide, pp. 53–60. New York: Plenum Press.

5. Yokoi, I., Edaki, A., Watanabe, Y., Shimizu, Y., Toda, H. & Mori, A. (1989): Effects of anticonvulsants on convulsive activity induced by 2-guanidinoethanol. In *Guanidines 2* eds. A. Mori, B.D. Cohen & H. Koide, pp. 169–181. New York: Plenum Press.

6. Yokoi, I., Itoh, T., Yufu, K., Akiyama, K., Satoh, M., Murakami, S., Kabuto, H. & Mori, A. (1991): Effect of 2-guanidinoethanol on levels of monoamines and their metabolites in the brain. *Neurochem. Res.* **16** 1155–1159.

Guanidino Compounds in Biology and Medicine, eds. by P.P. De Deyn, B. Marescau, V. Stalon and I.A. Qureshi. ©1992 John Libbey & Company Ltd., pp. 453–456.

Chapter 66

Effects of 5-guanidinovaleric acid on monoamine release in rat striatum

Hideaki KABUTO, Kazuo IWAYA, Isao YOKOI and Akitane MORI

Department of Neurochemistry, Institute for Neurobiology, Okayama University Medical School, 2-5-1 Shikata-cho, Okayama 700, Japan

Introduction

Many guanidino compounds have been shown to exist in the mammalian brain[3]. Most of these guanidino compounds have roles not only in general nitrogen metabolism, but also in specific physiological functions in the normal body and brain. It is well known, for example, that arginine is an intermediate in the urea cycle and that creatine phosphate is used in muscle and brain energy processes. Some of the guanidino compounds have the ability to induce seizures in experimental animals[6]. Moreover, endogenous guanidino compound levels are known to change before and after seizures in many kinds of epileptic animal brain[4]. Therefore it has been suggested that guanidino compounds may play a role in neurological disorders such as epilepsy.

In electrophysiological studies, it has been shown that 5-guanidinovaleric acid (GVA)-induced spike discharges on the electroencephalogram. Furthermore, spike discharges induced by GVA were antagonized by γ-aminobutyric acid (GABA), an inhibitory neurotransmitter in the mammalian central nervous system, and its agonist, muscimol, suggesting that GVA may be an endogenous GABA receptor antagonist[5].

Recently attention has been focused on the relationship between neuron systems, and it was reported that there were relationships between GABA and monoamine systems[1]. In the present study, we examined the effects of GVA on dopamine (DA), dihydroxyphenylacetic acid (DOPAC), homovanillic acid (HVA), serotonin (5-HT), and 5-hydroxyindolacetic acid (5-HIAA) in rat striatum by the method of a microdialysis technique combined with high performance liquid chromatography with an electrochemical detector (HPLC-ECD).

Materials and method

Sprague-Dawley rats weighing 250–350 were used. Stereotaxic surgery was performed under ether anaesthesia. The system used for intracerebral dialysis was similar to that described by Phebus *et al.*[2]. The dialysis probe and the microinfusion pump were hand made. After all preparations for tracheal intubation and artificial ventilation had been completed, rats were immobilized using succinylcholin chloride. The rats were than placed in a stereotaxic device. The dialysis probe inserted into the left striatum at the following coordinates: 2 mm anterior to the coronal suture and 3 mm on the left side of the sagittal suture, and 4 mm below the dura. The probe was perfused with normal saline at 2 μl/min and dialysate was collected at 20 min intervals in micro centrifuge tubes. In order to avoid the effects of anaesthesia, six or seven samples were collected prior to the administration of GVA. At least six samples were taken before GVA was added to the perfusion medium at the concentration of 1 and 10 mM. The administration of GVA was continued over the examination. The dialysate was injected and directly assayed for monoamines using HPLC-ECD immediately after the collection. Data were expressed as per cent changes from the values of a dialysate sample taken just prior to treatment. Statistical significance of differences between experimental conditions were tested using the U-test.

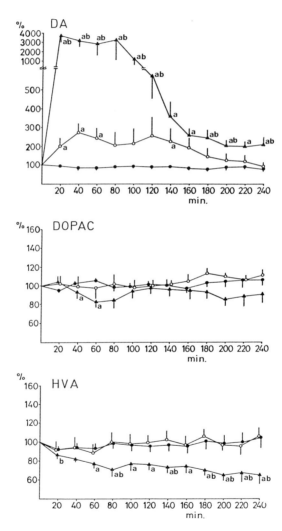

Fig. 1. The time-dependent changes (mean ± SEM) in the levels of DA, DOPAC and HVA after adminstration of GVA into the dialysate. ●: control; ○: GVA 1 mM; ▲: GVA 10 mM.
Data are expressed as per cent changes from the value of the sample just prior to the treatment. Statistical significance was tested by of U-test (P < 0 .05, a: between experimental and control groups; b: between 1 mM and 10 mM GVA groups) n = 6–12.

Results and discussion

Time-dependent changes in the extracellular levels of DA, DOPAC, HVA, 5-HT and 5-HIAA were monitored in the striatal dialysate after the injection of GVA.

DA and its metabolites

A significant increase in DA levels was observed in the samples between 0 and 60, and 120 and 140 min after the administration of GVA (1 mM), though there were no significant changes in DOPAC and HVA levels. A significant increase in DA levels was observed during the whole examination when GVA (10 mM) was administered. Especially within 2 h after the administration of GVA (10 mM), the increase was marked. A significant decrease in DOPAC levels was observed in the samples between 20 and 60 min after the administration of GVA (10 mM). A significant decrease in HVA levels was observed in the samples obtained at 40 min or later after GVA administration (Fig. 1).

5-HT and its metabolite

A significant increase in 5-HT levels was observed in all samples after the administration of GVA (1 mM, 10 mM). Though GVA (10 mM) had no effect on 5-HIAA levels, a significant increase was observed in the samples between 160 and 180, and 220 and 240 min after the administration of GVA (1 mM) (Fig. 2).

These results suggest that the administration of GVA into the rat striatum induced the significant increase of DA and 5-HT releases in a dose-dependent manner. Two h after the administration of GVA, the levels of DA tended to decrease. 5-HT remained at high levels over longer periods of time. A possible reason for this might be that the increase of DA release is only transient and the increase of 5-HT release is sustained while GVA is administered continuously (perhaps when GABA neurons

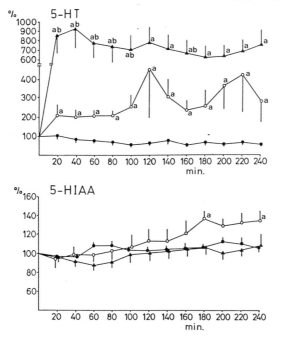

Fig. 2. The time-dependent changes (mean ± SEM) in the levels of 5-HT and 5-HIAA after administration of GVA into the dialysate. Symbols mean the same as in Fig. 1.

were inhibited continuously). The increase of 5-HIAA induced by GVA (1 mM) seems to be the result of the increase of 5-HT. The mechanism of the decreases of DOPAC and HVA is unclear. It might be a result of the reinforcement of flush out or the decrement of metabolism of DA. As a conclusion, GVA induced the unusual excitation of DA and 5-HT neurons.

References

1.   Kabuto, H., Yokoi, I. & Mori, A. (1988): Effects of muscimol and baclofen on levels of monoamines and their metabolites in the E1 mouse brain. *Neurochemical Res.* **13**, 1157–1161.

2.   Phebus, L.A., Perry, K.W., Clemens, J.A. & Fuller, R.W. (1986): Brain anoxia releases striatal dopamine in rats. *Life Sci.* **38**, 2447–2453.

3.   Robin, Y. & Marescau, B. (1985): Natural guanidino compounds. In *Guanidines*, eds. A. Mori, B.D. Cohen & A. Lowenthal, pp. 383–438. New York: Plenum Press.

4.   Wiechert, P., Marescau, B., De Deyn, P. & Lowenthal, A. (1986): Guanidino compounds in serum and brain of audiogenically sensitive rats. *Biomed. Biochim. Acta* **45**, 1339–1342.

5.   Yokoi, I., Tsuruta, K., Shiraga, H. & Mori, A. (1987): δ-Guanidinovaleric acid as an endogenous and specific GABA-receptor antagonist, electroencephalographic study. *Epilepsy Res.* **1**, 114–120.

6.   Yokoi, I., Edaki, A., Watanabe, Y., Shimizu, Y., Toda, H. & Mori, A. (1989): Effects of anticonvulsants on convulsive activity induced by 2-guanidinoethanol. In *Guanidines 2* eds. A. Mori, B.D. Cohen & H. Koide, pp. 169–181. New York: Plenum Press.

Guanidino Compounds in Biology and Medicine, eds. by P.P. De Deyn, B. Marescau, V. Stalon and I.A. Qureshi. ©1992 John Libbey & Company Ltd., pp. 457–459.

Chapter 67

Characterization of guanidinosuccinic acid-induced seizures

Y.Q. PEI, R. D'HOOGE, B. MARESCAU, I. POSSEMIERS, F. FRANCK
and P.P. DE DEYN

Laboratory of Neurochemistry, Born-Bunge Foundation, University of Antwerp (UIA), Universiteitsplein 1, B-2610 Wilrijk, Belgium

Summary

Convulsions induced by intraperitoneal and intracerebroventricular injections of guanidinosuccinic acid (GSA) were behaviourally assessed in Swiss mice. Mice (total n = 35) were injected intraperitoneally with increasing doses of GSA. Full-blown clonic or clonic-tonic convulsions appeared in a dose-dependent manner, with a median latency of about 25 min. With probit analysis, we determined the CD_{50} (and 95 per cent confidence limits) to be 363 (287–458) mg/kg. Often, clonic seizures would occur lasting 3–5 h. Some of the animals died from status epilepticus or tonic extension, depending on the dose used. The CD_{50} of GSA, injected intracerebroventricularly, was determined to be 5.1 (3.8–6.7) μg/mouse (total n = 30). Convulsive action was similar to that observed after intraperitoneal injection, only latency was much smaller. We also demonstrated, by simultaneous intracerebroventricular injection of subconvulsive doses, that GSA and *N*-methyl-D-asparate act synergistically.

Introduction

Guanidinosuccinic acid (GSA) is one of many naturally occurring guanidino compounds[7]. Possible physiological and pathophysiological actions of GSA in mammalian central nervous system (CNS) remain to be studied. It was reported that GSA inhibits GABA and glycine responses on mouse spinal cord neurons in cell culture, and it may play a role in the pathophysiology of uraemia-associated complications[2,3,8]. However, behavioural analysis of the actions of this putative endogenous excitatory compound is still highly incomplete.

This study deals with the characterization of convulsions induced by systemic and intracerebral administration of GSA to adult mice. Although many guanidino compounds are known to induce clonic and tonic convulsions experimentally[6], the convulsive action of GSA has not been reported as yet. We present the first assessment of behaviour induced by this compound, which is highly increased in serum, cerebrospinal fluid and brain of uraemic patients with neurological complications including epilepsy[3,4].

Materials and methods

Prior to their use in the experiments, male and female random-bred Swiss mice were kept under standard environmentally controlled conditions (12–12 h light–dark cycle, constant room temperature and humidity). For systemic administration of GSA, mice weighing 18–25 g and for intracerebral administration, 30–40 g, were used. GSA was purchased from Sigma Chemical Company St Louis, (MO, USA). For intraperitoneal (ip) administration, smooth suspensions of GSA in 30 per cent polyethylene glycol solution were used. Concentrations were always calculated in such a manner that each animal received 0.1 ml suspension /10 g body weight. Groups of five to 10 animals were injected ip with doses of GSA between 237–800 mg/kg.

For intracerebroventricular (icv) administration, the compound was dissolved in saline and delivered in a volume of 5 μl with a Hamilton microsyringe into the left lateral brain ventricle, according to a technique modified from Herman[5]. Briefly, the animal's scalp was cut and the exposed skull was pierced with a small stainless steel drill under local anaesthesia by 1 per cent lidocaine solution. The hole in the skull was made about 1 mm posterior to the coronal suture and 1 mm left to the sagittal suture. The animals were allowed at least 3 h rest for recovery and wearing off the local anaesthetic. An injection cannula was mounted on the microsyringe and fitted with a nylon stopper to ensure an injection depth of 3 mm from the surface of the skull. While the animals were restrained by hand, the 5 μl volume was administered at an injection rate of 1 μl/5 s and the cannula was kept in position for about 15 s. Post-mortem injection of black dye through the same hole in random control animals (n > 20) always revealed homogenous filling of the lateral brain ventricles. Groups of 5–10 animals were injected icv with doses between 1.56–12.5 μg GSA/mouse.

For ip as well as for icv administration, each series of injections commenced with a dose of maximal effect, thereafter, doses were decreased until zero effect and each mouse was used only once. Immediately after injection of GSA, the animals were placed in individual cylindrical plastic cages for observation of their behaviour. The character, onset time latency, evolution and duration of presumed epileptic activity were noted within a 1 h observation period.

Results

The convulsive response after ip injection of GSA suspension usually developed slowly, with a median latency of about 25 min, and went as follows: hyperventilation, hyperactivity or rearing preceded wild running and jumping up and down against the walls of the observation cages. This was followed by tremor of the head, mouth movements, mild myoclonus of the face, neck and forelimbs, with vertically raised tail. Subsequently, all muscles of the forepart of the body twitched. There followed typical generalized clonic seizures comprising automatic crawling or grasping movements of forelimbs and paws with head jerked backwards, hind legs broadly placed and toes spread. Then a typical status of clonic seizures (SCS) occurred, which lasted for 3–5 h. Often, the animals would foam at the mouth, some even bit their tongue, defecated and micturated, fell down but rose again or rolled around their axis. Eventually, some of the animals died from SCS or tonic extension depending on the dose used. Probit analysis yielded a CD_{50} (95 per cent confidence limits) of 363 (287–458) mg/kg (n = 35) for GSA after ip injection.

The convulsive response after icv administration of GSA was similar to that after ip injection, only the seizures appeared immediately and lasted for a shorter time, about 5–30 min, depending on the dose used. The CD_{50} was determined to be 5.1 (3.8–6.7) μg/mouse (n = 30). By simultaneous icv injection of subconvulsive doses of both drugs, we also demonstrated that GSA potentiates N-methyl-D-asparate (NMDA)-induced seizures. When 3.125 μg GSA/mouse (n = 20) or 36.8 ng NMDA/mouse (n = 20) were used alone, seizures were induced in only 10 per cent and 0 per cent of the animals respectively; when these two drugs were co-injected, seizures were induced in 87 per cent of the animals (n = 15).

Discussion

Many guanidino compounds are known convulsants. The results presented here add GSA to the list of convulsant guanidino compounds. Indeed, GSA induces clonic and tonic convulsions after ip and icv administration, and it does so in a dose-related manner. Although the potency of GSA after ip injection appeared to be very low (doses were in the mg and g/kg range), potency after icv injection was rather high (doses in the μg range, i.e. per kg body weight, more than 2500 times lower than after ip injection). This difference is probably due to the low solubility of GSA and its difficulty in getting across the blood–brain barrier.

Seizures induced by GSA were dose-dependent clonic and tonic seizures, the latter only occurring when large doses were used. After ip administration, GSA-induced clonic seizures have a slow onset, with a median latency of about 25 min. Wild jumping and running were the first symptoms of these seizures. This feature and the long-lasting GSA-induced SCS (continuous clonic convulsions during 3–5 h) are very similar to seizures induced by kainic acid and thus GSA-induced seizures could tentatively be characterized as limbic motor seizures[1] with high resistance to antiepileptic drugs (unpublished material).

Physiological and pathophysiological effects of GSA in the CNS are still very putative. In patients with renal insufficiency, guanidino compounds accumulate in serum, cerebrospinal fluid and brain, and especially GSA levels are highly increased in uraemia[4]. As uraemic patients often suffer from a large variety of neurological complications, including epilepsy, the convulsant action of GSA may be one of the underlying causes of these complications. The mechanism of GSA-induced seizures remains to be studied much further, but it may be related to GABA neurotransmission as suggested by De Deyn et al.[2,3], as well as to excitatory amino acid neurotransmission mediated by the NMDA receptor/ionophore complex.

Acknowledgements

Financial support was obtained from the University of Antwerp (UIA), Born-Bunge Foundation, NFWO grants N° 3.0044.92 and NDE58, and United Fund of Belgium. Yin-Quan Pei is a visiting professor from the Department of Pharmacology, Beijing Medical University, China, sponsored by the Belgian foundation of scientific research NFWO.

References

1. Ben-Ari, Y. (1985): Limbic seizures and brain damage produced by kainic acid. Mechanisms and relevance to human temporal epilepsy. *Neuroscience* **14**, 375–403.

2. De Deyn, P.P. & Macdonald, R.L. (1990): Guanidino compounds that are increased in cerebrospinal fluid and brain of uremic patients inhibit GABA and glycine responses on mouse neurons in cell culture. *Ann. Neurol.* **28**, 627–633.

3. De Deyn, P.P., Marescau, B. & Macdonald, R.L. (1990): Epilepsy and the GABA hypothesis. A brief review and some examples. *Acta Neurol. Belg.* **90**, 65–81.

4. De Deyn, P.P., Marescau, B., Cuykens, J., Van Gorp, L., Lowenthal, A. & De Potter, W.P. (1987): Guanidino compounds in serum and cerebrospinal fluid of non-dialyzed patients with renal insufficiency. *Clin. Chim. Acta* **167**, 81–88.

5. Herman, Z.S. (1975): Behavioural changes induced in conscious mice by intracerebroventricular injection of catecholamines, acetylcholine and S-hydroxytryptamine. *Br. J. Pharmacol.* **55**, 351–358.

6. Mori, A. (1987): Biochemistry and neurotoxicology of guanidino compounds. History and recent advances. *Pav. J. Biol. Sci.* **22**, 85–94.

7. Robin, Y. & Marescau, B. (1985): Natural guanidino compounds. In *Guanidines*, eds. A. Mori, B.D. Cohen & A. Lowenthal, pp. 383–438. New York: Plenum Press.

8. Stein, I.M., Cohen, B.D. & Horowitz, H.I. (1968): Guanidinosuccinic acid: The 'X' factor in uremic bleeding? *Clin. Res.* **16**, 397–399.

Guanidino Compounds in Biology and Medicine, eds. by P.P. De Deyn, B. Marescau, V. Stalon and I.A. Qureshi. ©1992 John Libbey & Company Ltd., pp. 461–463.

Chapter 68

Effect of ferric citrate and hyperoxygenation on guanidino compounds in mouse organs

R. EDAMATSU, S. WATANABE, T. ITOH, K. YUFU[1] and A. MORI

Department of Neurochemistry, Institute for Neurobiology and [1]Department of Anaesthesiology and Resuscitology, Okayama University Medical School, Shikata-cho 2-5-1, Okayama 700, Japan

Introduction

In the case of cerebral haemorrhage or other injuries in the brain, haem proteins are released in the interstitium. This occurs by the haemolysis and degradation of haemoglobin. Free iron ion is liberated from such haem proteins. It is already reported that excess free iron ion *in vivo* catalyses the formation of active oxygen species (AOS) and promotes peroxidation of neuronal cell membrane[1]. We have been interested in this process, because it is suggested to be the possible mechanism for the focus formation of traumatic epilepsy. It is also known that ferric citrate administered ip distributes in the liver or kidney, and promotes lipid peroxidation in the tissue. Meanwhile, it is reported that the AOS is also related in the generation of methylguanidine, which is a potent convulsant in animal models[3]. In our present study, we studied whether administration of ferric citrate or inhalation of pure oxygen can alter the production of methylguanidine in various mice organs. The change in the contents of other guanidino compounds was also investigated.

Abbreviations

Methylguanidine (MG), creatinine (CRN), arginine (Arg), γ-guanidinobutyric acid (GBA), α-guanidinoacetic acid (GAA), guanidine (G), *N*-acetyl-arginine (NAA), guanidinopropionic acid (GPA), guanidinoethanol (GEt), homoarginine (HArg) and superoxide dismutase (SOD).

Materials and methods

Six-weeks-old male ddY mice were used for all experiments. The mice were given mouse food (Oriental Yeast Co., Ltd., Tokyo) and water *ad libitum*. Ferric citrate (10 mg iron/kg) was administered to one group of mice every other day intraperitoneally. The other group of mice we exposed to pure oxygen for 1 week. All mice were sacrificed a week later. Blood was heparinized in a test tube which could be used for separating the serum. The kidney, brain, liver, pancreas and gastrocnemic muscle were removed quickly and kept at –80 °C until analysis. The serum was deproteinized by 30 per cent trichloroacetic acid solution. Other tissues were deproteinized with 1 per cent picric acid. The samples were kept at –20 °C until analysis. Guanidino compounds were quantified with a guanidino compounds analyser (G-520 JASCO Co., LTD., Tokyo).

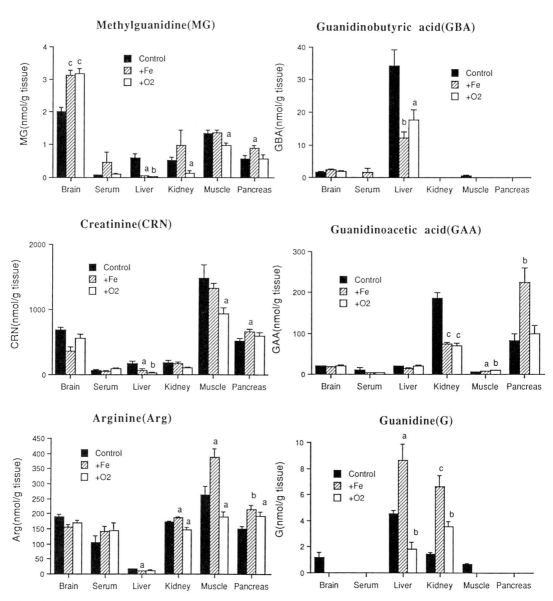

Fig. 1. Effect of ferric citrate and hyperoxygenation on MG, CRN and Arg. Each value represents mean ± SEM of eight to 11 mice.
(a) P < 0.05; (b) P < 0.01; (c) P < 0.001 vs control.

Fig. 2. Effect of ferric citrate and hyperoxygenation on GBA, GPA and G. Each value represents mean ± SEM of eight to 11 mice.
(a) P < 0.05; (b) P < 0.01; (c) P < 0.001 vs control.

Results

Effect of administration of ferric citrate

In the brain, NAA and MG were increased; CRN, Arg, G and HArg were decreased. In the serum, MG was increased; NAA and GPA were decreased. In the liver, G was increased; GPA, GBA, CRN,

N-acetylarginine(NAA)

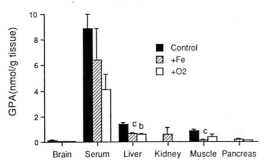

Guanidinopropionic acid(GPA)

Guanidinoethanol(GEt)

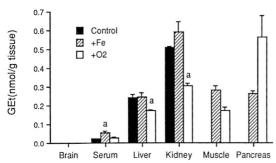

Fig. 3. Effect of ferric citrate and hyperoxygenation on NAA, GPA and GEt. Each vale represents mean ± SEM of eight to 11 mice.
(a) P < 0.05; (b) P < 0.01; (c) P < 0.001 vs control.

Arg and MG were decreased. In the kidney, G, MG and Arg were increased; GAA was decreased. In the pancreas, GAA, CRN, Arg and MG were increased; NAA was decreased. In the muscle, GAA and Arg were increased; GPA, G and GBA were decreased.

Effect of inhalation of pure oxygen

In the brain, MG was increased; HArg and G were decreased. In the serum, GEt was increased. In the liver, GPA, G, GBA, GEt, CRN and MG were decreased. In the kidney, G was increased; NAA, MG, GEt, GAA and Arg were decreased. In the pancreas, Arg was increased; NAA was decreased. In the muscle, GAA was increased; G, GBA, CRN, Arg and MG were decreased.

Discussion

Because ferric citrate and excessive oxygen enhanced peroxidation may occur, we expected an increase in MG level in the tissues of both the ferric citrate administered group and the pure oxygen inhalation group. The MG level increased in the brain, serum, kidney and pancreas of the ferric citrate treated mice, and in the brain of oxygen inhaling mice. In other organs, however, no significant change was observed. On the other hand, both in the liver of the ferric citrate administered group, and in the liver, kidney and muscle of the pure oxygen exposure group, decreased MG levels were observed. We already reported that SOD activity in the rat brain was increased under hyperoxygenation[2]. Therefore, this study suggests that SOD activity increased by the compensatory enhancement of reactive oxygen species scavenging system. The reason for the alternation of the other guanidino compounds still remains unknown.

References

1. Dunford, H.B. (1987): Free radicals in iron-containing systems. *Free Radical Biol. Med.* **3**, 405–421.

2. Mori, A., Hiramatsu, M., Yokoi, I. & Edamatsu, R. (1990): Biochemical pathogenesis of post-traumatic epilepsy. *Pav. J. Biol. Sci.* **25**, 54–62.

3. Willmore, L.J., Hurd, R.W. & Sypert, G.W. (1978): Epileptiform activity initiated by pial iontophoresis of ferrous and ferric chloride on cat cerebral cortex. *Brain Res.* **152**, 406–410.

Guanidino Compounds in Biology and Medicine, eds. by P.P. De Deyn, B. Marescau, V. Stalon and I.A. Qureshi. ©1992 John Libbey & Company Ltd., pp. 465–466.

Chapter 69

Effect of isobaric hyperoxygenation on guanidino compounds in rat brain

K. YUFU[1], T. ITOH, R. EDAMATSU and A. MORI

Department of Neurochemistry, Institute for Neurobiology, and [1]Department of Anaesthesiology and Resuscitology, Okayama University Medical School, Okayama 700, Japan

Introduction

O$_2$ inhalation therapy is widely performed in clinical practice, but oxygen toxicity is sometimes troublesome to clinicians. It is well known that oxygen free radical reactions are a main cause of oxygen toxicity[2] and major target organs are the respiratory system and the central nervous system. Some guanidino compounds are known as neurotoxins that cause convulsion[5], etc. Recently, it is reported by some authors that some guanidino compounds, especially methylguanidine, are related to oxygen free radical reactions[3,6]. Therefore, guanidino compounds may be involved in oxygen toxicity to the central nervous system. To study the relationship between their content in the central nervous system and hyperoxygenation, we performed this experiment.

Materials and methods

We used male Sprague-Dawley rats at 7 weeks of age. They were divided to three groups: a 12 h O$_2$ inhalation group, a 24 h O$_2$ inhalation group, and a control group. O$_2$ inhalation was performed in a $41 \times 26 \times 19$ cm box that had a 100 per cent O$_2$ flow at 3 l/min. O$_2$ concentration was monitored at the range of 90–95 per cent. The control group was kept in air. Both groups had free access to water and laboratory chow. After 12 h or 24 h of hyperoxygenation, rats were sacrificed by phlebotomy. Their brains were quickly removed and dissected on ice according to Inversen and Glowinski's method. We used the cerebral cortex for guanidino compounds' analysis. Each sample was homogenized with phosphate buffer (pH 7.8) and centrifuged at 3000 rpm, 15 min with 30 per cent trichloroacetic acid. Supernatant fluid was analysed with a high performance liquid chromatograph, JASCO G-520 (Japan Spectro-Scopic Co., Tokyo) with the phenanthrenequinone reaction. Statistical analysis was performed using the Student's *t*-test.

Results and discussion

Table 1 shows the guanidino compounds content in rat cortex. After 12 h of hyperoxygenation, only methylguanidine and creatinine was significantly increased. However, after 24 h, guanidino acetic acid, creatinine and methylguanidine were significantly increased. Moreover, guanidinoacetic acid and arginine were significantly increased between 12 h and 24 h of hyperoxygenation. No behavioural changes were observed, except that the rats seemed to be calm.

Table 1. Changes of guanidino compounds in rat cerebral cortex after 12 h or 24 h of isobaric hyperoxygenation. Each value represents mean ± SEM (nmol/g wet tissue) of six or seven rats

	Guanidinoacetic acid	Creatinine	Arginine	Methylguanidine
Control	11.72 ± 0.48	177.5 ± 9.0	676.7 ± 37.9	0.69 ± 0.05
12 h	12.06 ± 0.18	208.0 ± 2.8*	663.0 ± 8.7	0.90 ± 0.05*
24 h	13.25 ± 0.41*,**	215.9 ± 9.9*	710.6 ± 16.5**	1.02 ± 0.05*

*$P < 0.05$ *vs* control: **$P < 0.05$ *vs* 12 h of isobaric hyperoxygenation.

In this study, guanidino compounds showed an almost time-dependent increase. It is reported that the symptoms of oxygen toxicity to the central nervous system develop as a function of oxygen pressure × duration of hyperoxygenation[1]. Moreover, Kovachich *et al.* showed lipid peroxide level changes in the same manner in rat brain cortical slices[4]. Recently, in our preliminary study, we found that carbon-centred radical content in rat cerebral cortex increase in a time-dependent manner. Therefore, we suppose some oxygen free radical reactions may contribute to the changes of guanidino compounds in this study to some extent.

References

1. Dickens, F. (1962): The toxic effect of oxygen on nervous tissue. In *Neurochemistry*, eds. K.A.C. Elliott, I.H. Page & J.H. Quastel, pp. 851–869. Springfield: C.C. Thomas.

2. Gershman, R., Gilbert D.I., Nye S.W. *et al.* (1954): Oxygen poisoning and x-irradiation; a mechanism is common. *Science* **119**, 623–626.

3. Hiramatsu, M., Edamatsu, R., Kohno, M. & Mori, A. (1989): The reactivity of guanidino compounds with hydroxyl radicals. In *Guanidines 2*, eds. A. Mori, B.D. Cohen & H. Koide pp. 97–105.

4. Kovachich, G.B. & Mishra, O.P. (1980): Lipid peroxidation in rat brain cortical slices as measured by the thiobarbituric acid test *J. Neurochem.* **35**, 1449–1452.

5. Nagase, S., Aoyagi, K., Narita, M. & Tojo, S. (1986): Active oxygen in methylguanidine synthesis. *Nephron* **44**, 299–303.

6. Mori, A., Watanabe, Y. & Akagi, M. (1982): Guanidino compounds anomalies in epilepsy. In *Advances in epileptology, XIIIth Epilepsy Symposium*, eds. H. Akimoto, H. Kasamatsuri, M. Swino & A. Ward, New York: Raven Press.

Author Index

Subject Index